Polymer Fractionation

Contributors to this Volume

K. H. Altgelt
O. A. Battista
W. Burchard
H.-J. Cantow
Manfred J. R. Cantow
John H. Elliott
Alden H. Emery, Jr.
O. Fuchs
Hanswalter Giesekus
F. C. Goodrich
Maurice L. Huggins
Julian F. Johnson
Akira Kotera
H. W. McCormick
J. C. Moore
Hiroshi Okamoto
Roger S. Porter
W. Schmieder
J. Schurz
L. H. Tung

POLYMER FRACTIONATION

Edited by

MANFRED J. R. CANTOW

Richmond Laboratory
Chevron Research Company
Richmond, California

1967

ACADEMIC PRESS *New York · London*

COPYRIGHT © 1967, BY ACADEMIC PRESS INC.
ALL RIGHTS RESERVED.
NO PART OF THIS BOOK MAY BE REPRODUCED IN ANY FORM,
BY PHOTOSTAT, MICROFILM, OR ANY OTHER MEANS, WITHOUT
WRITTEN PERMISSION FROM THE PUBLISHERS.

ACADEMIC PRESS INC.
111 Fifth Avenue, New York, New York 10003

United Kingdom Edition published by
ACADEMIC PRESS INC. (LONDON) LTD.
Berkeley Square House, London W.1

LIBRARY OF CONGRESS CATALOG CARD NUMBER: 66:30024

PRINTED IN THE UNITED STATES OF AMERICA

List of Contributors

Numbers in parentheses indicate the pages on which the authors' contributions begin.

K. H. Altgelt, Chevron Research Company, Richmond, California (123)

O. A. Battista, Central Research Department, FMC Corporation, Princeton, New Jersey (307)

W. Burchard, Institut für Makromolekulare Chemie der Universität, Freiburg im Breisgau, Germany (285)

H.-J. Cantow, Lehrstuhl für Physikalische Chemie der Makromolekularen Substanzen, Institut für Makromolekulare Chemie der Universität, Freiburg im Breisgau, Germany (285)

Manfred J. R. Cantow, Chevron Research Company, Richmond, California (461)

John H. Elliott, Research Center, Hercules Incorporated, Wilmington, Delaware (67)

Alden H. Emery, Jr., School of Chemical Engineering, Purdue University, Lafayette, Indiana (181)

O. Fuchs, Abt. Kunststoff-Forschung, Farbwerke Hoechst AG., vormals Meister Lucius & Brüning, Frankfurt (M)-Hoechst, Germany (341)

Hanswalter Giesekus, Ingenieur-Abteilung Angewandte Physik, Farbenfabriken Bayer AG., Leverkusen, Germany (191)

F. C. Goodrich, Chevron Research Company, Richmond, California (415)*

Maurice L. Huggins, Stanford Research Institute, Menlo Park, California (1)

Julian F. Johnson, Chevron Research Company, Richmond, California (95)

Akira Kotera, Department of Chemistry, Tokyo Kyoiku University, Otsuka, Tokyo, Japan (43)

H. W. McCormick, Physical Research Laboratory, The Dow Chemical Company, Midland, Michigan (251).

J. C. Moore, The Dow Chemical Company, Freeport, Texas (123)

Hiroshi Okamoto, Electrical Communication Laboratory, Nippon Telegraph and Telephone Public Corporation, Musashino-Shi, Tokyo, Japan (1)

Roger S. Porter, Chevron Research Company, Richmond, California (95)†

---

* Present address: Department of Chemistry, Clarkson College of Technology, Potsdam, New York.

† Present address: Polymer Science and Engineering Program, University of Massachusetts, Amherst, Massachusetts.

W. Schmieder, Abt. Kunststoff-Forschung, Farbwerke Hoechst AG., vormals Meister Lucius & Brüning, Frankfurt (M)-Hoechst, Germany (341)

J. Schurz, Institut für Physikalische Chemie, Universität Graz, Graz, Austria (317)

L. H. Tung, The Dow Chemical Company, Midland, Michigan (379)

Preface

The physical properties of a given polymer type are to a large extent determined by the shape and width of its molecular weight distribution. Any study of the kinetics of polymerization and degradation of macromolecules is again greatly aided by the knowledge of the molecular spread of the sample. Thermodynamic and hydrodynamic solution behavior are also affected by the distribution. Nevertheless, only a disproportionately small percentage of investigations in polymer science is carried out on samples with a completely characterized molecular weight distribution. One reason for this may be the considerable additional labor which is required to obtain narrow fractions or to determine the distribution of a polymer. The present volume is intended to aid workers in the field in the selection of a fractionation method suitable for a particular case. Each chapter begins with a discussion of the theoretical background of the procedure. This is followed by a thorough description of instrumentation and experimental techniques. Several practical applications are presented in detail.

Chapters B.1–B.5 cover methods which yield sizable fractions in addition to distribution data on the whole polymer. Chapters C.1–C.5 are confined to analytical scale distribution methods only. The thermodynamics of polymer fractionation, the analysis of copolymers and mixtures, and the treatment of experimental data are presented in separate chapters. The deduction of kinetic information from a knowledge of the molecular weight distribution is treated in Chapter F. A final chapter tabulates suitable solvent–nonsolvent systems for the fractionation of a large variety of polymers by the various methods discussed elsewhere in the volume.

I want to express my deep gratitude to the authors of this volume for their efforts and their patience. I am greatly indebted to Dr. E. M. Barrall II for his assistance in matters of English construction in chapters by foreign authors. My thanks are also due to the management of Chevron Research Company for permission to undertake the editing of this treatise and for the use of facilities. The representatives of Academic Press have made every effort to be of assistance.

MANFRED J. R. CANTOW

Novato, California
October, 1966

Contents

LIST OF CONTRIBUTORS.. v
PREFACE.. vii

CHAPTER A. Theoretical Considerations
Maurice L. Huggins and Hiroshi Okamoto

I. Polydispersity of High Polymers....................................... 1
II. Phase Relations for Polydisperse Systems 5
III. Fractionation Theory.. 10
IV. Fractionation Efficiency... 27
    References.. 41

CHAPTER B.1. Fractional Precipitation
Akira Kotera

I. Nonsolvent Addition Method... 44
II. Fractionation by Solvent Evaporation................................. 58
III. Fractionation by Cooling.. 61
    References.. 65

CHAPTER B.2. Fractional Solution
John H. Elliott

I. Introduction... 67
II. Theoretical Considerations... 69
III. Experimental Methods.. 74
IV. Factors in Column Elution—Guiding Principles for Experimental Fractionation.. 82
V. Comparison between Fractional Solution Methods....................... 89
VI. Possible Areas for Future Research................................... 90
    References.. 92

CHAPTER B.3. Chromatographic Fractionation
Roger S. Porter and Julian F. Johnson

I. Introduction... 95
II. Apparatus.. 97
III. Column Operation.. 106
IV. Specific Fractionations.. 109
V. Theoretical Considerations and Comparison with Other Methods......... 109
VI. Preparatory-Scale Fractionation..................................... 118
    References.. 120

CHAPTER B.4. Gel Permeation Chromatography

K. H. Altgelt and J. C. Moore

I. Introduction	123
II. History of Gel Permeation Chromatography	127
III. The Theory of Gel Permeation Chromatography	130
IV. The Gels	147
V. Experimental Technique	158
VI. Evaluation of Data	169
References	173

CHAPTER B.5. Thermal Diffusion

Alden H. Emery, Jr.

I. History	181
II. Basic Theory	182
III. Studies of the Variables	183
IV. Polymer Fractionations	186
V. Methods of Fractionation	188
References	189

CHAPTER C.1. Turbidimetric Titration

Hanswalter Giesekus

I. Introduction	191
II. Outline of the Method and Range of Application	193
III. Elaboration of the Method	195
IV. Apparatus	199
V. Evaluation	211
VI. Application	228
References	246

CHAPTER C.2. Sedimentation

H. W. McCormick

I. Introduction	251
II. Theory	252
III. Experimental	255
IV. Sedimentation Velocity	258
V. Sedimentation Equilibrium	276
References	282

CHAPTER C.3. Isothermal Diffusion

W. Burchard and H.-J. Cantow

I. Introduction	285
II. Theory: The Differential Equation of Diffusion and Three Solutions	286
III. Experimental Methods for the Determination of Diffusion Constants	287
IV. Theory: Determination of Molecular Heterogeneity	289

V. Methods for the Determination of the Diffusion Coefficient Distribution....	301
VI. Experimental Results...	303
VII. Critical Remarks..	304
References..	305

CHAPTER C.4. Summative Fractionation

O. A. Battista

I. The Summative Method...	307
II. Mathematical Interpretation of Summative Data........................	309
III. Precipitating Power of Varying Compositions of Acetone–Water Precipitants...	311
References..	315

CHAPTER C.5. Rheological Methods

J. Schurz

I. Introduction to the Problem..	317
II. Parameter Methods...	318
III. Evaluation of Flow Curves..	321
IV. Evaluation of Relaxation Measurements...............................	332
V. Miscellaneous Methods...	335
References..	336

CHAPTER D. Chemical Inhomogeneity and its Determination

O. Fuchs and W. Schmieder

I. Causes for the Chemical Inhomogeneity of Macromolecular Compounds...	341
II. Determination of Chemical Inhomogeneity.............................	344
III. Quantitative Description of Chemical Inhomogeneity...................	356
IV. Literature Survey...	362
References..	371

CHAPTER E. Treatment of Data

L. H. Tung

I. Introduction..	379
II. Methods of Expressing Molecular Weight Distribution..................	380
III. Calculation of Molecular Weight Distribution from Fractionation Data....	387
IV. Calculation of Molecular Weight Distribution from Average Molecular Weight Measurements...	400
V. Comparison of Methods..	402
Appendix: Numerical Illustrations.....................................	405
References..	412

CHAPTER F. The Numerical Analysis and Kinetic Interpretation of Molecular Weight Distribution Data

F. C. Goodrich

I. Introduction	415
II. On the Prediction of Molecular Weight Distributions from Kinetic Schemes	418
III. Numerical Methods in the Handling of Molecular Weight Distribution Data	438
Appendix to Chapter F	452
References	459

CHAPTER G. Additional Methods of Fractionation

Manfred J. R. Cantow

Text	461
References	464
Appendix to Chapter G	465
References to Appendix	488

AUTHOR INDEX	499
SUBJECT INDEX	523

CHAPTER A

Theoretical Considerations

Maurice L. Huggins
STANFORD RESEARCH INSTITUTE, MENLO PARK, CALIFORNIA

AND

Hiroshi Okamoto
ELECTRICAL COMMUNICATION LABORATORY, NIPPON TELEGRAPH AND TELEPHONE
PUBLIC CORPORATION, MUSASHINO-SHI, TOKYO, JAPAN

I. Polydispersity of High Polymers 1
   A. Polydispersity and Fractionation 1
   B. Polydispersity and Polymer Properties 3
II. Phase Relations for Polydisperse Systems 5
   A. General Considerations 5
   B. Phase Relations for Polydisperse Polymer Solutions; Fractionation 6
   C. Phase Relations Involving a Crystalline Polymer Phase 9
III. Fractionation Theory 10
   A. The Flory–Huggins Theory of Polymer Solutions 10
   B. Equilibrium between Two Polymer Solutions 12
   C. Equilibrium between a Polymer Solution and a Gel 21
   D. Equilibrium between a Polymer Solution and a Crystalline or Semi-crystalline Polymer Phase 23
   E. Comparison of the Theoretical Solubility with Experiments 25
IV. Fractionation Efficiency 27
   A. General Considerations 27
   B. Definition of Fractionation Efficiency 29
   C. Dependence of Efficiency on Various Parameters 30
   D. Limitations and Extensions of the Theoretical Treatment 36
   E. Determination of Distribution Functions from Fractionation Data 39
   References 41

I. Polydispersity of High Polymers

A. POLYDISPERSITY AND FRACTIONATION

No synthetic high polymers are chemically pure substances in the strict sense. They are molecular mixtures of various components. This is also true for most natural high polymers.

Linear polymers, the molecules of which are composed of linearly connected identical units, are mixtures of linear homologous molecules of various chain lengths or molecular weights. Branched polymers, such as low-density polyethylene, have polydispersity with respect to molecular structure in addition to chain length polydispersity. In isotactic-block polymers, such as crystalline polypropylene, the lengths and the numbers of the sequences of d and l units are different for each molecule. Polydispersity with respect to chemical composition exists in copolymers, the molecules of which are composed of two or more chemically different types of structural units. The distribution of the units along a copolymer chain differs from molecule to molecule. A detailed discussion of these problems is presented in Chapter D of this book.

The polydispersity of linear homologous polymers can be defined by a *molecular weight distribution function*, $f(M_i)$, which gives the total mass of each molecular weight, M_i, in unit mass

$$\sum_i f(M_i) = 1 \tag{1}$$

Different experimental techniques measure different kinds of average molecular weights. The kinds of average which are most important in polymer chemistry are the following [1, 2]:

The *number average molecular weight*, $\overline{M}_n$, may be defined in either of the following ways:

$$\overline{M}_n = \frac{1}{\sum_i (1/M_i) f(M_i)} = \frac{\sum_i N_i M_i}{\sum_i N_i} \tag{2}$$

Here, N_i is the number of molecules of species i in the sample.

The *weight average molecular weight*, $\overline{M}_w$, is

$$\overline{M}_w = \sum_i M_i f(M_i) = \frac{\sum_i N_i M_i^2}{\sum_i N_i M_i} \tag{3}$$

The *z average* and the *(z + 1) average molecular weights* are,

$$\overline{M}_z = \frac{\sum_i M_i^2 f(M_i)}{\sum_i M_i f(M_i)} = \frac{\sum_i N_i M_i^3}{\sum_i N_i M_i^2} \tag{4}$$

and

$$\overline{M}_{z+1} = \frac{\sum_i M_i^3 f(M_i)}{\sum_i M_i^2 f(M_i)} = \frac{\sum_i N_i M_i^4}{\sum_i N_i M_i^3} \tag{5}$$

A. Theoretical Considerations

The *viscosity average molecular weight*, $\overline{M}_v$, is

$$\overline{M}_v = \left[\sum_i M_i^a f(M_i)\right]^{1/a} = \left(\frac{\sum_i N_i M_i^{a+1}}{\sum_i N_i M_i}\right)^{1/a} \tag{6}$$

where a is the exponent of the molecular weight in the intrinsic viscosity $[\eta]$ *versus* molecular weight relation

$$[\eta] = KM^a \tag{7}$$

The *integral distribution of molecular weights* $I(M_j)$ is the weight fraction of the molecules whose molecular weights are equal to or smaller than M_j.

$$I(M_j) = \sum_{i=1}^{j} f(M_i)$$

$$M_1 < M_2 < \ldots < M_j < \ldots \tag{8}$$

If a polymer is ideally monodisperse, the ratio $\overline{M}_w/\overline{M}_n$ is unity. This ratio, which increases in value as the polydispersity increases, is often used as a measure of the polydispersity of a polymer sample. Of course, for a fixed value of $\overline{M}_w/\overline{M}_n$, there can be an infinite number of different types of distribution.

Fractionation is carried out to reduce the polydispersity of high polymers, either for the purpose of evaluating the degree of polydispersity or for the preparation of samples of reduced polydispersity. The usual procedures for polymer fractionation involve the distribution of the polymer molecules between two phases. Fractionation results because of differences in the distribution for the molecules of different sizes or structures. This chapter deals with the theoretical considerations for polymer fractionation based on this principle.

B. Polydispersity and Polymer Properties

Before the detailed description of fractionation theories, we shall outline briefly some polymer properties which are closely related to the polydispersity. To deal with all aspects of this subject would be very difficult, so we shall limit our discussion to some of these which seem most important. We shall concern ourselves with homopolymers.

It has been found empirically that the viscosity of polymer melts is proportional to the 3.4 power of a certain type of average molecular weight, provided that that average lies above a critical value [3]. It is generally accepted that the type of average approximates the weight average $\overline{M}_w$. According to Bueche's theory [4], however, for a very wide distribution, the proper average is not $\overline{M}_w$ but a value somewhere between $\overline{M}_w$ and $\overline{M}_z$, with $\bar{M}_z$ a better approximation to the true average if $\bar{M}_w/\bar{M}_n > 2$. Care must be used in applying these relationships if there is any branching or if there are appreciable amounts of components of molecular weight lower than the critical value [4, 5].

Data obtained at low shear stress conform to the viscosity behavior just described. At high shear stress, the viscosity of a polymer melt decreases; in other words, it is non-Newtonian. The rate of viscosity reduction with the shear stress is enhanced by a broad molecular weight distribution [6–8].

Let us now consider a polymer melt in steady flow under a constant shear stress. If the stress is released at a time t, there is retarded elastic recovery. The steady state elastic compliance, which is the ratio of the recovered strain and the initial stress, is very sensitive to the higher molecular weight components of the polydisperse polymers. It was predicted theoretically [4] to be proportional to $\bar{M}_{z+1}\bar{M}_z/\bar{M}_w$, but experiments on polyisobutylene by Leaderman et al. [9] found it to be proportional, rather, to the square of $\bar{M}_{z+1}\bar{M}_z/\bar{M}_w$. At any rate, a polymer having a wide distribution of molecular weights shows a greater elastic effect than a homogeneous polymer with the same weight average molecular weight.

The retarded elastic recovery just described is one of the properties which are characteristic of viscoelastic materials such as polymer melts. The time-dependent phenomena of viscoelastic materials under the influence of an external force can be represented by a spectrum of relaxation times. The longer time part of that spectrum, for polymer melts, depends on the molecular weight and the molecular weight distribution [10, 11]. It is thus possible to estimate the distribution from viscoelastic data, a subject discussed in Chapter C.5 of this book.

The effects of the molecular weight distribution on mechanical properties have been examined for narrow and broad distribution polystyrene [12]. Tensile strength and elongation were found to depend upon a type of molecular weight average between the weight average and the number average, while the tensile or flux modulus was independent of both molecular weight and molecular weight distribution. Tensile properties of fractionated and unfractionated high-density polyethylene have been compared by Tung [13]. Rupture properties, such as total elongation, ultimate strength, and tensile impact strength, were found to be favored by high molecular weight and narrow distribution. On the other hand, yield strength and tensile moduli of high-density polyethylene were functions of the degree of crystallinity and independent of the molecular weight distribution.

The effects of molecular weight and its distribution on the deformation characteristics of polyethylene single crystals seem to be less important than the effects of defects in the crystals [14]. Crystal precipitation from dilute polyethylene solution has been studied morphologically for fractionated and unfractionated polyethylene [15, 16].

Spherulitic crystallization in crystalline high polymers has been attributed by Keith and Padden [17, 18] to polydispersity and high melt viscosity. Spherulites are formed by fibrous crystals radiating from their center. Continuing and profuse irregular branching of the radiating fibrous crystals leads to a spherulitic structure of relatively compact texture with radiating orientations of the crystalline subunits of each spherulite. The transformation from single crystal to spherulitic growth habit and the formation of fibrous crystalline structures were attributed to the effect of impurity segregation on the growing crystal faces. This effect also seems to play an important part in the irregular branching. Impurities considered to be effective in modifying the type of crystalline growth include low molecular weight molecules, branched molecules, and molecules of low tacticity. According to these considerations, polydispersity is thus one of the important factors in spherulitic crystallization.

In studies of solution properties of polymers, the effects of molecular weight distribution must be taken into consideration for properties related to intermolecular interactions, e.g., those which can be related to the magnitude of the second virial coefficient [19, 20]. In general, the effect of molecular weight distribution on these properties is negligible, except when very low polymer species are present.

II. Phase Relations for Polydisperse Systems

A. General Considerations

Theoretical considerations of fractionation based on different distributions of polymeric solutes of different molecular weight or molecular structure between two phases will be described in this and the following sections.

We first consider thermodynamic equilibria in a two-liquid-phase system containing a polydisperse polymer and either a single solvent or two solvents (one a good solvent and the other a poor solvent). A polymer solution, at first in a single phase, may separate into two phases as the temperature is lowered or as a poorer solvent is added. At equilibrium, the polymer distributions in the two phases differ from each other; i.e., one of the phases contains more of the higher molecular weight components than does the other phase. This fact is the basis for all two-liquid-phase fractionation processes.

In actual fractionations, thermodynamic equilibrium may in some cases be closely approached, but in other cases, a close approach to equilibrium distributions is very doubtful. Furthermore, in column elution fractionation, the surface of the column packing material may affect the fractionation efficiency. We shall consider primarily the ideal case of a system in complete equilibrium.

The thermodynamic state of a polymer solution can be described by the free energy of mixing ΔG. The solution is unstable and separates into two or more phases, if ΔG is not a monotonous function of the compositions. We consider a system of two phases, composed of n components $1, 2, \ldots$. For thermodynamic equilibrium, the chemical potential of mixing of the ith component $\Delta \bar{G}_i = \partial(\Delta G)/\partial n_i$ must have the same value in each phase, I and II.

$$\Delta \bar{G}_{1,\text{I}} = \Delta \bar{G}_{1,\text{II}}$$
$$\Delta \bar{G}_{2,\text{I}} = \Delta \bar{G}_{2,\text{II}}$$
$$\ldots \qquad \ldots \qquad (9)$$
$$\Delta \bar{G}_{n,\text{I}} = \Delta \bar{G}_{n,\text{II}}$$

Here, n_i is the number of molecules of the ith component.

We shall consider three types of two-phase equilibria: (1) equilibrium between two liquid solution phases, (2) equilibrium between a liquid solution and a gel, and (3) equilibrium between a liquid solution and a crystalline or semicrystalline solid phase. The first and second of these are frequently employed in separation on the basis of molecular weight. The third is of more importance if there are compositional and structural differences (including differences in stereoregularity) between the molecules.

B. Phase Relations for Polydisperse Polymer Solutions; Fractionation

Following Tompa [21], we consider a system composed of a solvent S and two homologous polymers P_1 and P_2 of different chain lengths. The phase diagram of such a system can be represented on a triangular diagram (Fig. 1). At first the amounts of P_1 and P_2 are considered to be independent. Binodials corresponding to several temperatures are shown on the diagram. When the system is held at a fixed temperature and represented by a point inside the binodial at that temperature, a single-phase solution is unstable and separates into two liquid phases (or, under certain circumstances, into one liquid and one gel phase). The points representing the compositions of the coexisting phases are on the binodial. The straight line joining each pair of these points is called a tie line. At a temperature for which the system

A. Theoretical Considerations

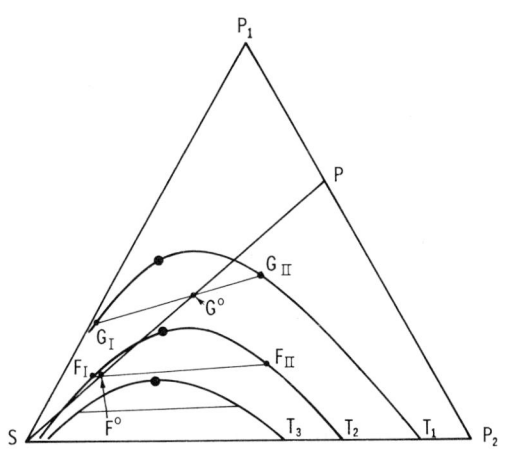

FIG. 1. A schematic phase diagram for a system consisting of a solvent (S) and two homologous polymers (P_1, P_2) of different chain lengths. The composition of the polymer mixture is indicated by P. Overall compositions of two solutions are represented by F^0 and G^0. F_I and F_{II} represent the compositions of the two coexisting phases, if the overall composition is F^0 and the temperature is T_2. Binodials for three temperatures (T_1, T_2, T_3) are shown. Critical points are indicated by small circles. This figure is adapted from one published by Tompa [21].

is completely miscible (for any relative amounts of P_1 and P_2), the binodial is absent. On lowering the temperature, the system becomes immiscible, in certain ranges of composition, as indicated by the appearance of a binodial. As the temperature is further decreased, the two-phase region gradually expands. (See the binodials for temperatures T_1, T_2, and T_3.) The limiting composition (at a given temperature), at which the two coexisting phases have identical compositions, is called a critical point. The relative amounts of P_1 and P_2 in a sample being fractionated are represented in the diagram by a point P on the line P_1P_2. Let the composition of a solution of this polydisperse polymer be represented by a point F^0 on the line SP. As long as the temperature is such that the binodial does not cross the line SP, a single-phase solution is stable. With decreasing temperature, the binodial reaches and crosses the line SP. The tie line through F^0 at the temperature represented by curve T_2 is $F_I F_{II}$. The ratio P_1/P_2 at the point F_{II} is markedly different from the overall ratio at F^0. In each step of a series of successive fractional precipitations, the concentrated phase is separated as a fraction. The ratio of volumes of the dilute and concentrated phases is given by $F^0 F_{II}/F_I F^0$. The remaining dilute solution, of composition F_I, is further fractionated at a lower temperature. In the case of successive fractional solution, if the composition of the solution is given by G^0, the dilute phase having composition G_I is separated as a fraction and the remaining solution has the composition represented by G_{II}.

In actual high polymers obtained by presently known methods, the number of polymer components is a tremendous number n, often approaching the number of mers in the component of highest molecular weight. The phase diagram for a solution of $n + 1$ components (a solvent and n polymer components) is a diagram in an n-dimensional polyhedron in n-dimensional space. Coexisting phases lie on an $(n - 1)$-dimensional binodial surface. The polymer of the given distribution is represented by a point P on the face of the polyhedron opposite the solvent corner S. The composition of the solution is represented by a point on the line PS. Fractionation theory must treat the distribution of the components between the two coexisting phases.

In the case of a system consisting of a good solvent, a poor solvent, and a polydisperse polymer, the composition of the solution of the given polymer is expressed by a point F^0 on a plane determined by the good solvent corner S, the poor solvent corner N, and the point P representing the composition of the given polymer. A simplified case in which the system is composed of a good solvent, a poor solvent, and two homologous polymers of different chain lengths is shown in Fig. 2.

The fractionation effect is not appreciable near the critical point where the two coexisting phases have similar compositions. The conditions deter-

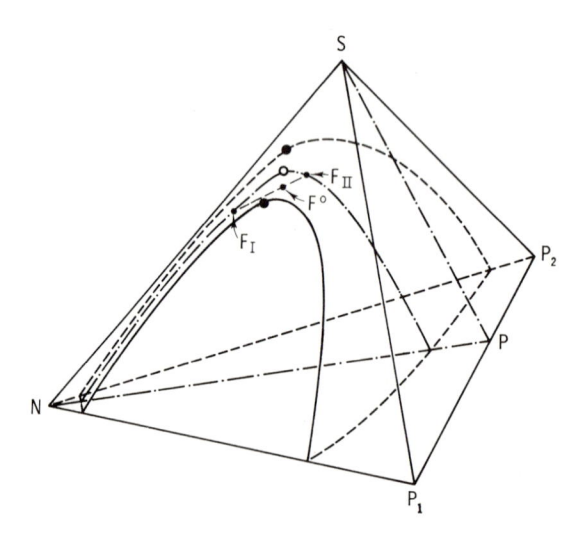

FIG. 2. A schematic phase diagram for a system: good solvent, S; poor solvent, N; polymer, P_1; polymer, P_2. The composition of a certain polymer mixture is given by the location of point P and that of its solution by F^0. F_I and F_{II} designate the compositions of the two coexisting phases. One binodial surface is shown. ○, Critical point for solutions of the polymer mixture of composition P; ●, critical points for solutions of polymers P_1 and P_2.

mining the critical points of an $n + 1$ component system, sometimes referred to in fractionation theory, have been given by Gibbs [22, 23].

For a system of $n + 1$ components under a constant pressure, the independent variables are the concentrations of the n components and the

$$\begin{vmatrix} \dfrac{\partial^2 \Delta G}{\partial n_1^2} & \dfrac{\partial^2 \Delta G}{\partial n_1 \partial n_2} & \cdots \\ \dfrac{\partial^2 \Delta G}{\partial n_1 \partial n_2} & \dfrac{\partial^2 \Delta G}{\partial n_2^2} & \cdots \\ \cdots & \cdots & \end{vmatrix} = U = 0 \tag{10}$$

$$\begin{vmatrix} \dfrac{\partial U}{\partial n_1} & \dfrac{\partial U}{\partial n_2} & \cdots \\ \dfrac{\partial^2 \Delta G}{\partial n_1 \partial n_2} & \dfrac{\partial^2 \Delta G}{\partial n_2^2} & \cdots \\ \cdots & \cdots & \end{vmatrix} = V = 0 \tag{11}$$

temperature. There are $n + 1$ independent variables for two equations; therefore, there remain $n - 1$ degrees of freedom for the critical phase. In the system of a single solvent and a polymer having a given distribution of n components, however, another restriction is required in addition to Eqs. (10) and (11). This restriction is that the concentrations of the polymer components in the solution must be proportional to their amounts in the original polymer. The number of the equations needed to specify this restriction is $n - 1$. As a result, there remains no degree of freedom for independent variables and the critical phase is represented by a single point for a (single solvent)–(polydisperse polymer) system.

In a system consisting of two solvents and a polydisperse polymer, one degree of freedom remains. One can properly speak of a "critical temperature" or "critical composition" only if the value of an additional parameter (e.g., the ratio of the amounts of the two solvents) is specified.

C. Phase Relations Involving a Crystalline Polymer Phase

The phase relations we have heretofore considered are for liquid–liquid (or liquid–gel) phase equilibria, applicable to solutions of polymers which do not crystallize on separation from solution. With some types of polymer, however, a solid crystalline phase may separate out from the liquid solution phase (Flory [2, p. 575]). Phase diagrams for typical systems, each consisting of a monodisperse crystalline polymer and a single solvent are shown in Fig. 3. From a single liquid phase, the crystalline phase separates at compositions and temperatures indicated by the lines *AB* and *DE*. The loop *BCD* is the boundary of the region within which a single phase is

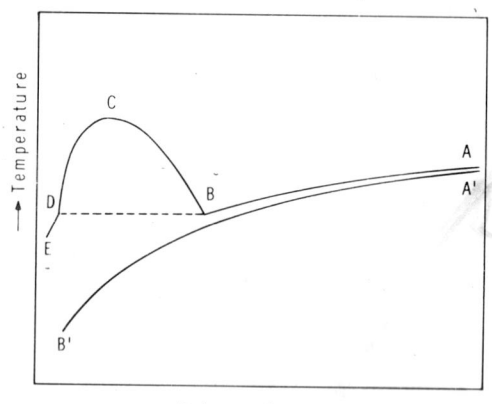

FIG. 3. Schematic phase diagrams for two solutions of a monodisperse polymer. The lower curve indicates the phase boundary between the liquid phase (above) and the crystalline phase (below), for a solution for which there is no two-liquid stability region. The upper curve is for a solution which is in equilibrium with a crystalline phase along the lines *AB* and *DE*, but which separates into two liquid phases if the point representing the temperature and overall composition is within the loop *BCD*. Point *C* is the critical point.

unstable. *C* designates the critical point. For a single good solvent, no two-liquid-phase region appears. The phase boundary between the crystalline phase and the solution phase is represented by the line *A'B'*. Whether a two-liquid-phase region exists or not for a given solvent is determined largely by its solvent power or, more precisely, by the relative solvent–polymer, solvent–solvent, and polymer–polymer interactions. A convenient measure of the "solvent power" is the interaction coefficient, χ^0, to be discussed below.

If crystalline polymers have polydispersity of molecular structure (branching, tacticity, etc.) the structural effect is usually minor in liquid–liquid or liquid–gel separations, but very important in the liquid–crystal separations. This subject will be dealt with in a later section.

III. Fractionation Theory

A. The Flory–Huggins Theory of Polymer Solutions

From appropriate models of polymer solutions, the free energy of mixing ΔG has been calculated in a number of ways. The phase relationships of the

A. Theoretical Considerations

polymer solutions can be derived from the analytical expression of ΔG by using the thermodynamical relations of Eq. (9). A general treatment is necessarily very involved mathematically, but for most purposes it suffices to use the relatively simple equations and concepts of the Flory–Huggins theory [2,24–30].

According to this theory, the partial molal Gibbs free energy of mixing of the solvent, $\Delta \bar{G}$, in a solution of linear polydisperse (in molecular size) polymer of a single type and a single solvent, is given by the equation

$$\Delta \bar{G}_1 = \mu_1 - \mu_1^0 = RT[\ln(1-\varphi) + (1-\bar{x}_n^{-1})\varphi + \chi^0 \varphi^2 \\ + \chi'\varphi^3 + \chi''\varphi^4 + \ldots] \qquad (12)$$

Here, μ_1 and μ_1^0 are the chemical potentials of the solvent, per mole, in the solution and in the pure solvent; R is the molal gas constant; T is the absolute temperature; $\bar{x}_n$ is defined as $\bar{V}_{n,x}/\bar{V}_1$, the ratio of the (number average) partial molal volume of the polymer to that of the solvent; φ is the volume fraction of polymer in the solution; and χ^0, χ', χ'', etc., are "interaction coefficients." These coefficients depend [24,29–31] on the types of solute and solvent and on the temperature, but not on the concentration, nor (appreciably) on the molecular weight of the polymer, provided it is high. To avoid complicated mathematics, we shall (like others who have dealt with this problem) at first (Section III,B,1) make the simplifying assumption that $\chi'\varphi^3$ and higher terms are negligible. Then, in Section III,B,2, we shall show the effect of including these higher terms.

The first two terms on the right side of Eq. (12) arise from the configurational entropy of mixing of the two components, polymer and solvent. The other terms allow for the change in the total energy (and enthalpy) of the system, when the components are mixed. This change results from changes in the relative intermolecular contact areas for solvent–solvent, polymer–polymer, and solvent–polymer contact. The later terms also include contributions resulting from the change from the randomness (hence the entropy) of orientation of each rigid segment in the polymer chain, relative to the preceding segment. The effects on energy and entropy, due to departures from perfect randomness of mixing in the solution, and some other usually neglected factors are also included.

For a given type of polymer, the values of the interaction coefficients χ^0, χ', etc., can be varied by varying the solvent or by varying the temperature. With regard to χ^0, the relationship

$$\chi^0 = a_\chi + (b_\chi/T) \qquad (13)$$

where a_χ and b_χ are constants, appears to hold sufficiently accurately for most purposes [2,24,30]. The two terms, a_χ and b_χ/T, can be roughly identified with the entropy and enthalpy contributions to χ^0, designated by Flory [2] as $\frac{1}{2} - \psi_1$ and κ_1, respectively.

Although the Flory–Huggins equations were originally derived for strictly linear homopolymers, they can also be used, with discretion, for branched polymers (with short branches) and for copolymers, provided $\bar{x}$ is appropriately averaged and χ^0, χ', etc., appropriately changed [32].

B. Equilibrium between Two Polymer Solutions

1. *A Polydisperse Homologous Polymer in a Single Solvent, Neglecting $\chi'\varphi^3$ and Higher Terms*

Using subscripts I and II to designate quantities referring to the more dilute and more concentrated phases, respectively, the chemical potentials of the solvent in these phases (neglecting the terms in the cube and higher powers of the concentration) are

$$\mu_{1,\text{I}} = \mu_1^0 + RT[\ln(1 - \varphi_\text{I}) + (1 - \bar{x}_{n,\text{I}}^{-1})\varphi_\text{I} + \chi^0 \varphi_\text{I}^2] \tag{14}$$

$$\mu_{1,\text{II}} = \mu_1^0 + RT[\ln(1 - \varphi_\text{II}) + (1 - \bar{x}_{n,\text{II}}^{-1})\varphi_\text{II} + \chi^0 \varphi_\text{II}^2] \tag{15}$$

The chemical potentials of the polymer species of relative size x in the two phases are

$$\mu_{x,\text{I}} = \mu_x^0 + RT[\ln \varphi_{x,\text{I}} - (x - 1) + x(1 - \bar{x}_{n,\text{I}}^{-1})\varphi_\text{I} + \chi^0 x(1 - \varphi_\text{I})^2] \tag{16}$$

$$\mu_{x,\text{II}} = \mu_x^0 + RT[\ln \varphi_{x,\text{II}} - (x - 1) + x(1 - \bar{x}_{n,\text{II}}^{-1})\varphi_\text{II} + \chi^0 x(1 - \varphi_\text{II})^2] \tag{17}$$

Equations for $\chi^0 = \chi_c^0$ and $\varphi_\text{I} = \varphi_\text{II} = \varphi_c$ at the critical point for phase separation can be deduced from Eqs. (10) and (11):

$$\chi_c^0 = \frac{1}{2}\left(1 + \frac{1}{\bar{x}_w^{1/2}}\right)^2 + \frac{(\bar{x}_z^{1/2} - \bar{x}_w^{1/2})^2}{2\bar{x}_w \cdot \bar{x}_z^{1/2}} \tag{18}$$

$$\varphi_c = \frac{1}{1 + (\bar{x}_w/\bar{x}_z^{1/2})} \tag{19}$$

Here $\bar{x}_w$ and $\bar{x}_z$ are the weight and z average of x, respectively [33]. For a monodisperse polymer these equations reduce to the following:

$$\chi_c^0 = \frac{1}{2}(1 + x^{-1/2})^2 \tag{20}$$

$$\varphi_c = 1/(1 + x^{1/2}) \tag{21}$$

The critical temperature is related [see Eq. (13)] to the critical value of χ^0 by the equation

$$T_c = b_\chi/(\chi_c^0 - a_\chi) \tag{22}$$

Note that if b_χ is negative, as it is for a solution of polymer in a good solvent, a finite (positive) critical temperature (and hence a region of stable

A. Theoretical Considerations

coexistence of two phases) can exist only if the magnitude of a_χ is greater than χ_c^0.

By equating the chemical potentials, we obtain the following relations between the concentrations of polymer species x in the two phases, $\varphi_{x,\text{I}}$ and $\varphi_{x,\text{II}}$

$$\varphi_{x,\text{I}}/\varphi_{x,\text{II}} = \exp[-\sigma_s^0 x] \tag{23}$$

$$-\sigma_s^0 = (1 - \bar{x}_{n,\text{II}}^{-1})\varphi_{\text{II}} - (1 - \bar{x}_{n,\text{I}}^{-1})\varphi_{\text{I}} + \chi^0[(1 - \varphi_{\text{II}})^2 - (1 - \varphi_{\text{I}})^2] \tag{24}$$

Equation (23) is one of the bases of fractionation theory. Because σ_s^0 is positive and not a function of x, the difference between the concentrations of the polymer species in the two phases progressively increases with increase of x. The polymer species of higher x are selectively transferred to the concentrated phase. For the simplified case of infinite molecular weight, Scott [34] calculated the values of σ_s^0 as a function of χ^0. It is seen in Fig. 4 that σ_s^0 increases with increase of χ^0, in other words, with the decrease of solvent power.

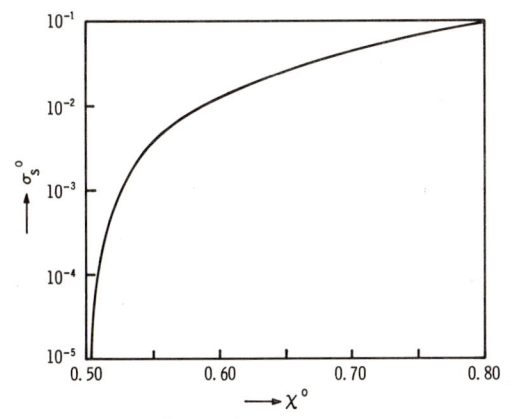

Fig. 4. The relation between σ_s^0 and χ^0 under the assumption of infinite molecular weight (from Scott [34]).

For a monodisperse polymer

$$\bar{x}_{n,\text{I}} = \bar{x}_{n,\text{II}} = x \tag{25}$$

and Eqs. (23) and (24) suffice to determine φ_{I} and φ_{II} for given values of x and χ^0. Figure 5 shows phase equilibrium curves obtained in this way. It can be shown that these curves are also very nearly correct for polydisperse polymers (with $x = \bar{x}_n^0$, the number average value of x), if the spread of molecular sizes is not too great.

The fraction of the total volume which is in phase II is related to the overall polymer concentration, φ^0, and the concentrations of polymer in the

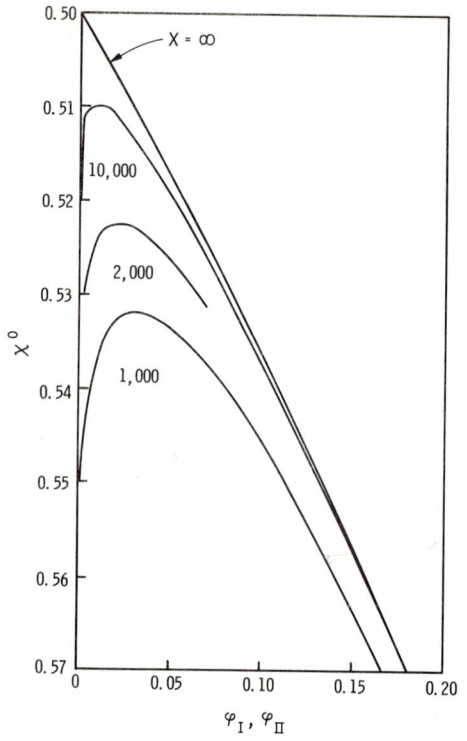

FIG. 5. Phase equilibrium curves for the values of x indicated. If a horizontal line, corresponding to a given value of the interaction coefficient χ^0, intersects one of these curves at two points, the abscissa at the left-hand intersection is φ_I, the polymer concentration in phase I, and the abscissa at the right-hand intersection is φ_{II}, the polymer concentration in phase II.

two phases according to the equation

$$\frac{V_{II}}{V^0} = 1 - \frac{V_I}{V^0} = \frac{\varphi^0 - \varphi_I}{\varphi_{II} - \varphi_I} \tag{26}$$

The fraction of polymer which is in phase II is

$$f_{II} = \frac{\varphi_{II}}{\varphi^0} \frac{V_{II}}{V^0} = \frac{\varphi_{II}}{\varphi^0}\left(\frac{\varphi^0 - \varphi_I}{\varphi_{II} - \varphi_I}\right) \tag{27}$$

Figure 6 shows how f_{II} varies with φ^0 for two χ^0 values, with x (or $\bar{x}^0$) = 1000.

The value of the overall concentration, φ^0, corresponding to a given value of f_{II} can be computed from the relationship

$$\varphi^0 = \frac{\varphi_I}{1 - [f_{II}(\varphi_{II} - \varphi_I)/\varphi_{II}]} \tag{28}$$

A. THEORETICAL CONSIDERATIONS

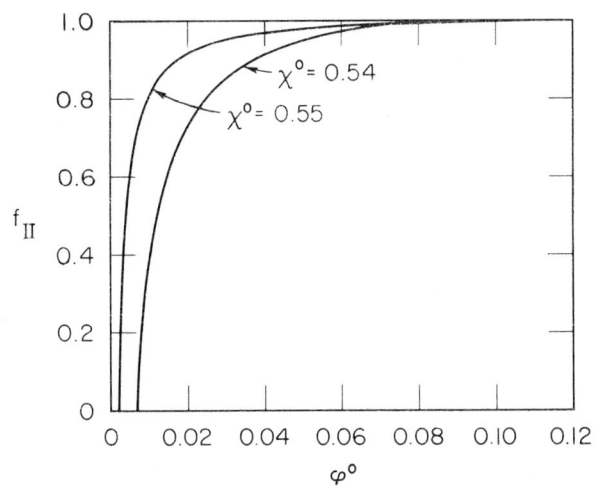

FIG. 6. The fraction of polymer in the more concentrated phase (II), as a function of the overall polymer concentration φ^0, for $x = 1000$ and the values of χ^0 indicated.

The φ_I values for this are obtainable from graphs like Fig. 5 or by computation, as described above.

For a mixture of polymers of the same type but different sizes, we can compute the fraction of polymer of a given size, x, in each phase by means of the following relationship, deduced from Eq. (23):

$$f_{xI} = 1 - f_{xII} = \frac{1}{1 + (V_{II}/V_I)\exp(-\sigma_s^0 x)} \qquad (29)$$

where σ_s^0 is given by Eq. (24). (If the difference between the average sizes of the polymer in the two phases is not very large, little error is introduced by putting $\bar{x}_{n,I} = \bar{x}_{n,II} = \bar{x}_n^0$.)

2. *A Polydisperse Homologous Polymer in a Single Solvent, Including Terms in Higher Powers of the Concentration*

In Section B,1, the effects of terms (in the expressions for the chemical potentials) in powers of the polymer concentration higher than the second have been neglected. Actually, these effects are far from negligible. Much of the reported poor agreement between experimental and theoretical phase diagrams [35], for example, can be attributed to this neglect. For quantitative treatment of fractionation equilibria and efficiencies, it is essential, in many cases at least, to include at least one higher concentration term.

Equations (14) and (15) should be expanded to the following:

$$\mu_{1,\text{I}} = \mu_1^0 + RT[\ln(1 - \varphi_\text{I}) + (1 - \bar{x}_{n,\text{I}}^{-1})\varphi_\text{I} + \chi^0\varphi_\text{I}^2 + \chi'\varphi_\text{I}^3 + \chi''\varphi_\text{I}^4 + \cdots] \tag{30}$$

$$\mu_{1,\text{II}} = \mu_1^0 + RT[\ln(1 - \varphi_\text{II}) + (1 - \bar{x}_{n,\text{II}}^{-1})\varphi_\text{II} + \chi^0\varphi_\text{II}^2 + \chi'\varphi_\text{II}^3 + \chi''\varphi_\text{II}^4 + \cdots] \tag{31}$$

Equations (16) and (17) should be replaced by

$$\mu_{x,\text{I}} = \mu_x^0 + RT[\ln \varphi_{x,\text{I}} - (x - 1) + x(1 - \bar{x}_{n,\text{I}}^{-1})\varphi_\text{I} + \chi^0 x(1 - 2\varphi_\text{I} + \varphi_\text{I}^2)$$
$$+ \tfrac{1}{2}\chi' x(1 - 3\varphi_\text{I}^2 + 2\varphi_\text{I}^3) + \tfrac{1}{3}\chi'' x(1 - 4\varphi_\text{I}^3 + 3\varphi_\text{I}^4) + \cdots] \tag{32}$$

$$\mu_{x,\text{II}} = \mu_x^0 + RT[\ln \varphi_{x,\text{II}} - (x - 1) + x(1 - \bar{x}_{n,\text{II}}^{-1})\varphi_\text{II} + \chi^0 x(1 - 2\varphi_\text{II} + \varphi_\text{II}^2)$$
$$+ \tfrac{1}{2}\chi' x(-3\varphi_\text{II}^2 + 2\varphi_\text{II}^3) + \tfrac{1}{3}\chi'' x(1 - 4\varphi_\text{II}^3 + 3\varphi_\text{II}^4) + \cdots] \tag{33}$$

The critical concentration is given, more precisely than by Eq. (19), by

$$\varphi_c = 1/(1 + f_c^{-1/2}) \tag{34}$$

with

$$f_c = (1/x) + 3\chi'_c\varphi_c^2 + 8\chi''_c\varphi_c^3 \tag{35}$$

For given values of χ'_c, χ''_c, etc., the critical value of χ_c^0 can be computed from the relationship

$$\chi_c^0 = \frac{1}{2(1 - \varphi_c)^2} - 3\chi'_c\varphi_c - 6\chi''_c\varphi_c^2 \cdots$$

$$= \frac{f_c(1 + f_c^{-1/2})^2}{2} - \frac{3\chi'_c}{1 + f_c^{-1/2}} - \frac{6\chi''_c}{(1 + f_c^{-1/2})^2} \cdots \tag{36}$$

The exponential relation between the concentrations of the two phases, Eq. (23), holds, but the σ_s^0 in that equation should be replaced by

$$\sigma_s = \sigma_s^0 + \chi'[(\tfrac{3}{2}\varphi_\text{I}^2 - \varphi_\text{I}^3) - (\tfrac{3}{2}\varphi_\text{II}^2 - \varphi_\text{II}^3)]$$
$$+ \chi''[(\tfrac{4}{3}\varphi_\text{I}^3 - \varphi_\text{I}^4) - (\tfrac{4}{3}\varphi_\text{II}^3 - \varphi_\text{II}^4)] + \cdots \tag{37}$$

To illustrate the nature and magnitude of the effect of including the terms $\chi'\varphi^3$ and $\chi''\varphi^4$ in the thermodynamic equations [Eq. (12), etc.], we show in Fig. 7 portions of the phase diagrams, with and without these terms, for $x = 2000$. For χ' and χ'' constant values of $\tfrac{1}{3}$ and 0.075 are assumed, and higher terms are assumed to be negligible. (These values of χ' and χ'' have been obtained by Krigbaum and Geymer [36] for the polystyrene–cyclohexane system. Similar values of χ' have been obtained by others for several other systems.)

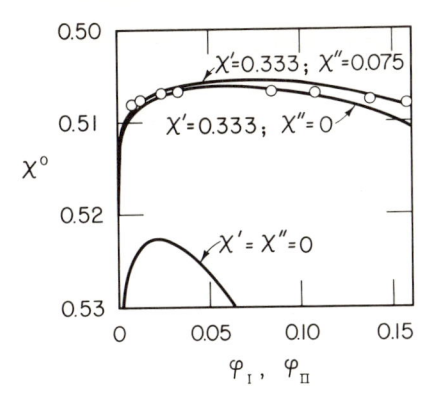

FIG. 7. Phase equilibrium curves for $x = 2000$, illustrating the effect of omitting or including the $\chi'\varphi^3$ and $\chi''\varphi^4$ terms. The small circles represent experimental data on polystyrene in cyclohexane, by Bergsnov-Hansen and Brady [37].

The inclusion of the additional terms changes the critical concentration φ_c and the critical value of χ^0 (hence the critical temperature) markedly and also spreads out the coexistence curve, in accordance with the observations of Shultz and Flory [35], Bergsnov-Hansen and Brady [37], and others (see Tompa [21, Section 7.7]).

3. A Polydisperse Polymer in a Mixture of Two Solvents

In many fractionations a mixture of a good solvent with a poor solvent is used. Different fractions can be separated by changing the proportions of the two solvents, instead of changing the temperature, as in the case of a single solvent. A precise theoretical treatment of the equilibrium relationships for two-solvent systems is very complicated [2,21,23]. We shall merely outline it here and shall neglect the higher concentration terms involving χ', χ'', etc. The principles involved are essentially the same as for a system consisting of a polydisperse polymer and a single solvent.

At first sight, one might consider it legitimate to treat a two-solvent system like a single-solvent system, using an interaction coefficient for the mixture computed from the proportions of the two solvents and the interaction coefficients for solutions of the polymer in each of them separately. It can readily be shown, however, that this is not justifiable [23]. The proportions of the two solvents in the two phases are, in general, very different, and the molecular weight distributions in these phases are therefore very different from what they would be if the solvent ratios were the same in both phases.

Let us represent quantities or functions characteristic of the better solvent, the poorer solvent, and the polymer by symbols with subscripts 1,

2, and 3, respectively, and let us use x_2 and x to represent partial molal volume ratios, as follows:

$$x_2 = \overline{V}_2/\overline{V}_1, \qquad x = \overline{V}_x/\overline{V}_1 \tag{38}$$

Then, for a solution of a polydisperse polymer in a mixture of the two solvents, the chemical potentials of the components are given by the equations

$$\mu_1 - \mu_1^0 = RT[\ln(1 - \varphi_2 - \varphi_3) + a_2\varphi_2 + a_3\varphi_3 + a_{22}\varphi_2^2 + a_{33}\varphi_3^2 + a_{23}\varphi_2\varphi_3] \tag{39}$$

$$\mu_2 - \mu_2^0 = RT[\ln \varphi_2 + b_0 + b_2\varphi_2 + b_3\varphi_3 + b_{22}\varphi_2^2 + b_{33}\varphi_3^2 + b_{23}\varphi_2\varphi_3] \tag{40}$$

$$\mu_x - \mu_x^0 = RT[\ln \varphi_x + c_0 + c_2\varphi_2 + c_3\varphi_3 + c_{22}\varphi_2^2 + c_{33}\varphi_3^2 + c_{23}\varphi_2\varphi_3] \tag{41}$$

$$a_2 = 1 - \frac{1}{x_2} \qquad a_3 = 1 - \frac{1}{\bar{x}_n} \qquad a_{22} = \chi_{12}^0$$

$$a_{33} = \chi_{13}^0 \qquad a_{23} = \chi_{12}^0 + \chi_{13}^0 - \frac{\chi_{23}^0}{x_2}$$

$$b_0 = 1 - x_2 + x_2\chi_{12}^0 \qquad\qquad b_2 = x_2 - 1 - 2x_2\chi_{12}^0$$

$$b_3 = x_2 - \frac{x_2}{\bar{x}_n} - x_2\chi_{12}^0 + \chi_{23}^0 - x_2\chi_{13}^0 \qquad b_{22} = x_2\chi_{12}^0 \tag{42}$$

$$b_{33} = x_2\chi_{13}^0 \qquad\qquad b_{23} = x_2\chi_{12}^0 - \chi_{23}^0 - x_2\chi_{13}^0$$

$$c_0 = 1 - x + x_2\chi_{13}^0 \qquad c_2 = x - \frac{x}{x_2} - x\chi_{13}^0 + \frac{x}{x_2}\chi_{23}^0 - x\chi_{12}^0$$

$$c_3 = x - \frac{x}{\bar{x}_n} - 2x\chi_{13}^0 \qquad c_{22} = x\chi_{12}^0$$

$$c_{33} = x\chi_{13}^0 \qquad c_{23} = x\chi_{13}^0 - \frac{x}{x_2}\chi_{23}^0 + x\chi_{12}^0$$

The χ^0's refer to the interaction coefficients for the pairs of components indicated by the subscripts. These relationships are the extended forms of those given by Flory [2, p. 549].

A. THEORETICAL CONSIDERATIONS

With the foregoing equations one can compute the chemical potentials as functions of the concentrations and constants characteristic of the individual components and interaction coefficients for pairs of the components. The interaction coefficients χ_{13} and χ_{23} are readily measured experimentally, and χ_{12} can be estimated from Hildebrand's solubility parameters [38]

$$\chi_{12} = (\overline{V}_1/RT)(\delta_1 - \delta_2)^2 \tag{43}$$

The critical concentration for sufficiently high molecular weight polymer is calculated [39] from Eqs. (10), (11), and (39)–(42) to be, approximately,

$$\varphi_c = C(\bar{x}_z^{1/2}/\bar{x}_w) \tag{44}$$

where the factor C is a complicated function of the various interaction coefficients. Numerical values of C have been given by Okamoto for various solvent mixtures [39].

It is interesting to compare the concentration of the critical point, at which no fractionation occurs, for a polymer solution in a single solvent Eq. (19) with that of the polymer solution in a solvent mixture Eq. (44). According to both equations, the concentration is proportional to $\bar{x}_z^{1/2}/\bar{x}_w$ for sufficiently high molecular weight. For a single-solvent system, the critical concentration is independent of the solvent used, provided the molal volume is not varied. With a mixture of solvents, the critical concentration is dependent on the solvent composition. This fact suggests a relationship between the fractionation efficiency and the solvent composition. No study of such a relationship has, to our knowledge, been reported.

The following relation between the concentrations of the polymer of size x in the two phases is derived by equating the chemical potentials [39,40].

$$\varphi_{x,\text{I}}/\varphi_{x,\text{II}} = \exp[-\sigma_m^0 x] \tag{45}$$

$$\sigma_m^0 = (1 - x_2^{-1} - \chi_{13}^0 + x_2^{-1}\chi_{23}^0 - \chi_{12}^0)(\varphi_{2,\text{II}} - \varphi_{2,\text{I}}) + (1 - 2\chi_{13}^0)$$
$$\times (\varphi_{3,\text{II}} - \varphi_{3,\text{I}}) + \chi_{12}^0(\varphi_{2,\text{II}}^2 - \varphi_{2,\text{I}}^2) + (\bar{x}_{n,\text{II}}^{-1}\varphi_{3,\text{II}} - \bar{x}_{n,\text{I}}^{-1}\varphi_{3,\text{I}})$$
$$+ \chi_{13}^0(\varphi_{3,\text{II}}^2 - \varphi_{3,\text{I}}^2) + (\chi_{13}^0 - x_2^{-1}\chi_{23}^0 + \chi_{12}^0)(\varphi_{2,\text{II}}\varphi_{3,\text{II}} - \varphi_{2,\text{I}}\varphi_{3,\text{I}}) \tag{46}$$

Equation (45), like the similar Eq. (23), can be considered a basic equation of fractionation theory.

If the coefficient χ_{12}^0, for interaction between the good and the poor solvent, is larger than $\frac{1}{2}[(1 + x_2^{-1/2})]^2$, the solvent mixture is not miscible for all concentrations. A mixture of polymer species will be distributed differently in the two phases, in accordance with Eqs. (45) and (46).

4. A Polydisperse Copolymer in a Single Solvent

Let us now consider solutions of a copolymer, the molecules of which are composed of mers of two types, distinguished by the subscripts a and b. Let there be n_a mers of type a and n_b mers of type b in a copolymer molecule. The ratio of the partial molal volume of the copolymer to that of the solvent is

$$x = (n_a v_a + n_b v_b)/v_1 \tag{47}$$

where v_a and v_b are the volume contributions of the two types of mers and v_1 is the volume of a solvent molecule. The values of n_a and n_b are spread over a range. Neglecting the higher concentration terms of various interaction coefficients, Kilb and Bueche [41] have treated the fractionation problem. The chemical potential of species x (composed of n_a a-mers and n_b b-mers) is

$$\mu_x - \mu_x^0 = RT[\ln \varphi_x + (1-x) + x(1-\bar{x}_n^{-1})\varphi + \frac{(\chi_a n_a v_a + \chi_b n_b v_b)}{v_1}$$
$$\times (1-\varphi)^2 + (\chi_a - \chi_b)\frac{(n_b v_b \varphi_a - n_a v_a \varphi_b)}{v_1}(1-\varphi) \tag{48}$$

where φ_x and φ are the volumes fractions of the species x and the copolymer as a whole, φ_a and φ_b are those of the two mers, χ_a and χ_b are the interaction coefficients between the solvent and the polymer molecules of a and b types, respectively. For equilibrium between the two solution phases,

$$\varphi_{x,\mathrm{I}}/\varphi_{x,\mathrm{II}} = \exp(-\sigma_a x_a - \sigma_b x_b] \tag{49}$$

where

$$x_a = (n_a v_a/v_1), \qquad x_b = (n_b v_b/v_1)$$

$$-\sigma_a = (1 - \bar{x}_{n,\mathrm{II}}^{-1})\varphi_{\mathrm{II}} - (1 - \bar{x}_{n,\mathrm{I}}^{-1})\varphi_{\mathrm{I}} + \chi_a[(1-\varphi_{\mathrm{II}})^2 - (1-\varphi_{\mathrm{I}})^2]$$
$$+ (\chi_b - \chi_a)[\varphi_{b,\mathrm{II}}(1-\varphi_{\mathrm{II}}) - \varphi_{b,\mathrm{I}}(1-\varphi_{\mathrm{I}})] \tag{50}$$

$$-\sigma_b = (1 - \bar{x}_{n,\mathrm{II}}^{-1})\varphi_{\mathrm{II}} - (1 - \bar{x}_{n,\mathrm{I}}^{-1})\varphi_{\mathrm{I}} + \chi_b[(1-\varphi_{\mathrm{II}})^2 - (1-\varphi_{\mathrm{I}})^2]$$
$$+ (\chi_a - \chi_b)[\varphi_{a,\mathrm{II}}(1-\varphi_{\mathrm{II}}) - \varphi_{a,\mathrm{I}}(1-\varphi_{\mathrm{I}})] \tag{51}$$

There is an infinite set of combinations x_a and x_b for a fixed value of $\sigma_a^0 x_a + \sigma_b^0 x_b$, hence the molecular size x is no longer the single factor which controls the fractionation. This is in marked contrast to the fractionation of homopolymers, using a single solvent.

In the foregoing treatment, the interactions between a given solvent and a copolymer molecule are assumed to be additive with respect to the numbers of the mers of the two types. The distribution of each mer along the molecular chain, e.g., a random or a blockwise distribution, is not taken into consideration. Precise treatment of these problems is much more complicated. For a discussion of these factors, a paper by Huggins [32] may be consulted. Further theoretical and experimental studies are both needed.

5. *A Mixture of Different Polymers in a Single Solvent or a Mixture of Two Solvents*

In the hypothetical case of a mixture of different types of polymers, with the molecules of each type all having the same molecular weight, the fractionation problem is the same as for fractionation of a mixture of low molecular weight compounds. For each polymer type, the size parameter x and the interaction coefficients have characteristic values. If these parameters are known, the chemical potentials for each type can be computed and the theory of phase separation and the distribution of the components between the phases can be developed in a manner analogous to that given above for separation on the basis of molecular size differences alone. Certain special cases have been dealt with by Scott [42], Flory [2], and Tompa [21]. A very small positive value of the interaction coefficient between polymer species of different types is sufficient to make a solution unstable over a large range of composition, causing it to separate into two phases.

If each type of polymer present has a spread of molecular sizes, the distributions of polymer and solvent molecules between the two phases, and so the fractionation behavior of the system, are complex functions of the x and χ values of the components. The solution of this problem, by methods analogous to those applied to a single polymer, is straightforward, though complicated.

C. Equilibrium between a Polymer Solution and a Gel

As the concentration of a polymer solution increases, the chains become more and more entangled with each other. The points of entanglement become effective cross-links, preventing independent motion of the individual molecules. (This effect may be dominant in fractionation by column elution methods.) Thermodynamically, the effect is the same as if actual chemical cross-links were formed; the average "molecular weight" and so the average value of x increases. The cross-links have the effect of changing phase II from a liquid to a semirigid gel.

If the cross-linking in a gel is not very extensive and the energy of cross-linking is merely the energy associated with replacing polymer–solvent contacts by polymer–polymer and solvent–solvent contacts, as in the solution, the chemical potentials of the components of the gel are essentially the same as for a solution of the same concentration. Equation (31) and (33), with x put equal to infinity, are applicable.

If some or all of the cross-links are strong, the elastic energy resulting from the swelling of the gel network may have a significant effect on the phase equilibria and the elastic properties. Flory [2, p. 577] has considered the case of swelling of a hypothetical model gel, containing a fixed number, v_e, of moles of cross-links in a volume (when the network is unswollen) V_{II}^0, with 4 chains connected together at each cross-link. Putting his result in our terminology and adding the term $\chi' \varphi_{II}^3$, we obtain for the chemical potential of the solvent in the gel

$$\mu_{1,II} = \mu_1^0 + RT\left[\ln(1 - \varphi_{II}) + \varphi_{II} + \chi^0 \varphi_{II}^2 + \chi' \varphi_{II}^3 + \cdots \right.$$
$$\left. + \bar{v}_1\left(\frac{v_e}{V_{II}^0}\right)\left(\varphi_{II}^{1/3} - \frac{\varphi_{II}}{2}\right)\right] \quad (52)$$

The corresponding equation for the chemical potential of species x in the gel is

$$\mu_{x,II} = \mu_x^0 + RT[\ln \varphi_{II} - (x - 1)(1 - \varphi_{II}) + \chi^0 x(1 - \varphi_{II})^2$$
$$+ \tfrac{1}{2}\{\chi' x\{1 - 3\varphi_{II}^2 + 2\varphi_{II}^3\} + \cdots - \tfrac{1}{2}x v_1(v_e/V_{II}^0)(1 - \varphi_{II})(2\varphi_{II}^{-2/3} - 1)] \quad (53)$$

In some gels used in fractionations, there may be both weak and strong cross-links. They will have fractionation properties intermediate between those derivable using Eqs. (52) and (53) and those obtained with similar equations omitting the last term in each. Only for large degrees of swelling and a high concentration of cross-links, however, would the elasticity terms be expected to have an appreciable effect on the fractionation behavior.

Treating the polymer molecules in the gel phase as all incorporated in an infinite network, as we have done, does not take account of the possible inclusion of some unincorporated polymer molecules within the pores of the gel. This neglect may also introduce some error, especially if some polymers of low molecular weight are present.

More thorough theoretical and experimental studies of gel–liquid equilibria in polymer systems are obviously desirable.

D. Equilibrium between a Polymer Solution and a Crystalline or Semicrystalline Polymer Phase

A second phase, separating out from a polymer solution, is sometimes a crystalline or semicrystalline solid phase. If so, the equilibrium relationships applicable to a system of two liquid phases or a liquid phase and a gel phase are not applicable. Fractionation may occur, but molecular weight will not be the principle factor determining the distribution equilibria. Departures from perfect regularity of structure, differing from molecule to molecule, will be much more important. Differences in the type or degree of branching, the type or degree of tacticity, or the sizes of the blocks in block copolymers, for example, produce differences in the ease with which molecules fit into a crystalline structure and so affect the phase distribution equilibria.

Let us first consider the equilibrium between a solution of a monodisperse polymer of regular (tactic) structure and a crystalline phase, containing no solvent. The chemical potential of the polymer in the crystalline phase is

$$\mu_{x,\text{cr}} = \mu_x^0 - (\Delta H_{m,x} - T_m \Delta S_{m,x}) \tag{54}$$

Here, $\Delta H_{m,x}$ and $\Delta S_{m,x}$ are the molal heat and entropy change on melting, respectively, and T_m is the (absolute) melting point. Using the relationship

$$T_m = \Delta H_{m,x}/\Delta S_{m,x} \tag{55}$$

we obtain

$$\mu_{x,\text{cr}} = \mu_x^0 - \Delta H_{m,x}[1 - (T/T_m)] \tag{56}$$

By equating the chemical potentials of the species in the solution and the crystalline phase, we deduce [16,27]

$$\frac{-\ln \varphi_1 + x - 1}{x} = \left(1 - \frac{1}{x}\right)\varphi_1 + \chi^0(1 - \varphi_1)^2 + \frac{\chi'}{2}(1 - 3\varphi_1^2 + 2\varphi_1^3)$$
$$+ \cdots + \frac{\Delta H_{m,x}}{RTx}\left(1 - \frac{T}{T_m}\right) \tag{57}$$

If the polymer molecules under consideration are perfectly linear, tactic, and sufficiently long, $\Delta H_{m,x}$ is quite closely proportional to the number of mers per molecule and to the molecular size parameter x [27,43] and T_m is also nearly independent of the size of the species [27,44]. Using these approximations, neglecting the higher terms in Eq. (57), and applying numerical values appropriate to a 0.1 wt % solution of polymethamer (linear polyethylene) in o-xylene, Wunderlich et al. [16] found T_m to be nearly independent of x provided the molecular weight is higher than 40,000. Therefore, for this system (and presumably for others as well), the phase separation between the solution (liquid) and the crystalline phase is not suitable for the fractionation of crystalline polymers on a molecular weight basis. Furthermore, in the actual polymer crystallizations, the mechanisms of crystal nucleation and growth and the difficulty of rearrangement of the molecules in the crystals must certainly make effective fractionation difficult or impossible.

Kawai [45] has extended Eq. (57) to polydisperse systems, using a simplified assumption with regard to the crystalline phase and considering the effects of molecular size, lamellae thickness, and the difficulties of attainment of thermodynamic equilibrium.

Actual high polymers of the usual types rarely, if ever, consist of molecules of perfectly regular structure and they rarely, if ever, separate out from solution as perfect crystals. Minor irregularities of molecular structure reduce the magnitude of $\Delta H_{m,x}$ and lower the melting point. The distribution of molecules of different degrees of irregularity between the solution phase and a crystalline phase will thus be different. Crystallization can thus be used, in principle, for structural fractionation.

With many polymeric systems, such as those of imperfect stereoregularity and copolymers in which sequences of one of the two or more types of mer are prone to aggregate into crystalline regions with other like sequences, the solid phase separating out from a solution contains both crystalline and amorphous regions. If the crystalline regions can be considered to be perfectly crystalline and if neither type of region contains solvent, the equilibrium behavior of the system can be roughly represented by Eq. (57), provided we insert a factor f_{cr}, to represent the degree of crystallinity, in the last term. This involves the approximation that the chemical potentials of the polymer molecules or parts of molecules in the noncrystalline parts of the solid phase are the same as in pure amorphous polymer.

In many solid polymers containing crystalline regions, the polymer chains in these regions are apparently frequently folded to give a lamellar type of structure [46,47]. The folding must obviously affect the free energy of the crystalline phase. If the molecules differ structurally, their abilities to conform to such a structure will vary, hence crystallization should produce fractionation on the basis of structure. Further discussion of the subject seems unwarranted at this time, because of the important differences of opinion existing with regard to the details of the structure of semicrystalline bulk polymers and even the factors affecting chain folding [46–50].

Assuming equilibrium conditions, whether a crystalline solid phase or a second liquid phase appears first, from a solution of a given solvent, as the temperature is lowered or the χ's otherwise altered, depends on which saturation limit is reached first (for any of the species present in the solution). If the spread of polymer sizes is small, it is probably sufficient to consider the polymer mixture, for this purpose, as if it consisted entirely of polymer having the average x value. From the equations given one can then (if the appropriate constants are known) determine the range of concentration and χ (or temperature) values within which a second liquid phase will appear first and the range within which a crystalline phase will first appear. (See Fig. 3.) In general, it is usually best, for a two-liquid-phase separation, to use a relatively poor solvent and, for a separation involving a liquid phase and a crystalline phase, to use a good solvent.

Several examples illustrating the above discussion may be cited. Fractionation phenomena during crystal precipitation from dilute linear polyethylene solution have been investigated by Kawai and Keller [51]. Very poor fractionation according to molecular weight was observed, as expected. Wijga et al. [52] fractionated polypropylene according to molecular weight, using a kerosene fraction (a good solvent) and butyl-carbitol (a poor solvent) at 150°C, where the liquid–liquid separation probably prevailed. They also fractionated the same polymer according to tacticity, using kerosene in the temperature range 30°–145°C, where the liquid–crystalline phase separation probably prevailed. Hawkins and Smith [53] fractionated polyethylene roughly on the basis of molecular structure, using a good solvent, xylene, in the temperature range 57°–104°C. In column elution fractionation, the requirements for selectivity according to molecular weight in depositing polypropylene on the column packing material were considered by Mendelson [54]. The abnormal reverse-order fractionation was observed when the polymer was deposited from a good solvent, but not when deposited from a mixture of a solvent and a poor solvent. In this experiment, the solvent mixture must have acted as a poor solvent and the polymer deposition must have been carried out according to molecular weight by the two-liquid-phase process.

An interesting procedure for fractionation by crystallization from a stirred solution has recently been reported by Koningsveld and Pennings [54a,54b]. The fractionation apparently results from the dependence of the interactions between polymer molecules, in a solution with a velocity gradient, on the sizes of the molecules. We shall not attempt to develop a quantitative treatment at this time.

E. Comparison of the Theoretical Solubility with Experiments

As we have seen, polymer solution theory, with or without the higher terms $\chi'\varphi^3$, etc., leads to relations (3), (37), and (45) between the concentrations of the species in the two coexisting phases. Equations (23) and (37) were derived for polymer solutions in a single solvent and Eq. (45) for solutions in a solvent mixture. It is interesting that all of these relations have similar exponential forms. Let R be the volume ratio of the dilute and the concentrated phase. The fraction of the species of size x in the concentrated phase is given by a function $K(R, \sigma, x)$ (Flory [2, p. 559])

$$K(R, \sigma, x) = \{1/[1 + R \exp(-\sigma x)]\}$$
$$\sigma = \sigma_s^0, \sigma_s, \quad \text{or} \quad \sigma_m^0 \tag{58}$$

The distribution function for the polymer in the concentrated phase is

$$f_{II} = (1/w)Kf(x) \tag{59}$$

where w is the weight fraction of the polymer in the concentrated phase.

The exponential solubility relation of Eq. (58) has often been used in fractionation problems [53–63]. An experiment to test this equation has been carried out by Okamoto and Sekikawa [64].

From Eq. (59)

$$w = 1 - R\sigma \int_0^\infty \frac{\exp[-\sigma x]}{\{1 + R \exp[-\sigma x]\}^2} I(x)\, dx \tag{60}$$

where $I(x)$ is the integral distribution of the original polymer. In Eq. (60), w, R, and $I(x)$ are measurable. If they are known, σ is the only unknown parameter. Solving Eq. (60) with respect to σ and substituting the solution into Eq. (59), they calculated the distribution of the polymer in the more concentrated phase. If the solubility relation is correct and there is no experimental error, the calculated and experimental distributions should coincide. Experiments on xylene, poly(ethylene glycol) of low molecular weight (400) and linear polyethylene showed small but nonnegligible discrepancies (Fig. 8).

The aim of this study was to check the simple solubility relation Eq. (45). No test of theoretical equations for σ_m^0 [e.g., Eq. (46)] was attempted.

Assuming a single effective interaction coefficient for a two-solvent system, Tung [65] found fair agreement between the simple theory and the results of fractionations which were made on the same system studied by Okamoto and Sekikawa. The effect on polymer species distribution of differences in solvent composition in the two phases was neglected, as were the effects of $\chi'\varphi^3$ and higher terms.

Bergsnov-Hansen and Brady [37] have measured phase equilibrium curves for solutions of polystyrene of narrow molecular weight distribution in cyclohexane, obtaining results which agree closely with the theoretical curve, *provided the $\chi'\varphi^3$ term is included.* (See Fig. 7.) (The magnitude of χ' was that reported by Krigbaum and Geymer [66].) Omission of this term leads to very poor agreement. Inclusion of still another term, $\chi''\varphi^4$, makes differences in the theoretical curves which are of the order of magnitude

A. Theoretical Considerations

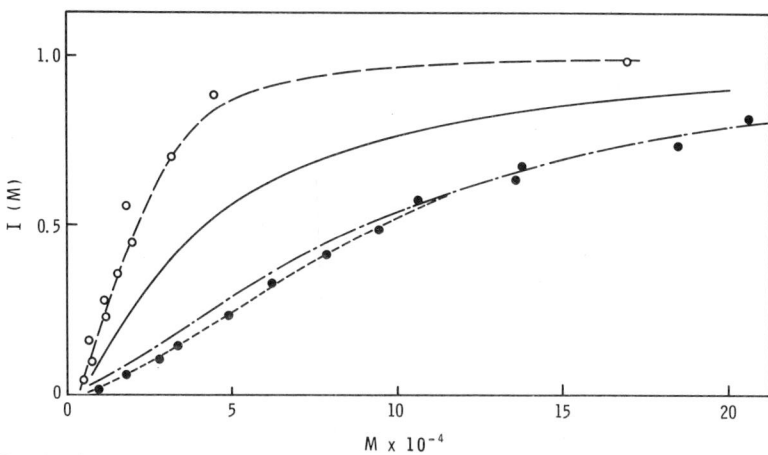

FIG. 8. The partitioning of the polydisperse polymer solution (xylene—poly(ethylene glycol) of molecular weight 400—linear polyethylene) between the two coexisting liquid phases: ———, the integral distribution of the unseparated polymer, – – – – –, the integral distribution of the concentrated phase calculated from the theoretical solubility relation, ● – – –, the experimental integral distribution of the polymer in the concentrated phase; ○ —— —, the experimental integral distribution of the polymer in the dilute phase (from Okamoto and Sekikawa [64]).

of the probable experimental error. On the basis of these results, one may state confidently that the primary reason for the poor agreement between theory and experiment, reported by Shultz and Flory and others, is the neglect of the higher terms in the theoretical equations. As a corollary, we can reasonably use the theoretical equations, with the higher terms, as a sound quantitative basis for deducing relations between the magnitudes of the parameters in liquid–liquid systems used for fractionations.

IV. Fractionation Efficiency

A. General Considerations

In this section we shall deal only with fractionation by molecular weight. In an ideal fractionation, the polymer species of molecular weights higher than a certain value are all in the more concentrated phase and those of lower molecular weight are all in the more dilute phase. Actual fractionations are all far from ideal. It is important to learn how to make actual fractionation approach the ideal as closely as possible.

The fraction of the species of size x in the concentrated phase is given by the function $K(R, \sigma, x)$ defined by Eq. (58). The function $K(R, \sigma, x)$ is $1/(1 + R)$ at $x = 0$ and gradually increases to unity as x increases. An inflection point exists at $x = (\ln R)/\sigma$, at which point K is $\frac{1}{2}$. The gradient $dK/d(\ln x)$ at the inflection point, which may be considered as a measure

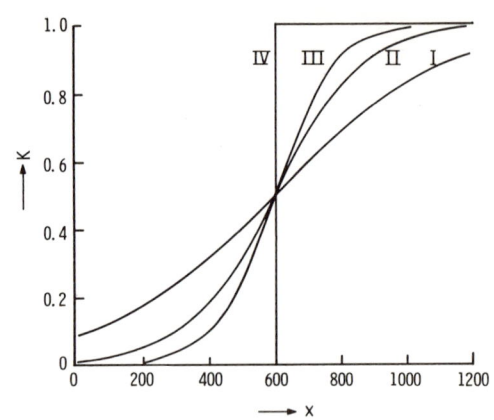

FIG. 9. The fractions of the species of size x in the concentrated phase $K(R, \sigma, x)$. The curves are given for various values of the volume ratio R of the dilute and the concentrated phase. Curve I for $R = 10$, Curve II for $R = 100$, Curve III for $R = 1000$. The values of σ are so chosen that every curve has the inflection point at the same x. An ideal separation curve, IV, is shown for comparison (from Schulz [55]).

of the sharpness of the fractionation, is $(\ln R)/4$. An ideal fractionation is a steplike function, as shown in Fig. 9. The function K is far from the ideal one but approaches the ideal as $\ln R$ increases. The curves in Fig. 9 illustrate this situation.

It is clear that large values of R and σ are necessary to obtain an effective separation at a certain molecular weight. The polymer concentration of the more concentrated phase becomes higher as σ increases. The higher the polymer concentration in the more concentrated phase, the smaller is the volume of that phase required to obtain a definite amount of fractionated polymer. The requirements for the values of R and σ are, therefore, not in contradiction with each other.

For a single-solvent solution, a rough relation between σ_s^0 and χ^0 is shown in Fig. 4. From this figure we can see that to obtain an effective fractionation, we should use a solvent with high χ^0 value unless the experimental technique becomes impractical. The value of σ or the concentration difference between the two phases is also increased by decrease of the overall polymer concentration. This can be qualitatively understood from the phase diagram of Fig. 1. (Consider the change of locations of the points F_1 and F_2 as the point F_0 moves along PS from the P side to the S corner.)

Flory has pointed out [2, p. 562] that the fractionation should be carried out at overall concentrations much less than the critical concentration. Theoretical relationships concerning the critical points, at which no fractionation effect occurs, were discussed in Section III,B. (It must be kept in mind that the experimental critical concentrations are often much higher than those predicted by the theory without the higher terms $\chi' \varphi^3$, etc.)

Numerical calculations of the fractionation efficiency, illustrative of the general considerations discussed above, are found in the following section.

B. Definition of Fractionation Efficiency

We shall deal only with single-solvent fractionation, but similar mathematical procedures should lead to corresponding results for two-solvent fractionations. For illustrative purposes we shall assume the Flory–Huggins dilute solution equations (neglecting $\chi'\varphi^3$ and higher terms) to hold rigorously, realizing that for many systems this assumption leads to errors of considerable magnitude.

To deal qualitatively with complex distributions of the different species in the original solution is quite impractical, but it is relatively easy to deal with mixtures of only two species, as we shall do, and the results we shall obtain can be expected to be applicable, with little error, to solutions of most polymer mixtures having the same (weight average)/(number average) ratio of molecular weights.

To be specific, we shall consider hypothetical systems in which there are equal numbers of two polymer species, having x values of $x_\alpha = \bar{x}_n^0(1 + a^0)$ and $x_\beta = \bar{x}_n^0(1 - a^0)$. Here, "$a^0$" can be designated as the *spread factor*. The (weight average)/(number average) ratio for this mixture is

$$\bar{x}_w^0/\bar{x}_n^0 = 1 + (a^0)^2 \tag{61}$$

For given $\bar{x}_n^0$ and χ^0, one can readily compute φ_I and φ_II, the polymer concentrations in the two phases. As an approximation (which can be shown to lead to negligible errors, unless the spread factor is large) we can put $\bar{x}_n = \bar{x}_n = \bar{x}_n^0$ in Eq. (24), and compute σ_s^0. The ratio R is

$$R = (\varphi_\text{II} - \varphi^0)/(\varphi^0 - \varphi_\text{II}) \tag{62}$$

Substitution of these values into Eq. (58) for the x value of one of the two species then yields the fraction of that species in each phase. Doing this also for the other species and averaging, one obtains $\bar{x}_{n,\text{I}}$ and $\bar{x}_{n,\text{II}}$, the number average polymer size in each of the two phases. For phase II, for example

$$\bar{x}_{n,\text{II}} = \frac{f_{\alpha,\text{II}} x_\alpha + f_{\beta,\text{II}} x_\beta}{f_{\alpha,\text{II}} + f_{\beta,\text{II}}} \tag{63}$$

where $f_{\alpha,\text{II}}$ and $f_{\beta,\text{II}}$ are number fractions of the molecules of sizes x_α and x_β, respectively.

A spread factor for the polymer in each phase, analogous to the spread factor in the original solution, can be computed from equations like Eq. (61). For example,

$$1 + a_\text{II}^2 = \bar{x}_{w,\text{II}}/\bar{x}_{n,\text{II}} \tag{64}$$

This is useful if the polymer in that phase is to undergo further fractionation.

The definition of a quantity to be called the "fractionation efficiency," for comparison of different combinations of χ^0 and φ^0 values, is somewhat arbitrary. It seems reasonable to choose a function proportional to the fraction of the total amount of polymer contained in the phase which is removed in a single fractionation step, and proportional to the fractional increase or decrease in x (or in molecular weight) relative to the original number average value. Our measure of efficiency will thus be proportional to $f_{\mathrm{II}}[(\bar{x}_{n,\mathrm{II}} - \bar{x}_n^0)/\bar{x}_n^0]$ if phase II is removed at each step, and to $f_{\mathrm{I}}[(\bar{x}_n^0 - \bar{x}_{n,\mathrm{I}})/\bar{x}_n^0]$ if phase I is being removed at each step. For perfect fractionation, with all of the high molecular weight component in phase II and all of the low molecular weight component in phase I,

$$f_{\mathrm{II}} = \tfrac{1}{2}(1 + a^0), \quad f_{\mathrm{I}} = \tfrac{1}{2}(1 - a^0), \text{ and } (\bar{x}_{n,\mathrm{II}} - \bar{x}_n^0)/\bar{x}_n^0 = (\bar{x}_n^0 - \bar{x}_{n,\mathrm{I}})/\bar{x}_n^0 = a^0 \quad (65)$$

Hence, to make the efficiencies equal to unity for perfect fractionation, we define

$$\varepsilon_{\mathrm{II}} = \frac{2 f_{\mathrm{II}} (\bar{x}_{n,\mathrm{II}} - \bar{x}_n^0)}{(1 + a^0) a^0 \bar{x}_n^0} \quad (66)$$

for removal of phase II at each fractionation step, and

$$\varepsilon_{\mathrm{I}} = \frac{2 f_{\mathrm{I}} (\bar{x}_n^0 - \bar{x}_{n,\mathrm{I}})}{(1 - a^0) a^0 \bar{x}_n^0} \quad (67)$$

for removal of phase I at each step. For a fractionation system in which, after separation at each step, both phases are repeatedly refractionated, it may be useful to define the efficiency as the average of $\varepsilon_{\mathrm{II}}$ and ε_{I}.

C. Dependence of Efficiency on Various Parameters

Calculations show that the differences between $\varepsilon_{\mathrm{II}}$ and ε_{I}, for given values of χ^0 and φ^0, are usually small, and the values of these variables giving maximum efficiencies are very nearly the same, especially when a^0 is small. We shall therefore present here only the results of calculations of $\varepsilon_{\mathrm{II}}$.

Calculation of $\varepsilon_{\mathrm{II}}$ for given $\bar{x}_n^0$, a^0, and χ^0, with variable φ^0, yields curves such as those of Fig. 10. The overall concentrations, φ^0, giving greatest efficiencies are relatively close to the concentrations, φ_{I}, in the more dilute phase. When χ^0 is not fairly close to the critical value for phase separation, however, these optimum concentrations are so low as to be impractical for most actual fractionations.

Figures 11, 12, and 13 show how the efficiencies at the optimum φ^0 values for each χ^0 depend on χ^0. The efficiency increases continuously as χ^0 increases, hence one can conclude that the χ^0 value should be as high

A. Theoretical Considerations

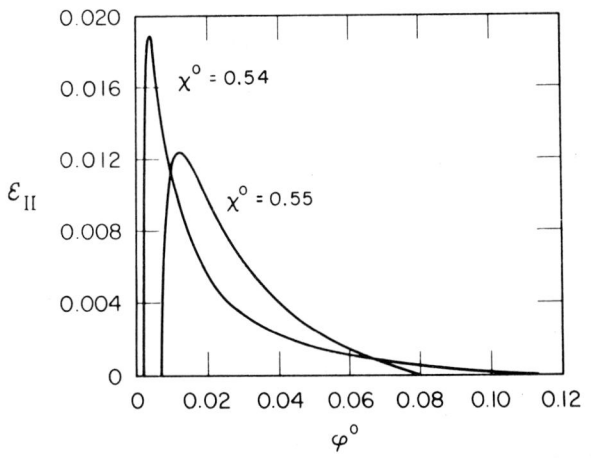

FIG. 10. Dependence of the efficiency on the overall concentration of polymer, for the values of χ^0 indicated, with $\bar{x}^0 = 1000$.

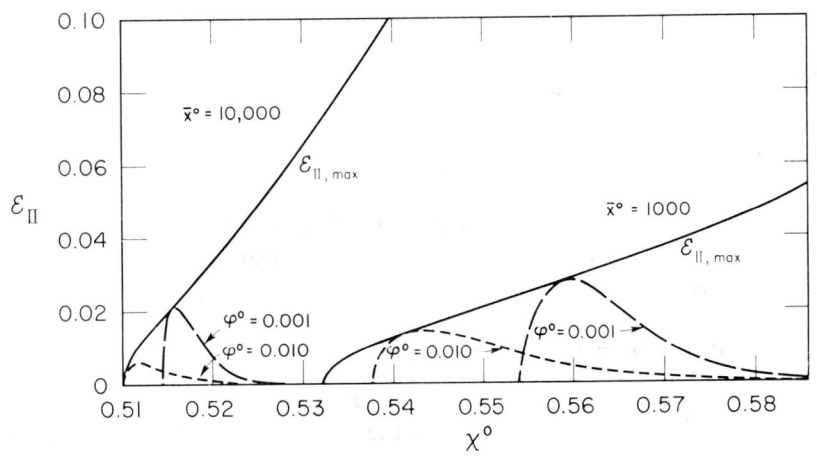

FIG. 11. Equilibrium efficiencies, as functions of the interaction coefficient $\chi = \chi^0$, for $\bar{x}^0$ values of 10,000 (curves on the left) and 1000 (curves on the right), with $a^0 = 0.01$.

as possible, *if* φ^0 can have the optimum (very dilute) value and *if* the properties of the more concentrated phase are such as to permit sufficiently rapid achievement of molecular weight distribution equilibrium.

These figures also show the dependence of the efficiency on χ^0, at a constant value of φ^0. If that value of φ^0 is the lowest concentration which is practical for the fractionation of interest (and if distribution equilibrium can be closely approached in the time available), the peak in the curve

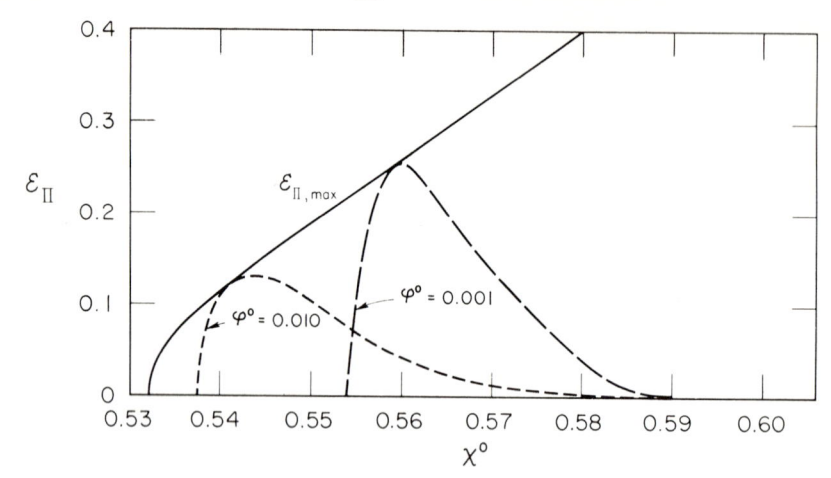

FIG. 12. Equilibrium efficiencies as functions of the interaction coefficient, for $\bar{x}^0 = 1000$ and $a^0 = 0.10$.

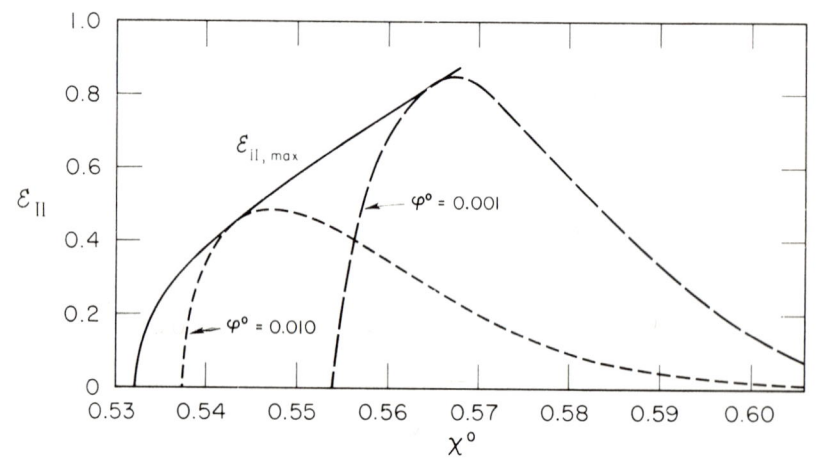

FIG. 13. Equilibrium efficiencies as functions of the interaction coefficient for $\bar{x}^0 = 1000$ and $a^0 = 0.50$.

indicates the best χ^0 value for the purpose and the maximum efficiency obtainable at that χ^0 value.

Figures 14, 15, and 16 show how the optimum φ^0 value varies with χ^0, for certain combinations of $\bar{x}_n^0$ and a^0. (In Fig. 16 reduced parameters are plotted; i.e., the concentration is divided by the critical concentration for phase separation, and the excess of χ^0 over its critical value for separation is plotted, rather than χ^0 itself.) Note the approximately rectilinear relationship

A. Theoretical Considerations

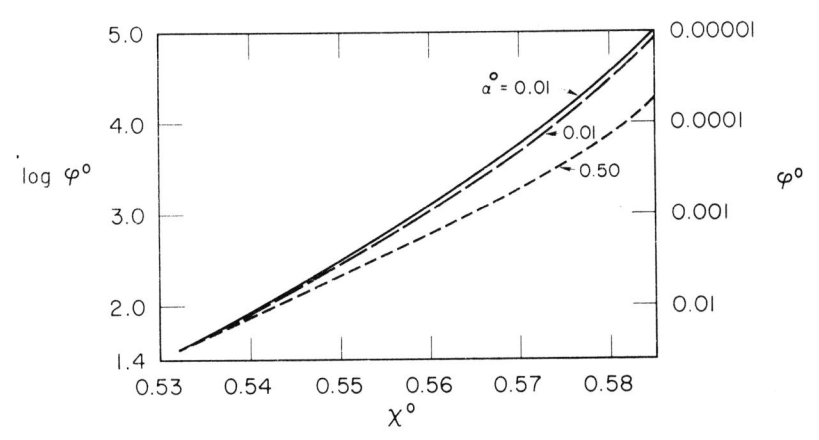

Fig. 14. Overall concentrations for maximum efficiency, ε_{II}, at each χ^0, as functions of χ^0, for $\bar{x}^0 = 1000$ and the values of a^0 shown.

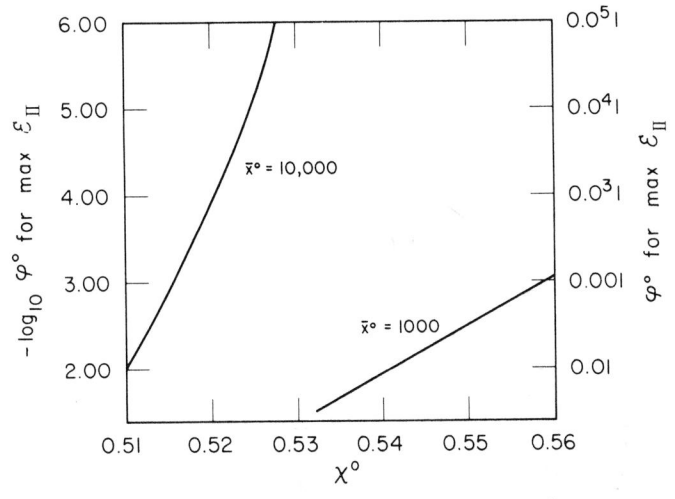

Fig. 15. Overall concentrations for maximum efficiency at each χ^0, as functions of χ^0, for $a^0 = 0.01$ and the values of $\bar{x}^0$ shown.

between χ^0 and the logarithm of φ^0 for optimum efficiency. This means that the optimum value of φ^0 decreases approximately exponentially with increase of χ^0.

Figure 17 shows the dependence of $(1 + a^0)\varepsilon_{II}$ on a^0 for certain values of χ^0. For small values of a^0, the efficiency is practically proportional to a^0. As would be expected, the greater the spread of molecular weights in the polymer being fractionated, the greater is the efficiency of separation.

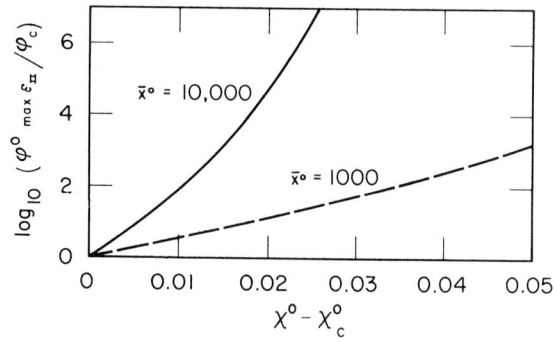

FIG. 16. Same as Fig. 15, except that the concentrations and interaction coefficients have both been normalized with respect to the critical values.

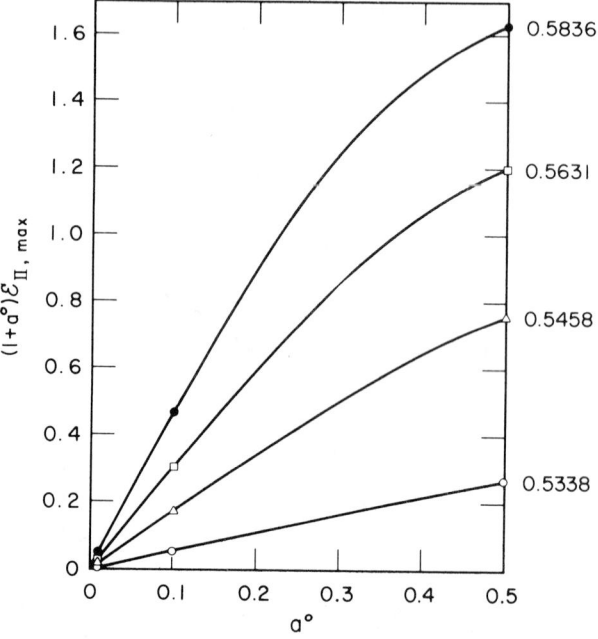

FIG. 17. Maximum efficiencies, times $(1 + a^0)$, as functions of a^0, for the interaction coefficients indicated. $\bar{x}^0 = 1000$.

The optimum value of χ^0 for constant φ^0 (0.01) and a^0 (0.01), and the efficiency at that optimum value are plotted against the reciprocal of $\bar{x}_n^0$ in Figs. 18 and 19, respectively. As expected, the maximum obtainable efficiency increases as the molecular weight decreases. For these φ^0 and a^0 values the maximum efficiency is approximately proportional to $(\bar{x}_n^0)^{-1/3}$, but the generality of this relationship has not yet been tested.

A. Theoretical Considerations

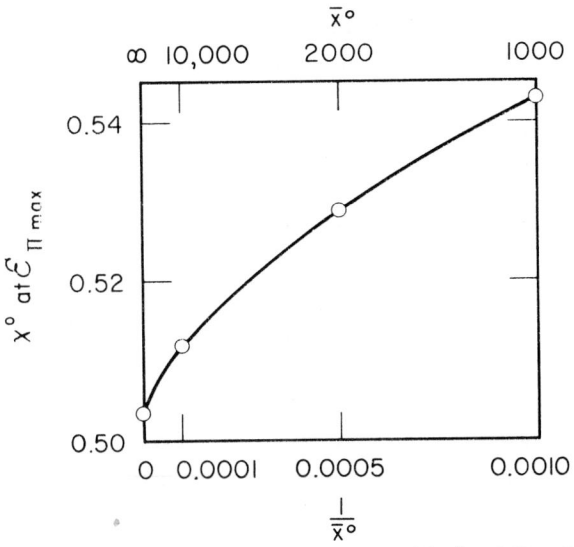

FIG. 18. Interaction coefficient for maximum efficiency, with $a^0 = 0.01$ and $\varphi^0 = 0.01$, as a function of the average initial value of x.

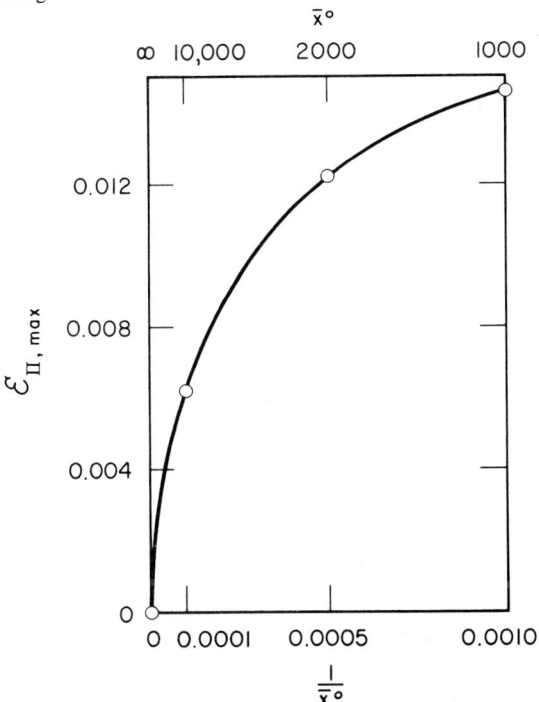

FIG. 19. Maximum efficiency, for $a^0 = 0.01$ and $\varphi^0 = 0.01$, as a function of the average initial value of x.

If the polymer in one of the phases (e.g., phase II) of a fractionation step is to undergo further fractionation, it is helpful to know the average size ($\bar{x}_{II}$), amount (f_{II}), and spread factor (a_{II}) of the polymer in that phase. The theoretical calculations which we have made for our simple polymer–solvent model system give these data. For example, Fig. 20 shows how f_{II} depends on χ^0, for $\bar{x}_n^0 = 1000$ and $\varphi^0 = 0.010$ and 0.001. Figures 21 and 22 show the dependence of $\bar{x}_{n,II}$ and a_{II}, respectively, on a^0, for $\bar{x}_n^0 = 1000$ and the χ^0 values giving maximum efficiency, at $\varphi^0 = 0.010$ and 0.001.

D. Limitations and Extensions of the Theoretical Treatment

It has been shown how the equilibrium efficiency of a fractionation step depends on the average molecular weight, the molecular weight spread, the overall concentration of polymer, and the polymer–solvent interaction coefficient, for systems for which the theoretical limiting law equation holds [i.e., for which Eqs. (14) and (15), with χ^0 constant, hold]. In most polymer

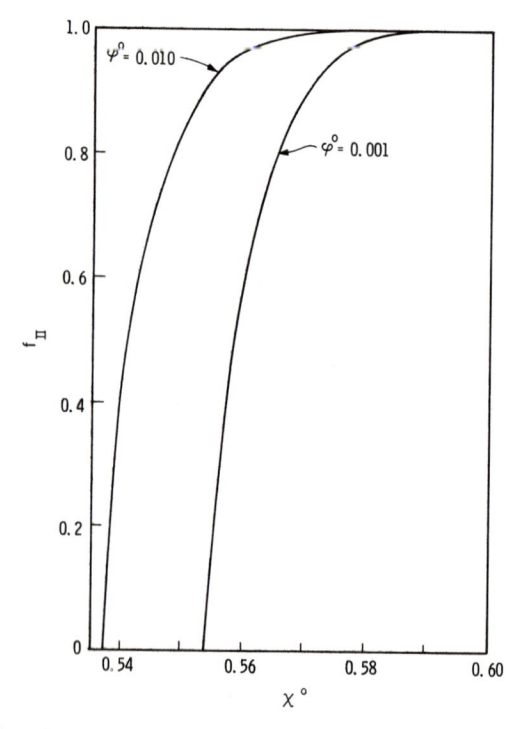

Fig. 20. Fraction of polymer in phase II, as a function of the interaction coefficient χ^0, for overall polymer concentrations of 0.010 and 0.001.

A. Theoretical Considerations

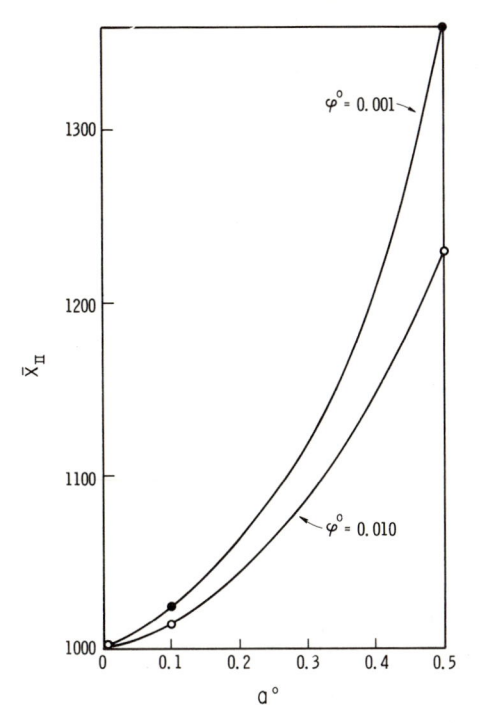

FIG. 21. Average polymer size in phase II, as a function of the initial spread factor, a^0, for overall polymer concentrations of 0.010 and 0.001, and the χ^0 values giving greatest efficiency at these concentrations.

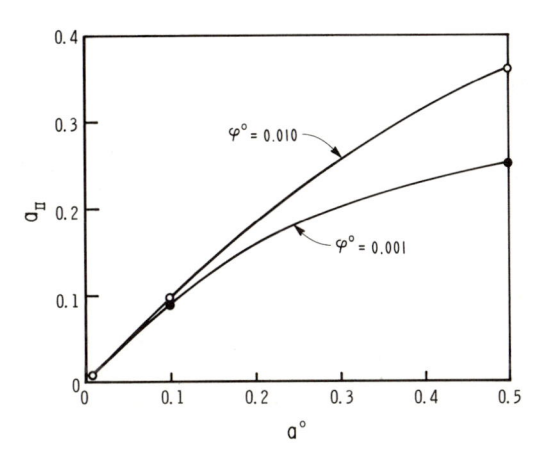

FIG. 22. Spread factor in phase II, as a function of the initial spread factor, a^0, for overall polymer concentrations of 0.010 and 0.001, and the χ^0 values giving greatest efficiency at these concentrations. $\bar{x}^0 = 1000$.

systems of interest, χ^0 is not accurately constant (or, equivalently, terms in higher powers of the concentration should be included), and the errors introduced by neglect of these departures from the limiting law are far from negligible. The efficiency results which have been presented must be considered merely qualitative and illustrative of what can be done, in a similar manner, using the more complex equations.

The efficiency calculations reported assume only a single solvent. At a given temperature the value of χ^0 is determined by the nature of the solvent and the polymer. To obtain a desired χ^0 (e.g., that for optimum efficiency for a specified overall concentration), one must choose a solvent having that χ^0 for the polymer being fractionated, or else vary the temperature. In general, χ^0 for a polymer–solvent pair is a function of temperature, conforming approximately to the relationship expressed by Eq. (13). An increase in temperature usually decreases χ^0.

As already noted, the distribution of polymer species between two coexistent phases can be altered by using a mixture of solvents and changing their relative proportions. For quantitative calculations of the polymer distributions, efficiencies, etc., it is then necessary to allow for the different proportions of the two solvents in the two phases. Some of the equations for this have been presented in Section III,B,3, but efficiency calculations have not yet been carried out.

The efficiency relationships reported in this chapter are far from quantitative, unless a reasonably close approach to equilibrium is achieved. Not only must the proportions of polymer and solvent in the two phases be close to the equilibrium values, but also the molecular weight distributions within the two phases must approximate the equilibrium distributions. To achieve this result in a fractionation step, it is obvious that the polymer molecules must be able to pass from one phase to the other with sufficient rapidity, relative to the time available, before the two phases are separated from each other. In a continuous process, in which the χ^0 value at a given location is being constantly changed, the approach to distribution equilibrium must be fast enough relative to the rate at which the equilibrium distribution is altered because of changing χ^0.

In general, the rate of approach to distribution equilibrium is expected to decrease as the viscosity increases, hence as the polymer concentration increases, but the dependence on concentration varies with the types of polymer and solvent and with the temperature. It seems likely that, as the concentration increases, there is a rather sudden decrease in rate of approach to equilibrium when a viscous solution changes to a gel, but we know of no quantitative data on this subject. Precise experimental data in this regard for a few typical systems would be very valuable. Then one might be able to predict the maximum concentration for phase II, which would give a sufficiently rapid approach to distribution equilibrium for the experimental setup under consideration. With the aid of the relationships given or exemplified in this chapter, one could then determine the optimum φ^0 and χ^0 values, for fractionation at a specified value of φ_1, or the optimum χ^0 value, for fractionation at a specified value of φ^0.

E. Determination of Distribution Functions from Fractionation Data

The foregoing discussions show that, even under optimum conditions, it is impossible to obtain a high degree of fractionation of a polymer mixture in a single fractionation step.

In a series of successive fractional precipitations, the solvent power of the solution is decreased, stepwise, by lowering the temperature or by adding poor solvent. In each step the concentrated phase is removed as a fraction. The molecular weight distributions of these fractions overlap each other because of the nonideal separation [54,55]. If w_i is the weight fraction of the polymer obtained in the ith fractionation step and $\overline{M}_i$ is its average molecular weight, the integral distribution $I(\overline{M}_i)$ [Eq. (1)] is usually assumed to be

$$I(\overline{M}_i) = 1 - \sum_{j=1}^{i-1} w_j - \tfrac{1}{2} w_i \tag{68}$$

The effect of overlapping of the distributions in the individual fractions is largely, but not entirely, compensated for by this procedure. This problem has been examined experimentally [67] and theoretically [55–57] by several authors. The values of R which are necessary to obtain a reliable distribution have been discussed [62,63]. An attempt has been made to improve the accuracies of the distribution by assuming an appropriate distribution for each fraction [68]. A method of estimating the distributions of the individual fractions has been proposed by Okamoto [64].

Spencer [69] has proposed the following procedure: To a series of samples of solutions of polymer in a good solvent, different amounts of a poor solvent are added. The gel phase produced in each case is separated and dried, the weight fraction of the polymer $w(s)$ and its weight average molecular weight $\overline{M}_w(s)$ are measured. Here, we designate by s a parameter indicating the relative amount of the good solvent and the poor solvent. If, in each precipitation, all the polymer of molecular weight higher than a given value, $M(s)$, but none of lower molecular weight, were in the more concentrated phase, then

$$w(s) = \int_{M(s)}^{\infty} f(M)\, dM \tag{69}$$

$$\overline{M}_w(s) = \frac{1}{w(s)} \int_{M(s)}^{\infty} M f(M)\, dM \tag{70}$$

$$M(s) = \frac{d[w(s)\overline{M}_w(s)]}{dw(s)} \tag{71}$$

The slope of the curve $w(s)\overline{M}_w(s)$ versus $w(s)$ gives the corresponding value of $M(s)$. A plot of $w(s)$ versus $M(s)$ gives the integral distribution curve.

The premise on which this procedure is based is certainly far from the truth, and as Billmeyer and Stockmayer [70] have shown, the distributions obtained thereby cannot be expected to be close to the true distributions. The method does, however, furnish a quick, inexpensive method for obtaining a parameter, comparable to our a^0, characterizing the spread of the distribution. Theoretical analyses of this problem have also been published by Matsumoto [58] and Broda et al. [59,60]. A detailed discussion may be found in Chapter C.4 of this book.

The theoretical equations of $w(s)$ and $M_w(s)$ are, from Eqs. (58) and (59),

$$w(s) = \int_0^{\infty} \frac{f(x)}{1 + R(s)\exp[-\sigma(s)x]}\, dx \tag{72}$$

$$\overline{M}_w(s) = \frac{1}{w(s)} \int_0^{\infty} \frac{x f(x)}{1 + R(s)\exp[-\sigma(s)x]}\, dx \tag{73}$$

In case $w(s)$, $\overline{M}_w(s)$, and $R(s)$ are known for every s, the above equations can be regarded as a simultaneous integral equation with respect to $f(x)$ and $\sigma(s)$. An approximate method for solving the equation was proposed by Okamoto [61,63].

Because of the nonideality of each separation step, to obtain moderately sharp fractions one must resort to repeated fractionations or to a continuous refractionation process. In a typical chromatographic procedure [71], a solvent mixture of continuously increasing solvent power is passed down a column of solid particles coated with the gel of the polymer to be fractionated. A temperature gradient is maintained, with the highest temperature at the top of the column. As the solvent mixture passes down the tube, it continually dissolves some of the polymer (preferentially the species of lower molecular weight) from the beads and precipitates some polymer (preferentially the species of higher molecular weight) on them. Since this process is equivalent to a series of fractional precipitations and fractional solutions, the fractions obtained from the solution leaving the column at the bottom have considerably narrower molecular weight spreads than fractions resulting from a single precipitation or solution.

Although the basic principles are simple, their precise quantitative application is very complex. Theoretical treatments, based on certain simplifications, have been given by Caplan [72] and Schulz *et al.* [73]. For a detailed discussion Chapters B.2 and L.3 may be consulted.

REFERENCES

1. International Union of Pure and Applied Chemistry, *J. Polymer Sci.* **8**, 257 (1952).
2. P. J. Flory, "Principles of Polymer Chemistry." Cornell Univ. Press, Ithaca, New York, 1953.
3. T. G. Fox, S. Gratch, and S. Loshack, *in* "Rheology" (F. R. Eirich, ed.), Vol. 1, p. 431. Academic Press, New York, 1956.
4. F. Bueche, "Physical Properties of Polymers." Wiley (Interscience), New York, 1962.
5. H. P. Schreiber and E. B. Bagley, *J. Polymer Sci.* **58**, 29 (1962).
6. J. F. Rudd, *J. Polymer Sci.* **44**, 459 (1960).
7. L. H. Tung, *J. Polymer Sci.* **46**, 409 (1960).
8. R. L. Ballman and R. H. M. Simon, *J. Polymer Sci.* **A2**, 3557 (1964).
9. H. Leaderman, R. G. Smith, and L. C. Williams, *J. Polymer Sci.* **36**, 233 (1959).
10. A. V. Tobolsky, "Properties and Structure of Polymers." Wiley, New York, 1960.
11. W. L. Peticolas, *J. Chem. Phys.* **39**, 3392 (1963).
12. H. W. McCormick, F. M. Brower, and L. Kin, *J. Polymer Sci.* **39**, 87 (1959).
13. L. H. Tung, *SPE (Soc. Plastics. Engrs.) J.* **14**, 25 (1958).
14. P. H. Geil, *J. Polymer Sci.* **A2**, 3813 (1964).
15. V. F. Holland and P. H. Lindenmeyer, *J. Polymer Sci.* **57**, 589 (1962).
16. B. Wunderlich, E. A. James, and T. W. Shu, *J. Polymer Sci.* **A2**, 2759 (1964).
17. H. D. Keith and P. J. Padden, *J. Appl. Phys.* **34**, 2409 (1963).
18. H. D. Keith, *J. Polymer Sci.* **A2**, 4339 (1964).
19. V. V. Varadaiah and V. S. R. Rao, *J. Polymer Sci.* **50**, 31 (1961).
20. H. Yamakawa and M. Kurata, *J. Chem. Phys.* **32**, 1852 (1960).
21. H. Tompa, "Polymer Solutions." Academic Press, New York, 1956.
22. J. W. Gibbs, "Thermodynamics," Vol. I of "Collected Works." Longmans, Green, New York, 1931.
23. R. L. Scott, *J. Chem. Phys.* **17**, 268 (1949).
24. M. L. Huggins, "Physical Chemistry of High Polymers." Wiley, New York, 1958.
25. M. L. Huggins, *J. Phys. Chem.* **46**, 151 (1942).
26. M. L. Huggins, *Ann. N. Y. Acad. Sci.* **43**, 1 (1942).
27. M. L. Huggins, *J. Am. Chem. Soc.* **64**, 1712 (1942).
28. P. J. Flory, *J. Chem. Phys.* **10**, 51 (1942).
29. M. L. Huggins, *J. Polymer Sci.* **16**, 209 (1955).
30. M. L. Huggins, *J. Am. Chem. Soc.* **86**, 3535 (1964).
31. S. H. Maron, *J. Polymer Sci.* **38**, 329 (1959).
32. M. L. Huggins, *J. Polymer Sci.* **C4**, 445 (1963).
33. W. H. Stockmayer, *J. Chem. Phys.* **17**, 588 (1949).
34. R. L. Scott, *J. Chem. Phys.* **13**, 178 (1945).
35. A. R. Schultz and P. J. Flory, *J. Am. Chem. Soc.* **74**, 4760 (1952).
36. W. R. Krigbaum and D. O. Geymer, *J. Am. Chem. Soc.* **81**, 1859 (1959).
37. B. Bergsnov-Hansen and A. P. Brady, unpublished research at Stanford Research Institute (1961).

38. J. H. Hildebrand and R. L. Scott, "The Solubility of Nonelectrolytes," 3rd ed. Reinhold, New York, 1950.
39. H. Okamoto, *J. Polymer Sci.* **33**, 507 (1958).
40. A. Münster, *J. Polymer Sci.* **5**, 333 (1950).
41. R. W. Kilb and A. B. Bueche, *J. Polymer Sci.* **28**, 285 (1958).
42. R. L. Scott, *J. Chem. Phys.* **17**, 279 (1949).
43. M. L. Huggins, *J. Phys. Chem.* **43**, 1083 (1939).
44. P. J. Flory and A. Vrij, *J. Am. Chem.* **85**, 3548 (1963).
45. T. Kawai, *J. Polymer Sci.* **B3**, 83 (1965).
46. L. Mandelkern, "Crystallization of Polymers." McGraw-Hill, New York, 1964.
47. P. H. Geil, "Polymer Single Crystals." Wiley (Interscience), New York, 1963.
48. M. J. Richardson, P. J. Flory, and J. B. Jackson, *Polymer* **4**, 221 (1963).
49. P. J. Flory, *Trans. Faraday Soc.* **51**, 848 (1955).
50. M. L. Huggins, *Makromol. Chem.* **92**, 260 (1966).
51. T. Kawai and A. Keller, *J. Polymer Sci.* **B2**, 333 (1964).
52. P. W. O. Wijga, J. van Schooten, and J. Boerma, *Makromol. Chem.* **36**, 115 (1960).
53. S. W. Hawkins and H. Smith, *J. Polymer Sci.* **28**, 341 (1958).
54. R. A. Mendelson, *J. Polymer Sci.* **A1**, 2361 (1963).
54a. R. Koningsveld and A. J. Pennings, *Rec. Trav. Chim.* **83**, 552 (1964).
54b. A. J. Pennings, *Preprint 216, Int. Symp. Macromol. Chem.* Prague, 1965.
55. G. V. Schulz, "Die Physik der Hochpolymeren" (H. A. Stuart, ed.), Vol. II, p. 726 *et seq.* Springer, Berlin, 1953.
56. E. V. Sayre, *J. Polymer Sci.* **10**, 175 (1953).
57. M. Matsumoto and Y. Ohyanagi, *Kobunshi Kagaku* **11**, 7 (1954).
58. M. Matsumoto, *Kobunshi Kagaku* **11**, 182 (1954).
59. A. Broda, B. Gawronska, T. Niwinska, and S. Polowinski, *J. Polymer Sci.* **29**, 183 (1958).
60. A. Broda, T. Niwinska, and S. Polowinski, *J. Polymer Sci.* **32**, 343 (1958).
61. H. Okamoto, *J. Polymer Sci.* **41**, 535 (1959).
62. H. Okamoto, *J. Polymer Sci.* **41**, 537 (1959).
63. H. Okamoto, *J. Phys. Soc. Japan* **14**, 1388 (1960).
64. H. Okamoto and K. Sekikawa, *J. Polymer Sci.* **55**, 597 (1961).
65. L. H. Tung, *J. Polymer Sci.* **61**, 449 (1962).
66. W. R. Krigbaum and D. O. Geymer, *J. Am. Chem. Soc.* **81**, 1859 (1959).
67. E. A. Haseley, *J. Polymer Sci.* **35**, 309 (1959).
68. G. Beall, *J. Polymer Sci.* **4**, 483 (1949).
69. R. S. Spencer, *J. Polymer Sci.* **3**, 606 (1948).
70. F. W. Billmeyer, Jr., and W. H. Stockmayer, *J. Polymer Sci.* **5**, 121 (1950).
71. C. A. Baker and R. J. P. Williams. *J. Chem. Soc.* p. 2352 (1956).
72. S. R. Caplan, *J. Polymer Sci.* **35**, 409 (1959).
73. G. V. Schulz, P. Deussen, and A. G. R. Scholz, *Makromol. Chem.* **69**, 47 (1963).

CHAPTER B.1

Fractional Precipitation

Akira Kotera
DEPARTMENT OF CHEMISTRY, TOKYO KYOIKU UNIVERSITY, OTSUKA, TOKYO, JAPAN

I. Nonsolvent Addition Method	44
A. General Procedure	44
B. Special Experimental Conditions	45
C. Additional Comments on Experimental Technique	56
II. Fractionation by Solvent Evaporation	58
A. General Procedure	58
B. Special Experimental Conditions	59
III. Fractionation by Cooling	61
A. General Procedure and Other Remarks	62
B. Solvent Systems for Semicrystalline Polymers	63
References	65

In this chapter we deal with the fractionation of polymeric substances by methods involving the fractional precipitation from solution of a series of fractions, the first member of which has the highest molecular weight and the succeeding ones progressively decreasing molecular weights. The precipitation is carried out by a stepwise decrease in the solvent power of the system. This may be achieved by any of the following three methods:

(1) Addition of nonsolvent (or precipitant)
(2) Elimination of solvent by evaporation
(3) Lowering the temperature of the system

These are the most widely used methods, and starting with the demonstration of the "polymer homologous series" by the school of Professor H. Staudinger [1,2] these techniques have been employed in almost all the fractionations carried out in the course of studies in polymer science. Even though new techniques are rapidly being developed, it is likely that those above will continue to be popular, because they are based upon very simple operations and do not require the use of elaborate or special equipment. In addition, they can be applied to small-scale experiments where the molecular weight distribution in small samples is to be examined, as well as to the large-scale separation of fractions needed for the study of various properties of high polymers [3].

Theoretically, the solvent power is conveniently expressed in terms of the "polymer–solvent interaction constant, χ." A solvent for a given polymer should have a χ value below 0.5, whereas a liquid having a χ value over 0.5 belongs to the category of nonsolvents. In this way, phase separation will occur when the χ value of the system is somewhat in excess of 0.5; the critical value χ_c will depend upon the molecular weight of the dissolved polymer molecules. If the size of a molecule is expressed in terms of its segment number, x, it has been shown by Flory [4] that

$$\chi_c = (\tfrac{1}{2}x)(1 + x^{1/2})^2 \sim \tfrac{1}{2} + 1/x^{1/2} \tag{1}$$

The first method cited above is based on the expectation that the gradual addition of a nonsolvent, with a large χ value, will cause the χ value of the system to successively exceed the χ_c value for each component; the component will then be precipitated in the order of decreasing value of the segment number x. Of course, the separation is not complete because the system is a multicomponent one.

The second method, fractionation through evaporation of the solvent, is based on the same principles but the technique is reversed: The concentration of nonsolvent, added at the beginning of the procedure, is increased by gradually expelling the more volatile solvent. The cooling method depends upon the temperature variation of the χ value of the system. Since χ with values of about 0.5 is known to have a negative temperature coefficient, the χ value will increase as the system is cooled.

I. Nonsolvent Addition Method

A. General Procedure

The usual procedure for fractional precipitation is as follows.

The polymer sample is dissolved in a suitable solvent (concentration ranges are discussed in Section 1,B,3) and an appropriate nonsolvent is added gradually to this solution at constant temperature. At the same time, the solution is subjected to vigorous agitation.

After a certain amount of nonsolvent has been added, the addition of one more drop of nonsolvent causes a turbidity which is not easily removed by agitation. At this stage the dropwise addition of nonsolvent is continued more carefully until the system reaches a point just short of the turbidity end point. The solution is then allowed to ripen for several hours, or longer if a good solvent has been used.

After ripening, the slow dropwise addition of a further small amount of nonsolvent is continued until the solution turns milky. The solution is then warmed several degrees to make it transparent again.[1] The amount of additional nonsolvent introduced at this stage should be such that the temperature increase required to remove the turbidity of the system is not too large, usually less than 5°C.

The solution is cooled gradually to the original temperature, with agitation, and it is kept at that temperature for several hours to settle the precipitate.[2] The precipitated phase is separated from the supernatant phase either (1) by decanting or syphoning off the latter, or (2) by withdrawing the former by means of a syringe, or drawing it off through a tap at the bottom of the vessel. The most suitable method will depend upon the nature of the precipitate, i.e., its fluidity, stickiness, etc.

The precipitated phase, which contains the fraction consisting of the highest molecular weight species, is collected by dissolving it in a small amount of solvent. The solution thus obtained is thrown into a large amount of nonsolvent to precipitate the fraction as a solid, which is then collected on a glass filter and dried to constant weight at an appropriate temperature in a vacuum.

The supernatant liquid is treated with a further volume of nonsolvent, using the same procedure described above, to obtain the next fraction. This procedure is repeated until a fairly large proportion of nonsolvent needs to be added in order to obtain the succeeding fraction. At this stage the solution is concentrated under reduced pressure, nonsolvent is added to it, and the precipitate is separated. This is usually the fraction before the last. The final fraction is obtained by evaporating the supernatant liquid to dryness in a vacuum.

B. Special Experimental Conditions

1. Special Equipment

As stated above, special equipment is not essential for carrying out these procedures. However, a few examples of such apparatus should be recommended, because their use is often convenient.

The two examples shown in Figs. 1 and 2 have been designed with reference to the literature [9–11] and used successfully in the author's laboratory. Figure 1 shows a pear-shaped flask with a small well at the bottom. If the precipitated phase is sufficiently fluid and does not stick to the wall of the vessel, the gel phase collects in the bottom of this well. With a graduated well, the volume of the gel phase may be measured. The

---

[1] It occasionally happens that cooling instead of warming is required to make the solution transparent. Poly-4-vinylpyridine in a methanol–toluene system is an example [5].

[2] In some cases, when the polymer is less dense than the solution, the gel phase floats on top of the sol phase. This has been observed to occur when the nonsolvent polyethylene glycol was added to toluene solutions of polyethylene [6], or of atactic or isotactic polystyrene [7,8].

volume of the gel phase will depend upon a number of factors, i.e., the concentration of the original solution, the magnitude and range of molecular weights in the parent polymer, the nature of the solvent–nonsolvent pair, and so on. However, an adequate volume for the well is usually found to be in the range of 1.0–1.5% of the total volume of the vessel. Some care needs to be exercised in choosing a shape for the well which will permit the gel

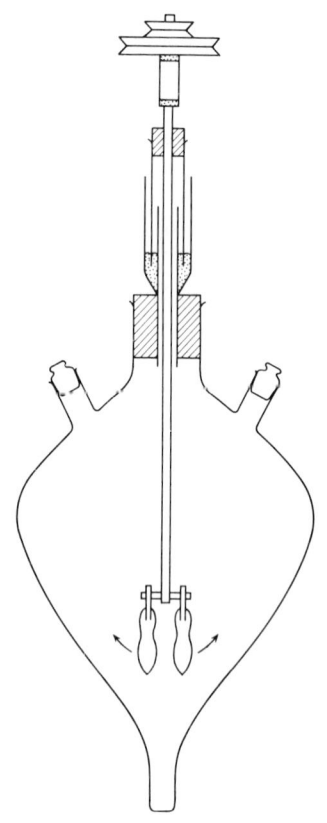

FIG. 1. Fractionation flask.

phase to be readily dispersed throughout the supernatant liquid when the solution is subjected to moderate agitation. Two of these flasks are required. In addition, some devices have been described which permit the gel phase to be separated from the supernatant liquid; their use appears to be convenient when the fluidity of the gel phase prevents the solution phase from being decanted [10,12].

Figure 2 shows a flask modified by the addition of a cock which enables the gel phase to be easily separated. The thermostat for this vessel needs to be refashioned with a hole in the bottom to permit access to the cock; this necessitates the use of a special cover to prevent heat losses via the cock.

If the sample is sensitive to oxidation, the vessel should be fitted with a number of side tubes for thermometer, tap funnel for introduction of the nonsolvent, gas inlet and outlet, etc. The whole system should be under an inert gas, e.g., nitrogen. In addition, it is helpful to add a small quantity of an antioxidant to the system. Other precautions must be taken according to the nature of the sample, for example, colored glass vessels must be used for a light-sensitive polymer.

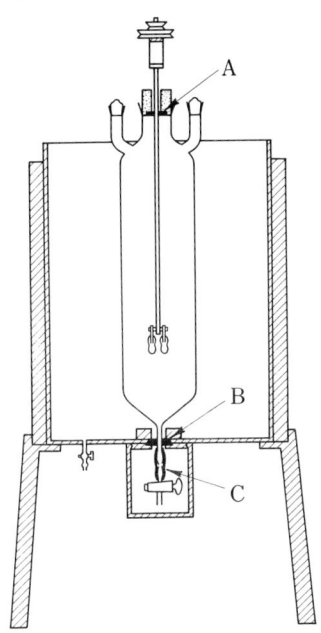

FIG. 2. Fractionation flask: A, oil seal; B, rubber packing; C, Teflon tube.

Recently, a number of solvent-resistant polymers have come into common use, and since these polymers, such as polyethylene, and polypropylene, are soluble only in certain hot solvents, fractionation must be carried out at an elevated temperature. Figure 3 shows some examples of equipment for this purpose. The one on the left, [Fig. 3(a)], is a fractionation vessel employed by Okamoto [13] for the evaporation fractionation of Kel-F polymer. As will be seen later, the same equipment may be used for the evaporation method and the nonsolvent addition method. The large flask A on the right-hand side of the figure is the main flask in which the precipitation takes place. After the gel phase has settled, it is transferred to the collector B (shown in the center of the figure) by means of mild suction applied at b and c. When the last portion of gel reaches e, the stop-cock E was opened, thus braking the vacuum. Solvent is added from D and the dissolved fraction in B is transferred to the receiver C on the left. The receiver can easily be taken out from the high-temperature thermostat in

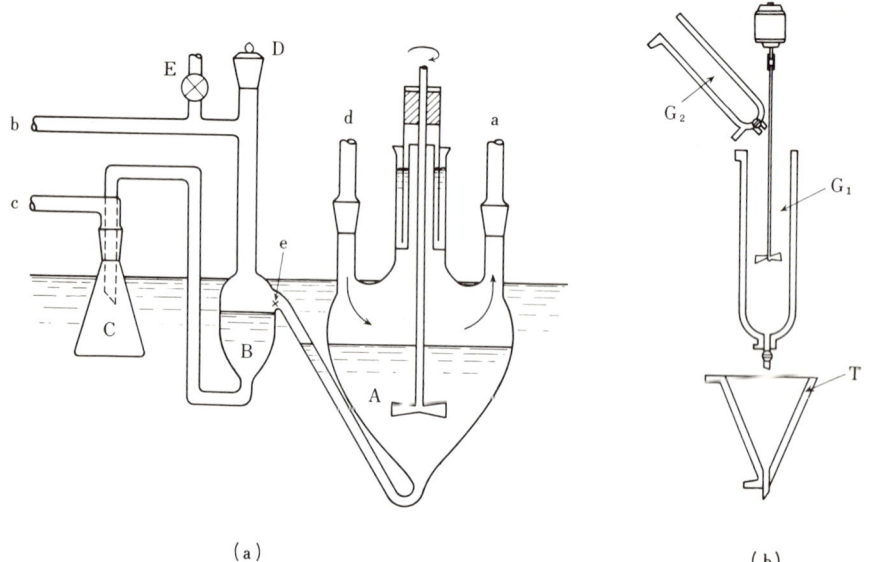

(a)   (b)

FIG. 3. Fractionation apparatus for use at elevated temperatures. (a) Okamoto [13]; (b) Überreiter [14]. See text for explanation.

which the whole equipment is housed. The air or nitrogen stream is maintained from d to a during the fractionation procedures.

Figure 3(b) shows an apparatus used by Überreiter et al. [14,15] for the fractionation of polyethylene and polyethylene terephthalate. All the glass vessels are of the Dewar type. Each vessel, G_1 or G_2, is covered with a watch glass between individual operations. The funnel T is employed to filter off the solid precipitate using filter paper.

2. Choice of Solvent and Nonsolvent

In the case of amorphous polymers, such as atactic polystyrene, or polyisobutylene, there is no essential limitation in the choice of solvent or nonsolvent. However, from the experimental point of view, consideration must be given to certain factors which are discussed below.

With regard to the selection of solvent, it is preferable to use a poor solvent, since only a small amount of nonsolvent will be required to produce the first fraction. In some cases, it is desirable to employ a mixture of solvent and nonsolvent in which the parent polymer is able to disperse.

B.1. Fractional Precipitation

For example, the Committee on Standards and Methods of Testing of the Division of Cellulose Chemistry of the American Chemical Society, recommends a mixture of acetone and water (91:9) as the solvent for cellulose nitrate to study the molecular distribution in parent cellulose samples [16].

Certainly, it is difficult to dissolve the samples in poor solvents, or in solvent–nonsolvent mixtures. However, if a good solvent is used to dissolve the polymer and a nonsolvent is added just past the cloud point, it is often observed that turbidity once produced will disappear when the solution is allowed to stand for several hours or overnight. This phenomenon is known as the "false cloud point" for which Boyer [9] has offered the following explanation: The polymer molecules, on being dissolved in a good solvent, might be expected to assume a rather extended form and are apparently subject to some degree of criss-crossing. When a nonsolvent is added to the system, the polymer molecules begin to form coils. As the proportion of nonsolvent is increased, the coils become tighter. However, if the nonsolvent is added too quickly, the desired stage (i.e., when the molecules are in separate tight coils which causes them to precipitate) is overshot due to the molecules not having had time to disengage from each other before coiling. A tangle of interwoven coils results. Therefore, it is necessary to allow the solution to stand for a period of time to give the molecules time to disentangle themselves as much as possible. In other words, the false cloud point is not the real equilibrium state, which requires some time to be reached. Boyer pointed out that it was sometimes necessary to wait more than 24 hours in the case of the polystyrene-benzene-methanol system, even when the cloud point had been approached very slowly. For this reason, it is much better to use a poor solvent.

With regard to the choice of a nonsolvent, it should first be noted that some solvent–nonsolvent pairs may actually produce a better solvent than the pure solvent alone. As pointed out by Gee [17], this may occur if the solvent and the nonsolvent have cohesive energy densities (CED) on either side of that of the polymer. Thus, their combination may result in a CED value equal to that of the polymer. In such a case, and, also, in the case of a nonsolvent with weak precipitation power, the large volume of nonsolvent required to start the fractionation may exceed the volume available in the apparatus.

On the other hand, it is very difficult to work with nonsolvents having a strong precipitation power. For example, if we choose dioxane as the solvent for the fractionation of polystyrene, and water as the nonsolvent, precipitation will occur with the addition of as little as 6% of water, and very little additional water is needed to precipitate the entire mass of polymer. Control of the fractionation is thus almost impossible. Further difficulties may arise in the succeeding operation. For example, during separation of the precipitation phase unavoidable evaporation may cause the precipitate to either redissolve or increase. The difficulty may be partially avoided by using a small proportion of solvent in the nonsolvent in order to decrease its precipitation power, or by application of the temperature cooling method described in Section III. It is usually advisable to look for a more suitable nonsolvent with moderate precipitation power.

One further requirement is to choose a solvent–nonsolvent system which produces an easily handled precipitate. In some cases, trouble in manipulating the precipitate will arise because of its stickiness or dryness. The difference in the densities of the supernatant liquid and the swollen precipitate should also be taken into account in order to minimize the settling time of the latter. By a process of trial and error, a suitable solvent–nonsolvent pair may be found which will reduce these difficulties.

Other factors to be considered are flammability, toxicity, vapor pressure, tendency to form peroxides, and so on.

It must be mentioned that some solvent–nonsolvent pairs have been found to be unsatisfactory because they cause "reverse-order fractionation," in which the low molecular weight polymers are precipitated before the higher ones. Usually, this is considered to be caused by the initial concentration being too high (see Section I,B,3). However, Morey and Tamblyn [18] found that reverse-order precipitation occurred during the use of an alkyl ether as the nonsolvent in the fractionation of cellulose acetate butyrate even when the initial concentration of the solution was dilute (0.035%).

In addition, Menčik [19] found that, in the fractionation of polyvinyl chloride, the well-known relation

$$\gamma = a/[\eta] + b \qquad (2)$$

(where a and b are constants) between the concentration of nonsolvent, γ, needed to bring a homogeneous polymer solution to the verge of precipitation, and the intrinsic viscosity $[\eta]$ of the solute, does not hold for the following solvent–nonsolvent pairs; tetrahydrofuran–amyl alcohol and cyclohexanone–n-butyl alcohol. An example of the anomalous relation for the former system is given in Fig. 4(a), where it is shown that the molecular species first precipitated is not that with the highest $[\eta]$, but the species with an intermediate molecular weight. In this case regular fractionation cannot be expected. Menčik showed further that in these abnormal systems the turbidity gradually increased from a point just before the real cloud point, whereas with the same sample in a normal system the turbidity was observed to rise sharply. This behavior is illustrated in Fig. 4(b). Menčik attributed this effect to association of the polyvinyl chloride molecules.

Fujisaki and Kobayashi [20] have found that the polyacrylonitrile–60% nitric acid–butanol system also shows the reverse-order effect, but the system obeys relation (2). Therefore, they have attributed the effect to the abnormal change of the polymer concentration in the precipitation phase.

At present there is no way of predicting with certainty when these anomalous effects will occur. To ensure a regular fractionation, it is therefore sometimes necessary to carry out preliminary experiments by making a small-scale trial, or by determining the turbidimetric behavior of the system to be used.

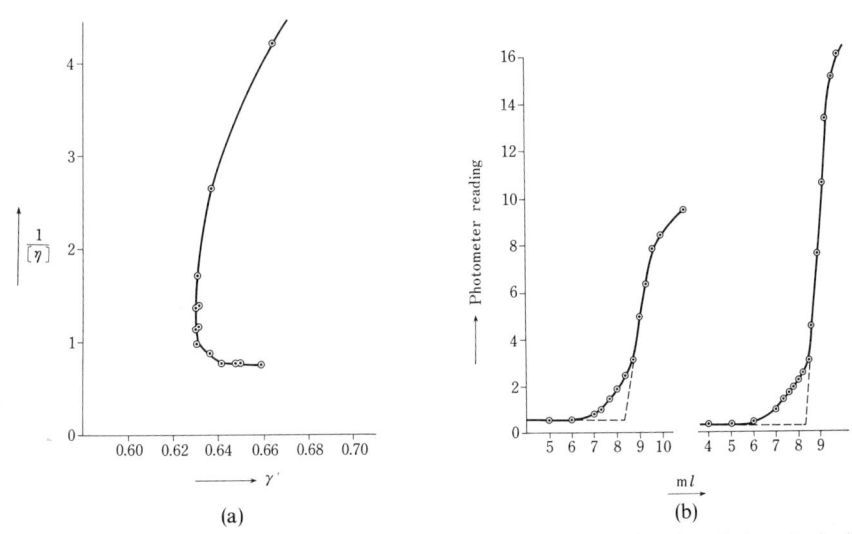

FIG. 4(a) Dependence of the precipitation point on the intrinsic viscosity of the polyvinyl chloride fraction in a tetrahydrofuran–amyl alcohol system (Menčik [19]). Measured at 22°C, initial concentration, $c = 0.2$ gm/100 ml. (b) Dependence of the intensity of scattered light on the volume of precipitant added. *Left:* the cyclohexanone–butyl alcohol system, a fraction of $[\eta] = 0.70$; $c = 0.20$ gm/100 ml. *Right:* the tetrahydrofuran–amyl alcohol system, a fraction of $[\eta] = 1.05$; $c = 0.20$ gm/100 ml. Initial volumes 5 ml, 22°C. The broken line shows the sharp break obtained in a normal system, using the same sample.

The behavior of semicrystalline polymers is discussed in some detail in Section III,B.

A few words about polyelectrolytes should be added. Although, they may be fractionated by employing the usual solvent systems, the same as with the normal nonelectrolyte polymers, it has been found that they may be effectively fractionated by a "salting-out" process as well. For example, McDowell [21] has fractionated alginate with $MnCl_2$ and $CaCl_2$. Marshall and Mock [22] carried out the fractionation of sodium poly-*p*-styrene sulfonate by dissolving it in $4 N$ aqueous NaI solution and adding a $9.1 N$ solution; however, only 42.5% of the dissolved polymer sample could be precipitated in this way.

In general, since most polymers have already been fractionated by prior investigators, it is advisable to begin by referring to Chapter G, where the reader will find some suitable solvent–nonsolvent systems.

3. Concentration

Flory [4] has shown that the fractionation efficiency, ε, increases in proportion to the logarithm of the ratio of the volume of the supernatant phase, V', to that of the precipitation phase, V, i.e.,

$$\varepsilon = (\tfrac{1}{4}) \ln(V'/V) \tag{3}$$

To increase the value of ε, it is necessary to reduce the value of V and to increase that of V'. Although the phase equilibrium theory for multicomponent systems is highly complicated, it is sufficient to realize that the former condition corresponds to not taking too large a volume of gel phase at a time, i.e., to an increase in the total number of fractions [23], and the latter condition to decreasing the initial concentration of the polymer solution. From a theoretical point of view, Scott [24] has recommended starting with a 1% solution and taking 10 fractions approximately equal in weight. Flory [4] has also suggested that the polymer concentration (expressed as the volume fraction) should not be more than some fraction of the inverse square root of the degree of polymerization (expressed as the segment number, x) of the polymer to be precipitated. Thus, for polymers of $\overline{M}_w = 10^6$, 10^5, and 10^4 (assuming that this corresponds to $x = 10^4$, 10^3, and 10^2), the initial concentration should not exceed say, if we apply $\tfrac{1}{4}$ as the value of the "fraction," 0.25%, 0.8%, and 2.5% for a given separation efficiency. Boyer [9] has pointed out, from preliminary experiments in the fractionation of polystyrene, that the initial concentration needs to be in the range of 0.1–0.25%.

Perhaps every worker in this field has had bitter experiences in which a series of fractions, particularly the first few fractions, were not obtained in the order of their intrinsic viscosities. This reverse-order fractionation has been shown by experiment to be the result of a high initial concentration; the possibility of its occurrence has also been deduced from theory [4,25,26].

Increasing the number of fractions decreases the quantity of each fraction. Thus, difficulty is experienced in determining the properties of each fraction due to its small size. The use of very dilute initial concentrations may cause various experimental difficulties, in particular a prolonged settling time for the precipitation phase.

A compromise is usually adopted in order to make the experiment not so troublesome; the number of fractions is around 10, and the initial concentration is adjusted to less than 1%, as Scott [24] has recommended.

4. Fractionation Schemes

The weakest point of the fractional precipitation method is that every fraction contains an appreciable amount of the lower molecular components. It is called the "tail effect" and is usually wrongly attributed to the inclusion of small molecules with the larger ones during precipitation of the latter, or else to incomplete separation of the gel phase. However, it is really an inherent characteristic of fractional precipitation [26,27]. Figure 5 shows the theoretical curves, calculated by Matsumoto from Flory's theory, for the distribution of molecular weights in each of the fractions obtained in (a) the fractionation by precipitation method, and (b) the fractionation by extraction method.

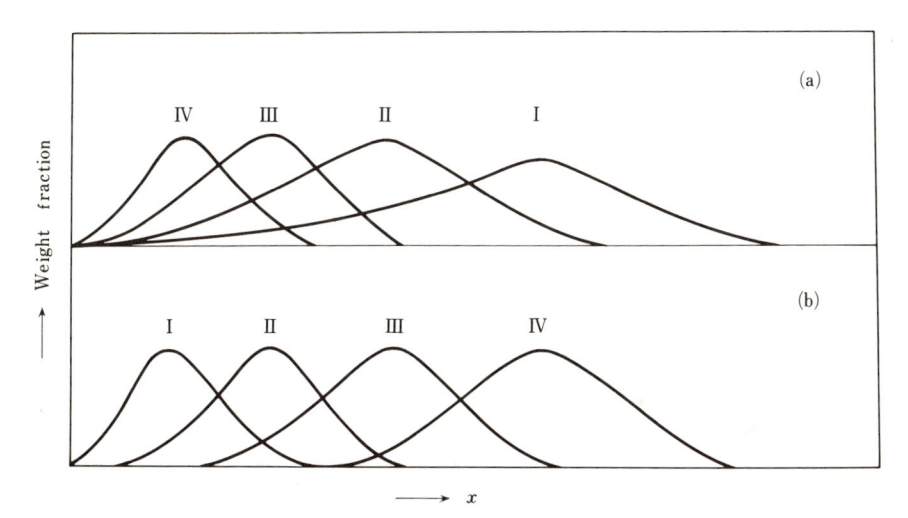

FIG. 5. Theoretical curves for the distribution of molecular weights in fractions [26,28]. (a) Fractional precipitation. (b) Fractional extraction.

To remedy this defect, Schulz [29] has proposed the refractionation of each fraction. The scheme for this procedure is shown in Fig. 6(b); the simple procedure described in Section I,A of this chapter is depicted in Fig. 6(a). As depicted in Fig. 6(a), the sample S_0 is first divided into the precipitated $F1$ fraction and the supernatant s_1 fraction, and s_1 is then divided into $F2$ and s_2, and so on.

In the case of the refractionation shown in Fig. 6(b), it is necessary to include an evaporation process to avoid expansion of the volume of the system. There are two popular methods of achieving this. The first is to evaporate the supernatant liquid of p_1 and dissolve the residue in s_1 with addition of a small amount of solvent, and the second is to evaporate the combined sol phase of s_1 and that of p_1 to the desired volume.

Another shortcoming in the simple procedure is that fractionation of the higher molecular weight portions of the sample is carried out at concentrations higher than those of the lower molecular weight portions. This

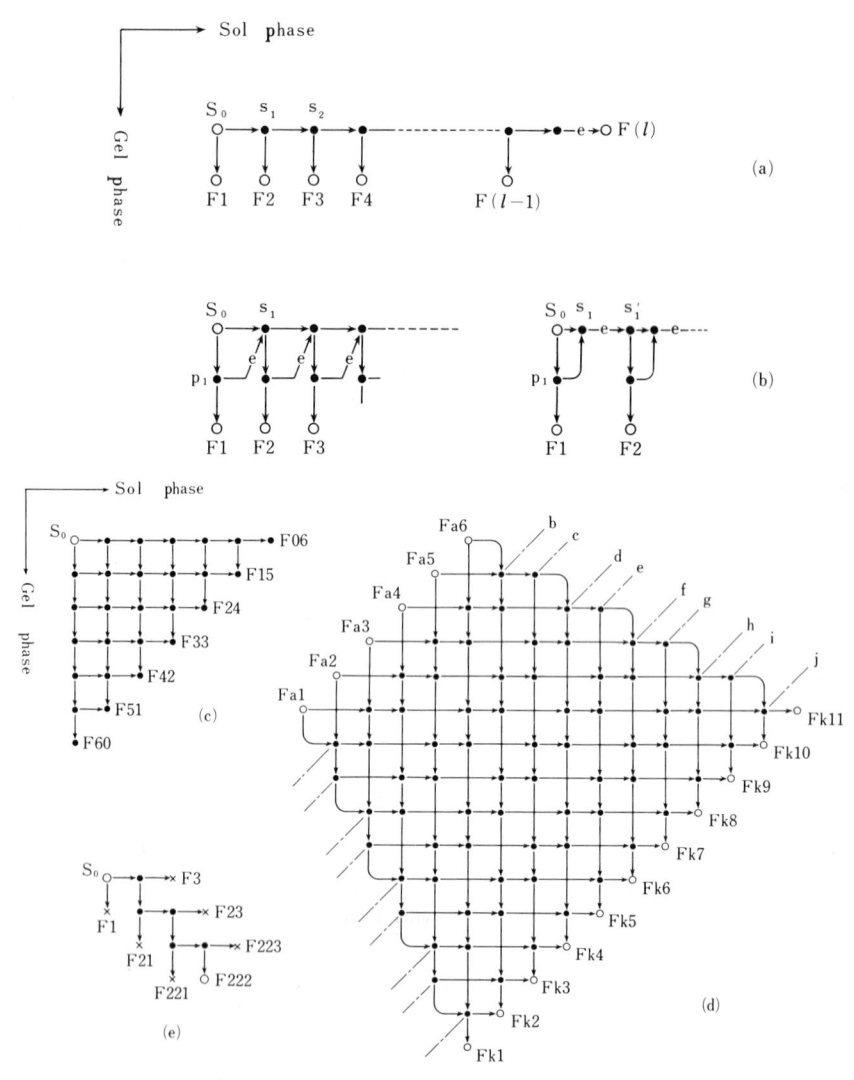

FIG. 6. Schemes for fractional precipitation. (a) Simple procedure. (b) Refractionation procedure (Schulz [29]; Flory [30]). (c) Triangle method (Meffroy-Biget [31]). (d) Scheme of Thurmond and Zimm [32]. (e) Scheme to homogenize a sample. See text for explanation.

is so, because the concentration of the sample is decreased, both by the successive addition of nonsolvent and by the separation of the preceding fractions. The fact that reverse-order precipitation, cited in Section B,3, appears in the first few fractions is chiefly due to this situation.

Attempts have been made to carry out the second steps of the refractionation, i.e., the fractionation of p_1 (and so on), under more dilute conditions. In this case, the volume of solution does not become so large, because the quantity of p_1 is only a fraction of that of S_0. For example, Flory [30] has fractionated samples of polyisobutylene starting with an approximately 1% benzene solution and using acetone as a nonsolvent. The first gel phase, after being separated, was diluted to 0.1% with benzene, and the same proportion of nonsolvent added to obtain the first fraction $F1$. The mother liquor was recombined and the acetone evaporated off almost completely by steam. The solution was then made up to the starting volume and the second fractionation obtained as before. As can be seen in this example, it is recommended that a solvent–nonsolvent system be used where both are sufficiently volatile and preferably with the latter having the higher vapor pressure.

The refractionation process is considered really effective in increasing the homogeneity of the fractions, but, unfortunately it is laborious and too time consuming. A compromise is often adopted by applying the refractionation steps only to the first few fractions to improve their homogeneity and to save time and labor.

The time-consuming nature of the fractional precipitation process is chiefly due to the need for attainment of quasi-equilibrium conditions throughout the whole process, i.e., careful addition of nonsolvent, a long waiting time just before the cloud point, and sufficient agitation time both during the equilibration of the gel and sol phases after the addition of an excess amount of nonsolvent, and during the cooling of the system which has been heated to redissolve the precipitate. A device for avoiding this time wastage is to fractionate quickly by sacrificing the long equilibration periods, and then to refractionate several more times in the same way.

Two examples of such schemes are shown, although various combinations of refractionation processes have been devised. Meffroy-Biget [31] has reported that the "triangle fractionation" of nitrocellulose [Fig 6(c)], in which the refractionation was repeated six times, required 3 weeks. Thurmond and Zimm [32] have fractionated branched polystyrene following the scheme, shown in Fig. 6(d), which was planned so that the lowest molecular weight component of fraction $Fa1$ would end up in the final fraction $Fk11$ of least molecular weight, and vice versa; fractions $Fa1$ to $Fa6$ were obtained beforehand by the simple fractional precipitation procedure. The whole procedure was said to have taken 5 weeks.

The main object in carrying out fractionation experiments is considered to be twofold. The first is to examine the molecular weight distribution in a given polymer sample, and the second is to obtain fractions of high homogeneity and having different (average) molecular weights. For the latter purpose, it is not always desirable to fractionate one sample. It is sometimes wiser to start with a number of samples having different average molecular weights and to extract their main component. Meyerhoff [33], for example, attempted the fractionation of a number of samples of polymethyl methacrylate into highly homogeneous fractions by means of a scheme [shown in Fig. 6(e)] in which he started with a variety of S_0's and collected $F222$'s. The scheme was so arranged that, for example, in the first fractionation stage, fractions $F1$ and $F3$ would amount to only 20–30% of the parent polymer sample, while the main portion appeared in fraction $F2$. Subdivision of $F1$ was by normal fractional precipitation, whereas the separation of $F3$ almost amounted to an extraction process since the amount of polymer in the gel phase was far greater than that in the sol phase. Other examples are the fractionation of polystyrene by Cantow *et al.* [34], of polyvinyl alcohol by Dailer *et al.* [35], and of isotactic polystyrene by Natta *et al.* [36]. In the last example the schedule was somewhat modified.

C. Additional Comments on Experimental Technique

Some of the factors which are of importance in choosing the optimum experimental conditions have been discussed in Section I,B. Here we deal with certain aspects of the experimental technique; these remarks may be regarded as a supplement to the description in Section I,A.

1. *Scale of Experiments*

This is decided by the quantity desired in each fraction and by the proposed number of fractions. For determination of the molecular distribution of a common polymer, it is possible to start with an amount less than 1 gm and to take more than 20 fractions. Providing the average molecular weight of the sample is sufficiently high, one may be content to carry out the molecular weight determination of each fraction by the so-called "one-point method" of viscometry. The largest amount of sample that can be handled in a laboratory is usually about 25 gm, when one uses a 5-liter flask and starts with a 1% solution. At the same time, allowance should be made for the volume of nonsolvent to be added to the system.

2. *Equipartitioning of Fractions*

One of the most difficult features in fractionation experiments is to adjust the conditions so that each fraction has practically the same weight. There is no straightforward method by which the amount of precipitate may be estimated with any certainty before it has settled. Turbidity measurements seemed to be useful, but they are not so simple to apply (see Section II,B,4). After phase separation, it is possible to roughly estimate the amount of precipitated polymer by measuring the volume of gel phase, since the swelling ratio may be found by preliminary experiments. Thus, Boyer [9] has found the swelling ratio for polystyrene to be about 12–15 in a benzene-methanol system. If the estimated quantity of precipitate is smaller than that desired, it may be increased by adding a further amount of nonsolvent, and if it is larger, solvent may be added to the system. However, some experience is necessary because the amount of nonsolvent or solvent to be added is not always proportionate. Furthermore, it is essential that the system be equilibrated once more, so that these adjustments tend to approximately double the experimental time.

3. *Temperature*

As the phase equilibrium is very sensitive to a temperature variation, it is imperative that the experiments be carried out at constant temperature. It is recommended that the temperature deviation be not larger than $\pm 0.05°C$, although this will depend upon the solvent–nonsolvent system, as well as on the magnitude and range of molecular weights in the sample. Furthermore, the working temperature should be as close as possible to the mean room temperature, so that the separation process may be carried out at the same temperature. This precaution is especially necessary when one wishes to use centrifugal separation.

4. *Treatment of the Precipitated Polymer*

As described in Section I,A, the gel phase is usually collected by dissolving it in solvent, and is then precipitated in a large amount of nonsolvent. However, it is better to evaporate the solution directly, especially in cases where the precipitated polymer sticks to the equipment. During the evaporation, nonsolvent must be added to precipitate the globular polymer before the formation of polymer film (if necessary, add more than one lot of nonsolvent!), otherwise solvent is included in the film and is very difficult to remove even in a vacuum at high temperatures. Of course, a freeze-drying technique is preferable, if there is a suitable solvent (e.g., the frozen benzene technique). It must be noted that the drying temperature must be over the glass-transition temperature of the polymer even when the drying is being done under vacuum.

5. *Some Special Devices to Assist in Gel-Phase Formation*

The successful fractionation of polyacrylonitrile is reported to require the use of certain devices. Kobayashi [37] has pointed out that the micro particles, formed by addition of benzene or toluene to the hydroxyacetonitrile solution of the polymer, are converted to micro gels with the help of ethyl alcohol at 35°C. He has also shown [38] that the addition of aqueous hydrochloric acid to the solvent (dimethyl formamide) is effective for the fractionation of the same polymer (nonsolvent: toluene). Mikhailov and Zelikman [39] have reported that fractionation occurs when the dimethyl formamide solution is heated to 60°C with the nonsolvent heptane. This procedure fractionates only about a half of the dissolved polyacrylonitrile polymer; the rest of the sample was obtained by the addition of a 2% $CaCl_2$ solution.

II. Fractionation by Solvent Evaporation

Since this method depends on a decrease in solvent power which is effected by evaporation of the solvent component, it is an alternative method for fractional precipitation. It differs from the nonsolvent addition method in the manner in which the composition of the solvent–nonsolvent system is altered.

Certain advantages are expected from this procedure:

(1) The volume of the system decreases as the fractionation proceeds.
(2) Continuous change of the solvent–nonsolvent composition is possible.
(3) Local concentration of the nonsolvent can be avoided.

The first item allows us to nearly double the scale of the experiment—about 90% of the total volume of the container can be utilized.

The second and third items make it possible to carry out the experiments at nearly quasi-equilibrium conditions. In practice the operations are easier to control than in the case of the nonsolvent addition method.

A. General Procedure

The polymer solution is introduced into a three-necked flask, placed in a thermostat and fitted with a powerful stirrer, and nonsolvent is added with vigorous agitation until the system almost reaches the cloud point. The solution is then left for a few hours. One neck of the flask is fitted with a gas inlet tube which is surrounded by a warm-water jacket and attached to a large $CaCl_2$ tube; through this inlet a stream of warm dry air is introduced and used to evaporate the solvent. Warm air is used since it serves to compensate for the heat losses, caused by vaporization of the solvent, preventing the surface of the liquid from cooling. The other neck is provided with a gas outlet tube, which is connected to a water pump through two traps, one of which is inverted to avoid a head current from the aspirator.

B.1. Fractional Precipitation

Some means of detecting the onset of turbidity should be used. For example, a suitable apparatus is one employing a phototube to measure the light transmitted by the solution. It is more convenient if some warning signal can be given when the turbidity increases beyond a given value.

Let the air stream begin to pass through the flask while the solution is continuously stirred. When the turbidimeter gives a signal, the air stream is slowed down and then stopped after the turbidity of the solution reaches some definite value. The solution is warmed several degrees to remove the turbidity and then allowed to cool down to the original temperature. After confirming that the turbidity is back to the value which it had before warming the solution, the stirring is stopped to allow the gel particles to settle [2]. The completeness of sedimentation is verified by the transparency of the solution with the aid of the turbidimeter. Then, the precipitated phase is separated from the supernatant liquid, and all the operations thereafter are entirely the same as the nonsolvent addition method described in Section I.

The chart of the turbidity of the solution, shown in Fig. 7, is taken from Kawahara's report [40] on the fractionation of polystyrene in a butanone-butanol mixture.

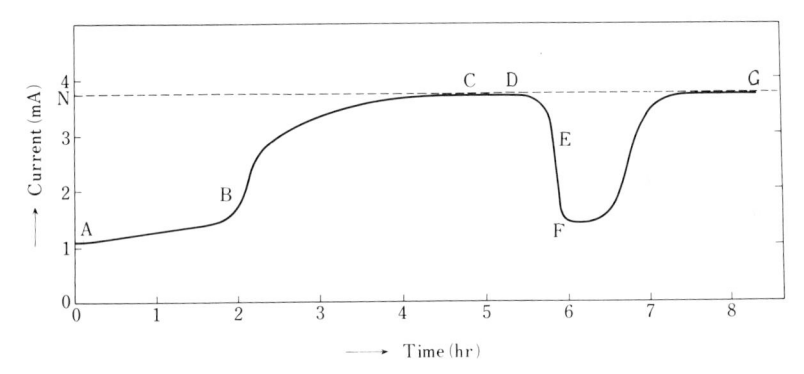

Fig. 7. Turbidity of solution in the course of fractionation experiment by evaporation (Kawahara [40]). A, Start of evaporation; B, rate of evaporation decreased; C, evaporation interrupted, while continuing the agitation; D, Solution warmed; F, solution cooled; G, agitation interrupted.

B. Special Experimental Conditions

1. *Equipment*

The equipment is similar to that used in the nonsolvent addition method, described in Section I,B,1, except that the fractionation flask should have at least two side necks to fit the gas inlet and outlet tubes.

2. Choice of Solvent–Nonsolvent System and Concentration of the Solution

It is absolutely necessary to choose a nonsolvent which is less volatile than the solvent. Otherwise, all the considerations for the selection of a solvent–nonsolvent system and also for the concentration of the solution are just the same as in the case of the previous discussion, i.e., Sections I,B,2 and I,B,3. For most common polymers, examples of fractionations by this method may be found in the literature. For polymers which are only soluble at high temperatures, for example, polyethylene or Kel-F, this method has been used in preference to the nonsolvent addition method.

3. The Experimental Scale and Fractionation Scheme

As stated above, it is possible to double the experimental scale using the same equipment, for example, 90 gm of sample may be fractionated at one time in a 10-liter flask. So it can be seen that this method is rather convenient for large-scale experiments.

Usually refractionation is not included in the experimental scheme when this method is used, perhaps because the refractionation procedure involves an increase in the volume of the system. As stated in Section I,B,4, if the scheme includes refractionation, then the usual evaporation step must also be included in order to reduce the volume of the system. It is necessary to choose a nonsolvent which can be distilled easily under reduced pressure but is still less volatile than the solvent.

4. Equipartitioning of Fractions

This procedure, when used in conjunction with a simple turbidimeter such as that described in Section II,A, is a popular choice for furnishing fractions of nearly equal weight. Table I is an example of the fractionation of a polystyrene sample by Kawahara [40] using a butanone–n-butanol system at 25°C. It must be noted that the turbidimeter failed to be of use for the last few fractions. In these, some additional time of evaporation was necessary after the turbidity had reached the set value. The extra time required had to be estimated roughly.

5. Temperature

It is always advantageous to work at a temperature near room temperature. An effort should be made to make the solvent evaporate moderately fast, by reducing the pressure in the system, increasing the gas flow velocity, and adjusting the position of the gas stream so that it is near the surface of the solution.

B.1. Fractional Precipitation

TABLE I
Data from a Fractionation by Solvent Evaporation[a]

Fraction number	Yield (grams)	Yield (%)	$[\eta]$	Huggins constant (k')	Average molecular weight ($M \times 10^{-4}$)
1	3.04	3.38	6.38	0.38	681
2	3.64	4.03	5.76	0.35	577[b]
3	2.73	3.03	5.91	0.38	603[b]
4	4.29	4.77	5.49	0.39	535
5	3.04	3.38	5.33	0.38	510
6	3.54	3.94	5.12	0.37	478
7	2.85	3.17	4.95	0.39	452
8	4.43	4.92	4.91	0.37	447
9	3.79	4.21	4.74	0.38	422
10	5.63	6.25	4.53	0.37	392
11	4.50	5.00	4.30	0.37	360
12	4.99	5.55	3.98	0.35	318
13	4.11	4.57	3.76	0.32	291
14	3.26	3.62	3.73	0.32	286
15	2.87	3.19	3.48	0.33	256
16	3.70	4.11	3.27	0.34	232
17	4.87	5.41	3.07	0.33	210
18	4.05	4.50	2.75	0.37	175
19	4.41	4.90	2.41	0.35	142
20	3.77	4.19	2.07	0.38	110
21	3.52	3.92	1.73	0.35	83
22	5.54	6.15	1.10	0.29	40
Residue	0.20	0.22			
Total	86.77	96.41			
Unfractionated sample			4.01	0.33	321

[a] Kawahara [40].
[b] Molecular weight reversed.

III. Fractionation by Cooling

This procedure depends upon a decrease in the temperature of the system causing a decrease in the solvent power. The merits of this procedure may be outlined as follows:

(1) The experiment is carried out in only one solvent system.
(2) The volume of the system is constant.
(3) Lack of uniformity in the medium can be avoided.
(4) Control of the fraction size can be adjusted more precisely.

Certain disadvantages are also experienced:

(1) Care must be taken that degradation of the polymer does not occur when high temperatures are involved.
(2) All the operations before the separation of the two phases, must be conducted at the same temperature as the precipitation.
(3) It is rather difficult to find a suitable solvent–nonsolvent system which will completely precipitate the polymer on cooling.

A complete fractionation by this method is not frequently made because of the disadvantages cited above. It regularly finds a partial application in experiments using the nonsolvent addition method [41–43].

It sometimes happens that the nonsolvent used is too powerful and that too small a quantity of it is needed to obtain the succeeding fraction. In such a case, it is preferable to obtain the succeeding fractions by cooling the system, because of the third and fourth merits cited above. This situation is most likely to arise in taking the first few fractions, where the molecular weight of the precipitated polymer is high. If possible, a nonsolvent of weak precipitation power should be employed.

A. General Procedure and Other Remarks

The procedure for this method is essentially very simple and does not need to be described here because it forms only a part of the procedures of the two methods discussed above.

The equipment shown in Fig. 2 or Fig. 3(a), for example, has been found to be suitable. The thermostat should be provided with a supplementary heater connected to a variable transformer and of sufficient power to maintain its temperature just below the desired one. Cooling below room temperature is usually not recommended because thermostating is difficult.

The cooling process should be sufficiently slow to approach the quasi-equilibrium condition. Usually several hours are necessary to lower the temperature of the system to the next stage. A stepwise decrease of the temperature of the thermostat is expedient, preferably under automatic control.

The size of fraction is determined by the temperature difference; its control is rather easier. It is always necessary to repeat the operation once more to reach an equilibrium. Hall [11] has recommended that after each fraction has been obtained, the system be reheated above the equilibrium temperature. This redissolves any precipitate unavoidably left on the wall of the fractionation vessel, before cooling to the next equilibrium temperature. If all of the fractions can not be obtained by cooling, the solution is warmed to the original temperature, and the same procedure repeated after a suitable amount of nonsolvent has been added to the solution [44]. Otherwise, the residual components have to be recovered and subjected to another method of fractionation.

For the solvent, Flory [4] has suggested that a mixed solvent system is preferable to a single solvent, in regard to the efficiency of fractionation. This has not yet been confirmed experimentally. However, in practice, it is very rare that a suitable one-solvent system is found.

B. Solvent Systems for Semicrystalline Polymers

The cooling procedure is particularly effective for "solvent-resistant polymers" such as polyethylene, polypropylene, polyfluoroolefins, and others, which are soluble in suitable solvents only at high temperatures. Most of these polymers belong to the group of so called "crystalline" polymers. We will now discuss as far as present experimental results permit solvent systems for their fractionation.

The phase equilibrium between a crystalline polymer and its solution is different from a two-liquid-phase equilibrium. A solid is stabilized below the liquid state by a certain amount of energy, which is determined by its heat of crystallization, ΔH_c, and equal to $\Delta H_c/RT$, where R is the gas constant and T, the absolute temperature.

In the case of a two-liquid-phase equilibrium, it is necessary to assume that the solubility of each polymer molecule depends only on its molecular weight. However, for a crystalling polymer, the relation between ΔH_c and the molecular weight must also be considered. Further complications may arise, because all of these polymers form only partially crystalline states. The degree of crystallization depends upon various factors. These are the molecular weight of the polymer, the solvent system applied, and so on. This is why the two-liquid-phase separation is desirable for the fractionation of such semicrystalline polymers.

The phase diagrams for these semicrystalline polymers have not yet been examined apart from the polyethylene–solvent system reported by Richards [45], and the poly(N,N'–sebacoylpiperazine)–solvent system reported by Flory *et al.* [46]. Only the former report will be discussed in some detail here.

Polyethylene is known as a typical polymer which may easily crystallize to a large extent. The results of Richards, which are of interest to us, are shown in Fig. 8(a) and (b). Figure 8(a) shows the border curves for three solvents, where A is the melting point of polyethylene, and curves $A B C$, and so on, represent the phase-separation temperature at given concentrations. It shows a prominent loop which represents the separation into two liquids and is similar to that found in amorphous polymer systems. This tells us the following: (1) Nitrobenzene has the widest concentration range, where the behavior of the system is similar to that of the amorphous polymer–solvent system. It is well known from other facts that nitrobenzene is the poorest of the three solvents for polyethylene. (2) Although xylene is

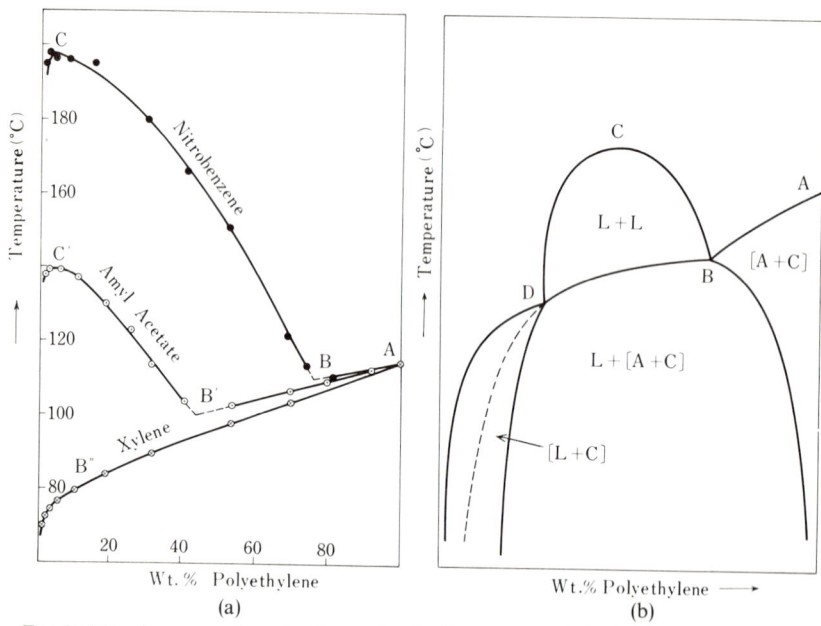

FIG. 8(a) Border curves for polyethylene in nitrobenzene, amyl alcohol, and xylene (Richards [44]). (b) Phase diagram for polyethylene–poor solvent systems showing the two-liquid region (not to scale) (Richards [44]). L, a free liquid or rubberlike amorphous phase; C, a free crystalline phase; A, a liquid phase in the form of an amorphous region of "solid" polyethylene surrounding the crystallites C. The bracket indicates that the pair of phases are not mechanically separable. The broken line indicates the disappearance of turbidity, which may occur at some temperature below that at which the last of the crystallites has dispersed. See text for further explanation.

classed as a good solvent, it behaves in an entirely different way. This may be said to be the normal behavior for a crystalline polymer solvent. (3) Amyl acetate is also known as a poor solvent, and has similar, but intermediate character to nitrobenzene. A similar diagram has been obtained in the case of the polyamide studied by Flory et al. [46], where m-cresol and o-nitrotoluene behave as good solvents, but diphenyl ether behaves like a poor one. It follows that the solvent system to be used for the fractionation of a semi-crystalline polymer by cooling must be as poor a solvent as possible.

Next, let us examine the phase diagram, Fig. 8(b) (it is not to scale). It can be seen in the figure that the region in which the two liquid phases can coexist has a low concentration limit, which is not seen in Fig. 8(a). This may be due to its location in a very dilute concentration range. There are two critical concentrations for the polyethylene-poor solvent system. Behavior as an amorphous polymer–solvent system lies between these two concentrations. In other words, if the solution is too dilute or too concentrated, no liquid–liquid separation is expected on cooling.

As has been frequently stated, the concentration of the system must of necessity decrease as fractional precipitation proceeds. It is possible that during the fractionation of semicrystalline polymers, in a poor solvent by cooling, the phase separation may change from liquid–liquid to liquid–crystalline phases, after several fractions have been separated, if the concentration of the system should decrease beyond the lower critical concentration. This has actually been observed to happen with a polyethylene–(amyl acetate + ~10% xylene) system in the author's laboratory [47].

In the literature, however, it has been reported that polyethylene glycol of low molecular weight is suitable for liquid–liquid phase separation in the fractionation of polyethylene [6,48,49] and isotatic polystyrene [8].

In general, it has been noticed that for the fractional precipitation of semicrystalline polymers by cooling, two conditions must be met to attain liquid–liquid phase separation, i.e., (1) the solvent must be as poor as possible, and (2) the concentration of polymer must be above the lower critical value of the system.

There are no precise experimental details for other cases, i.e., for the polymers less crystallizable than the polymers cited, or for the three-component phase relation for a semicrystalline polymer–solvent–nonsolvent system. At present nothing can be predicted except in the above-cited case of a highly crystalline polymer. Lacking this information, preliminary experiments must be carried out for any contemplated system to ensure regular fractionation.

REFERENCES

1. H. Staudinger, K. Frey, and W. Starck, *Chem. Ber.* **60**, 1782 (1927).
2. H. Staudinger, "Die hochmolekularen organischen Verbindungen," (Reprint) Springer, Berlin, 1960.
3. E. H. Merz and R. W. Raetz, *J. Polymer Sci.* **5**, 587 (1950).
4. P. J. Flory, "Principles of Polymer Chemistry," Chapter 13. Cornell Univ. Press, Ithaca, New York, 1953.
5. J. B. Berkowitz, M. Yamin, and R. M. Fuoss, *J. Polymer Sci.* **28**, 69 (1958).
6. M. L. Nicholas, *Compt. Rend.* **236**, 809 (1953).
7. K. Kawahara, *Kobunshi Kagaku* **18**, 687 (1961).
8. K. Kawahara and R. Okada, *J. Polymer Sci.* **56**, S7 (1962).

9. R. F. Boyer, *J. Polymer Sci.* **9**, 197 (1952).
10. C. Booth and L. R. Beason, *J. Polymer Sci.* **42**, 93 (1960).
11. R. W. Hall, in "Techniques of Polymer Characterization" (P. W. Allen, ed.), Chapter II, p. 19. Butterworth, London and Washington, D.C., 1959.
12. W. G. Harland, *J. Textile Inst., Trans.* **46**, 483 (1955).
13. H. Okamoto, *J. Polymer Sci.* **37**, 173 (1959).
14. K. Überreiter, H. J. Orthmann, and G. Sorge, *Makromol. Chem.* **8**, 21 (1952).
15. K. Überreiter and T. Götze, *Makromol. Chem.* **29**, 61 (1959).
16. R. L. Mitchell, *Ind. Eng. Chem.* **45**, 2526 (1953).
17. G. Gee, *Trans. Faraday Soc.* **40**, 468 (1944).
18. D. R. Morey and J. W. Tamblyn, *J. Phys. Chem.* **51**, 721 (1947).
19. Z. Menčik, *J. Polymer Sci.* **17**, 147 (1955).
20. Y. Fujisaki and H. Kobayashi, *Kobunshi Kagaku* **18**, 305 (1961).
21. R. H. McDowell, *Chem. & Ind. (London)* p. 1401 (1958).
22. C. A. Marshall and R. A. Mock, *J. Polymer Sci.* **17**, 591 (1955).
23. T. Kawai, *Kobunshi Kagaku* **12**, 63 (1955).
24. R. L. Scott, *J. Chem. Phys.* **13**, 178 (1945).
25. A. Münster, *J. Polymer Sci.* **5**, 333 (1950).
26. S. Matsumoto and Y. Ohyanagi, *Kobunshi Kagaku* **11**, 7 (1954).
27. R. L. Scott, *Ind. Eng. Chem.* **45**, 2532 (1953).
28. H. Takenaka, in "Jikken Kagaku Koza (Handbook of Experimental Chemistry)" (A. Kotera, ed.), Vol. VIII, Part I, Chapter II. Maruzen, Tokyo, 1956.
29. G. V. Schulz and A. Dinglinger, *Z. Physik. Chem.* **B43**, 47 (1939).
30. P. J. Flory, *J. Am. Chem. Soc.* **65**, 372 (1943).
31. A. M. Meffroy-Biget, *Bull. Soc. Chim. France* pp. 458 and 465 (1954); *Compt. Rend.* **240**, 1707 (1955).
32. C. D. Thurmond and B. H. Zimm, *J. Polymer Sci.* **8**, 477 (1952).
33. G. Meyerhoff, *Makromol. Chem.* **12**, 45 (1954).
34. M. Cantow, G. Meyerhoff, and G. V. Shulz, *Makromol. Chem.* **49**, 1 (1961).
35. K. Dialer, K. Vogler, and F. Patat, *Helv. Chim. Acta* **35**, 869 (1952).
36. G. Natta, F. Danusso, and G. Moraglio, *Makromol. Chem.* **20**, 37 (1956).
37. H. Kobayashi, *J. Polymer Sci.* **26**, 230 (1957).
38. H. Kobayashi and Y. Fujisaki, *J. Polymer Sci.* **B1**, 15 (1963).
39. N. V. Mikhailov and S. G. Zelikman, *Colloid J. (USSR) (English Transl.)* **18**, 715 (1956).
40. K. Kawahara, *Sen-i Kagaku Kenkyusho Nempo (Ann. Rept. Inst. Fiber Res. Japan)* **9**, 30 (1956).
41. L. H. Arond and H. P. Frank, *J. Phys. Chem.* **58**, 953 (1954).
42. S. N. Chinai and R. J. Samuels, *J. Polymer Sci.* **19**, 463 (1956); S. N. Chinai and R. A. Guzzi, *J. Polymer Sci.*, **21**, 417 (1956); S. N. Chinai, *J. Polymer Sci.*, **25**, 413 (1957).
43. W. H. Beattie and C. Booth, *J. Appl. Polymer Sci.* **7**, 507 (1963).
44. K. J. Ivin, H. A. Ende, and G. Meyerhoff, *Polymer* **3**, 129 (1962).
45. R. B. Richards, *Trans. Faraday Soc.* **42**, 10 (1946).
46. P. J. Flory, L. Mandelkern, and H. K. Hall, *J. Am. Chem. Soc.* **73**, 2532 (1951).
47. A. Kotera, T. Saito, K. Takamisawa, Y. Miyazawa, H. Nomura, T. Kamata, K. Yamaguchi, and H. Kawaguchi, *Rept. Progr. Polymer Phys. Japan* **3**, 58 (1960).
48. A. Nasini and C. Mussa, *Makromol. Chem.* **22**, 59 (1957).
49. H. Wesslau, *Makromol. Chem.* **26**, 96 and 102 (1958).

CHAPTER B.2

Fractional Solution

John H. Elliott
RESEARCH CENTER, HERCULES INCORPORATED, WILMINGTON, DELAWARE

I. Introduction 67
II. Theoretical Considerations 69
   A. Liquid-Phase Equilibria 69
   B. Diffusion 72
   C. Internal Criteria for a Fractionation 73
   D. Fraction Breadth 74
   E. Inhomogeneities in Composition and Structure 74
III. Experimental Methods 74
   A. Fractionation by Direct Extraction 74
   B. Fractionation by Film Extraction 76
   C. Fractionation by Coacervate Extraction 78
   D. Fractionation by Column Elution 79
IV. Factors in Column Elution—Guiding Principles for Experimental Fractionation 82
   A. Column Design 82
   B. Solvent–Nonsolvent Pair 83
   C. Polymer Support 84
   D. Method of Depositing the Polymer on the Support 85
   E. Rate of Elution 87
   F. Degradation during Fractionation 87
   G. Fraction Workup 88
   H. Reversals in Molecular Weight with Fraction Order 88
V. Comparison between Fractional Solution Methods 89
VI. Possible Areas for Future Research 90
   References 92

I. Introduction

   The almost explosive growth of polymer science over the last two decades has led to many advances in our ability to characterize polymers in some detail. Fractional solution techniques for polymer fractionation have played a significant role in these advances.

   In the following discussion, we shall be concerned primarily with fractionation by molecular weight. Our knowledge of fractionation by this variable is far more advanced than our ability to effect separations by other parameters such as small differences in composition, structure, or tacticity. For a detailed discussion of separations of the latter kind Chapter D may be consulted.

Emphasis will be placed on the scientifically and commercially important polyolefins, polyethylene and polypropylene. No apology is made for this choice of polymers, since space limitations make it impossible to consider a wide variety, and it seemed best to discuss this one important class in some depth. An extensive summary of polymer fractionations is given in Chapter G, and applications of fractional solution methods to other classes of polymers may be found there.

A fractionation by molecular weight may either be analytical or preparative. In an analytical fractionation, the goal is to obtain the molecular weight distribution curve of the polymer, which is then to be related to other characteristics of the sample, such as the method of synthesis or degradation, or the physical and mechanical properties. Isolation and characterization of individual fractions are done only to obtain the distribution curve and are time-consuming and burdensome parts of the experimental process. To simplify analytical fractionations, methods such as turbidimetric titration, sedimentation, and gel filtration, which do not isolate individual fractions, are being widely studied. These techniques are discussed in detail in other chapters. At our present state of knowledge, however, most of these methods must be calibrated by an analytical fractionation in which fractions are isolated and, at least, their molecular weights determined.

Preparative fractionation, on the other hand, is carried out to obtain significant quantities of a polymer having a narrow molecular weight distribution. (The distribution curve for the whole polymer is a by-product.) These narrow fractions are then used in other research; for example, the establishment of an intrinsic viscosity–molecular weight relationship, a study of physical properties as a function of molecular weight and molecular weight distribution, or as starting materials for chemical studies.

Fractionation by fractional solution lends itself to both analytical and preparative work. One of its greatest advantages is that it involves the handling and manipulation of far smaller quantities of solvent than fractional precipitation. For analytical work, small samples of the polymer (~ 1 gm) may be used to obtain the distribution curve and the technique is readily adapted to automation. For larger scale preparative fractionation, the relatively small quantities of solvent involved become a most important practical consideration. In addition, on theoretical grounds, one would expect the higher fractions obtained by fractional solution to be less contaminated with lower molecular weight polymer than the corresponding fractions obtained by precipitation.

Fractionation of a polymer by fractional solution involves preparing the polymer in an appropriate physical state and then extracting fractions of increasing molecular weight by use of a series of eluents of increasing

B.2. Fractional Solution

solvent power. In contrast to fractional precipitation, the lowest molecular weight fraction is the first and the highest is the last to be obtained. Many experimental arrangements have been employed. The finely divided polymer itself may be extracted or it may be deposited on a thin aluminum foil, or on a support (e.g., sand) in a column. A concentrated solution (coacervate) containing an appreciable quantity of polymer may be selectively extracted. Common to all of these methods is the preparation of the polymer in a form which is amenable to rapid extraction by the eluent. The immobilization of the polymer so as to permit this extraction with a minimum of mechanical manipulation is essential. Detailed consideration of the various methods will be given later. It seems appropriate, however, to point out at this time a differentiation between column techniques that has been made in the literature. When the polymer is eluted at a constant temperature, the process is called fractional solution or elution fractionation. On the other hand, when a temperature gradient is used alone or is superimposed on a solvent gradient, the process is called chromatographic fractionation. While this chapter is concerned with fractional solution, a comparison will be made with the chromatographic method since the differences between the two appear to be more semantic than real.

II. Theoretical Considerations

A. Liquid-Phase Equilibria

The present theoretical basis for fractionation, by way of solubility differences, is the Flory–Huggins statistical thermodynamic treatment of polymer solutions. While this theory and its implications have been discussed in detail in an earlier chapter, it is appropriate to restate certain of its consequences. It predicts that when a homogeneous polymer solution separates into two liquid phases as a result of a lowering of the solvent power of the solvent (due to a change in composition or temperature), the following will occur. A small quantity of a polymer-rich phase will separate, in equilibrium with a relatively large liquid phase of lower polymer concentration. If v'_x is the volume fraction of polymer of degree of polymerization, x, in the concentrated phase and v_x the volume fraction in the dilute phase, then

$$\ln(v'_x/v_x) = \sigma x \qquad (1)$$

Sigma is a function of the volume fractions and number average molecular weights of the polymer in each phase and also of χ_1, the dimensionless quantity characterizing the interaction energy of the solvent molecule with the polymer.

Theory does not permit the exact calculation of σ, and Flory [1] suggests that it be considered as an undetermined parameter. It is important to note, however, that it is a function of χ_1, the interaction coefficient. For a solution to separate into two phases the value of χ_{1c}, the critical value of χ_1, is

$$\chi_{1c} \cong \tfrac{1}{2} + x^{-1/2} \tag{2}$$

Values of χ_1 below this lead to homogeneous solutions and obviously no fractionation.

Fractionation by either precipitation or fractional solution, then is basically the adjustment of χ_1 so that two liquid phases are in equilibrium, removing one phase, and then adjusting χ_1 to obtain another set of equilibrium conditions. This adjustment of χ_1 can be obtained, in the case of fractional solution, using a solvent–nonsolvent pair, by increasing the solvent/nonsolvent ratio and thus lowering the value of χ_1. It can also be done by increasing the temperature, when a single solvent is used, since it has been found that

$$\chi_1 = A + B/T \tag{3}$$

where T is the absolute temperature and A and B are constants. Consideration of Eq. (1) shows first that a part of any polymer species is always present in each phase and, secondly that every species is more soluble in the concentrated than in the dilute phase, since σ is positive.

If we now let f_x be the fraction of the x-mer in the dilute phase and f'_x that in the concentrated, and $R = V'/V$, the ratio of the volumes of the two phases, then

$$f_x = 1/(1 + Re^{\sigma x}) \tag{4}$$

and

$$f'_x = Re^{\sigma x}/(1 + Re^{\sigma x})$$

giving

$$f_x/f'_x = 1/Re^{\sigma x} \tag{5}$$

This clearly shows that a low value of R, or fractional solution into a dilute solution, improves selectivity. Equation (5) has a further implication. Since σ is a function of χ_1 which in turn is a function of x, the ratios of the phases should be varied as a function of x for good fractionation. Flory [1] suggests that the maximum concentration should vary with $x^{-1/2}$. Pepper and Rutherford [2], in a thorough study of the large-scale fractionation of polystyrene by gradient elution, believe that the concentration of the polymer in the eluent is the most important single factor in fractionation efficiency. Schneider et al. [3] using their data have, at least qualitatively, confirmed this rule.

As shown in Eq. (2), χ_{1c} is a linear function of $x^{-1/2}$. Thus very small changes in χ_1, which require very careful control of temperature and eluent composition, are necessary to achieve good resolution in the high molecular weight region. This sensitivity to χ_1 is illustrated in Fig. 1, which shows the molecular weight of isotactic polypropylene fractions as a function of the composition of the eluent (temperature, 156°C) [4].

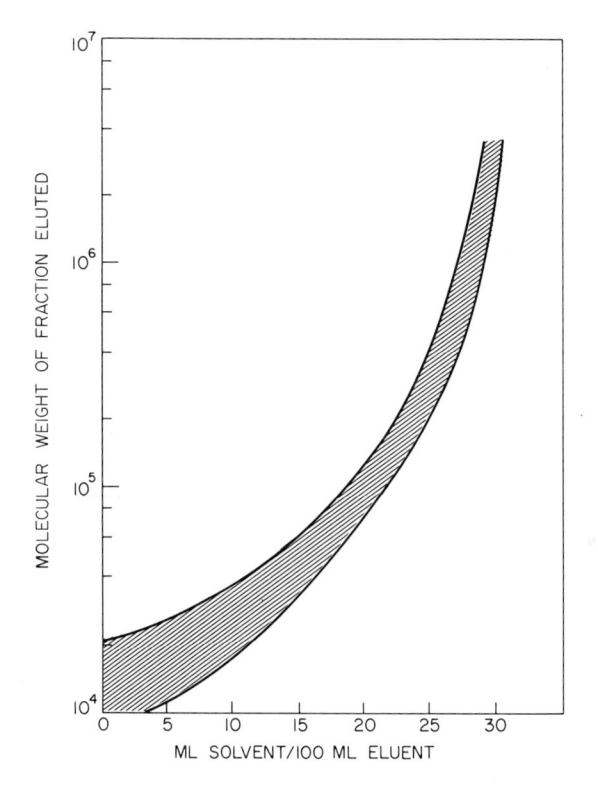

FIG. 1. Molecular weight of isotactic polypropylene fractions versus percent solvent in eluent (redrawn from data of Shyluk [4]).

Schulz et al. [5] have recently treated chromatographic fractionation in terms of stationary gel phase and a moving solution phase. General mass transport equations were derived, from which the equilibrium distribution of each component between the two phases could be formulated in terms of eluent composition, temperature, and degree of polymerization. This theoretical treatment permitted model calculations of column efficiency.

These considerations of phase equilibria do not permit quantitative predictions for any given fractionation due to our lack of detailed knowledge of the various parameters and to the complexity of the calculations. However, this limitation in no way detracts from the usefulness of the theory in establishing guiding principles.

B. Diffusion

It has been implicit in the previous discussion that equilibrium conditions prevail during a fractionation by fractional solution. This is not necessarily true, and indeed things are often arranged to take advantage of diffusion, rather than to establish true equilibrium. To illustrate this point, the method of deposition of the polymer on an inert support is often critical. Kenyon and Salyer [6] compared deposition of polyethylene by slow cooling (selective deposition) with simple solvent evaporation (nonselective deposition). The nonselective deposition gave poor resolution while the selective deposition resulted in good fractionation. Their work with polystyrene showed no difference with the method of deposition, which was attributed to its having a higher diffusion rate than polyethylene. Shyluk's [4] work with atactic and isotactic polypropylene brought out the importance of the deposition step very clearly. Figure 2 compares the integral distribution

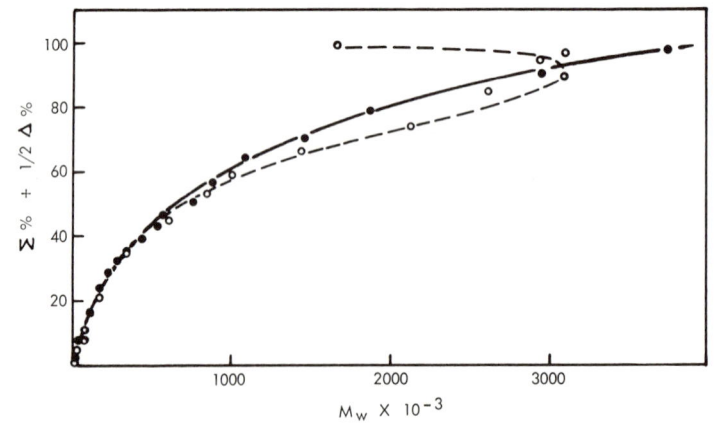

FIG. 2. Integral distribution curves for a sample of isotactic polypropylene deposited on sand by cooling from 126°C to room temperature in one hour (○) and in six hours (●) [4].

curves obtained on a sample of isotactic polypropylene when 1 hour and 6 hours were taken for deposition by cooling. The longer period gave a striking improvement in resolution and eliminated fraction-order reversal. Selectively depositing the higher molecular weight polymers first and the lower molecular weight last, greatly improved the resolution of the fractionation. Here advantage is taken of the low diffusion rates of polymers, and it is doubtful if true equilibrium of all polymer species between the two phases was ever achieved.

The slowness of polymer diffusion, coupled with the practical necessity of carrying out a fractionation within a reasonable time, accounts for the fact that the great majority of fractionations by fractional solution have been carried out by extraction of a liquid (or gel) phase. On the other hand, if equilibrium could be reached or closely approached in a reasonable time, fractionation involving a crystalline phase offers the possibility of achieving a high degree of selectivity. Meyer [7] showed that for normal hydrocarbons, the heat of fusion (610 cal/CH_2 group) is the critical parameter determining solubility and that a high degree of selectivity with molecular weight is to be expected. Measured solubilities of $C_{34}H_{70}$ and $C_{60}H_{122}$ in Decalin were in good agreement with the theory. As discussed in a later section, crystalline polymers have been fractionated by direct extraction and this is potentially a powerful technique. The operational difficulty is the time necessary to achieve any reasonable approach to equilibrium.

In fractionation by column elution, diffusion plays another and sometimes disconcerting role. As the polymer remaining on the column, toward the end of a fractionation, is of high molecular weight and in a highly swollen state, several things may happen. First, the column may "plug" due to the inability of the eluent to move through the swollen polymer which fills the interstices between the particles of the support. Second, low molecular weight material may be trapped inside this swollen gel, leading to molecular weight reversals with fraction number. These difficulties can be avoided by proper column design and polymer deposition techniques.

C. Internal Criteria for a Fractionation

There are, fortunately, two criteria which should always be used to establish that a given fractionation has been successful. First, the sum of the weights of the fractions should equal that of the original polymer. Second, since degradation, oxidation, and cross-linking during fractionation are ever present dangers, the weight average intrinsic viscosity of the fractions should be equal to that of the unfractionated polymer, i.e.,

$$[\eta] = \sum_i w_i [\eta]_i \tag{6}$$

where w_i is the weight fraction of the ith fraction, having an intrinsic viscosity $[\eta]_i$. Unless these two criteria are met, within experimental error, the fractionation cannot be considered successful.

D. Fraction Breadth

A brief word of caution about fraction breadth is in order. If a fraction has a weight average to number average molecular weight ratio ($\overline{M}_w/\overline{M}_n$) approaching unity, there is a strong tendency to regard it as essentially homogeneous. Such fractions are not homogeneous, and this fact must be considered in subsequent measurements. For example, a fraction, following a log-normal or Wesslau [8] distribution whose weight average molecular weight is 102,500 and having $\overline{M}_w/\overline{M}_n = 1.05$, will contain 5% by weight of polymer of molecular weight below 70,000 and 5% above 143,000.

Kenyon and Salyer [6] examined a polystyrene fraction, having $\overline{M}_w = 326,000$ and $\overline{M}_n = 278,000$ (ratio 1.17) in the ultracentrifuge and found species of molecular weight ranging from 38,000 to 740,000.

E. Inhomogeneities in Composition and Structure

The previous discussion has been concerned with the fractionation by molecular weight of a chemically homogeneous polymer. In many important instances this is not the case. There may be chemical differences between molecules such as variations in degree of substitution in the case of cellulose derivatives, or comonomer ratio in the case of copolymers. There may be steric differences, arising from branching, and in the case of polyolefins, atactic, isotactic, syndiotactic, or stereoblock copolymer species may be present. Such variations profoundly alter the behavior of a polymer and, as pointed out by Guzmán [9], generally require a fractionation by structure or composition and then one by molecular weight. An extensive discussion of these problems is presented in Chapter D.

III. Experimental Methods

A. Fractionation by Direct Extraction

As can be seen in earlier reviews [10], fractional solution, when used, was carried out by direct extraction of the polymer. After the pioneering work of Desreux and his colleagues [11–13], who developed the gradient elution method employing a column, direct extraction has been used only infrequently for fractionations by molecular weight. The difficulty is that the polymer particles become highly swollen during the extraction and equilibrium is very hard to attain [9,14]. Following the discovery of synthetic stereospecific polymers by Natta and his co-workers, however, direct extraction proved to be a very useful tool for carrying out a preliminary fractionation by structure, preparatory to subsequent fractionation by molecular weight. The technique is simple, involving merely the extraction of the finely divided polymer in either a flask or an extractor at an appropriate temperature. Krigbaum et al. [15] separated the atactic fraction of poly(1-butene) by extracting the whole polymer with boiling ethyl ether, the isotactic portion being insoluble under these conditions. Both fractions were then fractionated by molecular weight. Van Schooten et al. [16] carried out a large-scale fractionation (1000 gm) on polypropylene, separating the boiling-ether-soluble and boiling-heptane-soluble portions. The latter fraction, with the heptane insoluble residue, was further fractionated for a subsequent study of physical and mechanical properties. For an infrared investigation of polypropylene Luongo [17], prepared atactic polymer by extracting a sample, prepared by the Ziegler method, with acetone, and isotactic polymer by extracting a highly isotactic sample with heptane, then ether.

Natta and his co-workers [18–20] have made extensive use of extraction methods in their stereochemical characterization studies of polypropylene. The atactic portion is soluble in boiling ether while the isotactic is insoluble in boiling n-heptane. By successive extractions with a series of solvents of increasing boiling point, e.g., n-pentane, n-hexane, n-heptane, fractions of an intermediate degree of order and hence crystallinity are obtained. These are defined as stereoblock copolymers of atactic and either isotactic or syndiotactic polypropylene sequences. Further separation of these stereoblock copolymers was effected by an ingenious chromatographic technique [19] which will be discussed later (Section IV,C).

Keller and O'Connor [21] carried out a fractionation by molecular weight of a crystalline linear polyethylene (Marlex 50) using direct extraction in a Soxhlet. The polymer was extracted with trichloroethylene over the temperature range 46° to 86°C and with xylene from 84° to 122°C. The desired extraction temperature was obtained by varying the pressure in the system. Eleven fractions were obtained whose melting points increased monotonically with fraction number. The molecular weights of only a few fractions were determined but these also tended to increase with fraction number. These results are very intriguing and it is unfortunate that these fractions were not more completely characterized. The possibility of such a fractionation by molecular weight of a crystalline polymer is certainly implicit in Meyer's theoretical considerations and experimental results on the solubility of paraffins [7]. In the past, the long time thought to be necessary to achieve equilibrium in such a system has discouraged this approach. This work, however, strongly suggests that the problem should be reexamined.

Harrington and Zimm [21a] have developed an automatic apparatus for fractionation by direct extraction and have applied it to relatively monodisperse high molecular weight polystyrene ($\overline{M}_v = 3.0 \times 10^6$). Very good resolution was obtained using benzene-methanol mixtures of increasing solvent power; however, the elapsed time to fractionate a 1.4-gm sample was greater than 1 week. Effects due to adsorption are eliminated and, if equilibrium is attained, the fractionation is by solubility alone.

Nakajima and Fujiwara [21b] extracted polypropylene with a series of 17 normal aliphatic hydrocarbon fractions boiling over the temperature range 35°–130°C using a vapor-jacketed Soxhlet extractor. The tacticity of the fractions was measured by the infrared method of Luongo [17]. Fraction isotacticity, density, and melting point increased with the boiling point of the extracting solvent. Molecular weight was substantially constant for the first 0.2 integral weight fraction, then increased. Fraction melting points were treated by Flory's [21c] equation for random crystalline-amorphous copolymers. Positive deviations from the theoretical predictions suggest that stereoblock copolymers rather than random crystalline-amorphous copolymers were isolated.

Yamaguchi et al. [21d] extracted polypropylene stepwise in a jacketed Soxhlet with ethyl ether, n-pentane, n-hexane, n-heptane, toluene, and n-octane and compared the results with an elution column fractionation at 166°C using decahydronaphthalene–Butyl Carbitol mixtures. Their results demonstrate that the stepwise direct extraction fractionates primarily by tacticity, with only a minor molecular weight effect, while the column elution procedure fractionates by molecular weight, the tacticity of all these fractions being in a rather narrow range. Their work also indicates that at high tacticity levels, Luongo's tacticity parameter [17] is somewhat molecular-weight-dependent.

B. Fractionation by Film Extraction

Fuchs [22–24] developed an ingenious method of attaining equilibrium within a reasonable time in what is essentially a direct extraction fractionation. A 5–10 μ film of the polymer is deposited onto an aluminum foil by dipping the foil into the polymer solution. The film is then dried and cut up into pieces approximately 1 × 3 cm in size. For 500 to 800 mg of polymer, the surface area of the foil should be 600 to 1000 cm^2. The pieces of coated foil are then placed in an Erlenmeyer flask and extracted with 100 ml of a suitable solvent–nonsolvent mixture by gentle swirling. After equilibrium has been reached, the eluent is removed, and a second portion of somewhat greater solvent power added. This procedure is repeated until the fractionation is completed. The polymer dissolved in the various eluents is isolated and characterized in the usual ways. Fuchs has designed a jacketed flask [24] which permits operation at a controlled elevated temperature. This flask also contains a fritted glass disc, through which the extract is decanted, thus preventing contamination of the extract by bits of polymer which may have been dislodged from the aluminum foil.

Hall [14] gives a very detailed discussion of the application of this technique to the fractionation of polyvinyl acetate, which was thoroughly investigated by Fuchs [22,23]. This method has the advantages of rapidity and of requiring very simple apparatus. Fuchs [24] has used it successfully in the fractionation of a wide variety of polymers. Certain problems, however, arise in its use. In common with all fractional solution methods, the length of time to reach equilibrium in the extraction steps must be established experimentally, and this may be as long as an hour per step [14]. The optimum extent of drying of the polymer film on the foil involves a bit of art. It must be sufficiently dry to adhere to the foil but not so dry that the extraction steps are unduly long. Perhaps the greatest problem is the detachment of the polymer film from the foil during the course of the fractionation. This may be reduced by etching the foil before coating, but if it does occur, the highly swollen flakes of polymer are not efficiently extracted, giving rise to fraction-order reversals in the resulting integral distribution curve. Such an effect in the fractionation of a linear polyethylene is illustrated in Fig. 3. The better resolution of the column method

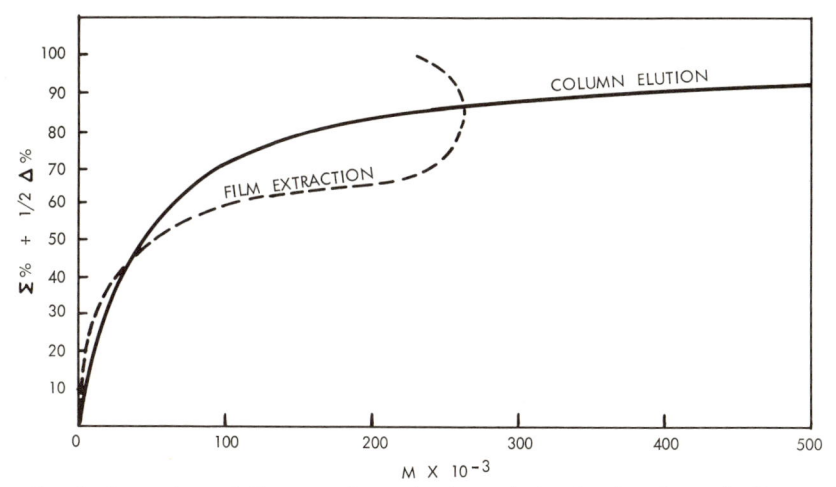

FIG. 3. Comparison of film extraction and column elution fractionations of a linear polyethylene (unpublished work, Hercules Incorporated).

in the higher molecular weight region is clear from these curves. In the extreme, these flakes of swollen polymer may plug the glass frit, through which the solution is decanted, and thus terminate the fractionation. These difficulties become greater as the molecular weight of the polymer increases, especially with polyethylene, which adheres poorly to the foil. It is for these reasons that the film extraction technique has been largely supplanted by column procedures.

C. Fractionation by Coacervate Extraction

In the current polymer literature, the word coacervate has been used to describe the polymer-rich liquid phase which separates upon addition of a nonsolvent to a polymer solution. This is a far narrower definition than that given by Bungenberg de Jong and Kruyt [25], who coined the word. They defined a coacervate as the fluid mixture which may occur in systems of lyophilic colloids and definitely imply the interaction of two colloids. In polymer fractionation, coacervate extraction is essentially the reverse of conventional fractional precipitation, and the so-called coacervate is nothing more or less than the polymer-rich phase that is present when liquid-phase separation occurs. In such a fractionation, nonsolvent is added to the solution until essentially all of the polymer is in the coacervate. The dilute polymer solution, containing the lowest molecular weight fraction, is removed and the polymer in it isolated. The coacervate is then extracted with a mixture having slightly greater solvent power. This procedure is repeated until the fractionation is completed. This technique has been used by Nasini and Mussa [26], Tung [27], and Nicolas [28] for high-pressure polyethylenes and by Davis and Tobias [29] and by Redlich et al. [30] for polypropylene. Other examples may be found in the tables of Chapter G.

The fractionation of polypropylene by Davis and Tobias [29] illustrates the method. It was carried out in the apparatus shown in Fig. 4. A temperature of 134°C was maintained in the jackets by refluxing isoamyl alcohol. A solution of 2 gm of polymer in 170 ml of xylene is placed in the stirred vessel and preheated polyethylene glycol of molecular weight about 200 added dropwise with vigorous stirring. When the liquid composition is about 65% xylene by volume, 90% to 100% of the polymer is in the concentrated coacervate phase which rises to the top when stirring is stopped. The system is allowed to stand overnight to ensure phase separation, then the lower layer containing the first fraction is withdrawn from the bottom of the apparatus. This is replaced by about 250 ml of a xylene–polyethylene glycol mixture of slightly greater solvent power. To have good mixing and to ensure the removal of any coacervate clinging to the walls, the hot xylene is added first, followed by dropwise addition of the preheated polyethylene glycol as above. The integral distribution curve found by this method was in satisfactory agreement with those obtained by fractional precipitation and by column elution. Despite the use of antioxidants and nitrogen blanketing, greater polymer degradation was observed in the case of coacervate extraction. This is attributed to the long times at elevated temperatures required by this procedure.

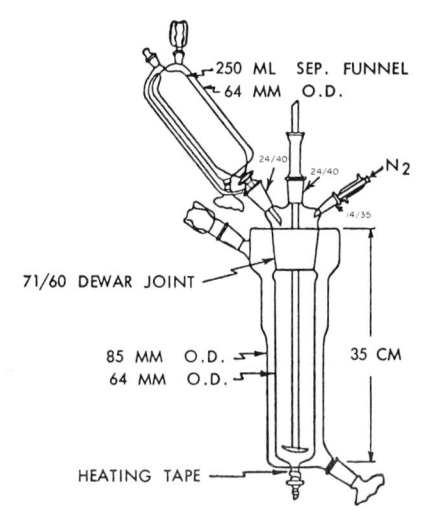

FIG. 4. Vapor-jacketed apparatus for fractionation by coacervate extraction [29].

The thermodynamics of phase relationships and coacervation, based on the Flory–Huggins theory, are discussed by Tompa [31] and Tompa and Bamford [32]. Redlich et al. [30], who fractionated a large sample (38 gm) of polypropylene, give a good discussion, based on solubility curves, of the proper selection of temperature, solvent, and nonsolvent for the formation of a liquid polymer-rich phase, or coacervate, in the case of a crystalline polymer.

Fractionation by coacervate extraction, being very similar to precipitation fractionation, basically possesses the same advantages and disadvantages, with the exception that the volumes of solvent handled at one time may be less in the case of coacervate extraction. When there is the possibility of a crystalline phase separating, this complication can be detected readily and avoided [26,28] by coacervate extraction. However, if the polymer–solvent system is sufficiently well characterized beforehand, the problem of the presence of a crystalline phase can be avoided in other fractionation systems.

D. Fractionation by Column Elution

The theoretical and experimental work of Desreux and his co-workers [11–13] about 15 years ago, constituted a major advance in our ability to characterize high polymers. Before their work, the fractionation of a polymer was generally an inordinately long and expensive procedure and could be considered only for samples of unusual interest. Their development of a fractional solution method in which the polymer, on the surface of a support in a column, is fractionated by elution, has made it possible to fractionate a polymer in a relatively short time. The method lends itself to automation and has been successfully scaled up [33] to fractionate a 50-gm sample into relatively sharp fractions, weighing 2–3 gm each. Of all the fractional solution methods, column elution is by far the most flexible and broadly applicable technique. While it has been used successfully with all types of polymers (see tables in Chapter G), we shall confine our discussion to the fractionation of isotactic polypropylene by Shyluk [4], since this work involved a thorough study of the variables which must be considered in a successful fractionation.

The fractionating column is shown in Fig. 5. The inner column contains sand (-40 to $+200$ mesh) to within 4–5 cm of the Trubore exit and is heated to 126°C with refluxing butyl acetate (reflux condenser not shown). A hot 1–2% p-xylene solution containing 1.2 gm of polypropylene is introduced into the top of the column and allowed to percolate through the top half of the sand bed. Flow is controlled by adjusting the concentration of the solution to achieve a suitable viscosity. The column is then cooled slowly (0.2°C/minute) to room temperature. The cooling rate is controlled by operating the heating system under partial vacuum, by closing the top of the reflux condenser and adjusting the power input to the boiler. After the system is cool, sand is added to the top of the column, a plug of glass wool placed on top, and the perforated disc clamped in the position shown, to prevent movement when the elution solvents are forced upward through the column. Atactic polymer is now removed by passing 200 ml of p-xylene at room temperature through the column. This fraction is isolated by partial evaporation of the xylene, then precipitated by the addition of a large volume of methanol. Since the extraction of the atactic portion is done at low temperature, no antioxidant is added to the extracting xylene. In all subsequent operations, all eluents contain 0.2% phenyl-β-naphthylamine and are sparged with nitrogen immediately before use, as shown in the figure. A suitable stopper is placed in the top of the column and nitrogen introduced through the hypodermic needle to provide an inert atmosphere over the polymer coated sand.

The p-xylene is now removed from the column by passing 250 ml of the nonsolvent (10% by volume of ethylene glycol in 2-butoxyethanol) through it. The column is then heated to the fractionation temperature of 156°C by refluxing cyclohexanone in the jacket. Suitable combinations of solvent (high-boiling petroleum fraction, b.p. 109°–114°/50 mm) and nonsolvent, with increasing solvent power are sparged in the vessel above the preheater, then preheated to 156°C and passed upward through the column at a rate of approximately 5 ml/minute.

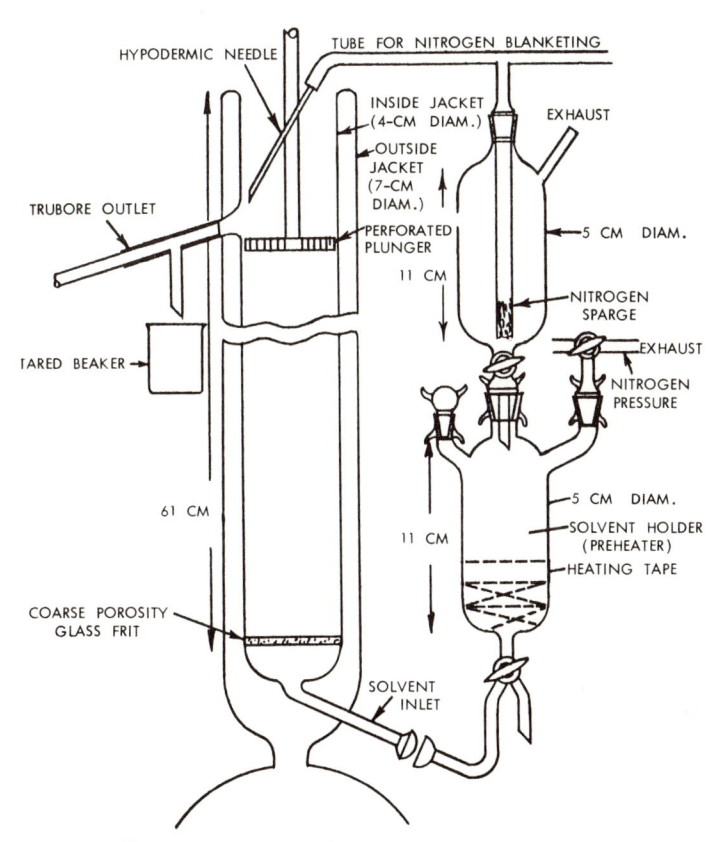

FIG. 5. Apparatus for fractionation by column elution [4].

Fractions are collected and precipitated in tared 150-ml beakers containing 35 ml of methanol and a few grams of Dry Ice. In general, 15 to 20 fractions are taken. The last fraction, which is eluted with 100% solvent, is collected in acetone, since this high-boiling hydrocarbon solvent is immiscible with methanol. The particle size of the precipitated polymer is increased by digesting for an hour on a steam bath. It is then allowed to settle overnight at room temperature. The solvent is decanted, the precipitated polymer washed several times with methanol to remove antioxidant, each washing being heated, cooled, and then decanted. The fractions are dried overnight in a vacuum oven at 65°C, cooled, and weighed.

The molecular weight of the fraction is then estimated by determining its specific viscosity at 135°C in Decalin using a polymer concentration of 0.1%.

Typical results are shown in the solid curve of Fig. 2. It can be seen that the resolution is quite good, the molecular weight of the highest fraction being several hundred times that of the lowest. The method is reproducible, selective, and relatively rapid. The actual fractionation takes about 6 hours; the elapsed time is 3–4 days. Nontechnical personnel can carry out the operation and the fractionation process may be automated.

IV. Factors in Column Elution—Guiding Principles for Experimental Fractionation

As is apparent from the description given above, there are a number of factors that must be considered in the fractionation of a polymer using the column elution technique. While these factors are discussed here in the framework of column fractionation, most of them are just as applicable to the other fractional solution methods. The emphasis on column techniques, however, seems appropriate in view of their broad area of applicability and of the magnitude of the current research effort on such methods.

A. COLUMN DESIGN

For fractionations, carried out primarily to obtain the molecular weight distribution, glass is the usual material of construction. For larger scale preparative columns, metal is generally used, although Henry [33] used a glass column about 5 feet long and having an internal diameter of 2.5 inches for the large-scale fractionation of polyethylene. Glass has the great advantage of transparency, which permits channeling or void formation to be observed. It is limited in strength, however, and if the fractionation must be carried out at elevated temperatures, cracks due to thermal shock in large columns are disastrous. Our own experience has been that large glass columns cannot be used safely above about 130°–140°C.

Good temperature control is essential in column fractionation. The absolute temperature is less important than its constancy. For fractionations at or near room temperature, circulating a constant temperature liquid through the jacket of the column is entirely satisfactory. At higher temperatures refluxing a liquid of appropriate boiling point in the jacket gives adequate temperature control.

In designing columns, particularly for preparative fractionations, dimensions should be selected with care. A column much over 6 to 7 feet long becomes unwieldy to set up, with its associated equipment, in the usual chemical laboratory. Too large a diameter leads to lateral temperature gradients and channeling, which markedly reduces the efficiency of the fractionation. Henry [33] obtained excellent resolution with polyethylene, using a column of about 2.5 inches diameter, but this probably is close to the upper limit. Cantow *et al.* [34,35] in designing columns for fractionation by the Baker–Williams [36], or chromatographic technique, used a 2.5-cm diameter, 1 meter long aluminum tube. They arranged six of these columns in parallel to fractionate larger samples.

A significant step forward in preparative fractionation has been made by Kenyon *et al.* [36a], who scaled up the method of Francis *et al.* [37] and Henry [33] to fractionate a 1-lb sample of polyethylene. The stainless steel column, packed with Celite, was 4 inches in diameter and 20 feet long. It could be operated over the temperature range 25°–137°C by means of an Aroclor heat exchanger. The polymer was selectively deposited on the Celite by programmed cooling. Eluents were pumped through the column using pressures up to 200 psi, 5 gal being used for each fraction. The integral distribution curve obtained from the large-scale fractionation of 1 lb of polymer agreed very well with a laboratory fractionation of 1.2 gm.

In the great majority of cases, the eluent is introduced at the top of the column and flows by gravity through the polymer-coated support. This is certainly the simplest method and is quite satisfactory for most fractionations. With certain systems, for example, isotactic polypropylene [4], the high molecular weight portion at the end of a fractionation tends to plug the frit at the bottom of the column. Reverse flow eliminates such plugging and, in addition, permits easier regulation of eluent flow and minimizes channeling and vapor entrapment. If difficulties are encountered with gravity flow, reverse flow should always be considered.

B. Solvent–Nonsolvent Pair

There are almost as many solvent–nonsolvent combinations reported, for a given polymer, as there are workers in the field. Details are given in Chapter G. When a polymer has not been fractionated before, it is necessary to carry out preliminary solubility experiments to establish the solvent–nonsolvent ratios and temperature where fractionation can be expected. After this has been done, the following considerations become pertinent.

As shown by Eq. (2) and Fig. 1, the eluent composition has a rather large range in the early stages of a fractionation, when low molecular weight material is being eluted, but this range becomes very small between successive fractions, at the high molecular weight end of the fractionation. In semiautomatic column fractionation, a logarithmic variation of eluent composition with volume run through the column is generally used [see, for example, 2,38]. This variation is readily accomplished by running the final eluent into a mixing vessel containing the original eluent at the same rate at which eluent is withdrawn from the vessel.

Gernert *et al.* [39] point out that there is no uniquely optimum gradient, as it changes with the molecular weight distribution of the polymer, and have described an ingenious apparatus which permits a wide variety of gradients to be obtained. Guillet *et al.* [38] found a linear gradient better than a logarithmic one for fractionating polyethylene, but their results appear somewhat inconclusive due to the possibility of gel band formation in their column. When investigating a new system, the logarithmic gradient is the obvious first choice on both theoretical and experimental grounds. If it proves to be unsatisfactory, then others can be tried.

The work of Horowitz on the high-resolution fractionation of isotactic polypropylene [40] points up the benefits of careful selection of solvent, nonsolvent, and operating temperature. If a relatively poor solvent (Butyl Cellosolve) is chosen, then variation in solvent power with percent nonsolvent (Butyl Carbitol) is much more gradual than if a good solvent is used. In addition, operation at 165°C, near the crystalline melting point of the polymer, minimizes difficulties due to possible crystallization. Using this technique, a sample was separated into 21 fractions, whose molecular weight increased monotonically from 6.9×10^3 to 2.77×10^6.

Another advantage of using a relatively poor solvent is that the precipitated polymer is not as highly swollen, which reduces its mobility and leads to better resolution [41].

C. Polymer Support

The purpose of the support is to immobilize the swollen gel of polymer in a thin film, so that it may be extracted readily by the eluent. For the fractionation of relatively small samples, fine sand and glass beads [see for example 4,6,34,37,40,42–45] are often used. Pepper and Rutherford [2] point out that for low molecular weight material, it is necessary to use a fine-grained support in order to minimize motion of the concentrated liquid phase. Kenyon and Salyer [6] found that a surface area of about 50 meter^2/gm of polymer was necessary for the successful fractionation of polyethylene.

When it is necessary to use larger polymer samples in a column fractionation, a support having a higher surface area per gram should be used. Henry [33], in scaling up the method of Francis *et al.* [37] to fractionate 50 gm of polyethylene, found it necessary to use Celite rather than sand as the support.

Metallic supports such as copper powder [2] have been used. The advantage of such a packing arises from the high thermal conductivity of metals and the consequent minimization of lateral thermal gradients. This packing permitted a faster rate of elution in the fractionation of polystyrene. Such metal surfaces are often catalytically active, however, and if they are used, the possibility of catalytic action on the polymer must be considered. This problem of interaction between the support and the polymer is not confined to metals. Recently, Rapp and Ingham [46] found degradation in the column fractionation of polyoxypropylene glycol–toluene diisocyanate polymers under chemically mild conditions. When the acid-washed glass beads, which they believe have a high inherent surface pH, were replaced with Haloport F or Fluoropak 80 (Teflon), there was considerably less degradation.

A review of polymer adsorption from solution has been given by Patat et al. [47]. However, the role of adsorption in fractionation is not clearly understood at this time. Certainly one would expect little or no adsorption of polyolefins on sand or glass. In the case of polystyrene, which is more polar, Krigbaum and Kurz [45] found that a temperature above the critical miscibility temperature was sometimes necessary to elute high molecular weight ($>3 \times 10^5$) fractions from sand. They attributed this to adsorption which increases both with molecular weight and with decrease in the particle size of the support. Schneider et al. [48] invoke adsorption as an explanation of comparable efficiencies in the fractionation of polystyrene by the gradient elution and chromatographic methods.

In the case of highly polar polymers and supports, adsorption is to be expected. Desreux and Oth [13] discuss the work of Brooks and Badger [49], who fractionated nitrocellulose, both as a function of molecular weight and of degree of substitution, by adsorption on starch.

Natta and his co-workers [19,20] have taken advantage of adsorption to separate stereoblock copolymers of atactic and isotactic or atactic and syndiotactic polypropylene. A column filled with highly isotactic polypropylene, or even better, isotactic polypropylene on a silica substrate, is used to adsorb the isotactic sequences in the stereoblock copolymers. The adsorbed polymer is then eluted, using isopropyl ether at various temperatures to effect a separation based on tacticity.

D. METHOD OF DEPOSITING THE POLYMER ON THE SUPPORT

In the case of analytical fractionations, involving only a gram or two of polymer, the simplest method of depositing the polymers on the support is to introduce a solution of the polymer into the packed column and to precipitate it by cooling [4,37]. With the larger diameter columns used in preparative fractionation, it is difficult to avoid channel formation and it is preferable to deposit the polymer on the support by stirring the support in a precipitating solution of the polymer. The coated material is then carefully packed into the column. Voids in the packing, which can lead to channeling during elution, must be avoided. It has been found helpful to partially fill the column with nonsolvent during this step.

The method by which the polymer is precipitated onto the support is one of the most critical experimental parameters in achieving a satisfactory fractionation. There is a considerable body of evidence which indicates that selective deposition of the polymer (e.g., by slow cooling at a controlled rate) effects a considerable degree of fractionation [4,6,29,41]; the higher molecular weight polymer being deposited first and the lower last, which is exactly what we want for an elution fractionation. This effect is strikingly illustrated in Shyluk's work on isotactic polypropylene [4] shown in Fig. 2. Slow precipitation (6 hours) gave a fractionation with good resolution and no reversals, in contrast to the results with rapid (1 hour) cooling.

Coating the support by solvent evaporation is not as effective in selectively depositing the polymer as slow precipitation and should be avoided if possible.

Davis and Tobias [29] and Horowitz [40] deposited polypropylene on the support from a solvent–nonsolvent mixture rather than from a good solvent [4]. Mendelson [50] found unsatisfactory fractionation of crystalline polypropylene when it was deposited from a good solvent and recommends deposition from a thermodynamically poor solvent, coupled with slow cooling.

If the Flory–Huggins theory is to apply, the precipitated phase must be a swollen liquid gel rather than crystalline polymer. Both Francis et al. [37] and Shyluk [4] showed by dilatometry that polyethylene and isotactic polypropylene were probably in a liquid phase under their fractionation conditions, albeit the crystalline melting points of both polymers are higher than the fractionation temperatures. Raff and Allison [51] and Nasini and Mussa [26] had difficulty in fractionating high-pressure polyethylene using the Desreux and Spiegels [12] fractional solution method and attributed these difficulties, as did Nicolas [28], to partial crystallinity in the gel phase under the conditions used. Schneider [41] in his review of column fractionation, also states that the precipitated phase must be amorphous. Crystallinity can be avoided by operating at higher temperatures and indeed Davis and Tobias [29] fractionated polypropylene at 170°C, which is almost the same as its crystalline melting point of about 173°C. Raising the temperature should be done cautiously, however, since degradation is more likely at higher temperatures and since the swollen polymer may become so fluid that it is not retained on the support [4].

If, on the other hand, we wish to fractionate by crystallinity as well as by molecular weight, then the precipitated phase must be crystalline.

Wijga et al. [52] found, in the case of polypropylene, that the gradient elution method gave excellent results in a fractionation by molecular weight. However, when the polymer was fractionated by increasing the temperature, using a single solvent, separation by both molecular weight and by crystallinity occurred.

E. RATE OF ELUTION

The rate of elution of a polymer from a fractionating column is an important variable in achieving good fractionation. As is true in so many other aspects of polymer fractionation, however, only guiding principles rather than absolute rules can be given. The concentration of the polymer in the solution leaving the column should be far below saturation. Flory's [1] suggestion that the maximum concentration should vary with (degree of polymerization)$^{-1/2}$ has been qualitatively confirmed [3]. Guillet et al. [38], using the chromatographic technique, found an optimum flow rate, and the same considerations hold for gradient elution. If the flow rate is too high, poor fractionation occurs, probably due to the low rate of diffusion. If the flow rate is too slow, back-diffusion may become a serious problem. The optimum flow rate must be determined for each system and it may be advantageous to change it during the fractionation. It is easier to control flow rates in reverse flow than in gravity flow columns [4]. Gernert et al. [39] describe a flow regulator which covers a wide range of rates and will compensate, in each 3-minute cycle, for adventitious rate changes as great as 10%. As a guide, elution rates of 2 to 6 ml/minute [4,37,38,41,42] have been satisfactory for the analytical fractionation (1–2 gm) of polymers and this is probably a good starting point when working on a new system. Suitable adjustments obviously must be made when operating on a different scale.

F. DEGRADATION DURING FRACTIONATION

As discussed earlier (Section II,C), one of the criteria of a good fractionation is that the weight average intrinsic viscosity of the fractions should be the same as that of the original polymer. When it is lower and the weight recovery is satisfactory, this is a clear indication that degradation has occurred. While temperature may be a factor, by far the most likely culprit is atmospheric oxygen, particularly when the fractionation must be carried out at high temperatures. Two steps can be taken to minimize oxidation. First, a suitable antioxidant should be present in all solutions and, second, operations, especially those at elevated temperatures, should be conducted in an inert atmosphere. This means sparging all solvents with nitrogen to remove dissolved oxygen, and protecting the surfaces of the column, solutions, and collecting vessels with an effective inert gas blanket. A gentle stream of nitrogen over an exposed surface is completely inadequate. Small gas exits are necessary to prevent back-diffusion of air and copious quantities of the blanketing gas must be used. When one remembers that one O_2 molecule can cause one chain break, the necessity for rigorous exclusion of air is apparent. Many fractionations have given meaningless or misleading results because of oxidative degradation. The necessity of taking all possible precautions cannot be overemphasized.

G. Fraction Workup

After the dilute solution of the fraction is obtained, it is necessary to isolate the polymer for further characterization. Often the solution can be run into a cold nonsolvent which precipitates it in a form that can be readily handled. As mentioned earlier, digesting on the steam bath is often effective in increasing the particle size of the precipitate in the case of crystalline polymers. In the case of amorphous or rubbery polymers that cannot be precipitated in an easily handled form, for example, atactic polypropylene [4], the polymer must be isolated by solvent evaporation. The workup should be carried out under the mildest feasible conditions of temperature and exposure to air, to minimize degradation. If the fractions are to be subsequently examined by either infrared or ultraviolet methods, all traces of antioxidant must be removed by careful washing with an appropriate solvent. At this stage, the sample is particularly susceptible to oxidation and the final drying should be done at a low temperature (50°–65°C) in a vacuum oven [4,44]. Each different polymer, due to its chemical nature, will require a different workup procedure; the guiding principle in all fraction workup is to use the mildest conditions possible.

H. Reversals in Molecular Weight with Fraction Order

In the column elution fractionation of polymers, it is not uncommon to find reversals in molecular weight with fraction order, particularly in the high molecular weight region. There are a number of causes for this effect. Francis et al. [37], in the fractionation of linear polyethylene, found examples of this due to operating at too high a temperature with inadequate protection against oxidation. Channeling in the column, aggravated by the high solution viscosity of the highest molecular weight fractions, is a common cause of fraction reversals. Careful packing of the column, adjustment of flow rates to maintain a low polymer concentration in solution [41], and the use of reverse flow [4] generally will rectify the situation.

Kenyon and Salyer [6] found, in the case of polyethylene fractionation, that if the temperature of the eluent entering the system differed from the operating temperature by 4°C, fraction reversal occurred. Cooper *et al.* [53] studied fraction reversal in the fractionation of polybutadiene using the chromatographic technique and experimentally demonstrated back-diffusion of low molecular weight polymer up the column. This was minimized by eliminating the temperature gradient and employing a very slow change in solvent power during the elution.

A very common cause of fraction reversal is nonselective deposition of the polymer on the support [4,6]. The importance of this step has been discussed earlier and the contrast between selective and nonselective deposition is clearly illustrated in Fig. 2. In any system where fraction reversals persist, selective deposition experiments are indicated.

V. Comparison between Fractional Solution Methods

In the writer's opinion, gradient elution, using a column, is far and away the most flexible and efficient of the molecular weight fractionation methods based on fractional solution. Direct extraction is of greatest value in preliminary separations based on differences in composition or structure, but little use has been made of this method in recent years for fractionation by molecular weight. The film extraction technique of Fuchs works very well in the lower molecular weight region, but difficulties have been observed with high molecular weight polymers (Fig. 2 and [14]).

The coacervation method is quite similar to the conventional precipitation method, and has many of its advantages and disadvantages. It is probably most useful for preparative scale work, albeit this involves handling relatively large quantities of solution, often at elevated temperatures. It is not readily adapted to automation on either the analytical or preparative scale.

Few direct comparisons between the various methods have been published. Nasini and Mussa [26] found the coacervation method superior to either the Desreux or the precipitation for the fractionation of polyethylene. Difficulties with the Desreux, however, may have been due to the presence of some crystalline polymer. Wijga *et al.* [52], on the other hand, found the column elution method to be quite satisfactory for crystalline polypropylene. Davis and Tobias [29] compared the precipitation, coacervate extraction, and column elution methods on the same polymer. They found less degradation with the column elution method, together with a high degree of reproducibility.

More comparisons have been made between the gradient elution or Desreux and the chromatographic or Baker–Williams methods. Guillet *et al.* [38] compared the original Desreux and Spiegels method [12] for polyethylene, using toluene at increasing temperatures as the eluent, with a chromatographic procedure, employing both solvent and temperature gradients. Ultracentrifuge studies showed that the fractions from the latter were sharper [54].

Perhaps the most thorough comparison was carried out by Schneider et al. [48], who fractionated polystyrene by both methods. They expected the Baker–Williams to give higher resolution since they, as well as others, had considered the gradient elution method as a single-stage process, and the Baker–Williams as multistage [36,55]. They found comparable resolution, however, with both methods.

Schneider [41] has made a critical review of the various important parameters in both gradient elution and chromatographic fractionation. Despite the fact that in later comparisons, using polystyrene, the Baker–Williams method gave a higher molecular weight fraction at the end, results by both methods are quite comparable. This similarity is very surprising if the gradient elution is really a single-stage method and the chromatographic is multistage, due to successive precipitations and dissolvings as a fraction is subjected to both a thermal and solvent power gradient while moving through the column. Certainly selective precipitation, commonly used in the gradient elution method, but not ordinarily used in the chromatographic, is an important factor, and may be a partial explanation for the similar results by both methods.

The gradient elution method is probably not strictly a single-stage process. As fresh eluent advances, it picks up some of the poorer solvent previously in the column by diffusion from the gel layer itself, the stationary layer at the surface, and side pockets between the packing particles. These effects could readily lead to some precipitation and redissolving of a fraction while it is passing through the column. Conversely, the chromatographic method may not have as many stages as one would expect. The experimental time scale may not be long enough to really achieve many precipitation and dissolution steps. In addition, it has been shown that a thermal gradient is conducive to back-diffusion of eluted polymer [53] and this would drastically reduce the number of effective stages. Further research is needed to resolve these questions.

VI. Possible Areas for Future Research

This brief review of fractional solution methods leaves a number of unanswered questions which suggest areas for additional research. Many more will undoubtedly occur to the reader, but the following seem worthy of some consideration.

In comparing the gradient elution method with the chromatographic, one would expect far better resolution with the latter method. While some published comparisons do show an advantage in fraction sharpness for this procedure, it is not of the magnitude that one would expect. Based on what we know experimentally, it appears unlikely that the gradient elution is merely a single-stage process and doubt is cast on the assumption that the chromatographic is really many stage. A possible way of obtaining basic information about what is really going on in the column, under operating conditions, might be to add a small quantity of a sharp, radioactively tagged, fraction to a wide distribution polymer and follow it during the process.

Direct extraction is a simple and powerful method of effecting separations by composition and structure. There are at least indications that it may be used to effect separations by molecular weight in the case of a crystalline polymer [21]. This technique appears to warrant further investigation to establish its potentialities and limitations. Is the time necessary to achieve equilibrium, or a practical approach to it, as long as we suspect?

Most of this discussion of fractional solution techniques has been concerned with fractionation by molecular weight. It often happens, however, that fractionation by ability to crystallize, by tactic sequences, by composition, or by degree of long-chain branching may be as or more important to a particular investigation than separation by molecular weight. It would appear that the work of Natta and his co-workers [18–20], involving a preliminary separation by direct extraction, followed by a separation based upon degree of adsorption on a suitable chromatographic material, has a high probability of success for the above type of fractionations and deserves systematic study.

In preparative-scale fractionation, narrow fractions of 5 to 10 gm are about the upper limit with present column equipment, both gradient elution [33] and chromatographic [35]. If extensive physical testing or determination of mechanical properties and chemical characteristics is required, special small-scale methods must often be developed in order to conserve the valuable sample. Practical means of obtaining larger quantities of sharp fractions would simplify many investigations. One group believes that conventional chemical engineering techniques can be adapted to preparative fractionation and success here would be a significant advance in polymer research.

Koningsveld and Pennings [56] have published preliminary work on the fractionation of linear polyethylene by crystallization from p-xylene at 85°–90°C. Their results indicate that very narrow molecular weight fractions can be obtained. Fractionation by ability to crystallize appears to warrant further study.

ACKNOWLEDGMENTS

The writer is indebted to his colleagues, Dr. H. M. Spurlin and Mrs. Shirley Shyluk, for many helpful discussions during the preparation of this chapter.

REFERENCES

1. P. J. Flory, "Principles of Polymer Chemistry," Chapters VIII-3 and XIII. Cornell Univ. Press, Ithaca, New York, 1953.
2. P. C. Pepper and P. P. Rutherford, *J. Appl. Polymer Sci.* **2**, 100 (1959).
3. N. S. Schneider, J. D. Loconti, and L. G. Holmes, *J. Appl. Polymer Sci.* **3**, 251 (1960).
4. S. Shyluk, *J. Polymer Sci.* **62**, 317 (1962).
5. G. V. Schulz, P. Deussen, and A. G. R. Scholz, *Makromol. Chem.* **69**, 47 (1963).
6. A. S. Kenyon and I. O. Salyer, *J. Polymer Sci.* **43**, 427 (1960).
7. K. H. Meyer, "Natural and Synthetic High Polymers," 2nd ed., p. 712. Wiley (Interscience), New York, 1950.
8. H. Wesslau, *Makromol. Chem.* **20**, 111 (1956).
9. G. M. Guzmán, *in* "Progress in High Polymers" (J. C. Robb and F. W. Peaker, eds.), Vol. I, p. 172. Academic Press, New York, 1961.
10. L. H. Cragg and H. Hammerschlag, *Chem. Rev.* **39**, 79 (1946).
11. V. Desreux, *Rec. Trav. Chim.* **68**, 789 (1949).
12. V. Desreux and M. C. Spiegels, *Bull. Soc. Chim. Belges* **59**, 476 (1950).
13. V. Desreux and A. Oth, *Chem. Weekblad* **48**, 247 (1952).
14. R. W. Hall, *in* "Techniques of Polymer Characterization" (P. W. Allen, ed.), Chapter 2, p. 38. Butterworth, London and Washington, D.C., 1959.
15. W. R. Krigbaum, J. E. Kurz, and P. Smith, *J. Phys. Chem.* **65**, 1984 (1961).
16. J. van Schooten, H. van Hoorn, and J. Boerma, *Polymer* **2**, 161 (1961).
17. J. P. Luongo, *J. Appl. Polymer Sci.* **3**, 302 (1960).
18. G. Natta, G. Mazzanti, and P. Longi, *Chim. Ind. (Milan)* **40**, 183 (1958).
19. G. Natta, M. Pegoraro, and M. Peraldo, *Ric. Sci. Suppl.* **28**, 1473 (1958).
20. G. Natta, I. Pasquon, P. Corradini, M. Peraldo, M. Pegoraro, and A. Zambelli, *Atti. Acad. Nazl. Lincei, Rend., Classe Sci. Fis., Mat. Nat.* [8] **28**, 539 (1960).
21. A. Keller and A. O'Connor, *Polymer* **1**, 163 (1960)
21a. R. E. Harrington and B. H. Zimm, *Abstr. Am. Chem. Soc. 149th Meeting, Detroit, 1965* (*Div. Polymer Chem. Preprints*) Vol. 6, p. 346.
21b. A. Nakajima and H. Fujiwara, *Bull. Chem. Soc. Japan* **37**, 909 (1964).
21c. P. J. Flory, *J. Chem. Phys.* **17**, 223 (1949).
21d. K. Yamaguchi, H. Kojima, and A. Takahashi, *Intern. Chem. Eng.* **5**, 169 (1965).
22. O. Fuchs, *Makromol. Chem.* **5**, 245 (1950).
23. O. Fuchs, *Makromol. Chem.* **7**, 259 (1951).
24. O. Fuchs, *Z. Electrochem.* **60**, 229 (1956).
25. H. G. Bungenberg de Jong and H. R. Kruyt, *Kolloid-Z.* **50**, 39 (1930).
26. A. Nasini and C. Mussa, *Makromol. Chem.* **22**, 59 (1957).
27. L. H. Tung, *J. Polymer Sci.* **20**, 495 (1956).
28. L. Nicolas, *Compt. Rend.* **236**, 809 (1953).
29. T. E. Davis and R. L. Tobias, *J. Polymer Sci.* **50**, 227 (1961).
30. O. Redlich, A. L. Jacobson, and W. H. McFadden, *J. Polymer Sci.* **A1**, 393 (1963).
31. H. Tompa, *Trans. Faraday Soc.* **45**, 1142 (1949).
32. H. Tompa and C. H. Bamford, *Trans. Faraday Soc.* **46**, 310 (1950).
33. P. M. Henry, *J. Polymer Sci.* **36**, 3 (1959).
34. M. J. R. Cantow, R. S. Porter, and J. F. Johnson, *Nature* **192**, 752 (1961).

B.2. Fractional Solution

35. M. J. R. Cantow, R. S. Porter, and J. F. Johnson, *J. Polymer Sci.* **C1**, 187 (1963).
36. C. A. Baker and R. J. P. Williams, *J. Chem. Soc.* p. 2352 (1956).
36a. A. S. Kenyon, I. O. Salyer, J. E. Kurz, and D. R. Brown, *Abstr. Am. Chem. Soc. 148th Meeting, Chicago, 1964* (*Div. Polymer Chem. Preprints*) Vol. 5, p. 638.
37. P. S. Francis, R. C. Cooke, Jr., and J. H. Elliott, *J. Polymer Sci.* **31**, 453 (1958).
38. J. E. Guillet, R. L. Combs, D. F. Slonaker, and H. W. Coover, Jr., *J. Polymer Sci.* **47**, 307 (1960).
39. J. F. Gernert, M. J. R. Cantow, R. S. Porter, and J. F. Johnson, *J. Polymer Sci.* **C1**, 195 (1963).
40. R. H. Horowitz, *Abstr. Am. Chem. Soc. 145th Meeting, New York, 1963* (*Div. Polymer Chem. Preprints*), Vol. 4, p. 689.
41. N. S. Schneider, *Anal. Chem.* **33**, 1829 (1961).
42. H. Hirooka, H. Kanda, and K. Nakaguchi, *J. Polymer Sci.* **B1**, 701 (1963).
43. R. L. Combs, J. E. Guillet, D. F. Slonaker, J. T. Summers, and H. W. Coover, Jr., *Abstr. Am. Chem. Soc. 140th Meeting, Chicago, 1961* (*Div. Org. Coatings Plastics Preprints*) Vol. 21, No. 2, p. 249.
44. P. Parrini, F. Sebastiano, and G. Messina, *Makromol. Chem.* **38**, 27 (1960).
45. W. R. Krigbaum and J. E. Kurz, *J. Polymer Sci.* **41**, 275 (1959).
46. N. S. Rapp and J. D. Ingham, *J. Polymer Sci.* **A2**, 689 (1964).
47. F. Patat, E. Killmann, and C. Schliebener, *Fortschr. Hochpolymer.-Forsch.* **3**, 332 (1964).
48. N. S. Schneider, J. D. Loconti, and L. G. Holmes, *J. Appl. Polymer Sci.* **5**, 354 (1961).
49. M. C. Brooks and R. M. Badger, *J. Am. Chem. Soc.* **72**, 4384 (1950).
50. R. A. Mendelson, *J. Polymer Sci.* **A1**, 2361 (1963).
51. R. A. V. Raff and J. B. Allison, "Polyethylene," p. 197. Wiley (Interscience), New York, 1956.
52. P. W. O. Wijga, J. van Schooten, and J. Boerma, *Makromol. Chem.* **36**, 115 (1960).
53. W. Cooper, G. Vaughan, and J. Yardley, *J. Polymer Sci.* **59**, S2 (1962).
54. L. D. Moore, Jr., G. R. Greear, and J. O. Sharp, *J. Polymer Sci.* **59**, 339 (1962).
55. J. L. Jungnickel and F. T. Weiss, *J. Polymer Sci.* **49**, 437 (1961).
56. R. Koningsveld and A. J. Pennings, *Rec. Trav. Chim.* **83**, 552 (1964).

CHAPTER B.3

Chromatographic Fractionation

Roger S. Porter* and Julian F. Johnson
CHEVRON RESEARCH COMPANY, RICHMOND, CALIFORNIA

I. Introduction 95
II. Apparatus 97
   A. Column Construction and Temperature Control 97
   B. Column Support Materials 98
   C. Solvent Gradient Production 98
   D. Flow Control 102
   E. Fraction Collection and Evaluation 103
III. Column Operation 106
   A. Sample Size and Preparation 106
   B. Choice of Nonsolvent–Solvent Pair and Gradient 107
   C. Selection of Volumes and Flow Rates 108
IV. Specific Fractionations 109
V. Theoretical Considerations and Comparison with Other Methods 109
   A. Theories Involving Temperature Gradients 109
   B. Determination of Fractionation Efficiency 113
   C. The Effect of the Temperature Gradient—Comparison of the Chromatographic Method with Other Fractionation Techniques 114
VI. Preparatory-Scale Fractionation 118
References 120

I. Introduction

In 1956 Baker and Williams [1] introduced a chromatographic procedure for the molecular weight fractionation of high polymers. A schematic diagram of an improved apparatus of this type is shown in Fig. 1. The essential elements are a mixing device to provide a solvent gradient, a column packed with support material, heaters to produce a linear temperature gradient in the column, and a sample collecting device. The samples, in this case [1] several polystyrenes, were coated on a small amount of glass beads by evaporation from a good solvent; the beads were then added to the top of the column as a slurry in the poor solvent. The solvent gradient ranged from 100% ethanol to 100% methyl ethyl ketone with a logarithmic change in solvent composition with time. The column temperature was 60°C at the top of the column and 10°C at the bottom. Molecular weight distributions were determined by collecting eluted

* Present address Polymer Science and Engineering Program, University of Massachusetts, Amherst, Mass.

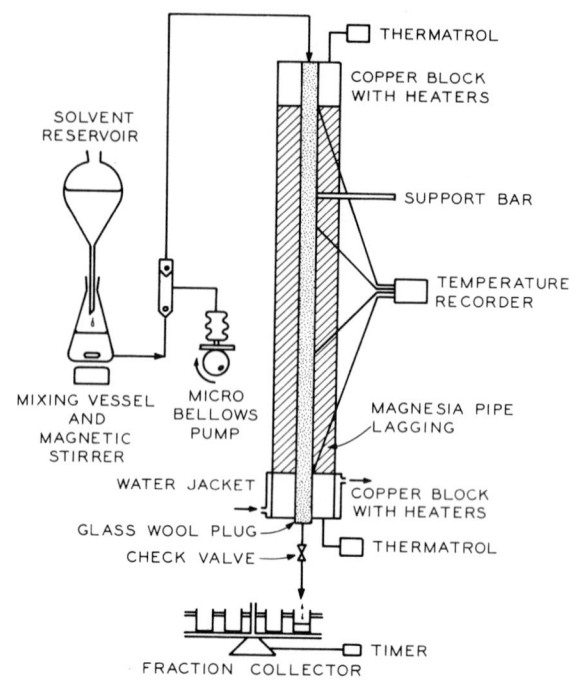

FIG. 1. Chromatographic fractionation apparatus [3].

fractions, recovering the polymer, measuring the amount and molecular weight of polymer in each fraction, and constructing integral and differential molecular weight distribution curves from the data. Fractionation of the polystyrene was good, as judged by refractionation and comparison with theoretical distributions. Baker and Williams attributed the mechanism of separation to a multistage fractional precipitation. This model postulated some of the polymer dissolving in the highest temperature portion of the column and traveling as a saturated solution to a cooler portion where, assuming a positive temperature coefficient of solubility, it would precipitate. Reequilibration of the precipitated polymer with the mobile liquid phase would occur as a higher proportion of good solvent became present. These steps would occur progressively throughout the column length until polymer emerged from the end of the column as a saturated solution at the temperature at the bottom of the column. Because of the postulated precipitation mechanism, the method has been referred to as "precipitation chromatography."

In the period since the introduction of this method, considerable controversy has arisen as to whether the temperature gradient improves the fractionation over isothermal gradient elution methods, treated in

Chapter B.2, or indeed reduces the efficiency. The mechanism of separation has been explained by a variety of different models. These conflicting viewpoints will be treated in detail in Section V.

Regardless of separation mechanism, the Baker-Williams method has been widely applied to successfully fractionate a number of polymer systems. These are tabulated in Section IV. Sections II and III are concerned with apparatus and experimental technique for analytical-scale columns where the objective is usually to determine molecular weight distribution. Larger scale columns designed to isolate fractions of sufficient size for other subsequent tests are discussed in Section VI.

II. Apparatus

A. Column Construction and Temperature Control

The original Baker-Williams columns, as cited previously, were glass, 35 cm long and OD 2.4 cm, surrounded by a cylindrical aluminum block heated by a 60-watt electrical spiral heater and cooled by a coil with circulating tap water at the bottom. Better heat transfer is obtained by use of metal columns with heavy walls. Thus, Jungnickel and Weiss [2] used copper pipe with $\frac{1}{8}$-inch walls and established linear temperature gradients over a 90-cm length by heating and cooling at opposite ends of the column. The column was insulated with 2.5–5 cm of magnesia pipe lagging. A similar arrangement was used by Flowers et al. [3], as shown in Fig. 1. They used temperature controllers with resistance thermometer, sensing elements to regulate the temperatures of the upper and lower heating elements. Use of temperature regulators, while not mandatory, is helpful because elapsed times on fractionations are relatively long and use of a variable voltage transformer may result in undesirable temperature fluctuations. An alternate method giving good control is use of a variable voltage transformer supplied by a voltage regulator [4]. Similarly, if subambient temperatures are desired, refrigerated liquid from a regulated bath may be supplied [5]. A different approach has been employed by Krigbaum and Kurz [6]. This consists of wrapping a glass column with nichrome wire in such a way as to obtain an approximately linear temperature gradient. Aluminum has relatively good heat conductivity and, because of its light weight, permits use of bigger wall thicknesses [7].

If reactivity with the column is a problem, for example, catalytic degradation by a metal column, glass columns inside metal columns may be used, although the heat transfer properties are not as good. Copper columns may be gold-plated to avoid reactions [2].

Column dimensions reported have ranged from internal diameters of 0.5 to 5 cm and lengths from 10 to 150 cm. Commonly used columns have diameters of about 2.5 cm and lengths of 50 cm. No systematic studies of the effect of column dimensions have been made, as detailed in Section V,B.

Complete column descriptions including detailed drawings for machining of metal columns are given by Flowers et al. [3], and constructional details with suitable connections for glass columns can be found in the work of Schneider et al. [4].

B. Column Support Materials

The column support is intended to serve as an inert material with enough surface area to support the polymer as a thin solvent-swollen film. The most commonly used material is small glass beads [1–3,7]. Size ranges of about 40–70 μ are often employed, although other ranges would undoubtedly work. The beads must be cleaned before use. Contaminants include both organic materials and metals. A satisfactory cleaning procedure is repeated washing with hot concentrated hydrochloric acid until the supernatant liquid is no longer yellow, followed in succession by washing with hot nitric acid, water, and a volatile solvent such as acetone. Coarse-grained Ottawa sand has also served as a support for fractionation [6]. Copper powder, although attractive from the view of thermal conductivity, introduces the possibility of catalytic reaction [8].

A different and interesting approach to the selection of a column support material has been utilized by Vaughan and Green [9]. They suggested that use of cross-linked polystyrene beads might permit the depositing of thicker films. It is necessary to have sufficient cross-linking so that the polymer cannot penetrate the matrix; otherwise fractionation by gel permeation occurs. Gel permeation produces fractions in inverse order to column fractionation; that is, the higher molecular weight fractions elute first, so the two methods would work in opposite directions, resulting in little or no fractionation. The method is discussed in detail in Chapter B.5. Good fractionation was obtained on polystyrene using cross-linked polystyrene beads containing 2% divinylbenzene, with a diameter of 1000 μ.

C. Solvent Gradient Production

The choice of solvent gradient is a crucial variable because it markedly influences the degree of separation. There is no uniquely optimum gradient, as this changes with the molecular weight distribution of the specific polymer to be fractionated. A widely used gradient is the logarithmic type recommended by Alm et al. [10] and others [11–13] for the general case of gradient elution chromatography. A common way to produce such a gradient is to have a mixing vessel with a good stirrer initially filled with a nonsolvent and to replace the volume removed from the mixing chamber into the column by addition of solvent at an equal rate. Equations (1) to (5) (below) relate the various dimensions. Other gradients may be produced by various geometrical shapes of mixing vessels containing fixed or moving baffles [14,15]. An example is shown in Fig. 2. A solvent gradient controller

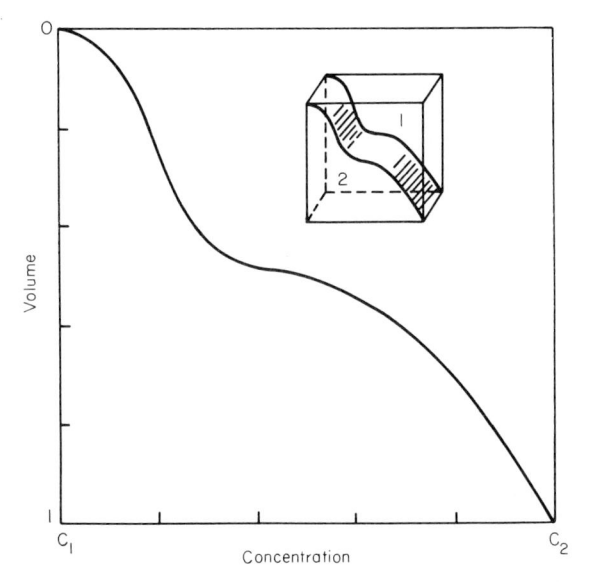

FIG. 2. Solvent gradient produced by rectangular tank with baffles [14].

capable of producing a wide variety of gradients is shown in Fig. 3 [16]. The stirred mixing chamber contains two thermistor probes. One near the bottom of the mixing chamber is always in liquid; the other is located at the vapor–liquid interface. A Wheatstone bridge contains the two probes as arms. If both probes are in liquid, the bridge is balanced and no signal is produced. If the liquid level is below the interface probe, the probe is surrounded by vapor which has a lower thermal conductivity than the liquid. Therefore, a sharp rise in temperature takes place, causing a marked reduction in resistance of the thermistor. The unbalanced bridge provides a signal to actuate a pump, adding liquid from the reservoir until the level reaches the reference thermistor. Automotive fuel pumps with the rubber sealing gasket replaced by Teflon are convenient and inexpensive. An alternate solution is gravity feed controlled by a solenoid valve.

The interface probe may be used in a fixed position. This can be varied to change the initial volume in the mixing chamber. If the probe does not move, a logarithmic change in gradient with time results. Other gradients

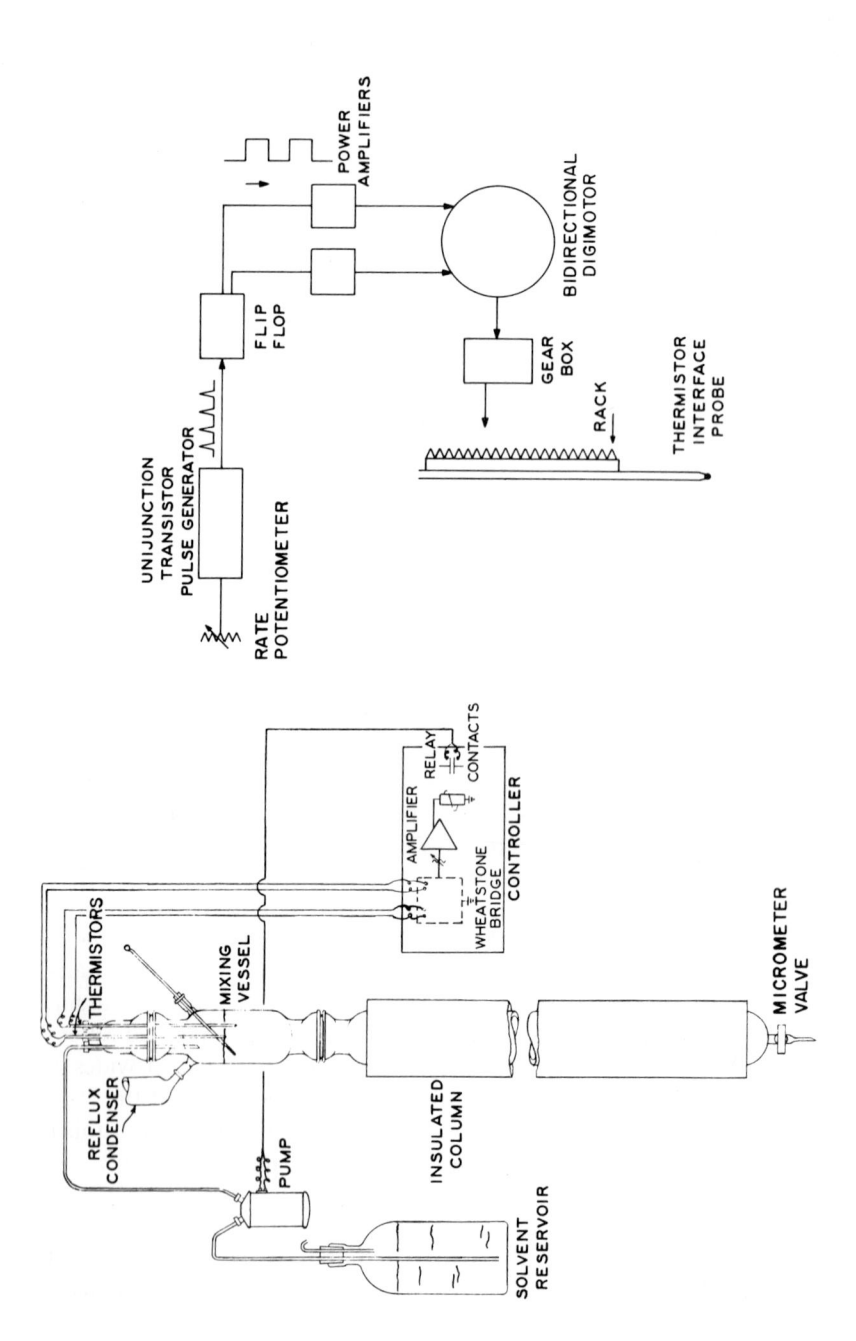

Fig. 3. Controller for producing wide range of solvent gradients [16]

B.3. Chromatographic Fractionation

may be produced by moving the probe either up or down. Figure 3 shows the schematic diagram of a motion controller. A gear train and gear rack to move the probe is actuated by a bidirectional Digimotor. The motor is driven by a flip-flop stage through power amplifiers. A unijunction pulse generator determines the rate of input pulses. If a linear potentiometer is employed, the output is such that the rate of motion of the probe will be linear. Speeds of 2.5×10^{-3} to 2.5 mm/hour have been employed, although others could be used. It is not necessary to use a linear potentiometer. Use of nonlinear potentiometers to produce logarithmic, sine, or other rates of change of the probe, are possible.

The parameters that affect the solvent gradient may be calculated as shown in Eqs. (1)–(5), using the following symbols:

φ_0 Volume fraction of liquid B in mixing chamber at time $t = 0$
V Volume of liquids A + B in mixing chamber at any time t
φ Volume fraction of B in mixing chamber at any time t
r Volume of A + B withdrawn per unit time
x Volume of A + B added per unit time
C_B Volume fraction of B in added mixture
k Motion of probe rate per unit time, assumed positive for upward and negative for downward motion
a Radius of mixing chamber
h_0 Height of liquid in mixing chamber at time $t = 0$
V_0 Volume of A + B in mixing chamber at time $t = 0$

It follows that

$$d\varphi V/dt = C_B x - \varphi r$$

$$dV/dt = x - r$$

and combining we obtain

$$d\varphi/dt = x(C_B - \varphi)/V_0 + (x - r)t$$

If we then let

$$\alpha = x/(x - r) \quad \text{and} \quad \beta = V_0/(x - r) \qquad (1), (2)$$

solving yields

$$\varphi = C_B - \frac{(C_B - \varphi_0)}{(1 + t/\beta)^\alpha} \qquad (3)$$

If the rate of input x is controlled by the probe moving at a velocity k, then Eqs. (1) and (2) read

$$\alpha = 1 + (r/\pi a^2 k) \quad \text{and} \quad \beta = h_0/k \qquad (1a), (2a)$$

The proper velocity k of the probe for any desired volume x to be added per unit time is given by

$$k = (x - r)/\pi a^2 \tag{4}$$

If the probe is not moving, i.e., for the case of $x = r$, Eq. (3) reduces to the familiar logarithmic gradient

$$\varphi = C_B - (C_B - \varphi_0)e^{-rt/V_0} \tag{5}$$

Equation (3) shows that in the limiting case of probe motion downward where $x = 0$, the concentration of the mixture entering the column does not change. This permits keeping the concentration of polymer in the eluting liquid from becoming too high for a polymer with a relatively narrow distribution. Obviously, with the probe motion upward the gradient is more rapid than logarithmic and conversely [17].

D. Flow Control

Fractionation columns are usually operated in a vertical position with flow from top to bottom. The pressure gradient to induce flow may be supplied by gravity and controlled by a needle valve [1,8]. Another method is to use a heated capillary to control flow [18]. A preferred method of ensuring constant solvent flow rate is the use of a constant volume pump [2,7]. One satisfactory arrangement includes a microbellows pump operating in conjunction with a check valve at the column exit to give a constant backpressure. Use of the check valve reduces flow fluctuations due to the sensitivity of the pump to pressure head [2]. Weakley et al. [19] utilized upward flow of solvent and pointed out that the inverted column would have the advantage that polymer gel, if it flowed at all, would flow into a region of the column with better solvent properties at higher temperature because the gel is more dense than the polymer solution with which it is in equilibrium. If the column were in the normal position, flow of polymer gel into a cooler region with lower solvent power could cause the column to become blocked. The inverted column configuration using gravity flow requires degassing of the solvents before they enter the column. A vented, heated chamber prior to the column entrance is sufficient [19]. Other arrangements for degassing have been described [4].

A different type of flow controller is shown in Fig. 4 [16]. It uses two photoelectric liquid level detectors [20]. It operates as follows: Initially the liquid level is below the lower sensing element. As the liquid crosses this sensing element, a timing circuit is started. The liquid continues to rise in the calibrated tube. Three cases are possible. If the liquid passes the upper sensing device in the preset timing period, $\pm 1\%$, the valve position is not changed. If the timing period elapses before the level reaches the senser,

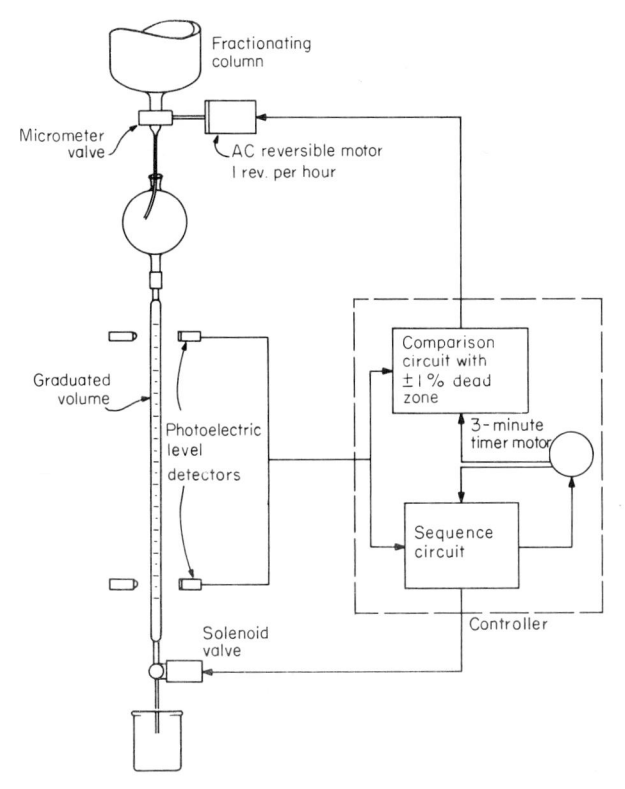

FIG. 4. Wide range flow controller [16].

the driving motor to the valve operates, opening the valve until the level reaches the senser. Should the liquid level actuate the detector before the timing period is ended, the valve closes for the length of time between the signal and the end of the timing period. This provides a valve adjustment proportional to the difference between the actual and required flow rate. At the end of the timing period or upon a signal from the upper photoelectric senser, whichever comes later, the solenoid at the bottom opens, empties the calibrated tube, and after a delay period the recycle repeats.

In the particular control cited the dead-time zone was $\pm 1\%$; the time cycle, 3 minutes. This unit controlled flows of 0.1–300 ml per 3 minutes and permitted a maximum change of 10% in flow rate for each cycle.

E. FRACTION COLLECTION AND EVALUATION

In order to determine the molecular weight distribution of the sample it is necessary to collect a number of fractions, recover the polymer, and determine the amount and molecular weight of polymer in each fraction. From these data integral and differential molecular weight distribution curves may be constructed. The number of fractions collected is a matter of balancing information gained versus labor required. The definition of the distribution curve obviously increases as the number of fractions collected and evaluated. However fraction evaluation is laborious, usually accounting for three-fourths or more of the total effort, so there is a practical limit to the number of fractions to be collected. Each polymer represents a different case, but usually a minimum of about 10 fractions is necessary to justify the fractionation with the most efficient number, probably in the 15–25 range.

There are numerous commercially available fraction collectors for collecting the column eluant on a time basis [18]. Jungnickel and Weiss [2] cite one example and used the collector for taking fractions from two or three columns simultaneously. Fractions may be collected on a volume basis by use of a syphon which periodically discharges a fixed amount into a receiver [19]. The receiver must be changed, and if this is done on a timed signal, variations in flow rate can cause difficulties such as two fractions in one receiver or syphon discharge during the interval when receivers are being changed. Cooper *et al.* [21] used a metal float in the syphon chamber with a coil wound around the syphon chamber to produce a signal from the change in inductance to actuate motion of a turntable that changed the receivers. This eliminated the difficulties cited. Their article gives a complete circuit diagram for a fraction collecting apparatus to allow simultaneous operation of two columns. A similar device using a float to operate mercury contacts was employed by Baker and Williams [1]. Hulme and McLeod [5] actuated a turntable on a signal from the change in dielectric constant in a condenser located on the discharge arm of the collecting syphon.

When collecting fractions by use of a syphon, care must be taken to avoid evaporation of solvent in the syphon. The polymer solution is generally saturated as it emerges, so evaporation will cause precipitation or separation of a second phase. One method of minimizing this is to connect a small solvent reservoir to the syphon so that the air space in the syphon is always saturated with solvent [1]. Thermostating the syphon at a temperature higher than that of the column exit reduces the chance of precipitation occurring.

Caplan [22] collected fractions continuously with a paper strip collector. The collector used a spreader to apply the column effluent in a thin uniform manner across the paper. This method is particularly suited to dyed or labeled polymer. A colorimetric or tracer method of analysis can result in an essentially continuous monitoring of the column results.

The selection of the method for fraction recovery depends upon polymer type. Addition of the polymer solution to a relatively large volume of cold nonsolvent may precipitate the polymer and permit collection by filtration [18]. Often, however, the polymer must be collected by solvent evaporation. If this is necessary, great care must be taken to use the lowest temperature possible and to exclude air to reduce degradation to a minimum.

Evaluation of polymer fractions usually involves determination of molecular weight by one of the established methods such as solution viscosity, ultracentrifugation, light scattering, and osmometry. Other measurements, often spectroscopic, to detect fractionation by chemical type in contrast to separation by molecular weight may be employed.

For repeated fractionation of polymers of the same chemical type a considerable amount of effort may be saved by establishing fixed conditions of temperature and solvent gradient, after which a constant relationship exists between molecular weight of eluted polymer and effluent volume. Then only the fraction weight need be determined. This procedure was suggested by Caplan [22]. Figure 5 shows a typical double logarithmic plot of solvent composition versus molecular weight for polyisobutylene [7].

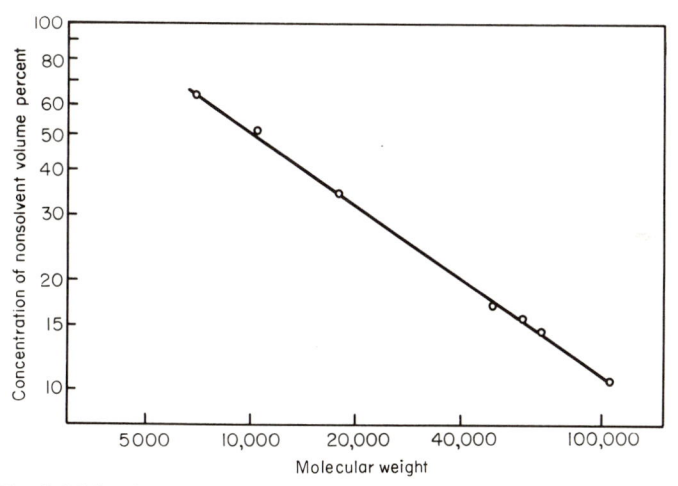

FIG. 5. Molecular weight versus eluant composition for polyisobutylene [7].

The curve was obtained from a fractionation by plotting the solvent composition, determined from refractive index measurements, versus solution viscosity molecular weights for the specific fraction. Subsequently numerous fractionations over a period of several months were spot checked on random fractions and showed excellent agreement in molecular weights determined from eluant composition and solution viscosities.

III. Column Operation

A. SAMPLE SIZE AND PREPARATION

Several methods have been used to coat the polymer on glass beads for introduction into the top of the column. In the original procedure Baker and Williams [1] added 300 mg of polymer dissolved in 10 ml of methyl ethyl ketone, the good solvent in the fractionation, to 30 gm of glass beads and evaporated the solvent using a stream of hot air. The beads were added to the top of the column in a slurry in the poor solvent. They occupied approximately the same length and location as the upper heater. Similar procedures were followed by Cooper et al. [23] and by Jungnickel and Weiss [2]. Schneider et al. [24] also prepared the polymer by deposition on beads by evaporation of solvent. They emphasized that stirring is necessary during the drying to avoid or minimize the formation of clumps. To further reduce the clumps the sample was passed through a No. 30 sieve. Microscopic examination suggested that most of the polymer was deposited in the inner space between the beads rather than as a thin film.

An interesting variation of this coating technique for elastomer fractionation is described by Hulme and McLeod [5]. They found that the evaporation procedure did not work as the mixture with elastomer formed balls rather than discrete particles. They used a highly absorbent solid, calcined diatomaceous earth, trade name Chromosorb. This material has a relatively high surface area and retained a considerable amount of the polymer even under prolonged, severe extraction conditions. This difficulty was overcome by a preliminary impregnation with a high molecular weight elastomer of the same composition as that to be fractionated. This treated Chromosorb was then extracted in the column by carrying out a blank run using the same eluants to be used in the fractionation. After this treatment, the recovered absorbents were used successfully many times for fractionations. The recovery of elastomer samples from the pretreated adsorbent was 97–100%.

An alternative procedure involves slowly cooling the polymer solution in the presence of glass beads until it is all precipitated [23]. Presumably this could improve fractionation as the polymer should be deposited onto the beads in descending order of molecular weight. This has been shown to be an important consideration in obtaining good results from elution fractionation [25]. It should improve column fractionation because the order of depositing should aid in the fractionation. A similar procedure involving the cooling of the polymer solution in place on the column was described by Guillet et al. [26], although apparently they filled the entire column with solution. This then would more nearly resemble a combination of elution and chromatographic fractionation rather than either procedure separately.

Sample sizes for analytical fractionations range from less than a milligram [22] up to several grams. Sample size is primarily controlled by the concentration of polymer in the eluant. Thus a larger sample of a broad molecular weight distribution may be fractionated with the same efficiency that could handle only a much smaller sample of a narrow distribution. Fractionation efficiency generally decreases with increasing sample size; see, for example, Guillet et al. [18]. No fixed rule can be evoked for polymer size selection, but, in general, samples should be kept as small as possible, compatible with the need to recover sufficient polymer for characterization in enough fractions to adequately describe the molecular weight distribution. The amount of polymer that can be fractionated decreases with increasing molecular weight. Flory [27] has suggested that the maximum concentration for efficient fractionation should be inversely proportional to the square root of molecular weight. At least qualitative confirmation of this has been experimentally determined from chromatographic fractionation [28].

B. Choice of Nonsolvent–Solvent Pair and Gradient

A wide variety of nonsolvent–solvent systems has been used for fractionating various polymers (see Section IV). There are no universal rules on selection. Properties such as boiling point and chemical stability must, of course, be compatible with the temperature employed and the polymer to be fractionated.

Hulme and McLeod [5] have described a method for selecting nonsolvent–solvent combinations. A dilute solution, about 0.5 gm polymer per 100 ml, is prepared using a known "good" solvent. Aliquots are titrated to the cloud point with a selection of different nonsolvents at room temperature. The nonsolvent chosen for the fractionation is the one that requires the largest addition to reach the cloud point. The solvent is chosen by making a series of solutions, again at about 0.5 gm per 100 ml, in various solvents and titrating with a known nonsolvent to the cloud point. The solvent which requires the least nonsolvent to cause precipitation is considered the most desirable. A dilute solution in the selected solvent is then titrated with the selected nonsolvent at the high and low temperatures of the column. This fixes the composition range over which the pair will effect solution of the polymer. This will permit selection of the initial composition in the gradient mixing chamber, θ_0, and the composition of the solvent added, C_B, see Eqs. (1)–(5), Section II,A.

The previously described method of selection of the nonsolvent–solvent pair was designed to give the widest composition limits of solubility. Pepper and Rutherford [8] argue that this consideration is not relevant since for any two compositional limits the change may be made as slowly as desired by regulating the choice of gradient itself. There is no conclusive evidence to support either viewpoint. For convenience, however, the wider compositional limits make preparation of the solution easier and require less precision in their preparation. There are theoretical grounds [22] to support the use of the logarithmic gradient for polymers with a broad molecular weight distribution. This arises from the usual solubility relationships and provides for an approximately equal resolution over the molecular weight range. Because it is convenient to maintain such a gradient with simple apparatus, it has been widely used. Other gradients have also been employed —a linear one, for example—in fractionation of polyethylene [18]. It was stated that the linear gradient gave better results than the logarithmic, although the results were somewhat ambiguous. Again from solubility relationships it would be expected that linear gradients would be preferable for comparatively narrow molecular weight distributions.

C. Selection of Volumes and Flow Rates

The choice of the volume of solvent–nonsolvent to be used to elute polymer in the course of the fractionation is very important, as has been demonstrated repeatedly. The solvent volume, if too low, will produce a poor fractionation. Recovery of polymer fractions becomes more difficult, however, if the solutions are extremely dilute.

Pepper and Rutherford [8] chose 0.5% as the upper permissible concentration limit of polymer in eluant. Using this value, and from various theoretical considerations, they concluded that the volume of the eluant in ml should be at least 200 times the polymer weight in grams. For a logarithmic gradient this would require that the mixing chamber volume be at least 100 times that of the polymer weight. Schneider et al. [24,28], using the suggestions of Flory [27] that the amount fractionated successfully should be proportional to the inverse square root of molecular weight, determined this factor, that is, concentration times the square root of molecular weight for a number of fractionations. The value of $cM^{1/2}$ at its maximum, that is, in the fraction containing the most polymer, for successful fractionations usually was below about 400. Although this value may be somewhat too high and is not defined definitely, it still suggests that values much above 300 should not be employed. This would imply considerably greater volumes than those recommended by Pepper and Rutherford for high molecular weight samples. Usually an elution volume of 500 times the weight of the polymer, assuming arbitrarily a polymer density of 1, is sufficient, although only experiment will show if this is true for the higher molecular weights. Concentrations here are referred to in weight fraction.

The selected flow rate is limited at the faster end by failure to maintain equilibrium, thus resulting in poor fractionation, and at the lower end by back diffusion, again causing the loss of resolution in fractionation. The optimum flow rate depends on column parameters. One study, that of Guillet et al. [18], indicated that optimum rates for the specific column used was about 3–6 ml/minute with poorer results observed at 1.2 and 10.6 ml/minute. For the average analytical column with sample sizes of approximately 1 gm, flow rates are usually of the order of 2–6 ml/minute.

IV. Specific Fractionations

Table I lists details of a number of polymer fractionations carried out by the chromatographic technique.

V. Theoretical Considerations and Comparison with Other Methods

As cited in the introduction, a considerable body of contradictory literature has arisen on the effect, if any, of the thermal gradient in chromatographic fractionation. These views range from evidence that contrary to increasing efficiency the temperature gradient actually harms it, to the opposite view that the temperature gradient is essential in order to achieve efficient fractionation. This section will attempt to present equitably the pertinent references and viewpoints from theory and comparisons with other methods. Discussion will be confined to the case of temperature gradient; Chapter B.2 treats the theory of elution fractionation without temperature gradient. Other variations exist, for example a stepwise temperature gradient without a solvent gradient [36]. Therefore, any such alternate theoretical models will be omitted in the following section.

A. Theories Involving Temperature Gradients

There are two principal models that have resulted in equations predicting an effect on fractionation due to the presence of both a temperature gradient and a solvent gradient. The first of these is due to Caplan [22]. For the case of amorphous polymers, Caplan cites experimental evidence that the polymer phase diagram is an asymmetric binodal with a critical point very close to the solvent ordinate. Therefore, he postulated that a dilute solution of polymer upon cooling crosses the binodal and that a very viscous or gel phase is precipitated out in equilibrium with a much larger volume of essentially pure solvent. The model assumed that a dilute solution of this type existed in every zone in the column that contained polymer. Then following the treatment of Baker and Williams, gel will be precipitated at a temperature which, introducing the Flory temperature θ for a polymer

TABLE I

Specific Fractionations

	Polystyrene	Polystyrene	Polystyrene	Polystyrene	Polyisobutylene	Polyisobutylene	Polymethyl-methacrylate	Polymethyl-methacrylate	Polyethylene
Molecular weight, unfractionated	$\bar{M}_n$ 1.08 × 10^5 3.50 × 10^5	$\bar{M}_n$ 7.25 × 10^5	$\bar{M}_w$ = 250,000	$\bar{M}_w$ = 1.63 × 10^5 ("Monodisperse")	$\bar{M}_n$ = 980–40,000	$\bar{M}_v$ = 1 × 10^6	—	$\bar{M}_v$ 5–60 × 10^3	$\bar{M}_v$ 63,400; 69,300
Column length (cm)	35	35	94	35	90	100	—	35	115
Inside diameter (cm)	2	2	2.5	4	2.5	4	—	2	3.7
Column support	Glass beads, $D = 0.1$ mm	Glass beads, $D = 0.1$ mm	Glass beads, $D = 200$–325 mesh	Glass beads	Glass beads	Glass beads, $D = 0.1$ mm chromosorb	—	Glass beads	Glass beads, $D = 100$ μ
Nonsolvent	Ethanol	50% Ethanol, 50% methyl ethyl ketone	50% Ethanol, 50% methyl ethyl ketone	20% Ethanol, 80% methyl ethyl ketone	Methyl ethyl ketone	n-Propanol	Methyl alcohol	Cyclohexane	2-Butoxy-ethanol
Solvent	Methyl ethyl ketone	10% Ethanol, 90% methyl ethyl ketone	15% Ethanol, 85% methyl ethyl ketone	Methyl ethyl ketone	Benzene	Mixed xylenes	—	Methyl ethyl ketone plus ethanol	1,2,3,4-Tetra-hydro-naphtha-lene
Solvent gradient	Logarithmic	Logarithmic	Logarithmic	Logarithmic	Logarithmic	Logarithmic	Logarithmic	Logarithmic	Linear and logarithmic
Temperature gradient (°C)	60–10	60–10	61–24	70–40	90–40	60–10	55–15	60–10	160–110
Flow rate (ml/hr)	5	5	12	—	20	10	—	—	180–360
Sample size (gm)	0.300	0.300	1	0.8	10	0.30	<1	—	2–5
Method of coating support	Evaporation	Evaporation	Evaporation	Evaporation	Evaporation	Evaporation and precipitation	—	Evaporation	Evaporation
Amount of support sample coated on (gm)	30	30	—	—	50	—	—	—	15–30
Number of fractions	~50	~50	22	13	15–20	9	—	20–40	—
Molecular weight range of fractions ($\bar{M}_v$)	(0.4–10) × 10^5	(0.2–1.7) × 10^6	(1.0–68.0) × 10^4	(1.4–3.2) × 10^5	(1–18) × 10^3	(10–200) × 10^4	—	—	(0.1–2.5) × 10^5
Reference	1	1	2	4,33	34	35	30	19	18

TABLE 1—continued

SPECIFIC FRACTIONATIONS

	Polypropylene	Polyester (hexanediol succinic acid)	cis-1,4-Polybutadiene	Polyalkane (C_{10-18} α Olefin)	Polysarcosines	Polypeptide	Polyisobutylene–styrene	Copolymer natural rubber–methyl methacrylate
Molecular weight, unfractionated	—	—	—	$\bar{M}_w = 450{,}000$	$\bar{M}_n$ (1.5, 5.0, 11.8) $\times 10^3$	—	$\bar{M}_n$ 78,000	—
Column length (cm)	115	90	30	94	30	35	50	35
Inside diameter (cm)	3.7	3.8	2.4	1.9	0.3	2	3.6	2.4
Column support	Glass beads, $D = 100\,\mu$	Glass beads, $D = 75\,\mu$	Chromosorb	Glass beads, $D = 200$–325 mesh	Glass beads, $D = 0.1$ mm	Glass beads, $D = 0.1$ mm	Glass beads, $D = 100$–$160\,\mu$	Glass beads, $D = 0.1$ mm
Nonsolvent	90/10 Butyl Carbitol/tetralin	20% MEK[a]; 80% cyclohexane	Several pairs	Ethanol	Dioxane	Cyclohexane	Isopropanol	80/20 Petroleum naptha/benzene
Solvent	70/30 Tetralin/Butyl Carbitol	60% MEK[a]; 40% cyclohexane	Several pairs	Benzene	Water	20% Methanol, 80% ethanol	Cyclohexane	25/75 Petroleum naphtha/benzene
Solvent gradient,	Logarithmic	Logarithmic	Logarithmic	Logarithmic	Logarithmic	Logarithmic	Logarithmic	Logarithmic
Temperature gradient (°C)	180–140	60–25	60–10	60–20	65–15	60–10	180–140	50–18
Flow rate (ml/hr)	180–360	180–240	125	12	1.5	5	11–13	20
Sample size (gm)	1–1.5	4–5	0.3–0.6	1	0.0001	5	2	—
Method of coating support	Evaporation	Evaporation	Evaporation	Evaporation	Evaporation	Evaporation	Precipitation	—
Amount of support sample coated on (gm)	200	200–250	—	—	0.1	—	—	—
Number of fractions	13	13	—	15	Continuous	20	40	25
Molecular weight range of fractions ($\bar{M}_v$)	—	(6–70) $\times 10^3$	—	(0.04–177) $\times 10^4$	—	—	—	(0.1–10) $\times 10^5$
Reference	26	32	5	2	22	29	31	21

[a] Methylethylketone.

of infinite chain length, would correspond to this temperature θ [37]. A better solvent gradient would dissolve the gel and reprecipitate at lower temperature. The volume of the eluant in the column at any time was considered small compared to the volume used in the entire solvent gradient. Therefore, the difference between solvent composition at the two ends of the column was ignored. From this, Caplan derived Eq. (6)

$$M = \left[\frac{KT_2}{\sigma e^{-V/G} + \tau - T_2}\right]^2 \quad (6)$$

where M is the molecular weight of the polymer species; T_2, the temperature of the bottom of the column; τ, σ, parameters describing the solvent–nonsolvent system (τ may be equated with the Flory temperature of the solvent); G, the volume of the mixing chamber, considered constant; and V, the volume of the solvent at any time which has passed any given point since the start of the fractionation. Using various methods of estimating the parameters in the equation, Caplan found good agreement for the case of a fractionation of polystyrene. This would indicate a definite temperature gradient effect. The difficulty in precise evaluation lies in the inability to experimentally determine the necessary constants σ and τ with precision.

Schulz et al. [38] used a different model in developing a theory of chromatographic or precipitation chromatography. Their model was based on derivation and solving of transport equations for the exchange process between a gel layer at rest and a sol layer in motion for the case of both the solubility and temperature gradient. They considered that the overall fractionation could be separated into two steps—an introductory process during which a continuous gel phase is formed on the support, and then the principal process, during which fractionation takes place due to the exchange distribution between the sol phase and the gel phase. In the first process it was considered that portions of the polymer were dissolved on the upper heated end of the column and redeposited on the support material as a gel further down in the column at a lower temperature. After a certain amount of solvent had passed through, a resulting continuous, finely distributed gel layer has been formed with a rough preliminary fractionation by molecular weight. In the principal process every component of the polymer at any location is subjected to a characteristic distribution equilibrium between the two phases depending only on the molecular weight. Equilibrium was considered to be obtained rapidly in comparison with a slow process of flow. Therefore, the combination of solvent and temperature gradient had two effects. Each molecular weight moves through the column with a characteristic velocity, and, because of the temperature gradient, the leading edge of the component moves more slowly than the trailing edge;

that is, it is compressed together in an increasingly smaller zone on the column.

Using this model it was shown that theoretically extremely high resolution could be obtained. In practice, of course, this is limited by the finite velocity for equilibrium.

These two theories, involving different models, both predict a pronounced temperature effect on the degree of fractionation.

B. Determination of Fractionation Efficiency

Criteria for evaluating fractionation efficiency are necessary in order to compare methods of fractionation and determine which is more effective. Unfortunately, in the case of polymer fractionation, these are not easy to determine.

Fractionations must be internally consistent in all cases. This can be determined easily. The recovery of polymer fractions should equal the polymer charged. To show lack of degradation during the fractionation the intrinsic viscosity of the original polymer should equal the summation of the weight fraction times the intrinsic viscosity for each fraction. In addition, it is to be expected that, in the absence of chemical heterogeneity, there should be a regular progression of molecular weight in order of fractions collected.

Even with all internal factors consistent, a quantitative comparison of fractionation efficiency is not available. A frequently used method is to compare distributions from repetitive fractionations. This is important to show the reproducibility of the method, but it does not define column efficiency.

What is desired is the distribution within each fraction. This may be obtained, for example, by refractionation of fractions, e.g., Baker and Williams [1], Hansen and Sather [32], and Henry [39]. This is laborious and suffers from the limitation that if the same fractionation procedure is used it may not necessarily resolve the fraction under study. Secondary methods such as determination of weight and number average molecular weight, $\overline{M}_w$ and $\overline{M}_n$, and examination of their ratio are useful, but only for fairly broad fractions. This is because of limitations in precision for ratios of $\overline{M}_w/\overline{M}_n$ of less than 1.03. Other methods of determining molecular weight distribution may be more significant indications of dispersion, for example, ultracentrifugation patterns [40] or possibly turbidimetric titrations.

An additional method is to compare fractionation distribution with distributions known from the method of polymerization [1,19].

It is also desirable to examine the fraction distributions in a concentration versus molecular weight plot. An example of this is given by Schneider and

Holmes [33], who used the chromatographic method to determine the actual distribution in a "monodisperse" polystyrene of the Szwarc type. Unfortunately, monodisperse samples have not been established. Another possibility would be to use tagged samples for the ease of determining the molecular weight distribution curve. However, these would also not be monodisperse.

C. The Effect of the Temperature Gradient—Comparison of the Chromatographic Method with Other Fractionation Techniques

Flowers et al. [3] have developed a stepwise continuous solvent gradient for use with elution chromatography. This is essentially a procedure for very carefully controlling the solvent gradient in the high molecular weight region where small changes in solvent power have large effects on the molecular weight solubility curve. In their work they carefully compared fractionation of a copolymer of 1-octadecene and 1-dodecene with and without a thermal gradient. The results are shown in Fig. 6. The backlash,

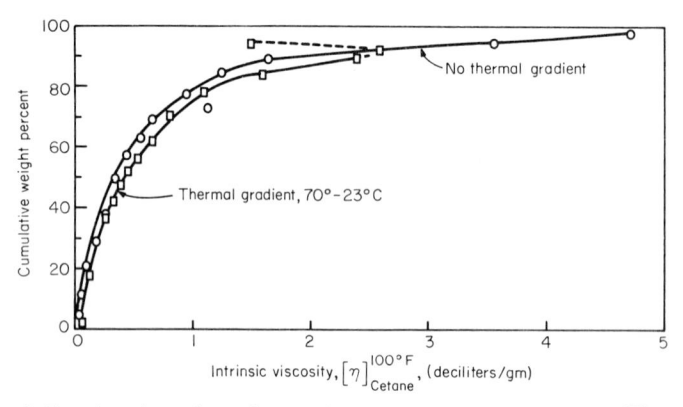

FIG. 6. Fractionation of copolymer of 1-octadecene and 1-dodecene, $\bar{M}_w$ 710,000, with and without thermal gradient [7].

that is, reversal in molecular weight, increases with fraction order with a thermal gradient and failure to resolve high molecular weight portions is evident. The fractionation is clearly better without the thermal gradient. The isothermal fractionation was carried out at 23°C, the lower temperature in the chromatographic column. One possible explanation of this effect advanced by Flowers et al. is that the actual fractionation mechanism is not that of selective precipitation but that an additional effect could occur in the exchange of lower molecular weight polymer on the support material for higher molecular weight polymer in the saturated solution as it flows through lower, as yet uneluted zones, of polymer-coated beads. This mechanism would require rapid equilibration. They found support for this by results of flask fractionations which showed equilibration to be achieved in less than 10 minutes.

Another comparison of elution fractionation with the thermal gradient method is shown in Fig. 7 from the same reference. Here the fractionation results on the copolymer, which has a considerably lower average molecular weight than that in Fig. 5, agree well although the elution fractionation is on a 10-gm sample in comparison to a 1-gm sample used in the chromatographic procedure. Also shown in Fig. 7 is a refractionation of the two highest molecular weight fractions obtained by the elution method.

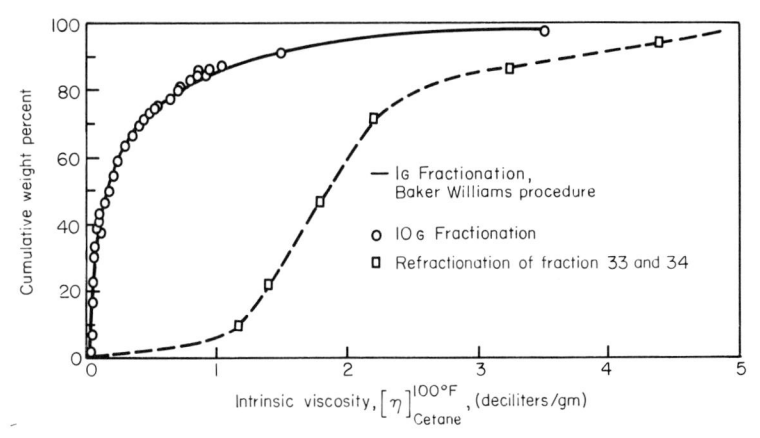

FIG. 7. Fractionation of copolymer of 1-octadecene and 1-dodecene, $\overline{M}_w$ 270,000 with and without thermal gradient [3].

Schneider et al. [24,41] also compared fractionation by the thermal gradient and elution methods. Two samples of polystyrene with a viscosity average of 5×10^5 and 3.8×10^6 were fractionated. The thermal gradient method produced better fractionation in terms of resolution and reproducibility. However, the difference was not as great as would be expected if the chromatographic method were a multistep process and the elution fractionation [42] a single-stage process. Schneider et al. concluded that possibly the enhanced resolution of the elution was due to adsorption on the beads and that this also contributes to the fractionation by the chromatographic method. Other evidence indicates significant effects due to adsorption [43].

Cooper et al. [23] compared the gradient elution fractionation of a polybutadiene, molecular weight about 2.7×10^5, with that of the chromatographic method. They found no appreciable differences with or without temperature gradient. They concluded that the most important factor controlling the efficiency of fractionation by either method was the establishment of the proper solvent gradient. This is in accord with the conclusions of others [23]. However, because of the large variations in solvent gradient and flow rate employed, a direct comparison of the two methods is not conclusive. Variations in the flow rate and solvent gradient tend to mask subtle differences between the two procedures.

Guillet et al. [26] compare the chromatographic method with that of the partial precipitation method of Tung using a high-density polyethylene. The agreement between the two methods was good, as is to be expected. Further, Guillet examined the distribution in a fraction from a column designed to fractionate large samples. He found an $\overline{M}_w/\overline{M}_n$ ratio of 1.04 by analysis using an ultracentrifuge.

A comparison of a fractionation of polyethylene with and without temperature gradient is shown in Fig. 8 [18]. Here the results using the temperature gradient are clearly better, particularly in the high molecular

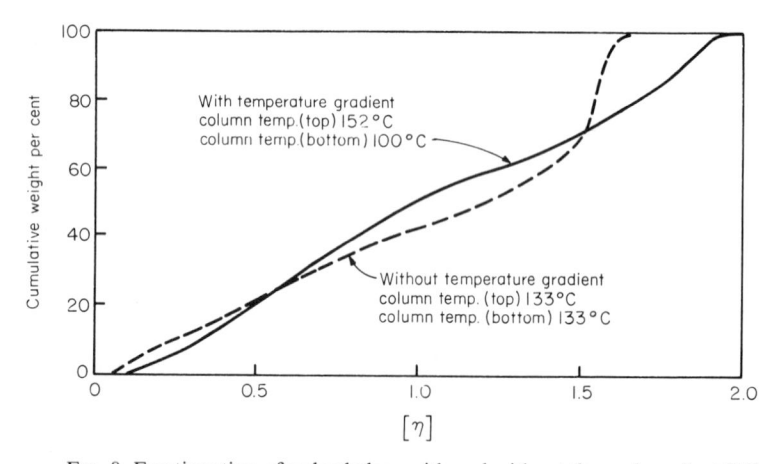

FIG. 8. Fractionation of polyethylene with and without thermal gradient [18].

weight region, than data obtained with no temperature gradient. One argument that might limit this conclusion is that the isothermal fractionation was made at 133°C, a temperature intermediate to those with the gradient, 152°–100°C. Usually extraction fractionation improves with lower

B.3. CHROMATOGRAPHIC FRACTIONATION

temperatures; thus it is possible that at 100°C the isothermal fractionation would have been as good as that with the thermal gradient.

Weakley et al. [19] fractionated polymethylmethacrylate with a $\overline{M}_w$ of about 130,000 by the chromatographic method and by elution fractionation without a temperature gradient. They determined $\overline{M}_n$ by osmotic pressure and $\overline{M}_w$ by solution viscosity measurements. For the chromatographic fractionation thirty-seven fractions had a $\overline{M}_w/\overline{M}_n$ ratio between 1.05 and 1.15. The ratio was independent of molecular weight. The elution fractionation had a similar $\overline{M}_w/\overline{M}_n$ of 1.05 to 1.15 for twenty-seven fractions. In this method the ratio increased with molecular weight so it could be concluded that the chromatographic method was better for the higher molecular weight range.

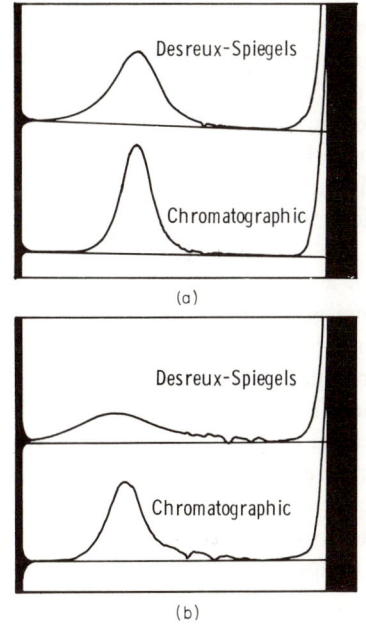

FIG. 9. Comparison of dispersion of polyethylene fractions obtained from fractionation with (a) and without (b) temperature gradient [18].

An interesting comparison of the distribution of fractions of polyethylene prepared by the elution and chromatographic methods in shown in Fig. 9 (Moore et al. [40]). The upper portion of Fig. 9 shows the ultracentrifuge pattern for two fractions of polyethylene of the same molecular weight, the upper curve being that prepared by elution techniques, the lower curve from a chromatographic fractionation. The starting polymer in this case had a relatively narrow distribution [18]. Moore *et al.* [40] further showed an even bigger difference in fractions prepared by the two methods from a broad distribution polymer as shown in the lower part of Fig. 9. They concluded that the chromatographic method was definitely superior to that of the elution technique. This is certainly justified by the data shown. However, some reservations must exist because of the larger sample size, 10 gm versus 2 gm, for the extraction method.

Further evidence that the temperature gradient is nonessential is cited by Vaughan and Green [9], who postulate that the mechanism of fractionation is that the polymer dissolves in the solvent–nonsolvent mixture but that precipitation does not occur. Instead, only those molecules which can stay in solution move into the cooler regions of the column. Thus, the technique is essentially an extraction method. If this mechanism is correct, the length of column and the temperature gradient would be relatively unimportant and the controlling factors would be the rate of change of the solvent gradient, the rate of flow, and the exit temperature of the column. In the discussion following Vaughan and Green's paper [9], F. D. Hartley cites use of a column without temperature gradient for a wide variety of polymers—vinyls, polyesters, and polyethers—with good results, although no temperature gradient was employed.

It is apparent from the literature that no firm conclusions can be drawn on the effect, if any, of a temperature gradient. To conclusively determine the effect of the thermal gradient, extremely well controlled conditions are required. An obvious method would be to study the effect of column length. To date, the time-consuming nature of fractionation has resulted in little work in this field. There is no exact theory to predict the selection of the various variables involved in fractionation. Experiments tend to be essentially empirical. Therefore, successfully demonstrated fractionation procedures are most often used rather than spending time in proving which, if any, of the conditions are unnecessary or less than optimum. In this aspect, the use of the temperature gradient appears to be justified.

VI. Preparatory-Scale Fractionation

In analytical fractionation only sufficient sample is used to provide enough material for characterization, usually by molecular weight. There are many cases, however, where it is desirable to prepare comparatively large amounts of narrow distribution fractions. There has been relatively little work reported in this field. One apparatus using the chromatographic principle is shown in Fig. 10. This apparatus has the same principle as the analytical-scale chromatographic fractionation. The increase in sample and,

B.3. Chromatographic Fractionation

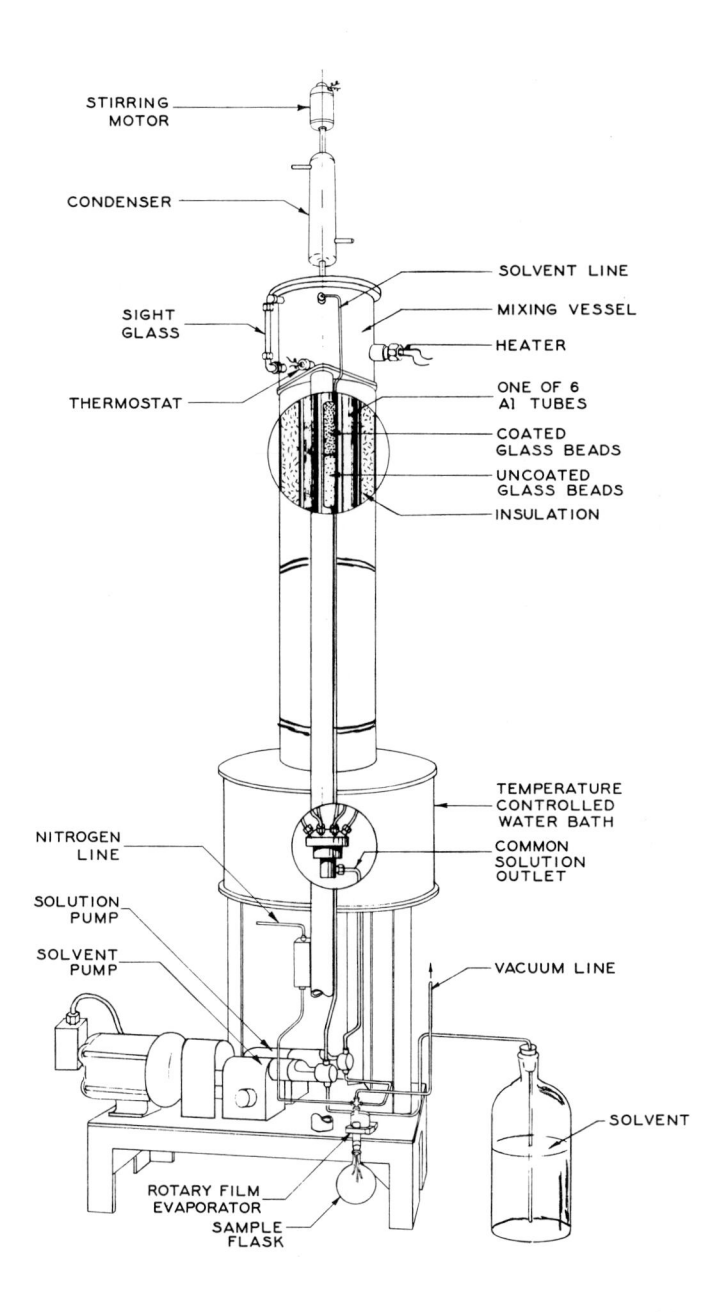

Fig. 10. Preparatory-scale fractionation apparatus, multi-column unit [7].

therefore, fraction size, is obtained by operating six columns in parallel. The choice of the parallel method was dictated by the difficulty encountered in simple scale up by increasing the diameter of the column. Experience in distillation, liquid–solid chromatography, and gas–liquid chromatography, has shown that above some maximum diameter usually a sharp decrease in efficiency results due to channeling and/or lateral temperature gradients. The apparatus shown has been used on a variety of polymers in the 35–100 gm sample charge range [7,44,45]. Two fractions from the large-scale apparatus were refractionated using a single analytical-scale column. The $\overline{M}_w/\overline{M}_n$ of the two fractions were 1.020 and 1.013, respectively, indicating that very good fractionation had been obtained. These results are comparable to the breadth of distribution obtained in analytical-scale columns.

The time required both in terms of actual operating time and elapsed time are comparable for the multicolumn apparatus and for analytical-scale instruments. Drawbacks inherent in large-scale fractionation are the large volumes of solvents that must be handled and the necessarily large size of the apparatus [46].

Acknowledgments

The authors are pleased to acknowledge the courtesy of the *Journal of Polymer Science* published by Interscience Publishers, a division of John Wiley & Sons, Inc., for permission to reprint Figs. 1 and 3 through 10, and also to *Analytical Chemistry* for Fig. 2.

REFERENCES

1. C. A. Baker and R. J. P. Williams, *J. Chem. Soc.* p. 2352 (1956).
2. J. L. Jungnickel and F. T. Weiss, *J. Polymer Sci.* **49**, 437 (1961).
3. D. L. Flowers, W. A. Hewett, and R. D. Mullineaux, *J. Polymer Sci.* **A2**, 2305 (1964).
4. N. S. Schneider, L. G. Holmes, C. F. Mijal, and J. D. Loconti, *J. Polymer Sci.* **37**, 551 (1959).
5. J. M. Hulme and L. A. McLeod, *Polymer* **3**, 153 (1962).
6. W. R. Krigbaum and J. E. Kurz, *J. Polymer Sci.* **41**, 275 (1962).
7. M. J. R. Cantow, R. S. Porter, and J. F. Johnson, *J. Polymer Sci.* **C1**, 187 (1963).
8. D. C. Pepper and P. P. Rutherford, *J. Appl. Polymer Sci.* **2**, 100 (1959).
9. M. F. Vaughan and J. H. S. Green, *Soc. Chem. Ind. (London), Monograph* **17**, 81 (1963).
10. R. S. Alm, R. J. P. Williams, and A. Tiselius, *Acta Chem. Scand.* **6**, 826 (1952).
11. A. Cherkin, F. E. Martinez, and M. S. Dunn, *J. Am. Chem. Soc.* **75**, 1244 (1953).
12. L. M. Marshall, K. O. Donaldson, and F. Friedberg, *Anal. Chem.* **24**, 773 (1952).
13. G. Meyerhoff and J. Romatowski, *Makromol. Chem.* **74**, 222 (1964).
14. R. M. Bock and Nan-Sing Ling, *Anal. Chem.* **26**, 1543 (1954).
15. K. O. Donaldson, V. J. Tulane, and L. M. Marshall, *Anal. Chem.* **24**, 185 (1952).
16. J. F. Gernert, M. J. R. Cantow, R. S. Porter, and J. F. Johnson, *J. Polymer Sci.* **C1**, 195 (1963).
17. T. K. Lakshmanan and S. Lieberman, *Arch. Biochem. Biophys.* **45**, 235 (1953).

18. J. E. Guillet, R. L. Combs, D. F. Slonaker, and H. W. Coover, Jr., *J. Polymer Sci.* **47**, 307 (1960).
19. T. J. R. Weakley, R. J. P. Williams, and J. D. Wilson, *J. Chem. Soc.* p. 3963 (1960).
20. R. E. Jentoft and A. A. Carlstrom, *Chemist-Analyst* **50**, 116 (1961).
21. W. Cooper, G. Vaughan, and R. W. Madden, *J. Appl. Polymer Sci.* **1**, 329 (1959).
22. S. R. Caplan, *J. Polymer Sci.* **35**, 409 (1959).
23. W. Cooper, G. Vaughan, and J. Yardley, *J. Polymer Sci.* **59**, S2 (1962).
24. N. S. Schneider, J. D. Loconti, and L. G. Holmes, *J. Appl. Polymer Sci.* **5**, 354 (1961).
25. A. S. Kenyon and I. O. Salyer, *J. Polymer Sci.* **43**, 427 (1960).
26. J. E. Guillet, R. L. Combs, D. F. Slonaker, J. T. Summers, and H. W. Coover, Jr., *SPE* (*Soc. Plastics Engrs.*) *Trans.* **2**, 164 (1964).
27. P. J. Flory, "Principles of Polymer Chemistry," p. 341. Cornell Univ. Press, Ithaca, New York, 1952.
28. N. S. Schneider, J. D. Loconti, and L. G. Holmes, *J. Appl. Polymer Sci.* **3**, 251 (1960).
29. N. T. Pope, T. J. R. Weakley, and R. J. P. Williams, *J. Chem. Soc.* p. 3442 (1959).
30. J. V. Ch'ien and S. N. Chu, *K'o Hsueh T'ung Pao* p. 525 (1959); see *Chem. Abstr.* **54**, 7216 b (1960).
31. A. Chapiro, P. Cordier, J. Jozefowicz, and J. Sebban-Danon, *J. Polymer Sci.* **C4**, 491 (1963).
32. C. M. Hansen and G. A. Sather, *J. Appl. Polymer Sci.* **8**, 2479 (1964).
33. N. S. Schneider and L. G. Holmes, *J. Polymer Sci.* **38**, 552 (1959).
34. W. F. Haddon, Jr., R. S. Porter, and J. F. Johnson, *J. Appl. Polymer Sci.* **8**, 1371 (1964).
35. C. J. Panton, P. H. Plesch, and P. P. Rutherford, *J. Chem. Soc.* p. 2586 (1964).
36. S. W. Hawkins and H. Smith, *J. Polymer Sci.* **28**, 341 (1958).
37. J. B. Kinsinger and R. A. Wessling, *J. Am. Chem. Soc.* **81**, 2908 (1959).
38. G. V. Schulz, P. Deussen, and A. G. R. Scholz, *Makromol. Chem.* **69**, 47 (1963).
39. P. M. Henry, *J. Polymer Sci.* **36**, 3 (1959).
40. L. D. Moore, Jr., A. Greear, and J. O. Sharp, *J. Polymer Sci.* **59**, 339 (1962).
41. N. S. Schneider, *Anal. Chem.* **33**, 1829 (1961).
42. V. Desreux and M. C. Spiegels, *Bull. Soc. Chim. Belges* **59**, 476 (1950).
43. R. A. Mendelson, *J. Polymer Sci.* **A1**, 2361 (1963).
44. M. J. R. Cantow, R. S. Porter, and J. F. Johnson, *Nature* **192**, 752 (1961).
45. M. J. R. Cantow, R. S. Porter, and J. F. Johnson, *J. Appl. Polymer Sci.* **8**, 2963 (1964).
46. A. S. Kenyon, I. O. Salyer, J. E. Kurz, and D. R. Brown, *J. Polymer Sci.*, **C8**, 205 (1965).

CHAPTER B.4

Gel Permeation Chromatography

K. H. Altgelt
CHEVRON RESEARCH COMPANY, RICHMOND, CALIFORNIA

AND

J. C. Moore
THE DOW CHEMICAL COMPANY, FREEPORT, TEXAS

I. Introduction..123
   A. Special Features of Gel Permeation Chromatography.............124
   B. The Principle of Gel Permeation Chromatography................125
   C. Applications of Gel Permeation Chromatography.................126
II. History of Gel Permeation Chromatography.......................127
   A. Gel Permeation Chromatography in Aqueous Solutions............127
   B. Gel Permeation Chromatography in Nonaqueous Systems...........129
III. The Theory of Gel Permeation Chromatography....................130
   A. The Mechanism...130
   B. The Distribution Coefficient K_d..............................138
   C. The Theoretical Plate Concept.................................141
   D. Assessment of Optimal Conditions for Gel Permeation Chromatography 143
   E. Side Effects..145
IV. The Gels...147
   A. General Considerations..147
   B. Gels Used in Aqueous Systems..................................149
   C. Gels Used in Nonaqueous Systems...............................153
V. Experimental Technique..158
   A. Analytical Separations..159
   B. Preparative Separations.......................................166
   C. Special Column Techniques.....................................169
VI. Evaluation of Data...169
   A. The Elution Curve...169
   B. Relations between Elution Volume and Molecular Weight..........171
   C. Computing the Molecular Weight Distribution Curve.............172
   References..173

I. Introduction

This chapter was written primarily for the polymer chemist who is concerned with molecular weight distributions or with obtaining fractions for further work. It is our aim to acquaint the polymer chemist with a powerful new method that is convenient, fast, mild, adaptable to analytical, as well as preparative, separations, and which, last but not least, separates solutes according to molecular size.

The protein chemists and the biochemists, who contributed so much to this method which they generally call Gel Filtration, may be disappointed that so many important applications to their work have only been mentioned in passing or are completely neglected. However, they will find the most important aspects of Gel Permeation Chromatography (GPC) described here: its potential, a variety of applications, its theory, latest instrumental designs, and evaluation of data. For scientists working with aqueous solutions, there are already excellent detailed surveys of this field, viz., Flodin's booklet, "Dextran Gels and Their Applications in Gel Filtration" [1], from which parts of our discussion were extracted, the review articles by Porath and Flodin [2] and by Tiselius et al. [3], the very commendable treatment of GPC in biochemistry by Morris and Morris [4], and more recently the comprehensive reviews by Determann [5] and by Gelotte [6]. AB Pharmacia in Uppsala, Sweden, publishes brief descriptions of the theory, experimental technique, and applications in "Gel Filtration" brochures [7] which are revised from time to time. Another of its publications [8] describes its Sephadex gels with properties and applications. Perhaps most important, this company offers a free literature service on GPC in two forms, (1) a booklet [9] containing the references and (2) a card system with abstracts of papers on GPC. The last one, in particular, is very useful for looking up special information. The references in Section I,C are largely based on these abstracts.

A. Special Features of Gel Permeation Chromatography

Gel permeation chromatography is a column fractionation method based on the molecular sieve effect. In principle it has been known since the early 1950's, but only after Porath and Flodin rediscovered and exploited it vigorously was it recognized by the general scientific world. Porath and Flodin published their first paper on gel filtration [10], as they called it, in May, 1959. From this date until January, 1964, more than 300 papers have been published on the new method.

This large number of contributions demonstrates the general interest for the advantages of GPC, some of which were recognized at once while others are still being uncovered.

Its most obvious and most important feature is that GPC separates particles by size; i.e., under suitable conditions, the separation is not affected by the chemical nature of the components to be separated. This is unique since most of the established fractionation methods are based on solubility, which is a function of both molecular weight and structure.

The other methods, like sedimentation, molecular and thermodiffusion, are complex and depend also on more than one variable.

With GPC many separations were made possible for the first time, particularly in the various fields of natural products, where separations by size only are especially important. Thus, its earlier development was advanced chiefly by biochemists and other categories of scientists working with biological materials.

Extension of GPC to the synthetic polymers was slower. Specialized gel structures had to be developed. There were many advantages to impel such development. Being a column technique, GPC is convenient and versatile. It can be modified in many ways to fit almost all requirements such as exclusion of oxygen, scaling up and down, and automation. It is cheap since the gel can be used over and over again. The fractionation volume is comparatively small. And GPC is rapid: Small-scale separations can be performed in 10 minutes; large-scale runs may take about 6 hours. Even before one run is completely finished, the column is ready for the next one. Thus, semicontinuous separations, going on for weeks automatically, can be achieved [8,11]. Perhaps the most important application has recently been published (see Section I,C). On this basis, molecular weight distributions of natural and synthetic polymers can be determined in a few hours, with high accuracy, and in the range from 18 (water) to several million molecular weight.

All these facts considered, GPC seems to be one of the most promising fractionation techniques available today.

B. The Principle of Gel Permeation Chromatography

A column is packed with small gel particles which have pores of variable size. Solvent fills the interstitial space as well as all the pores of the gel. The sample is dissolved, introduced into the column, and eluted, all with the same solvent. Small solute molecules diffuse freely into and through the gel pores; they permeate the gel. Some species may be so large that they cannot enter the gel and thus are completely excluded; others are excluded from the smaller pores only.

Since solvent flow occurs only in the space between the gel particles, the excluded larger molecules are flushed through the column first. Smaller molecules are delayed. Under ordinary conditions, the permeation or diffusion of solute into the gel is sufficiently faster than the elution rate. Therefore, diffusion equilibrium is assumed. Separation is then based on the different volumes inside the gel particles which are accessible to solute molecules of different size: Totally excluded molecules are eluted at a volume equal to the interstitial or void volume, V_0; other molecules are eluted at a volume composed of the interstitial volume and that part of the internal volume, V_i, which is available to them. The elution volume, V_e, of a molecular species is then: $V_e = V_0 + K_d V_i$. K_d is the volumetric distribution coefficient.

Because separation is here achieved by the different permeability of the gel particles, the name gel permeation chromatography was chosen. Porath and Flodin called the method "gel filtration." This name, however, implies filtration through a plate or a short column and thus appears to be misleading. Also, on filtration the larger particles are retained and the small ones go through the filter. In contrast, the name "gel permeation chromatography" seems to reflect more truly the basis of separation.

C. Applications of Gel Permeation Chromatography

GPC can serve three main purposes: group separations, including separations of large molecules from small ones, fractionation of polymers or oligomers, and molecular weight determinations.

Group separations can be achieved in short times and under mild conditions. Examples are:

Isolation, or at least enrichment, of high molecular weight compounds from natural or reaction mixtures [10,12–57], e.g., from enzyme digests [23,25,30,31,36].

Purification, particularly of high molecular weight compounds from low molecular weight contaminants [10,19,21,25,28,51,58–92a]. If a low molecular weight compound is to be purified, it can sometimes be complexed [93] and then separated from its admixtures.

Separation of two or more main products in a mixture such as enzymes and cofactors [94] of different proteins present in venoms or sera [13,17,95–100], of polymer additives from mineral oil [101], and a great number of other diverse mixtures [10–12,35,38,43,46,48,49,53,54,61,83,96,97,100,102–126].

Desalting and change of solvent (e.g., buffer) [10,16,26,30,60,67,92a,96,104, 127–130]. GPC is, in many cases, better than dialysis since it works faster, under milder conditions, and with higher throughput. Various clinical tests were made feasible or were at least greatly simplified by GPC [24,25,34,44, 55,68,69] because of the quick separation of macromolecules from salt or other compounds of low molecular weight.

Fractionations have been achieved on samples ranging from 18 MW to more than 1,000,000 MW. In aqueous solutions many diverse compounds have been separated: Oligo- and polysaccharides [92a,127,128,131–134], peptides [23,33,95,102,103,113,135,136], proteins [38,50,73,86,96–100,102–104,107,113,137–143], enzymes [71,104,105,123,144,145], nucleic acids and nucleotides [40,70,71,125,146–149], enzyme digests [23,25,30,31,36,81,124, 148,150–153] or chemically degraded natural macromolecules [132,144], even beer constituents [106] and coffee extracts [117]. In nonaqueous solutions fractions of polystyrene [154–162], polybutene [163], polypropylene glycol [157], and other polymers [156,157] as well as macromolecular natural products [163] were reported. Also small molecules in the molecular weight range of 50–1200 were separated [164–166].

One of the most important applications of GPC is the determination of molecular weights [157,167–175] and molecular weight distributions [157]. A column must be calibrated with substances of known molecular weight. Once a calibration curve has been established for a certain species, molecular weights of unknown samples of similar shape and structure can be obtained from their elution volumes. In comparison to ultracentrifugal determinations of this kind, GPC is more convenient and it can be performed at much lower concentrations, making extrapolation to zero concentration more accurate or even unnecessary. Also, GPC can be adapted to automation. A complete molecular weight distribution of an ordinary synthetic polymer can be obtained in 2–5 hours by a moderately skilled technician. Because GPC can be conducted at extremely low concentrations, it lends itself to studies of dissociation effects [173,174].

II. History of Gel Permeation Chromatography

A. GEL PERMEATION CHROMATOGRAPHY IN AQUEOUS SOLUTIONS

In 1925 Ungerer [176] observed a separation of small ions on account of their different sizes when he studied their adsorption on clay. This was the first step in the development of molecular sieves or ion exclusion, from which later GPC has evolved.

Wiegner [177] and, shortly after him, Cernescu [178] demonstrated the gradual exclusion of ammonium ions with increasing degrees of methylation from zeolite and permutites. In all cases the smallest ions were adsorbed to a greater extent because they would permeate the pores of the adsorbents more completely whereas the larger ones were excluded partly and thus did not find so many active sites to be adsorbed on.

Claesson and Claesson [179–181] applied this principle to the fractionation of nonionic polymers on columns packed with nonionic adsorbents. They obtained good results with nitrocellulose, neoprene, methyl methacrylate, and other polymers on charcoal, aluminum hydroxide, and calcium carboxide. Again, the larger species were excluded and, therefore, adsorbed to a lesser extent than the smaller ones.

Deuel [182,183] was one of the early pioneers of GPC. He separated natural polymers, such as pectins and algins, from their monomers, galacturonic and manuronic acid; fractionated polydisperse samples, such as enzyme digests of pectin acids or polyphosphoric acids of varying chain length; and prepared several gels from pectins by cross-linking them with formaldehyde and epichlorohydrin. At this early time Deuel had already suggested that this technique be used for the determination of molecular size and for rapid dialysis.

In 1953 Wheaton and Bauman [184] wrote an extensive paper on GPC in which they described the procedure and part of the theory of this method as they are still valid today. In their work on ion exclusion, Wheaton and Bauman had noticed that neutral compounds, too, could be separated on ion exchangers and even on neutral gels and that no adsorption was required for this separation. They studied the new phenomenon carefully on polystyrene gels and recommended it for analytical separations. However, these gels were permeable only to relatively small molecules.

Two years later Lindquist and Storgårds [185] reported separations of amino acids and peptides on starch columns and explained them in terms of a molecular sieve effect. In the same year, 1955, Lathe and Ruthven [167,186] fractionated mixtures of different proteins and smaller molecules on starch columns. They were the first ones to fractionate a series of solutes with molecular weights covering a wide range and to interpret these separations in terms of restricted permeations according to the varying solute sizes. They found, further, that the molecular weight range of fractionation could be extended by swelling the starch granules and thus increasing their pore size. Using columns of swollen starch, Lathe and Ruthven [167] determined the molecular weights of insulin (6000) and myoglobin (35,000). On the same columns they achieved separation of a mixture of amylopectin (molecular weight 1,000,000), globulin (150,000), hemoglobin (67,000), and compounds of lower molecular weight.

The great upsurge of GPC, as a generally accepted fractionation method, was induced by a paper by Porath and Flodin [10] in 1959 in which they described newly developed dextran gels with useful ranges of porosity and fair rigidity. The authors demonstrated convincingly the fractionation of a mixture of glucose and two dextran fractions of 1000 MW and 20,000 MW, respectively. Furthermore, they showed the convenient and fast separation of salt from serum proteins and pointed out the advantages of this technique as compared with dialysis. Cross-linked polyvinyl alcohols were also named as suitable gels for fractionation.

In short sequence, a number of papers from the same group in Uppsala followed the first one, describing fractionations of polypeptides [12], proteins [12,13,59,102], peptides [59,102], amino acids [95,103], poly- and oligosaccharides [131]. Other papers from this laboratory dealt with various applications of GPC such as purification of hormones [58] and enzymes [59] and separations of macromolecules from molecules of low molecular weight [102] and with special problems such as sorption effects [42], effects of flow rate, gel particle size, volume and viscosity of the sample [187], and automation [11].

Pharmacia, in Uppsala, Sweden, who produces dextran gels and sells them under the name of "Sephadex," started a very effective advertising campaign with brochures [7,8], free samples of Sephadex, and, perhaps most important, a comprehensive and free literature service [9]. It might well have been because of this well-organized endeavor that GPC caught on so rapidly.

More and more applications were found, chiefly in separation, isolation, and purification of biological materials but also in studies of complex formation [93] and other chemical reactions [21,30,47,62,65,111,114,188].

Investigations of drug binding to protein [47] and to plasma [114] and of protein binding to corticoids [111] and to small ions [188] were aided by GPC. In some cases the protein-bound compounds were simply separated on Sephadex gels from the unbound materials; in other cases different degrees of binding could be discriminated by using the plasma [114] or the ion solution [188] as eluents and analyzing the eluates for protein concentration.

Besides Sephadex, other gels were used for fractionating biological materials of high molecular weights: agar [138], gelatine [138], polyvinyl ethyl carbitol [140], polyvinyl pyrrolidone [140], and polyacrylamide [5,109,140,151,189,190]. Although most of them achieved good separation in their molecular weight range, most were less rigid than Sephadex and, therefore, did not find such wide acceptance. Only polyacrylamide gels were recently made commercially available and currently seem to be very promising [190]. They are highly resistant to bacterial growth [5,190] and have different sorption properties from polydextrans [149]. We shall come back to these gels in Section IV,B.

B. Gel Permeation Chromatography in Nonaqueous Systems

As already mentioned, only a few publications dealt with GPC in nonaqueous media [101,154–166]. One reason for the apparent lack of interest in this field was that no suitable gels seemed to be available.

Vaughan [154–156] found only poor fractionation of polymers on cross-linked polystyrenes, although he indicated the possibility of better fractionation. Brewer [101,164,165] reported excellent separation of several oligoisoprenes reaching up to 1200 and a polybutene sample of 18,000 MW on lightly cross-linked rubber. However, he obtained little or no fractionation of the polybutene sample itself.

Altgelt [191] achieved fractionations of several polybutenes on rubber gel. Up to molecular weights of 15,000, the separation was as good as with the Baker–Williams method; beyond, however, fractionation was nil.

For chemists working with synthetic polymers, the molecular weight range below 15,000 is generally of less interest. Therefore, GPC did not seem to offer any incentive to them. In addition, a study by Cortis-Jones [166] seemed to indicate that the chemical structure of the solute molecules had a grave effect on their elution volumes. This would have meant that, in nonaqueous systems, GPC did not fractionate by size only. However, the study had been made on polar substances of very low molecular weight where hydrogen bonding and shape can affect drastically the pore volume available to the permeating molecules. Brewer [165] found similar effects in the molecular weight range below 800, but not above. At high molecular weights and with suitable solvents polarity plays a very minor role in most cases.

In 1962 Vaughan [155] published the efficiencies of a wide variety of gels and of other porous particles for the fractionation of polystyrene. While most materials performed rather poorly, an expanded silica gel gave results comparable with regular precipitation fractionation. This was the first time that successful GPC of a polymer of high molecular weight in an organic solvent had been reported.

Later in the same year, Moore [157,192] reported excellent fractionation of polystyrene samples in a molecular weight range of 700–1,000,000. He had prepared a series of polystyrene gels which, in contrast to previous ones, were highly cross-linked and, at the same time, highly porous. This combination of properties was achieved by polymerization in solvent mixtures which were good solvents for the monomers but marginal as solvents for the polymers. These gels were rigid, packed easily without clogging, and could be produced with almost any pore size.

Moore undertook to design an analytical instrument for routine repetitive operation. By combining the principle of GPC with the experimental procedures of gas chromatography, he constructed a machine that, after calibration, produces a molecular weight distribution in a few hours. The molecular weight range can be adjusted according to need by utilizing gels of different pore sizes. Maley [159] has described the commercial version of this gel permeation chromatograph which is built by Waters Associates, Inc., Framingham, Massachusetts.

III. The Theory of Gel Permeation Chromatography

A. The Mechanism

The theory of GPC is still in a preliminary stage, but it provides a basis for the evaluation of experimental work and for certain predictions. In the introduction we discussed briefly the principle of GPC as it is seen by most workers in this field. In this section we shall go into more detail.

A gel particle consists of a network of more or less heavily cross-linked polymer chains which are themselves more or less aggregated into strands with random joinings. Electron micrographs of heavily cross-linked gel sections show that the "pores" are not tubular passages, but are merely the interconnected apertures of the void spaces between chains or the strands and clusters of chains.

In lightly cross-linked gels the network is flexible, and the structure collapses, i.e., the internal volume is squeezed out when the gel is dried or is suspended in a nonsolvent. Clearly the pore size of a lightly cross-linked gel is affected greatly by the solvent power of the solvent and even by the compatibility with a solute if this is present in appreciable concentration.

Heavily cross-linked gels swell to a much lower degree and their structure is rather impervious to the properties of solvent and of solute. This is one of their virtues; another one is their overall rigidity, which prevents clogging in columns under pressure.

The mechanism of GPC is the same in lightly and heavily cross-linked gels although some practical aspects may be vastly different. The gel particles in a column are suspended in solvent. The channels between the gel particles are much larger than the pores inside the gels; therefore, solvent flow occurs only in the interstitial space. Solute molecules permeate the gel pores as far as their size permits and move practically without restriction in the solvent contained in the gel. Only very close to the network strands, where the gel segment density is high, the diffusion rate drops sharply [1].

Under normal conditions molecules comparable in size to solvent molecules will distribute through the entire pore volume. Bigger molecules are excluded from the denser parts of the network, but they can diffuse freely through the more open passages. The larger a solute molecule is, the fewer apertures suitable for its size it will find. Finally, there may be molecules which are so big that they are completely excluded from the gel.

However, of the large pores in the gel, only a few will be open to the outside with their full width; most of them will be connected by apertures of more or less drastically reduced size. Therefore, the larger solute molecules admitted to the gel will reside only in the outer pores, as illustrated in Fig. 1.

At this point we shall be satisfied with this observation and not pursue it further or question it. It follows that each solute species has a certain volume at its disposal which it can occupy. This volume is the sum of the void volume V_0 and of the accessible part of the pore volume. An amount exactly equal to this volume has to pass the column for a given solute to

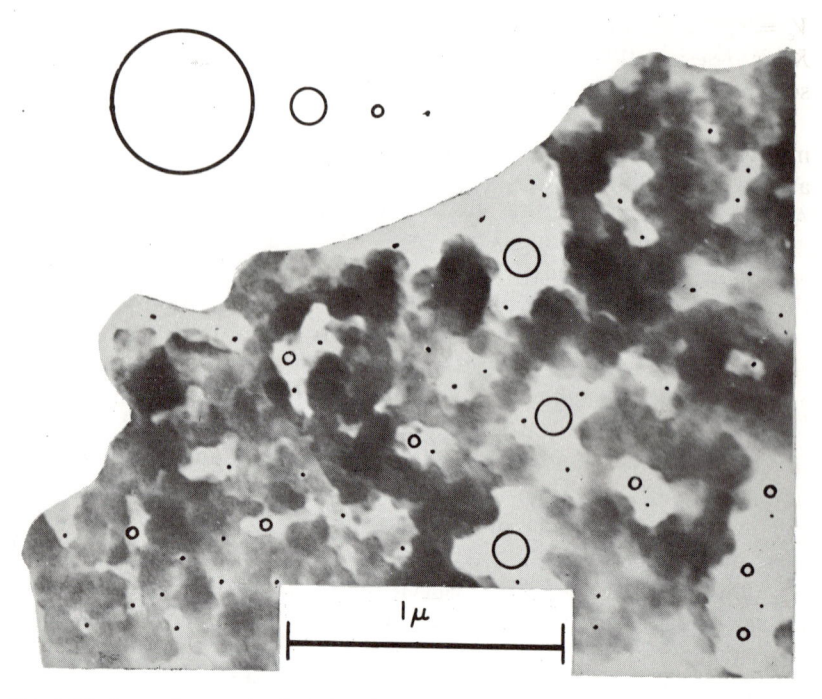

Fig. 1. Electron micrograph of gel showing network structure. The circles are drawn to the average radii of gyration for solute molecules with $K_d = 0$, 0.3, 0.5, respectively. The small molecules can diffuse into all the chambers; the intermediate-sized ones stay in the outer shell of the gel particle, and the large ones are completely excluded. (Gel N, see Figs. 4 and 5.)

be eluted:

$$V_e = V_0 + K_d V_i \tag{1}$$

As already stated, K_d is the volumetric distribution coefficient between the total internal volume V_i and that part of it which is accessible to a given solute, $V_{i,\text{acc}}$. As a consequence

$$K_d = V_{i,\text{acc}}/V_i \tag{2}$$

K_d can be calculated from the readily available volumes V_e, V_0, and V_i

$$K_d = (V_e - V_0)/V_i \tag{3}$$

It is effectively synonymous to the partition coefficient known from regular chromatography and will be discussed further under Section III,B. Here it shall suffice to repeat that for totally excluded solutes $K_d = 0$ and, hence,

$V_e = V_0$; whereas, for very small molecules, $K_d = 1$ and $V_e = V_0 + V_i$. K_d is independent of column size and geometry, and, within a given gel–solvent system, a solute species is determined by its K_d value.

Porath [193] was the first who developed a theory which relates the molecular size of a solute, or its molecular weight, with K_d. In his model he assumed all pores to be of equal size and of the same equal conical shape. A solute molecule could enter such a pore only up to a depth where its diameter was equal to the cone diameter. From the volume of the whole cone and of the part accessible to the solute molecule, Porath derived the equation

$$K_d = k \frac{V_{i,\text{acc}}}{V_i} = k\left(1 - \frac{2R}{A}\right)^3 \qquad (4)$$

R is the effective hydrodynamic radius of the solute molecule; A, the cone diameter; and k a proportionality coefficient introduced because of the inadequacy of the model.

Setting

$$A^3 \sim R_s - \alpha$$

and, for flexible chain molecules,

$$R \sim M^{1/2}$$

leads to

$$K_d = k\left[1 - k_1 \frac{M^{1/2}}{(R_s - \alpha)^{1/3}}\right]^3 \qquad (5)$$

R_s is the solvent regain; α is a part of R_s which is incorporated in the swollen gel structure and is not exchangeable for solute. Plots of K_d vs. $M^{3/2}$ yielded linear curves [133,162] and seemed to confirm the theory. However, as Laurent and Killander [194] pointed out, certain proteins also seemed to follow Eq. (6), although they were not threadlike molecules and therefore did not fit the model underlying relation (5). The assumption of conical gel pores which determines Eq. (4) appears grossly over-simplified in view of Fig. 1. Porath himself [194a] prefers now Laurent and Killander's theory [194]. A contributing factor to the apparent agreement between this model and experimental results may be the limited range of molecular weights used for testing a relatively insensitive function such as Eq. (6).

Laurent and Killander's theory [194] is based on a model consisting of a network of straight rigid rods which are infinitely long and distributed at random in the gel. The available volume for spherical particles in such a system was calculated by Ogston [195]. A different distribution coefficient K_{av} is introduced which is related to K_d by

$$K_{av} = K_d \frac{V_i}{V_i + V_{GM}} \qquad (6)$$

where V_{GM} is the volume taken by the gel matrix. This K_{av} was found to be

$$K_{av} = \exp[-\pi L(r_s + r_r)^2] \qquad (7)$$

L is the concentration of rods in the system, r_s and r_r are the radii of spherical solute particles and of the gel rods, respectively.

This model can only be an approximation. But with a suitable choice of L (to get a best fit) and of r_r (equals $7 \cdot 10^{-8}$ cm throughout) Laurent and Killander find that Eq. (9) fits experimental data taken from the literature as well as their own [194].

Until recently these were the only attempts to relate theoretically solute molecular weight with K_d or with the elution volume. Mostly empirical relations are used, such as a plot of the logarithm of molecular weight versus the elution volume, which gives straight curves over wide ranges (see Fig. 5). As yet, no theoretical foundation was given for this type of plot. Brewer [164] found a linear relation between the molecular weight and the logarithm of the retention volume $V_e - V_0$, which might serve as a confirmation of Eq. (9).

The mechanism of GPC as described in this section is based on the assumption of diffusional equilibrium; that is, the solute molecules are assumed to have time enough to distribute themselves among the interstitial space and the available gel pores. Experimental conditions, i.e., gel particle size and flow rate, are usually chosen such as to fulfill this condition reasonably well. The time it takes a solute zone to pass by a gel particle is generally much greater than the half-equilibrium time for diffusion of the solute molecules into the gel pores.

As an example, let us assume a column of 125-cm length and 1.5-cm width packed with a highly cross-linked gel. The total volume is close to 200 cm³, and $V_0 + V_i$ is about 150 cm³. At a flow rate of 1 ml/minute, the smallest solute species is eluted after 150 minutes. If we assume a zone width of only 1 cm, it would take this zone $(150/125) \times 60 = 72$ seconds to proceed down the column by 1 cm, i.e., by its own width. Actually, half this time, i.e., 36 seconds, should be compared with the diffusion time of the solute since only in the first half of the zone is the solute concentration rising and causing the molecules to diffuse into the gel. After half the zone has gone by, the outside concentration decreases, and the solute molecules start diffusing out again.

Flodin [1] estimated half-times of diffusion equilibrium for different gel bead sizes from an equation which was derived by Vermeulen [196]

$$t_{0.5} = 0.030(r_0^2/\overline{D}) \qquad (8)$$

Recently, Vink [197] obtained the same type of formula by a different treatment. Only the proportionality coefficient is slightly higher,

$$t_{0.5} = 0.039(r_0^2/\bar{D}) \tag{9}$$

$t_{0.5}$, measured in seconds, is the half-time of a solute species for self-diffusion through spherical particles of radius r_0 cm; $\bar{D}$, in cm² second⁻¹, is the diffusion coefficient in the gel. In the following example, which was given by Flodin [1], $\bar{D}$ is assumed to be equal to the regular diffusion coefficient D. With $D = 10^{-6}$, the half-equilibrium time for bovin albumin is found to be 3 seconds for gel particles of 0.1-mm radius and 0.03 second for 0.01-mm radius if diffusion takes place into the very center of the particle. For very small solute molecules with $D = 10^{-5}$, the half-times would have been 0.3 and 0.003 seconds, respectively.

Since the larger solute molecules, which can enter the gel pores, will have access only or mainly to the outer shell of the gel particles, their travelling distance is smaller than the particle radius r_0. Hence, under the assumption of free diffusion their diffusion time will be of the same order or even less than that of the small molecules which can penetrate the whole gel. Of course, this will be the case only with gels of suitable and not of excessive pore size.

On the other hand, the diffusion constant $\bar{D}$ in the gel is smaller than that in the absence of the gel matrix for several reasons. The gel pores are not straight cylindrical openings but randomly bent or kinked and of varying size and shape. A diffusing molecule thus encounters many obstacles on its way which delay it. Also, its mobility is reduced by interactions with the wall at spots where the pore diameter is only slightly larger than the molecule diameter. For our estimate of diffusion equilibrium, we shall therefore from here on assume $D \approx 1/10 D$. Readers who would like to go into more detail of this aspect are referred to Helfferich's book, "Ion Exchange" [198].

Vink [197] derived a simple equation for the concentration equilibration of a solute diffusing into a gel

$$\Delta C = \Delta C_0 \exp\left[-\frac{2\bar{D}t}{V}\left(\frac{1}{V} + \frac{1}{V_e}\right)\right] \tag{10}$$

ΔC is the difference of concentrations inside and outside the gel at time t; ΔC_0 is this concentration difference at time 0; V and V_e are the "per area volumes" of the gel and the external solution, respectively. In our case we extended time. For a spherical gel particle, $V = \tfrac{1}{3}r$. Thus, we obtain

$$\frac{\Delta C}{\Delta C_0} = \exp\left(-\frac{18\bar{D}t}{r^2}\right) \qquad (11)$$

For small molecules with $D = 10^{-5}$, i.e., $\bar{D} = 10^{-6}$, and for $r = 0.01$ cm and $t = 36$ seconds, a deviation from equilibrium of less than 0.2% is computed from Eq. (11).

Helfferich [198] quotes a different equation for the concentration difference of a solute diffusing out of a gel particle

$$\frac{\Delta Q(t)}{\Delta Q_\infty} = 1 - \frac{6}{\pi^2} \sum_{n=1}^{\infty} \frac{1}{n^2} \exp\left(-\frac{\bar{D}t\pi^2 n^2}{r^2}\right) \qquad (12)$$

$\Delta Q(t)$ is the difference of solute amounts inside and outside the gel at time t; ΔQ_∞ is this difference after equilibration. From tabulated values of Function (12), we find 97% equilibrium attained after 30 seconds and 99.6% after 50 seconds for our example. From these calculations we conclude that under normal circumstances, equilibrium conditions are reasonably closely attained in GPC.

Equations (8) and (9) illustrate how important it is to keep the gel particles small since the diffusion time increases with the square of the particle diameter. The newer gels are all available in rather small sizes, viz., around or smaller than 200 US mesh which corresponds to about 75-μ diameter. Since the gels are produced by bead polymerization, they are round and cause no flow impediments as did bulk polymerizates which had been ground to the same size.

Recently Laurent and Laurent built an electrical analog to the gel permeation process [198a]. With this computer the effect of nonequilibrium conditions on the elution profile of a sample can be predicted. Conversely, it might be possible to determine the approach to equilibrium on a column by comparing experimental with theoretical elution curves. This, of course, could only be done with a column on which zone broadening due to uneven packing is negligible compared to that by the partitioning of solute between the internal and the interstitial volume. The extent to which zone broadening is caused by the one or the other effect can also be studied now with Laurent and Laurent's electrical analog.

The mechanism of GPC as outlined here seems to be confirmed in full by the experiment. In most instances, change of flow rate does not affect the elution volume, indicating that the system is at or very near equilibrium. Yet, the picture given above is only very crude. In Fig. 1 we show solute molecules which can diffuse through all the pores, even beyond the constrictions. Then, there are others which are so large that they can enter certain pores, but only in the outer shell of the gel particle. However, there must certainly be molecules of intermediate size which can pass through the pore constrictions though at a sharply reduced rate because of the interaction with the walls. Craig [198b] has clearly shown that the escape rates of solutes in differential diffusion through membranes are not sharply different when the membrane pores are considerably larger than the escaping solute molecules. However, the escape rate becomes a sensitive measure of molecular size for molecules only a little smaller than the pores. It is apparent that differential diffusion and GPC are closely related. The diffusion rate of spheres in smooth capillaries with increasing ratio of sphere to capillary radius, a/r, is given by

$$D_R = D_0 \left(1 - \frac{a}{r}\right)^2 \left[1 - 2.104\left(\frac{a}{r}\right) + 2.09\left(\frac{a}{r}\right)^3 - 0.95\left(\frac{a}{r}\right)^5\right] \quad (13)$$

This equation was derived by Ackers and Steere [199] from a similar formula by Renkin [200]. Ackers and Steere showed that their equation describes very well restricted diffusion of hemoglobin and other proteins through membranes of agar gels, i.e., through irregular pores of real systems. On this basis it should be possible to develop a more general and comprehensive theory of GPC than the present ones.

In a recent paper [201] Ackers states that in more highly swollen gels such as Sephadex 200 separation is based on restricted diffusion while only in the denser gels—Sephadex 75 and 100—is it solely or mainly based on the accessible internal gel volumes. Using Eq. (13) and equating D_R/D_0 with K_d he arrives at

$$\frac{V_e - V_0}{V_i} = \left(1 - \frac{a}{r}\right)^2 \left[1 - 2.104\left(\frac{a}{r}\right) + 2.09\left(\frac{a}{r}\right)^3 - 0.95\left(\frac{a}{r}\right)^5\right] \quad (13a)$$

For nonspherical diffusing molecules a is the Stokes radius. Since the relation between a and the molecular weight can be determined by measurement of the intrinsic viscosity, sedimentation, or diffusion, Eq. (13a) correlates column data with molecular weight. By means of a computer, Ackers calculated a table of a/r values for a number of $(V_e - V_0)/V_i$ values; this makes the use of relation (13a) more convenient.

Busse [201a] and Goldsmith and Mason [202] called attention to another effect which contributes to separation in gel columns, if only to a negligible extent. In any system of molecules or particles flowing through capillaries or channels such as the interstices in a column bed, the larger particles near the walls extend toward the center of the capillary further than the small ones. On the average they seem to concentrate in the capillary core where flow is fastest and are therefore flushed out faster than the small particles which can reside closer to the capillary walls.

B. The Distribution Coefficient K_d

In chromatography the partition coefficient K is defined as the ratio of the solute concentration in the stationary phase to the concentration in the mobile phase. In GPC the "stationary phase" is the internal volume, V_i, and the "mobile" phase is the void volume, V_0. Therefore,

$$K = \frac{C \text{ (stat)}}{C \text{ (mobile)}} = K_d = \frac{C \text{ (internal volume)}}{C \text{ (void volume)}} = \frac{C_i}{C_v} = \frac{m_i}{V_i} \cdot \frac{V_0}{m_0} \quad (14)$$

The concentration in the stationary phase is taken as the weight (mass) of solute inside the gel, m_i, per total internal volume, V_i. At equilibrium the ratio of m_i to the accessible internal volume, $V_{i,\text{acc}}$, must be equal to the outer concentration

$$\frac{m_i}{V_{i,\text{acc}}} = C_v = \frac{m_0}{V_0} \quad (15)$$

This, inserted into Eq. (14), leads immediately to

$$K_d = \frac{V_{i,\text{acc}}}{V_i} \quad (16)$$

The distribution coefficient of a certain solute species is thus equal to that fraction of the internal volume which is accessible to the solute.

Naturally, K_d can never become greater than 1. In other words, should the elution volume of a species be greater than $V_0 + V_i$, some other effect has come into play. Generally, adsorption is the explanation for these cases; sometimes, partitioning occurs if mixed solvents are used or if the solvent is chemically very different from the gel.

In GPC, a substance is characterized by its K_d value just as in regular chromatography. K_d is independent of the column dimensions and can be employed with advantage to compare GPC runs on different columns. It is, however, dependent on anything that might change the pore size of the gel, i.e., the kind of gel, solvent, and temperature.

An important practical application of K_d is in finding the right conditions for a given task, e.g., a proper gel or the right column dimensions.

Let us first consider finding a proper gel. Without knowledge of K_d this can be time consuming, particularly in cases when the usual gels like Sephadex, Bio-Gel P, or polystyrene cannot be used. Time and effort can be saved by a series of simple concurrent diffusion equilibrium experiments. A row of vials, containing the solution and each a different gel, is set up.

B.4. GEL PERMEATION CHROMATOGRAPHY

From the difference of solute concentrations in the original solution and in the supernatant after the addition of solvent-swollen gel, K_d can be computed by the equation [184]

$$K_d = \frac{V'_0}{V_i} \frac{C_{original} - C_{supernatant}}{C_{supernatant}} \qquad (17)$$

V'_0 is the original volution volume, and V_i is the internal volume of the added gel.

For the case in which the solution is contacted with dry gel, another equation is used

$$K_d = 1 - \frac{V'_0 + V_i}{V_i} \cdot \frac{C_{supernatant} - C_{original}}{C_{supernatant}} \qquad (18)$$

The former method reflects the true mechanism in the gel column better and gives more accurate results. The latter method is very simple experimentally, It can lead to wrong absolute values of K_d because some solvent is taken up by the swelling gel, but it is used advantageously for relative determinations [191].

The internal volume of the gel, V_i, can be obtained by several methods. The easiest is to take it from the mixing ratio of solvent and monomer during polymerization [158]. If the polymerization is complete and the density of the monomer is assumed to be the same as that of the polymer, the mixing ratio of solvent and monomer will be the same in the gel as it was in the reaction mixture. Then V_i equals this mixing ratio times the volume of the swollen gel. This is true for highly cross-linked gels only [203], since soft gels swell to different degrees in different solvents.

In a better approximation the change of density during polymerization can be taken into account.

With soft gels V_i is usually calculated from the solvent regain, R_s, the solvent density, ρ_s, and from the weight of the dry gel, a, by the relation

$$V_i = aR_s/\rho_s \qquad (19)$$

R_s, in turn, is determined in a separate experiment. Granath and Flodin [133] describe an elegant modification of Pepper's [204] method. After allowing the gel to swell in water for 24 hours they transfer about 10 ml "... into a weighted adapter which consists of a tube with a dense screen on the bottom. The adapter is placed in a centrifuge tube, and the liquid in the void space is centrifuged down through the filter at 1000–2000 rpm for 20 minutes (radius 15 cm). The adapter with its contents is then weighed and the contents transferred to a beaker and dried to constant weight at 105°C. The water regain is expressed as grams of water imbibed per gram dry gel."

This method can be adapted to organic solvents. However, it will not work with heavily cross-linked polystyrenes of large pore size since these lose part of their internal solvent in the adapter even without centrifuging. In this case a tube is filled with swollen gel and just enough solvent to fill the void volume. From the total weight of gel plus solvent, W_t, the weight of the dry gel, a, the solvent density, ρ_s, and V_0, R_s is calculated by

$$\frac{W_t - V_0 \rho_s}{a} - 1 = R_s \tag{20}$$

Another way to obtain V_i is given by Eq. (21)

$$V_i = (V_t - V_0)\frac{d}{\rho_s}\frac{R_s}{1 + R_s} \tag{21}$$

Here all the quantities are easily measured. V_t is the total volume of the column and is composed of the void volume, the internal (liquid) volume, V_i, and the volume of the dry gel matrix, V_g

$$V_t = V_0 + V_i + V_g \tag{22}$$

d is the density of the swollen gel. The other designations were given before.

Equation (21) is readily derived starting out from the volume of the swollen gel, V_{swg}, in a column bed. V_{swg} can be expressed by the ratio of its weight, W_{swg}, and its density, d, and is composed of V_i and V_g

$$V_{swg} = W_{swg}/d = V_i + V_g \tag{23}$$

With

$$W_{swg} = aR_s + a = a(1 + R_s) \tag{24}$$

and Eq. (22), we obtain

$$(V_t - V_0)d = a(1 + R_s) \tag{25}$$

Equation (19) divided by Eq. (25) yields the Relation (21).

The void volume, V_0, is usually 30–35% of the total column volume, V_t. It, too, can be determined by various methods. Solute molecules, which are large enough to be totally excluded from the gel, will be eluted at V_0 (e.g., india ink or Pharmacia's Blue Dextran in aqueous systems). Also, an equation similar to Eq. (21) can be derived for V_0

$$V_0 = V_t - (a/d)(1 + R_s) \tag{26}$$

or together with Eq. (20)

$$V_0 = \frac{V_t - W_t/d}{1 + \rho_s/d} \tag{27}$$

Again, all variables are easily measured. The density of the swollen gel, d, and the solvent regain, R_s, are, of course, dependent on the kind of gel, solvent, and temperature but independent of the column dimensions. For a given gel–solvent system they need be determined only once.

The K_d values of two molecular species to be separated can be used to estimate the needed column size. According to Flodin [1], the volume of a sample solution, V_{ss}, should be smaller than the internal gel volume multiplied by the difference of K_d values.

$$V_{ss} < V_i(K_d' - K_d'')$$

K_d' refers to one solute, K_d'' to the other. For group separations this estimate can be very useful, as is discussed in more detail under Section III,D.

In ordinary synthetic polymers and other polydisperse samples, the difference of K_d values of neighboring fractions is so small that one has to find a compromise between resolution and practical column dimensions. The sample should always be as small as possible.

C. The Theoretical Plate Concept

In actual runs, each solute zone broadens while it is passing down the column because of nonideal packing and local nonequilibrium conditions and because of the statistical distribution of solute molecules between gel pores and interstitial volume. In a zone of one solute species only, the concentration follows approximately a Gaussian distribution [205], and, consequently, a Gaussian elution curve is obtained. The number of theoretical plates N of the column is equal to the square of the ratio between the peak elution volume V_e and the standard deviation σ of the elution curve. Glueckauf's formula [205] used the width β of the curve at $1/e$ of its maximum height to obtain a value of N little affected by the tailing often observed in gas chromatography. Several international committees [206,207] have preferred to use the more convenient base line width w between lines drawn tangent to the curve at its inflection points. For Gaussian curves the expressions are identical [207]

$$N = \left(\frac{V_e}{\sigma}\right)^2 = 8\left(\frac{V_e}{\beta}\right)^2 = \left(\frac{4V_e}{w}\right)^2 \tag{28}$$

The height equivalent of a theoretical plate, $HETP$, is the ratio of column length, l, and the number of theoretical plates, N

$$HETP = l/N \tag{29}$$

Both N and $HETP$ are used to characterize the separation qualities of a column. For other expressions of column efficiencies see, e.g., Ettre's review [208].

The number of theoretical plates defines column efficiency for a specific peak. The peak resolution defines the overlapping of peaks and is a function of two factors, viz., of column efficiency and of the separation factor. The separation factor is reflected in the relative retention of a species and is determined by $V_{i,acc}$ in GPC, by solubility in liquid–liquid chromatography, etc. Column efficiency in terms of theoretical plates and separation factor are independent from each other.

The theoretical plate concept was derived for well-defined curves representing one or a few species of molecules only. For polymolecular systems, such as solutions of synthetic polymers, narrow fractions are often used, and lower limits of N are obtained. The theoretical plate concept has proved valuable for characterizing the efficiency of a column and for testing the effects of gel type, particle mesh size, solvent, viscosity, and flow rate on this efficiency.

Flodin [1,187] studied these effects quite thoroughly on Sephadex columns. Some of his results are collected in Table I. Clearly, the greatest effects are obtained by changing the gel particle size as is predicted by Eqs. (8) and (9). Generally small particle size and low flow rate decrease the *HETP*, i.e., they improve the separation. Anything that furthers equilibration will result in better separation. The gel particle diameter, however, is of major importance since its square is proportional to the equilibration time. The flow rate affects the equilibration rate only in a linear manner. Working at elevated temperature also aids the equilibration and thus the separation [199].

TABLE I

HETP VALUES OBTAINED ON SEPHADEX COLUMNS UNDER VARIOUS CONDITIONS

Gel type	Sieve fraction (mesh)	Solute	Column dimensions (cm × cm)	Flow rate (ml/hour)	N	*HETP* (cm)
G-25	50–80	Uridylic acid	2 × 65	24	900	0.072
G-25	50–80	Uridylic acid	2 × 65	51	430	0.15
G-25	50–80	Uridylic acid	2 × 65	190	120	0.55
G-25	140–200	Uridylic acid	2 × 65	51	3100	0.021
G-25	140–200	Uridylic acid	2 × 65	10	6500	0.010
G-25	200–400	Glucose	4.5 × 132	30	3300	0.040
G-25	200–400	Cellubiose	4.5 × 132	30	3400	0.039
G-25	200–400	Cellotetraose	4.5 × 132	30	4000	0.033
G-200	100–200	Serum albumin	4 × 42	75	68	0.62
G-200	100–200	Serum albumin	4 × 42	50	88	0.48
G-200	100–200	Serum albumin	4 × 42	25	180	0.23
G-200	200–270	Serum albumin	4 × 42	25	525	0.08

D. Assessment of Optimal Conditions for Gel Permeation Chromatography

We have seen that the resolving power of a GPC column is aided by small gel particle size, long columns, slow flow rate, and other factors. Some of these variables increase the duration of a run as they improve the separation. On the other hand, speed of operation is one of the advantages of GPC which one does not want to abandon. Therefore, it often pays to assess the optimal conditions for certain separations of a desired degree of resolution.

In simple cases of group separations, e.g., rather short columns will give satisfactory separations. Glueckauf [205] developed a method to calculate the column length for a desired degree of separation of two monodisperse solutes. In his model, Glueckauf considers two different samples of equal amount and with Gaussian concentration distributions in their respective bands in the column. In Fig. 2 the ratio of elution volumes V''_e/V'_e is plotted in four curves versus the number of theoretical plates necessary to achieve 90%, 99%, 99.9% or 99.99% separation of the bands [205]. The values for V''_e/V'_e are readily computed from

$$\frac{V''_e}{V'_e} = \frac{\alpha + K''_d}{\alpha + K_d} \tag{30}$$

with $\alpha = V_0/V_i$.

Two examples given by Flodin [1] will illustrate the time saving this method affords: The fraction, α, is here taken as 0.30. For removing salt from protein, one will choose a gel of small pore size for which K'_d (protein) = 0 and K''_d (salt) $\sim$ 0.9. From Eq. (30) we find for this system

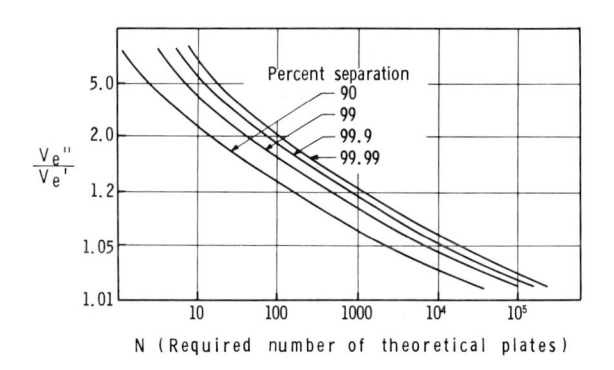

Fig. 2. The ratio of elution volumes for calculation of column length for a desired separation of two monodisperse solutes, by Glueckauf's method. Reproduced from Glueckauf [205].

$V''_e/V'_e = 4$. From Fig. 2 we read that a 99% separation requires 15 theoretical plates. According to Table I, a column of little more than 1-cm length containing Sephadex G-25 will be sufficient for this purpose.

For such a case the desired sample size may be quite large, and this will then become the controlling factor. Short columns have the double advantage of affording quick separations and of causing little zone spreading and thus little dilution of sample.

Much longer columns are required if solute mixtures of narrowly spaced K_d values are to be fractionated. For separating sucrose from glucose ($K_d \sim 0.7$ and 0.8, respectively, on Sephadex G-25) with 99% efficiency, the number of plates necessary is 2000. With a gel of 200–400 mesh size, a column of 60-cm length must be employed. According to Porath, the best separations are obtained on gels with pore sizes just large enough to admit the solutes.

With polydisperse samples, the problem is different; here we want to fractionate across a certain molecular weight range. This fractionation takes place within a rather small elution volume, viz., between V_0 and $V_0 + V_i$. We know there are gels with wide molecular weight ranges of separability and others with narrower ones; and, in order not to lose

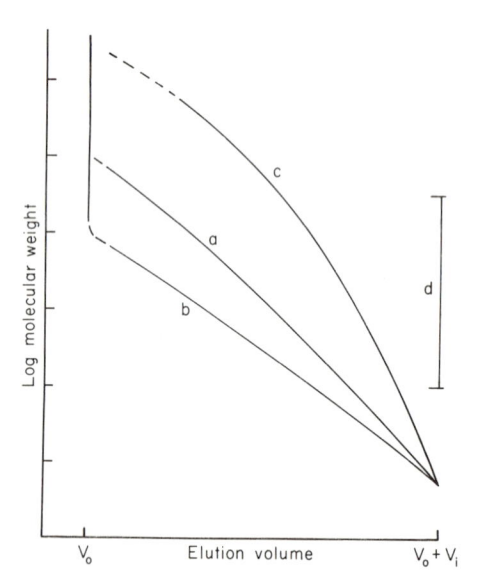

FIG. 3. The elution volumes of narrow cuts as a function of the average molecular weight of the cut, with gels of different average pore sizes. Curve a, proper pore size for the samples to be fractionated; b, pore size too small; c, pore size too large; d, molecular weight range of sample to be fractionated.

efficiency, we will choose a gel that separates the desired molecular weight range just within its total internal volume. This is illustrated in Fig. 3, curve a, where the logarithm of the molecular weight is plotted versus the elution volume. Under ideal conditions, a linear curve is obtained over the whole molecular weight range.

Often the curve in such a plot comes out rather flat with bends at the low or the high end of the elution volume, as in Fig. 3, curve b. This means the pore size range and, hence, the resolving power of the column is too small for the sample. If the curve is too steep, as in Fig. 3, c, the resolving power again is insufficient but this time because the pore sizes and their range are too large. In many cases a composite column of gels with different pore size ranges offers an advantageous solution. A useful discussion of such problems was worked out by Waters [209].

E. Side Effects

If we wish to observe the rather small differences in elution volume due to GPC then other effects which are easily capable of producing large differences must be substantially absent. Three such interfering effects are adsorption, partitioning, and incompatibility.

1. *Adsorption*

Adsorption usually occurs with more-polar solutes in less-polar solvents. In a homologous series it becomes stronger with increasing chain length, i.e., the larger molecules are retained to a greater extent than smaller ones. Early descriptions of a reversal of this order, in the case of high molecular weight polymers, were based on experiments in which the larger molecules had been excluded from the pores of the adsorbent [179–183,210–213], i.e., on GPC effects rather than on true adsorption. Addition of small amounts of a polar additive to the solvent usually prevents or reduces adsorption.

There are situations where combination of GPC with adsorption is beneficial [21,95,102,129], as Porath [102] first pointed out. A discussion of this aspect, however, would go beyond the scope of this chapter.

Gelotte [42] studied extensively the effects of sorption of a wide variety of compounds on Sephadex gels. He found an ion exchange mechanism caused by a few residual acidic groups present in the polydextran. This finding was confirmed by Miranda *et al.* [214]. With aromatic and heterocyclic solutes also, true adsorption takes place, particularly with salt-free water as eluent. Small amounts of salt presumably cover the active groups and thus reduce adsorption [42].

2. Partitioning

Since a gel can absorb a like component from a solvent mixture, a rigid porous gel with a suitable solvent mixture can function as a very flexible medium for liquid partition chromatography. Even like solute components can be absorbed by the gel from a solution if the solvent is very different in nature from the gel and the solute. Therefore, single solvents of solubility parameters similar to that of the gel are preferred in GPC. The addition of small amounts ($\leqslant 5\%$) of mismatched solvent, e.g., to prevent adsorption, usually has no detrimental effect.

Again, by prudent choice of different solvents, one can take advantage of this phenomenon and achieve separations first by size and then by polarity.

3. Incompatibility

Another interference with GPC occurs in cases where the solute is chemically very different from the gel and is excluded by incompatibility. Altgelt [191] observed this phenomenon when he tried to fractionate a solution of 10% polyvinyl acetate in benzene on a rubber gel. Despite the low molecular weight of the sample (2000), no fractionation took place under these conditions. Only at a much lower concentration (0.2%) was fractionation achieved.

The phenomenon of incompatibility, which is well known from the physical chemistry of polymer solutions [215–219], is the phase separation which occurs on mixing the solutions of two different polymers in the same solvent. It is found in systems of moderate or stronger positive interaction energies between the solutes. The entropy of mixing of two polymers with each other is very small because of the small number of molecules involved and can usually not overcome an even moderate positive heat of mixing. A positive chemical potential results, and the two polymers separate rather than mix. Dobry and Boyer-Kawenoki [215] found phase separation in 32 out of 35 polymer pairs studied at low and moderate concentrations; i.e., the effect is quite common. Recently Langhammer and Nestler discussed this effect in GPC in more detail [219a].

With highly swollen gels incompatibility leads to fairly strong exclusion. On heavily cross-linked and therefore only light swollen gels the effect is smaller but still discernible. Altgelt observed partial exclusion of polyvinyl acetate and of asphaltenes of relatively low molecular weight (1000–20,000) also from heavily cross-linked polystyrene gels [163, 248].

The effect of incompatibility works directionally like GPC in that the larger molecules are excluded to a higher degree than smaller ones. However, it should be avoided, at least in cases of mixed polymers, since it is sensitive to chemical structure. It can distinctly be diminished by working at very low concentrations and at high temperature.

Cases of incompatibility are readily detected by comparing their elution volume–molecular weight relation (see Section VI,B) with that of ordinary compounds. Excluded samples are eluted earlier than others of the same molecular weight. Experimental tests for this effect are described under Section V,A,5.

IV. The Gels

A. General Considerations

A great variety of gels has been used or tried for GPC. They are generally polymers of varying degrees of cross-linking and usually swell in the solvents for which they are made. Examples are polydextrans for aqueous solutions and polystyrenes for nonaqueous solutions. In contrast to the general views in this matter, the swelling was found not to be essential; the permeability (or porosity) is the important factor. Vaughan [155] demonstrated in an extensive study of various gels and other porous materials that an expanded silica gel (Santocel A from Monsanto) gave excellent fractionation of polystyrene in benzene. As a hydrophilic substance, silica gel certainly does not swell in benzene. Another example is the recently developed porous glass [219b].

Aside from permeability, the rigidity of a gel determines its usefulness in practice. Soft particles do not pack well and tend to clog columns. This means high degrees of swelling are actually undesirable. High permeability and low swelling do not contradict each other necessarily since it has been possible to prepare highly cross-linked gels of high porosity. The permeability of a heavily cross-linked gel is built into it by the diluent which must be present during cross-linking. If the diluent is well matched to the gel substance, increasing the amount of it is equivalent to increasing the swelling of the gel; the network becomes more tenuous. However, when the composition of the diluent is changed, it becomes possible to make the gel more rugged so that it becomes more permeable but not more tenuous.

Diluents have often been present during the formation of a gel but without full recognition of their effects on the gel network. Attempts to make very highly permeable gels by going to extremely low cross-linking or very great dilution have usually been disappointing. This has followed from a too literal view of gel structure in terms of average cross-linking density. For some time, lightly cross-linked networks have usefully been described in terms of the average length of the linear chain segment. Perhaps because attention was thus focused on the apparent direct relation between segment length and permeability, the other route to permeability has largely been ignored. Mikes [220] first noticed the decisive effect of a diluent present during cross-linking. Flodin [1] compared this effect with the influence of cross-linker content and chain length of his dextran "monomer" and found it is of importance. Millar et al. [203] and Alfrey and Lloyd [221,222] studied it in detail theoretically and experimentally. While one essential point is the *presence of diluent* during polymerization, the other is the *solvent match* of the diluent with the gel substance.

Synthetic ion exchangers have long been known to be heteroporous gels, but their permeabilities have been relatively low [184,223–225]. Highly permeable ion exchange resins have recently appeared [226–229]. In these resins, to open up the gel structure, a relatively small amount of diluent was used, a solvent compatible with the gel monomers but a precipitant for the polymer. Combining these two contributions, we have the basis for a more detailed understanding of gel structure.

A rigid gel is a specialized structure. An ion exchange resin needs to be rigid and permeable, but it also needs a maximum of ion capacity, that is, of mass and, therefore, a minimum of diluent. Much more diluent and a full range of internal structures should be available for the gel designed for a permeation column. Utilizing as variables the amount and composition of this diluent in highly cross-linked polystyrene gels, Moore [157] has shown that a wide variety of gel structures can be made. These were produced in fine bead form and were adequately strong and rigid to permit high elution rates in packed columns.

Electron micrographs in Fig. 4 of very thin sections of these gels show clear structural details in all but the most finely structured gels. A typical gel is thus shown to consist of randomly interconnected strands of granules where the granules seem quite uniform in size. The void spaces around the structures are highly nonuniform in size and shape. A proportionality is qualitatively evident between the solvent match of gel and diluent during polymerization and the size range of the gel structure. The better the solvent match between the gel and the diluent in which it was prepared, the smaller the unit granule of the gel structure and the smaller the size range of the voids in it. The permeability range of such a gel was also in accord with this picture: The coarser the structure, the higher the permeability limit shown. The preparation of several of these gels will be detailed in Section IV,C, and the mechanisms that determine gel structure will be discussed in connection with these examples. It will be seen that the permeability of the gel proceeds not only from the space around the granular structure of the gel but also from fine pores in the granules themselves.

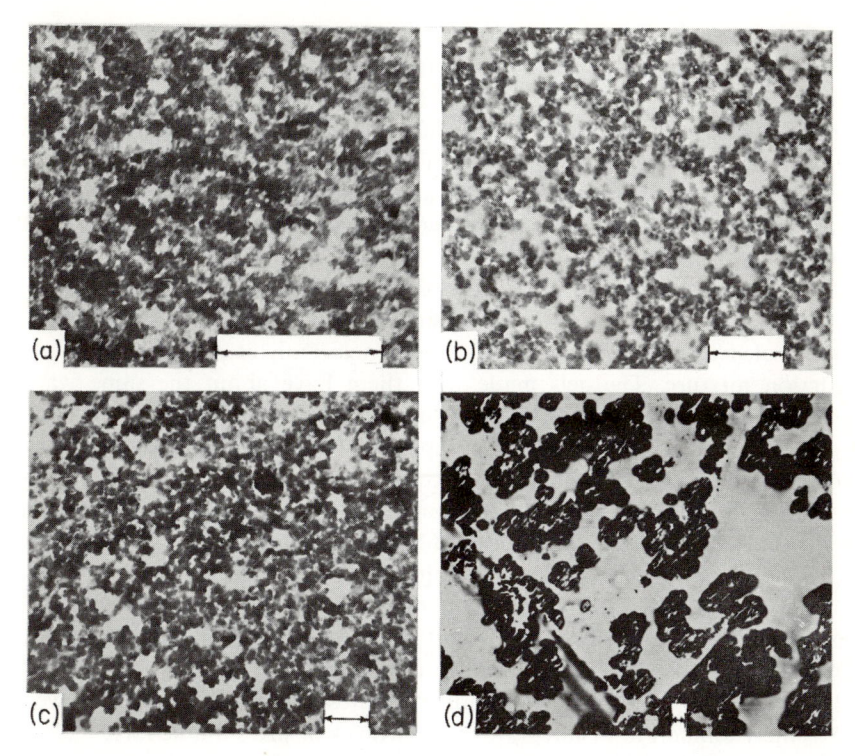

FIG. 4. Electron micrographs show structures of thin sections of some highly cross-linked polystyrene gels. (a) Gel C′, (b) Gel M, (c) Gel N, (d) Gel O. Magnification is shown by the white bar, 1 μ wide in each case. For compositions see Tables IV and V and Fig. 5.

B. Gels Used in Aqueous Systems

Two kinds of gels have been offered commercially to date: Sephadex, a group of cross-linked dextrans produced by Pharmacia in Uppsala, Sweden, and Bio-Gel P, produced by Bio Rad Laboratories in Richmond, California. We shall first describe briefly the already well-known Sephadex and later the recently introduced (1964) Bio-Gel P.

Eight grades of Sephadex are offered with different degrees of cross-linking and resulting different degrees of permeability. They are standardized in terms of the molecular weights of soluble dextran fractions which are just excluded from the gel particles. In Table II the eight grades are listed with some of their physical data. Since the exclusion limits are based on separations of the threadlike dextrans, they will be different for more extended or more compact molecules as indicated in Table IIA.

Flodin describes the preparation of a variety of dextran gels in detail [1]; we shall only outline the general aspects. Dextrin of $10-300 \times 10^6$ molecular weight is made by biosynthesis, then partially hydrolyzed to about 40,000 molecular weight and fractionated. Suitable fractions ($M_w/M_n = 2$–3) are cross-linked with epichlorohydrin (ECH) in the presence of NaOH. The cross-linking was first done in bulk and later in beads. The most porous samples, G-100 and G-200, are rather soft and can only be used in bead form.

Flodin [1] tried several other diepoxides as cross-linking agents but does not comment in detail on their properties. Some aliphatic α, β-diepoxides are labeled as useful because they are nonelectrolytes and soluble in the reaction mixture. The gels made with them have about the same water regain as dextrin ECH gels obtained under similar conditions.

Gels of higher water regain are prepared from starch and from hydroxyethyl and alkyl hydroxyethyl ethers of dextran or other polysaccharides. Starch gels are subject to oxidation and are more difficult to obtain free from ionized groups. Gels made of β-hydroxypropyl ether of dextran and ECH swell in water and in ethyl alcohol.

Sucrose and Sorbitol were polymerized with ECH to gels with rather high water regain. Polyvinyl alcohol did not give satisfactory gels with ECH [1] (see also Porath [12]); with diepoxides, porous gels of very high elasticity were obtained. Bulk polymerizates of this composition were extremely difficult to disintegrate into particles of size suitable for GPC [1].

Lea and Sehon [140] cross-linked three monomers, polyacryl amide, polyvinylethyl carbitol, and polyvinyl pyrrolidone with N,N'-methylenebis(acrylamide). They used moderately high cross-linker contents with 9%, 16%, and 16%, respectively, and high dilution ratios with 5%, 15%, and 15% monomer in water, water + alcohol, and water, respectively, as solvent. The polymerizations were carried out in bulk and initiated with dimethylaminopropionitrile and with ammonium persulfate. In tests with mixtures of human serum proteins, lysozyme, and components of ragweed pollen, the gels were found to have rather large pore size. The proteins and several pigments were at least partially fractionated, whereas, lipid was too big and was excluded from the gel.

Morris and Morris [4] recommend polyacrylamide gels for several reasons: They can be prepared with a wide range of pore size, and they do not shrink with increased ionic strength like the dextrans with low degrees of cross-linking. The permeability, P, can be varied by the concentration, C, of the acrylamide monomer present in the polymerization mixture independently of the cross-linker concentration according to the equation

$$P = K/C^{1/2} \tag{31}$$

B.4. GEL PERMEATION CHROMATOGRAPHY

TABLE II
AVAILABLE SEPHADEX TYPES AND THEIR PROPERTIES[a]

Type	Optimal molecular weight range[b]	Approx. exclusion limit (molecular wt.)	Water regain (g/g dry gel)	Hydrated bed volume (ml/gm)	Wet density (gm/ml)	Particle size (μ)
G-10	up to 700	700	1.0 ± 0.1	2–3	—	40–120
G-15	up to 1,500	1,500	1.5 ± 0.1	2.5–3.5	—	40–120
G-25	160–5,000	5,000	2.5 ± 0.2	5	1.11	
Fine						20–80
Coarse						100–300
G-50	1,000–7,000	10,000	5.0 ± 0.3	10	1.06	
Fine						20–80
Coarse						100–300
G-75	1,000–13,000	50,000	7.5 ± 0.5	12–15	1.03	40–120
G-100	—	100,000	10.0 ± 1.0	15–20	—	40–120
G-150	—	150,000	15.0 ± 1.5	20–30	—	40–120
G-200	—	200,000	20.0 ± 2.0	30–40	—	40–120

[a] Reference [8].
[b] According to a study by Granath and Flodin with soluble dextran fractions on cross-linked dextrans similar to the commercially available Sephadex samples.

TABLE IIA
FRACTIONATION RANGES OF SEPHADEX FOR VARIOUS SOLUTES[a]

Type	Dextrans (molecular wt.)	Peptides and globular proteins (molecular wt.)
G-10	700[b]	—
G-15	1500[b]	—
G-25	100– 5,000	5,000
G-50	500– 10,000	10,000
G-75	1,000– 50,000	3,000– 70,000
G-100	1,000–100,000	4,000–150,000
G-150	1,000–150,000	5,000–400,000
G-200	1,000–200,000	5,000–800,000

[a] Pharmacia, 1966.
[b] Polyethylene glycols instead of dextrans.

where K is a constant. Stable gels can be obtained over a concentration range of 3–30% monomer. Preliminary results showed that fractionations of proteins and nucleic acids up to a molecular weight of at least 68,000 (hemoglobin) could be achieved. The retention volumes of model proteins decreased in a regular manner with increasing monomer initial concentration [4].

Equation (31) reflects the change of permeability as a consequence of dilution during polymerization, as we discussed in some detail under Section IV,A.

Hjertén [189] described in detail the preparation of polyacrylamide gels which he and his co-workers used for various separations and purifications. Hjertén and Mosbach [108] achieved partial separations of lactase (molecular weight 57,000) from carbonic anhydrase (molecular weight 31,000), further separations of phycoerythrin, phycocyanin, hemoglobin, and cytochrome c on gels containing 3% or 5% polyacrylamide. Boman and Hjertén [149] separated microsomal from soluble ribonucleic acid and from adenylic acid.

Very recently, polyacrylamide gels were offered commercially by the Bio-Rad Laboratories in Richmond, California. Some properties of the eight grades available are listed in Table III. The advantages claimed by the manufacturers are fractionating power extending to higher molecular weights, absence of ionic groups, and, consequently, minimized adsorption effects.

The exclusion limits of the gels were based on experiments with a variety of proteins (cytochrome c, α-chymotrypsinogen, carbonic anhydrase, egg albumin, bovin serum albumin, aldolase) and refer to the molecular weight of a fictitious protein which would give an R_f value of 0.9. Corresponding to the general definition of R_f, this quantity is here

$$R_f = V_0/V_e$$

The swelling is claimed to be negligibly affected by changes in ionic strength, Bio-Gel P-100 decreasing 2% in going from 0 to 0.4 M phosphate buffer[1] [190]. The gels are insoluble in water, salt solutions, and common organic solvents and can be used within the pH range of 2–11.

Natural gels which have been used for GPC are starch [167,186], gelatine [190a], and agar [138,170,171]. They all have in common that they are only lightly cross-linked and are soft and exert great resistance to solvent flow in a column. Further, they contain groups which give rise to ion exchange and adsorption. Therefore, for most purposes the natural gels were abandoned in favor of the more easily controlled artificial gels.

---

[1] Compare the section on solvents, Section V,A,3a.

TABLE III

Available Bio-Gel P Types and Their Properties[a]

Type	Approx. exclusion limit (molecular wt.)	Hydrated bed volume (ml/gm dry gel)	Particle size (wet mesh)
Bio-Gel P-10	10,000	11	50–150
Bio-Gel P-20	20,000	11	50–150
Bio-Gel P-30	30,000	12	50–150
Bio-Gel P-60	60,000	16	50–150
Bio-Gel P-100	100,000	25	50–150
Bio-Gel P-150	150,000	34	50–150
Bio-Gel P-200	200,000	45	50–150
Bio-Gel P-300	300,000	68	50–150

[a] Reference [190].

On the other hand, agar seems to have unusually large pores. Polson [138] achieved protein fractionations of up to 6.6×10^6 molecular weight using granulated agar as gel medium. Steere and Ackers [170,171] even separated viruses and cell components on agar gels. Polson [138] found a simple relation between the agar concentration, c, in the gel and the molecular diameter, d, of a solute just admitted by the gel, $c = k/d$, with the constant $k = 70$. This is the equation of a hyperbola; and, thus, the greatest change of d (i.e., of pore size) per unit change of agar concentration would be expected at low agar concentrations, c, and correspondingly, of gels with large pore size.

Haller's porous glass [219b,219c] and some of its applications are discussed in the following section. Here, only his separations of tobacco mosaic virus from tobacco ring spot virus and of southern bean mosaic virus from bovine serum albumin need be mentioned.

C. Gels Used in Nonaqueous Systems

By far the most important gels to date for fractionating in nonaqueous solution are the heavily cross-linked polystyrenes [157,158,163]. The only other rigid gels reported so far in the literature are "Santocel A" [155], an expanded silica gel produced by Monsanto Chemicals and the recently reported "gels" of porous glass [219b,219c]. Santocel A gives excellent results with polystyrene [155] and, probably, with other nonpolar polymers, but it adsorbs more polar compounds [191]. Also, it is only available in irregular grains and offers appreciable resistance to flow in a column even after sizing.

Tempered alkali borosilicate glasses can be leached with acid and alkali to yield systems of quite uniform pore size in the range of 170–2500 Å [219b,219c]. Because of the narrow pore-size distribution, porous glass particles lend themselves uniquely to fundamental studies of the GPC mechanism (230) as well as to special separations. For general use, combinations of different porosities can be applied. On porous glass, artificial mixtures of plant viruses and bovine serum albumin [230a] as well as polystyrene and polyisobutene samples with molecular weights between 10,000 and 3,400,000 [230b] were separated.

Glass particles combine the advantages of being completely rigid and nonswelling with inertness to corrosive solvents. Thus, especially reproducible packings can be made; and contaminated columns can be cleaned with hot nitric acid. These advantages, together with the adjustable uniform pore size, make porous glass an extremely promising packing material for GPC. The only disadvantage may be adsorption which is more prominent in aqueous than in organic solvents. So far this point was not discussed in any detail in the literature. In general, little information on glass "gels" is available, but this is certain to change in the near future.

In summer 1966, porous glass in the form of crushed particles will be available from Bio-Rad Laboratories [219c].

Recently Pharmacia started marketing Sephadex LH-20, a partially alkylated Sephadex G25-type gel which swells with polar organic solvents, including chloroform, but not with hydrocarbons, such as benzene and others. Separation of lipids [230b], polyethylene glycols, various glycerol esters, and low-molecular-weight polystyrenes [230c] were reported.

Lightly cross-linked gels such as rubber [164,165], polystyrene [154,155], polyvinyl alcohol [155], and others [155] are either too tenuous or they permit only fractionation of material of moderate or small size. Rubber, e.g., gives very good separation from 40 up to 15,000 molecular weight [191], but beyond this limit fractionation is poor. Vaughan [155] did not determine the limits for his gels. He used a polystyrene of wide distribution containing a top fraction of about 10^6 molecular weight and compared only the viscosities of the first and the last fractions obtained in a GPC run. In these terms he found moderate fractionation on polystyrene beads (lightly cross-linked) and on cellulose, fair fractionation on low-density polyethylene and on cellulose acetate, and poor fractionation on high-density polyethylene and on polyvinyl alcohol. The poor results with polyvinyl alcohol and with polyethylene were certainly caused, to some extent, by the incompatibility, or "negative adsorption," of polystyrene with these gels (see Section III,E). In high-density polyethylene, probably the crystallinity limited drastically the available internal volume.

The heavily cross-linked polystyrene gels are so useful not only because they combine permeability with ruggedness but also because they can be tailored according to need. In the following paragraphs their preparation and their properties are discussed in greater detail.

Recipes and procedures for preparing polystyrene gel beads by suspension polymerization are abundant in the literature on ion exchange resins [203,231–237]. In these, droplets of the oil phase containing the monomers and a polymerization catalyst are suspended with continuous agitation and heating in an aqueous (continuous) phase containing a protective colloid. The ratio of aqueous phase to oil phase is usually between 4:1 and 1:1. The droplets congeal in 1 or 2 hours at 60–80°C, and polymerization should be essentially complete in 20 hours. The colloid is then removed as much as possible by thorough washing. The size range and uniformity of the beads are determined mainly by the agitation and the colloidal agent, but the nature and amount of diluent present with the monomers also affects the droplet size. As colloids, many materials have been used: sodium polyacrylate, polyvinyl alcohol, methyl cellulose, soluble starch, gelatine, finely divided alkaline earth phosphates, silicates, and carbonates. Finer beads require more energetic agitation and more of the coating agent. For the present purpose, beads 10 to 100 μ in diameter are suitable, and narrower fractions within this range are preferred.

In gel column practice, the advantages of higher resolving power per foot of column and, hence, shorter columns and elution times were associated with gel particles of small diameter. Lower resistance to liquid flow was found with spherical beads of very narrow size range. This is in agreement with the findings of Hamilton [238] in ion exchange chromatography. In that work, he also showed a convenient hydraulic classifier for making narrow size fractions. The slurry of ion exchange resin beads was placed in a separatory funnel, and consecutive size fractions were floated up and out by a series of increasing flow rates of water, admitted at the bottom of the funnel. Since unsulfonated polystyrene beads are not wetted by water, it was necessary in adapting this procedure to use a compatible solvent, such as xylene or diethylbenzene, for the elutriation. In this way fractions were readily prepared with 80% of the particles within a range of 20% plus and minus the average diameter. Satisfactory columns were obtained with such fractions.

To arrive at a set of formulations for rigid column packings of varied permeabilities, it was necessary to consider the interrelated effects of cross-linker and diluent on gel structure. Sufficient rigidity was obtained with 8–12% cross-linker in the monomer–diluent mixture. Different permeabilities in the low range were obtained by using a well-matched diluent and varying the amount of it. The change of gel structure across such a series, for example, with toluene replacing styrene, must be like the different stages of polymerization in an undiluted copolymer. The gel structure first formed is quite tenuous, with very little fine structure. Thus, the fractionating range of a gel made with 80% toluene was largely near its upper limit. At 60% toluene the structure was considerably filled in, its size range well balanced, its upper limit sharply reduced. The toluene-swollen gel made with 8% divinylbenzene and 92% styrene, no diluent, showed its porosity mostly in the low range, its calibration curve concave upward to a low limit. In this context see also the paper of Millar et al. [203].

Higher permeability ranges were readily obtained with diluents less well matched to the gel structure, differing from it in solubility parameter or in hydrogen bonding [239,240]. In this case, as each polymer chain is propagated, its structure will tend to shrink, being less than fully solvated by the liquid phase. Also, the newly formed polymer will tend to extract and absorb monomer molecules out of the liquid phase. The result is a more rugged, open gel structure rather than a well-dispersed one. A gel bead of this kind can be visualized as a cluster of grapes where each grape itself has a network structure. The very fine pores are chiefly available to solvent, while the larger pores between the "grapes" represent the gross internal gel volume. The actions leading this macroreticular structure, as Kunin et al. called it [240–242], proceed from the same molecular interaction properties more familiar in other contexts; the contraction or expansion of a polymer in different solvents leading to different solution viscosities, the partitioning of a solute between two unlike phases. Gordon [243] has shown that samples of fluorene and phenanthrene were eluted promptly by toluene without appreciable separation from a column packed with very permeable polystyrene beads; but, when hexane was used as the flowing solvent, these aromatic molecules were sorbed in the gel and eluted much later with a fair separation between them.

These mechanisms that affect gel structure are subject to delicate control and subtle variation. For example, if the monomers are styrene and divinylbenzene and the diluent is cyclohexane, a poor solvent but not a precipitant, the solubility parameter (polarity) of the liquid phase changes gradually as the monomers are removed from it by the polymerization. The more diluent in the starting mixture, the less change during the process. But, if the diluent is a mixture of a good solvent and a precipitant, for example, diethylbenzene and isoamyl alcohol, such that its average solubility parameter is the same as that of cyclohexane, the effect is different. During the polymerization, the good solvent is differentially sorbed by the growing solid phase so that the unit granules of the gel have a diluent different from that outside them, creating a wider diversity of structures. Also, since the polymerization is being carried on in droplets suspended in water, if the precipitant part of the diluent tends to absorb a little water as the other components leave it, the structural differentiation will be augmented. In the extreme case, precipitation and syneresis may take place before the gel is set, and the internal structure is lost.

These effects are illustrated in Tables IV and V and Figs. 4 and 5. A series of these gel bead preparations were packed into columns; and samples of narrow-range polystyrenes and polypropylene glycols were eluted with compatible solvents such as toluene, perchloroethylene, or tetrahydrofuran. The peak elution volumes were plotted against the average chain length of the sample substance, thus calibrating the column's separating range. The variety of gel structures is evident as is also the lack of linear polymers suitable for calibrating the upper range of the more permeable columns.

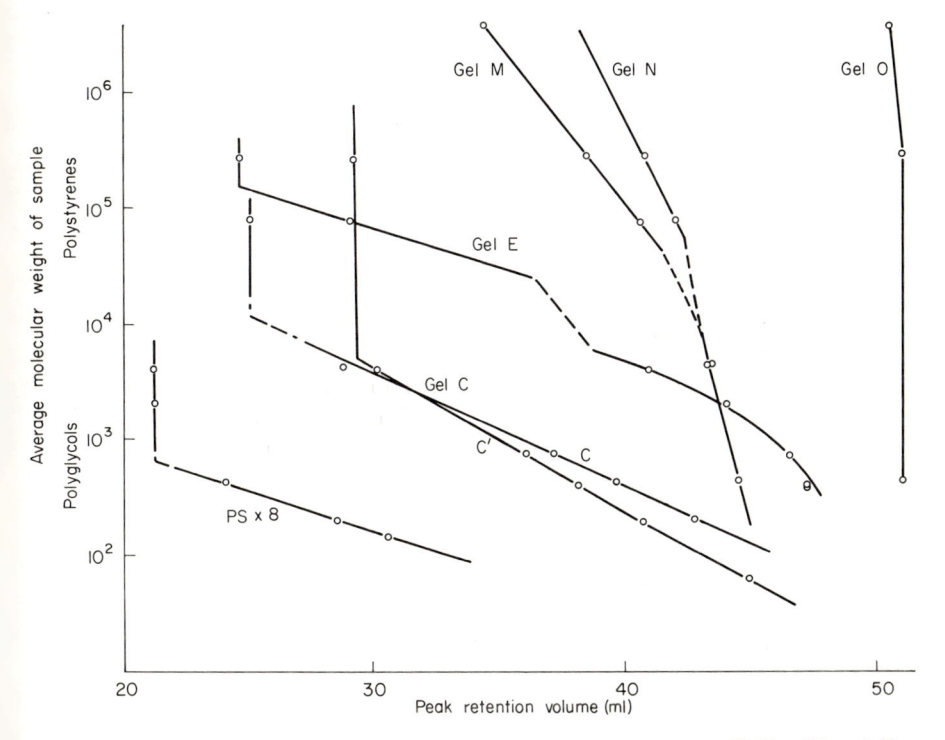

Fig. 5. Calibration curves for seven gels. For their compositions see Tables IV and V, except for C′ which was made by polymerizing 33 parts of divinylbenzene plus 27 parts of ethylvinylbenzene (60 parts 55% DVB) plus 40 parts of diethylbenzene. All columns were 120 cm × 7.8-mm ID, eluting solvent was tetrahydrofuran, 1 ml/minute, 25°C. Reproduced from Moore [157].

TABLE IV

PERMEABILITIES OF STYRENE GELS WITH VARYING PROPORTIONS OF TOLUENE

Gel	Styrene (wt. %)	Divinyl benzene (wt. %)	Toluene (wt. %)	Molecular weight Permeability limit	Notes
PSX8	92	8	0	1,000	
PSX4	96	4	0	1,700	
PSX1	99	1	0	3,500	Rubbery
PSX0.1	99.9	0.1	0		Too soft to pack
A	79.1	4.2	16.7	2,500	
B	65.7	5.7	28.6	~7,000	
C	30	10	60	7,000	
D	24.8	2.5	72.7		Too soft to pack
E	9	11	80	250,000	

TABLE V

PERMEABILITIES OF GELS WITH VARIOUS DILUENTS, ALL MADE FROM 30% STYRENE, 10% DIVINYLBENZENE, 60% DILUENT[a]

Gel	Diluents (parts/100 parts of gel)	Molecular weight permeability limit
C	60 Toluene	7×10^3
F	30 Toluene, 30 diethylbenzene	1.5×10^4
G	60 Diethylbenzene	1.2×10^4
H	45 Toluene, 15 n-dodecane	1×10^5
I	30 Toluene, 30 n-dodecane	3×10^5
J	15 Toluene, 45 n-dodecane	2×10^6
K	10 Toluene, 50 n-dodecane	$<2 \times 10^3$
L	40 Diethylbenzene, 20 isoamyl alcohol	$\sim 3.6 \times 10^3$
M	20 Diethylbenzene, 40 isoamyl alcohol	$\sim 8 \times 10^6$
N	13.3 Diethylbenzene, 46.7 isoamyl alcohol	$\sim 10^{10}$
O	60 Isoamyl alcohol	Extremely high

[a] "Styrene" is a mixture of styrene and ethylvinylbenzene.

V. Experimental Technique

Here we must distinguish between analytical and preparative procedures. For analytical purposes, utmost resolving power is expected of a column with rather small samples, whereas, for preparative separations, a compromise has to be found between accuracy and, usually, large sample

B.4. GEL PERMEATION CHROMATOGRAPHY

size. Since column techniques are very well developed and described in several fields of chromatography [5,244,245], we shall confine our comment to the aspects peculiar to GPC and to recent refinements.

A. ANALYTICAL SEPARATIONS

1. *Column Design*

The design of a separation system will reflect the characteristics of the process and the requirements to be met. The salient point of design for a GPC system is the relatively small eluate volume within which the entire separation appears. Where a polymer is to be fractionated, a very long range of molecular sizes is usually present in a continuous but not necessarily simple distribution. It has been shown that gels can be made with a sufficiently wide range of internal structures to differentiate molecules over this whole range, but the separation factors available within such a distribution are therefore small. If there are special features or dissymmetries in the distribution, then a column of high resolution, a large number of theoretical plates, will be required to reveal them.

Several factors in the design and operation of a gel column system are important for its ability to resolve the components of a polymer sample. The better the design, the more freedom remains for the operational factors such as elution time per sample.

As the gel preparation has drawn on the developments of ion exchange chromatography, so new column and system design are indebted to gas chromatography as, e.g., in the system developed by Moore and improved by Waters.

With a rigid packing, unaltered by the elution of a sample or by liquid flow rates, long stainless steel columns which are tightly packed and sealed to low-volume inlet and outlet fittings can be used. This enables sample addition without interrupting the flow of the mobile phase and is consistent with a continuous-flow device for detecting the eluted components. The resulting steady base line conditions permit the use of a highly sensitive detector with small internal volume. With column diameter reduced in proportion to detector needs, the plate count is increased. A further increase in resolution is available by reducing the diameter of the gel particles. The increase in back-pressure is minimized by using closely sized spherical particles and is not too great for standard loop-type sample valves.

With a continuous differential refractometer as detector, several measures are necessary to reduce base line noise to a low level. Pump pulsations are effectively damped to keep pressure waves out of the flat-walled refractometer cell. A mixing chamber after the pump makes gradual any changes in the solvent, and a reference column is used with flow adjusted so that the remaining changes would be received simultaneously, as far as possible, in the two sides of the cell.

To prevent radial viscosity changes in the column and to permit a reproducible calibration of its permeability range, the columns are held at constant temperature, and the eluate is measured in an accurate volumetric fraction collector held at the same temperature.

In order to keep the dead volume down, Moore and Waters devised special end fittings. One kind, which is satisfactory with gel particles larger than 50 μ, is made by drilling short sections of rod with a 15° taper to meet the $\frac{1}{16}$-inch OD by 0.020-inch ID transfer tubing and pressing a small wad of fine glass wool into the apex of the taper. A more convenient end fitting trouble-free down at least to 15 μ was designed by J. L. Waters; the packing is supported on a thin flat disk of felted stainless steel fibers, backed with a plate finely grooved radially and concentrically, which leads the liquid flow to or from the 0.020-inch ID connecting tubing. The commercially available column system of this type is shown in Figs. 6 and 7. More details can be found in a paper by Maley [159].

Porath and Bennich [246] applied to GPC the principle of recycling which had been employed before in gas chromatography [247]. They use a relatively short column and recycle the eluate several times until the separation is satisfactory as indicated by a continuously recording detector, here a UV analyzer. The advantage of this method is the possibility of drawing off parts of the mixture which have been separated sufficiently, while the

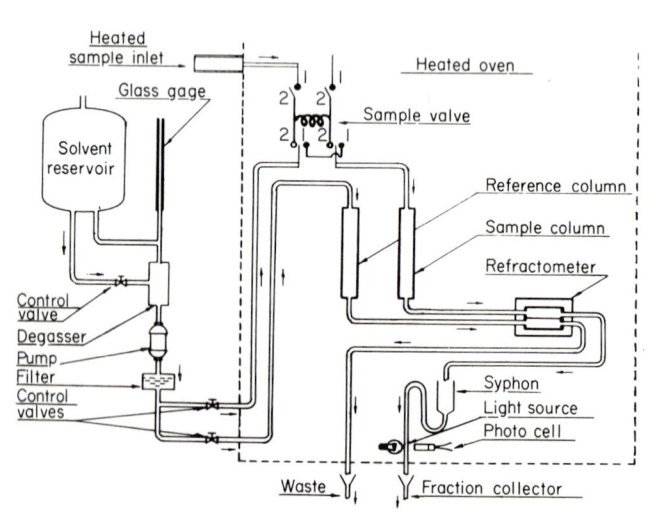

FIG. 6. Waters liquid chromatography assembly. Reproduced from Maley [159].

B.4. Gel Permeation Chromatography

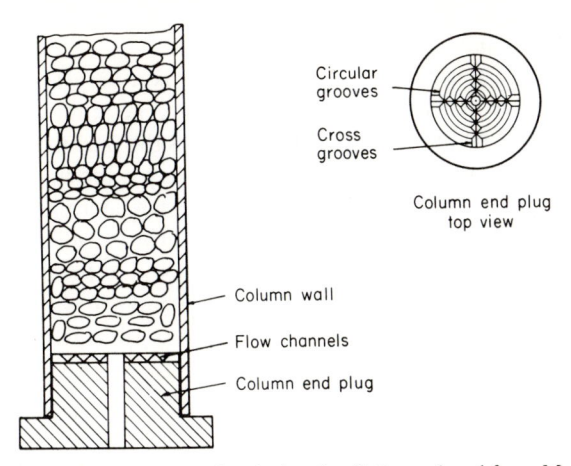

Fig. 7. Column cross section and end plug detail. Reproduced from Maley [159].

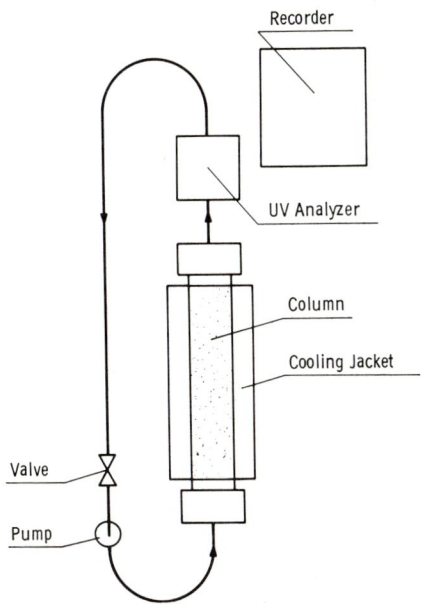

Fig. 8. Schematic diagram showing the principle of recycling chromatography. Reproduced from Porath and Bennich [246].

rest is submitted to longer and more extensive fractionation. Figure 8 illustrates the principle of this arrangement. The crucial point in making this apparatus work is to keep the liquid volume outside the gel packing as low as possible, since it is passed through many times during fractionation.

2. Packing of Columns

Arrangement of the packing in the column is a very real design factor. Unlike the practice in general chromatography, gel columns must be packed wet, and care must be taken that air never enters the packing. The gel should have a rather narrow size distribution; particularly, the fine and the coarse particles should be removed. The fine particles tend to obstruct the flow and are removed by stirring a suspension vigorously, allowing it to settle for a few minutes, and then decanting the turbid supernatant. This procedure may have to be repeated several times. Coarse particles disturb the pattern of the packed gel and reduce the resolving power. They may easily be sieved out dry. Better packings are made with narrow fractions of gel bead sizes, e.g., as obtained by Hamilton's floating technique [238] or by wet sieving.

Two packing techniques are recommended by Flodin [1,187] and by Pharmacia [7]. In one, the column is extended by a tube of the same width and about the same length, and a funnel is put on top. Both column and extension tube must be perfectly vertical. The assembly is filled with solvent up into the conical part of the funnel, and, under gentle stirring, the gel slurry is added. As the gel settles slowly, the column outlet is slowly opened until a uniform flow is obtained. A rising horizontal surface of gel in the tube indicates uniformity in packing.

In the other, better, method, a homogeneous slurry of swollen gel is poured into the solvent-filled column in a steady stream. When a layer a few centimeters thick has formed, the bottom valve is opened and solvent withdrawn at an even flow. The rest of the suspension is added continuously until the column is evenly packed. Both methods give good results except that they can be subject to gradual settling. Moore finds that consolidation with a plunger or vibration decreases the column efficiency markedly, apparently by creating channeling problems. It may be important to use high gel–solvent ratios [248]. At high particle concentrations convection currents caused by the sedimentation of the particles are restricted to smaller regions in the column. Therefore, more random packings are obtained. Rotating the columns during addition of the slurry further helps avoid uneven gel distribution [163,249]. Beling [250] and Rothstein [251] point out that more uniform packings are obtained by agitating the liquid in the entire column in an alternating fashion. Rothstein [251] used a serpentine stirrer which reached into the column and was attached to a reversing motor programmed to change the direction of rotation at 2-minute intervals.

According to Moore, with very rigid packings the best columns are made by pressure filtration at a relatively fast flow rate, as high as 15 ml per minute and per cm^2 cross section. The packing is suspended in several volumes of liquid, a mixture of perchloroethylene and toluene or tetrahydrofuran for polystyrene gels, to maintain a constant slurry density. For packings not excessively cross-linked, some methanol is added to oppose the slight swelling effect of the solvent mixture, otherwise a slightly better solvent match to polystyrene than either solvent alone.

The ease of extruding the gel packing from a column with the end fitting removed makes it evident that column back-pressures in packing and in operating are acting almost undiminished to compress the gel beads at the column outlet. Moore therefore arbitrarily limited column lengths to 4 feet. These sections can be packed and tested individually, then connected in series to any desired length. The packing is readily recovered from a column accidentally plugged or channeled, and successfully reused in most instances. Barring such incidents, service lives of many months and sometimes years have been experienced without loss of resolving power from well-packed columns of this design.

3. *Solvents*

To have a separation according to molecular size, the eluting solvent must match the gel substance closely enough so that sample components are not partitioned between the two according to solubilities in the two phases. The solvent must therefore be quite close in solvent nature to the gel substance. Further, it should be stable, dissolve the sample, aid and not interfere with the solute detector system, and perhaps allow the sample to be recovered readily from the separated fractions.

a. *Aqueous Solvents.* For separations on dextran and other hydrophilic gels, pure water, a wide variety of aqueous buffer solutions, and mixtures of water with organic solvents [104,252] have been used. Three properties of an aqueous solvent can be varied: the solvent power, the ionic strength, and the pH. According to Porath [12], an increase of ionic strength from 0.01 to 0.2 has little effect on the retention volumes of amino acids and proteins on Sephadex columns. However, $0.5 M$ Na^+ ions markedly reduced the retention volume of lysozyme. At higher ionic strength association–dissociation and denaturation effects can be expected. Also, dextran gels shrink in solutions of high ionic strength [12]. Moderate ionic strength and pH do not seem to affect the pore size of the gels markedly. Thus the pH was found not to influence the separation of amino acids, whereas the effect on proteins was variable [214]. This was to be expected, since polyelectrolytes experience expansion or contraction depending on their degree of ionization, while pH changes will not affect the shape of a small charged molecule.

Addition of organic solvents to the eluent dehydrate and shrink the gel.

Special effects can be achieved with such mixed solvents, probably because of partitioning.

b. *Hydrophobic Solvents.* With styrene-divinylbenzene gel columns, the following solvents have been found useful: methylene chloride, chloroform, tetrahydrofuran, methyl ethyl ketone, benzene, toluene, perchloroethylene, cyclohexane, N-methyl pyrrolidone, dimethylformamide, m-cresol, tetrahydronaphthalene, o-dichlorobenzene, 1,2,4-trichlorobenzene. Small additions of other compounds have proved useful in certain cases and are not detected by the eluate monitor when present equally in sample solvent, eluting, and reference streams. Thus, antioxidants may be used to protect the solvent or the sample. Another case is a polymer which contains weakly polar groups, which may effectively be covered with a small concentration of an organic acid or base with a sufficiently nonpolar "tail."

4. Temperature

Many separations may be conducted satisfactorily in gel columns at ambient temperature or below, but an elevated temperature lowers liquid viscosity and aids diffusion rates, yielding more resolution or shorter elution time. To avoid trapping air bubbles in the column, the solvent must be deaerated above the column temperature–exit pressure conditions. At a linear flow rate of 4 ft/hour in columns of 7.75-mm ID, based on empty column cross section, plate counts of 1000 to 2000 per foot have frequently been demonstrated at 120°C with column packings in the 30–50 μ diameter size range.

5. Sample Load

The sample load in a gel column affects the resolution of its components in three interrelated ways, as in other column separations. First, the volume of the sample itself is added arithmetically to the volume of the elution peak the column is capable of producing for each component. Then, the concentration of each component affects its own peak width. This effect becomes stronger as molecular size increases, and secondarily affects the small-molecule components in proportion to the time they spend in the overloaded zone. A different limitation is imposed on sample concentration by the viscosity of a polymer solution, called "viscous fingering" in some studies of fluid flow through packed beds. The more abrupt the viscosity change at the rear boundary of the sample zone, the less stable that boundary becomes. The extra pressure drop caused by sample viscosity then permits the solvent to push through at some weak point, and the liquid velocity profile becomes very uneven until considerable spreadout has occurred. In a study with hemoglobin and salt as solutes in the presence of high molecular weight dextrans, Flodin [1] found good separation at viscosities below 4 centipoises. Above, the resolution was impaired, and at 11.8 centipoises the zone profiles were badly distorted and drawn out.

The difference between these two limitations on sample load is shown by repeating a given sample elution with half the load, once with the concentration reduced and the original volume, then again with the original concentration but half the volume. Viscous fingering, if present, will be markedly reduced by lowering the concentration but not by reducing only the sample volume. The overloading of the diffusional mechanism will be relieved equally in both cases, so that peaks will be narrower and earlier near their tops. Overloading affects the first rise point very little, and delays the return to base line only when it is severe, but the peak shape is a sensitive indicator. Also, in any broad distribution, some single compound of small size may be included in a sample as a marker, and its elution peak compared with a similar peak obtained with the polymer absent.

Two other observations have been useful in checking the validity of an elution curve. A polar mismatch of sample to gel, i.e., incompatibility, is indicated when the elution curves of a series of decreasing loads show first rise points successively later, a marked advance coupled with a decreasing slope. With a normal separation below the overload point, decreasing the sample by half should narrow the peak only a very little at its base line intersections, and should not change its shape appreciably. The other test consists of rerunning collected cuts. This will be discussed in the section on the evaluation of data.

6. *Monitoring of Column Effluent*

Several excellent recent treatments of this subject are available. We will therefore only mention a few of the more important methods, most of which can be applied to automatic recording.

a. *Refractometer.* The continuous differential refractometer has been developed to extremely high sensitivity by J. L. Waters and is currently the method of choice for gel columns where a single eluting solvent is used with a column 0.5 cm or more in diameter. It is capable of showing polymer concentration changes down to 2 ppm with a cell volume slightly under 0.10 ml, making possible short elution times for fractionations of very high polymers. A microcell with only 0.010-ml volume has been developed by Waters for this instrument. Extensive experience with it is lacking, but peaks as narrow as 0.25 ml should be followed well by it.

b. *Ultraviolet Photometer.* This instrument is widely used for concentration measurements of protein solutions and those of other biological substances. In its simplest form, it is an inexpensive instrument and yet capable of high sensitivity. However, it can only be used for solutes which absorb light in the ultraviolet region, i.e., essentially for aromatic compounds, and is useless for many synthetic polymers.

c. *Evaporation.* The techniques of monitoring liquid column separations by evaporation are currently in a state of rapid change. Still a method of choice for some preparative separations, the Cahn Electro-balance technique using 2-ml Teflon cups and weighing up to 100 samples/hour to microgram accuracy may reduce considerably the time required for this method, perhaps causing it to be reconsidered for some analytical separations also. The smallest microcolumns may use this method with the flame ionization detector by using a moving wire or chain to carry the column effluent through a drying zone and into the flame [253–255].

d. *Other Methods.* Determination of the electrolytic conductance and of the dielectric constant also belongs to the standard methods of biochemists though they are not employed to the same extent as the ones previously mentioned. Colorimetry of color reactions, such as that of ninhydrin with amino acids and peptides, also is still in wide use.

B. Preparative Separations

1. *Column Design*

Whereas analytical columns are small and usually have a very high ratio of length to width, in preparative work large volumes are required which can best be obtained with wide columns and length-to-width ratios of 10–20. Glass columns of 20-cm diameter have been constructed and work very satisfactorily for desalting [7] as well as for fractionations. Their design is rather simple. They consist of a glass tube with a shallow, funnel-like bottom, which is fitted with a metering stopcock[2] or a valve. For organic solvents, all stopcocks which come into contact with solvent should be made of Teflon. The conical part of the column bottom is filled with glass beads of 0.5–1 mm size which may rest on a small screen put right above the outlet or on a wad of glass wool. Better are designs such as are offered by Pharmacia [256], where the conical part is very flat and the dead volume correspondingly small.

The column head can be very simple or may contain fittings for sample addition, a solvent reservoir, and nitrogen in- and outlets. In cases where the sample is denser than the solvent, the solution is often injected by means of a motor-driven syringe right above the gel surface where it spreads by

---

[2] These are stopcocks fitted with a needle valve for accurate and easy flow control.

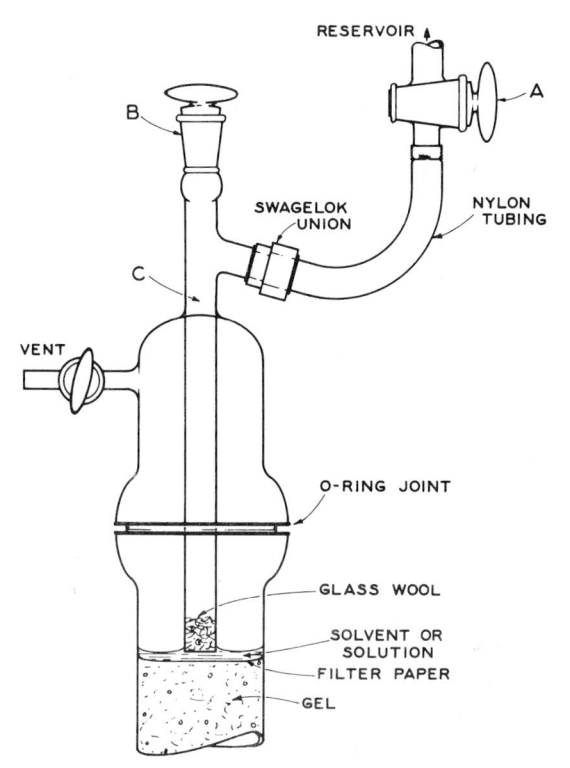

FIG. 9. A simple, versatile column top. For work under exclusion of air, a nitrogen inlet and a drainage stopcock can be included. Reproduced from Altgelt [163].

itself. A versatile top that gives good service under all conditions is shown in Fig. 9.

The top of the gel packing is best protected with a snugly fitting filter paper or a dense nylon net in order to avoid disturbance due to rapid solvent or sample flow. A filter paper, especially when held in place by a wire screen, should never be forced into the packing and jammed as, then, uneven flow results.

Other more elaborate columns have been described by Broman and Kjellin [257] and by Rothstein [251]. They were specifically designed for reverse flow, which mitigates strongly bed compaction usually encountered with highly swollen gels.[3]

Pharmacia developed adjustable adapters for reverse flow to be used with both aqueous and organic solvents. Columns fitted with such adapters

---

[3] One of these columns is commercially available from Future Plastics, Inc., 152 Columbia Street, Cambridge 39, Massachusetts.

at both ends have very little dead volume and can be operated under regular-, as well as reverse-flow, conditions. They are convenient and lend themselves to a number of special applications such as recycling or combinations of several columns.

Similar columns are available from Bio-Rad Laboratories [190].

Preparative columns may be fitted with warming or cooling jackets; but, generally, this will be unnecessary as long as strong draft or other sudden or large temperature changes are avoided. It is one of the advantages of GPC that it can be performed with very simple and inexpensive equipment.

As in gas chromatography, an intermediate design is frequently desirable. If a column of analytical size or somewhat larger is equipped for repetitive sample injection and automatic fraction collection, a creditable amount of each fraction is soon accumulated.[4]

2. General Procedure

Preparative gel columns are packed and run the same way as analytical columns. After prolonged service, flow may become obstructed. This usually results from fine deposits in the upper parts of the gel and can be corrected by replacing the upper 3–5 cm of packing by fresh gel. Sometimes, back-flushing the column with solvent, i.e., pumping solvent from the bottom up to the top, will restore the previous good flow conditions.

With respect to solvents and temperature, there is no difference between analytical and preparative procedures. The sample load will be greater in preparative runs than in analytical ones, where its lower limit is only determined by the sensitivity of the detector. Often, a compromise has to be found between maximum throughput and resolution. If only modest fractionation is required, the column can be overloaded considerably as long as viscosities beyond 3 centipoises are avoided. Especially, if only separation of large molecules from small ones is desired and a gel is employed which excludes the large molecules, sample volumes as high as $\frac{1}{4}-\frac{1}{5}$ of the column can be run. Even if some tailing occurs, the separation or fractionation may still be sufficient to serve the purpose.

The same detectors as in analytical setups can be used to monitor preparative separations although here the fractions are collected and usually evaporated and weighed. There are several devices for evaporation of small and large liquid volumes. Probably the most convenient and a very satisfactory procedure for evaporating a great number of cuts is to place the sample bottles or vials into a vacuum oven and apply vacuum while blowing filtered nitrogen or air through the chamber. By a simple modification,

---

[4] Equipment of this design has been offered by Waters Associates, Inc.

all vacuum ovens can be changed in such a way as to obtain a fanned air stream across the whole area, thus providing for fast evaporation.

C. Special Column Techniques

It is beyond the scope of this chapter to describe all the special techniques that have been or could be applied to GPC. We shall merely call to the readers' attention three modifications which can be useful in certain instances.

In some cases where adsorption occurs, advantage can be taken of this fact by eluting with different solvents [42,104]. Here, the techniques of GPC and regular chromatography are combined. First, the inert part of the sample is fractionated with a nonpolar solvent; then, the adsorbed part is gradually desorbed by eluting with solvent of increasing polarity. Several combinations of the two methods are possible.

Porath [86] suggested the method of "zone precipitation," a variation of the precipitation–dissolution process which is believed to be the basis of the Baker–Williams fractionation method [258]. (Compare Chapters B.2 and B.3.) A gradually decreasing concentration gradient of a nonsolvent in a solvent is fed into the column. A sample moving through such a column at a faster rate than the solvent–nonsolvent mixture will soon encounter a region where it is precipitated. However, since the following eluent has a greater solvent power, the sample is redissolved and proceeds down the column until it is precipitated again. In solute mixtures the different components will precipitate in different parts of the column according to their solubilities and thus be separated. Porath [86] suggested this technique for finding suitable conditions for crystal formation of proteins.

Thin layer chromatography is a technique of ever-rising importance for quick analyses of high resolution. Gel permeation, too, can be modified for this technique as Determann [113] and Johansson and Rymo [120,259] have demonstrated.

VI. Evaluation of Data

The evaluation of GPC data, in terms of molecular weight, requires determining three properties of the system: its validity, its calibration, and its resolution. These are interrelated, and they must finally be established by direct evidence. Thereafter, indirect evidence may attest that these properties have not changed. For a preliminary study, some indirect evidence has been found useful.

A. The Elution Curve

The plot of solute concentration of a fraction versus its elution volume is called the elution curve. For a monodisperse sample the elution curve obtained with a well-packed and well-operated column has the shape of a Gaussian curve; with pauci- or polydisperse samples, complex curves are obtained. When the components of a sample are separated sufficiently in the column, the shape of an elution curve is quite revealing and makes possible crude predictions of the size distribution of the sample.

Often the molecular weights of selected fractions are measured, and, thus, a column is calibrated for a particular solute. It is frequently desirable to interpret elution curves where no direct study of calibration and resolution has been made on the samples in question. If the column was not overloaded and the characteristic elution volumes of a series of curves have been shown to be reproducible, then a calibration is possible. It seems reasonable to assume that the exclusion of randomly coiled molecules in very dilute solution from the gel is a function of their hydrodynamic diameter. A calibration may be calculated on this basis from the elution volumes of a series of known samples of another polymer. Since many polymers are quite similar in their coil extension, as measured by elution

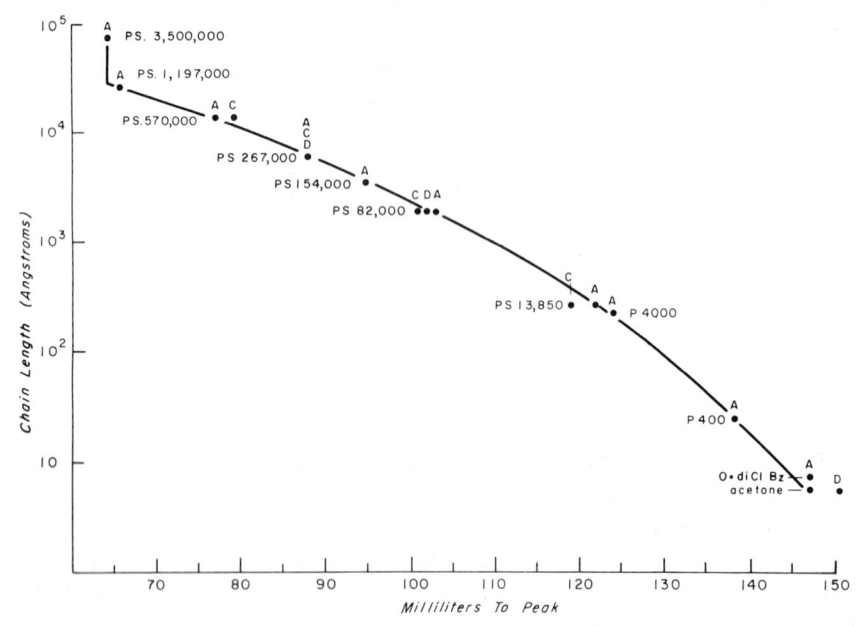

FIG. 10. Correlation of molecular chain length with elution volume for two polymers in good solvents. Narrow range polystyrenes and polypropylene glycols, molecular weights as shown. Columns 12 feet by 0.305 inches, Gel 843–845. I: A, 2170 pl, Toluene 90°C. II: C, 2790 pl, THF 55°C; D, 3200 pl, Tetralin 125°C. Reproduced from Moore and Hendrickson [158].

through a gel column as well as by their intrinsic viscosities, a correlation in terms of chain length is often acceptable as a preliminary estimate.

Such a correlation is shown in Fig. 10 for several polystyrenes and polypropylene glycols in three different solvents, each at a different temperature, and in two different columns [157]. Three points for polyvinyl chloride also fell close to this line in one of the solvents but were not run in the others. Simply comparing the elution curves of a polymer or copolymer in a column, eluted with one solvent, against those obtained with another solvent, may be quite informative. An agreement is frequently obtained, showing that solvent differences are not a factor. If the curves are shifted to different elution volumes, then a column may be indirectly calibrated for one solvent if it was calibrated for the other one. If considerable deviations are found between the two solvents, and curve shapes are altered, then polarity effects may be interfering with a size separation, or structural differences within the components of a sample may be suspected.

B. Relations between the Elution Volume and Molecular Weight

The direct calibration of a gel column system for a given polymer or copolymer requires first a series of narrow fractions of the substance, either made by some other method or generated by the gel column itself, if necessary, by repeated sampling and collection. These fractions must then be run on the gel column, and their molecular weights must be evaluated by one or more other methods. Since molecular weight determinations give different types of average values, unless the cut is extremely narrow in its distribution, the average obtained will be different from the molecular weight of the most abundant species, as indicated by the peak of the gel column elution curve of the cut. In other words, often the average molecular weight of a cut and its peak molecular weight are not the same. If the resolving power of the column is high, the elution curves of the cuts may be used in computing weight and number average molecular weights with an approximate calibration curve from which the needed correction is then available.

If the logarithm of the molecular weight of sufficiently narrow fractions is plotted versus the elution volumes on a column, almost linear curves such as shown in Fig. 5 (calibration curves) are obtained. Polymers of different structure usually yield slightly different calibration curves in the same column. Also, one polymer run in different solvents often gives a series of calibration curves which come together at their low molecular weight end. This behavior can be expected since polymers are more extended in good solvents than in poor ones, and these differences in hydrodynamic volume are diminished with decreasing molecular weight.

C. COMPUTING THE MOLECULAR WEIGHT DISTRIBUTION CURVE

It will be found relatively easy to set up gel columns showing 10,000–20,000 theoretical plates on samples of small pure species. These columns will then yield elution curves which, for relative purposes, are usable directly with peak calibration curves for the determination of the molecular weight distribution. With a linear detector, such as the continuous differential refractometer, the recorded elution curve may be continuously integrated by a recorder attachment. This ignores the obvious fact that the elution of any sample through a packed column will be accompanied by some longitudinal diffusion. On the other hand, it may also be true that even the uncorrected gel column data are more accurate than previously available data, and they may be adequate to the need for comparative data without correction.

It is to be hoped that the precaution of rerunning a gel column cut from the side of an elution curve will be taken before accepting peak calibrations as absolutely accurate for anything other than peaks. Even without extensive computation, it is evident that such a cut receives more material from the diffusional spread of the components on the peak side of the cut than it does from those on the low side. When rerun as a sample, its elution curve will be broadened by these contributions and also by their diffusional spread during the rerun. If this elution curve is narrow compared to the whole distribution from which it came and displaced in peak location only slightly toward the high side of the distribution, then further resolution or computation may not be required.

The series of elution curves from rerun cuts from a distribution made in the same column under the same conditions offers a set of data from which the relative distribution of the molecular species present can be derived. The resolution of the column and its response to load are readily observed with low molecular weight pure samples. Starting with this and the original elution curve, the composition of each cut may be calculated; similarly, the elution curve of the cut rerun through the same system will follow. By iterative methods, the assumptions of original distribution and the added diffusional spread due to increased molecular size should be corrected so that, when agreement is reached between the calculated and observed rerun curves, then the original distribution and the column's response to it also are known in detail.

A number of interesting computations concerning GPC were recently published. Berger and Shultz [260] calculated theoretical elution curves of linear polymers based on the assumption of a linear relation between $\log M$ and V_e and on three different molecular distribution functions. The molecular weights of the GPC peaks were $M_n < M_{gpc} \leq M_w$.

Rodriguez and Clark [261] described a simple method for the quick determination of M_n, M_w and M_z from GPC elution curves. They approximate a distribution by a triangle with the corners given by M_L, M_0, and M_H, the lowest, peak, and highest molecular weight, respectively.

Pickett, Cantow, and Johnson [262] devised a computer program which evaluates analytical as well as preparative GPC separations. Integral and differential distributions are tabulated and plotted as curves and as histograms. Number-, viscosity-, weight-, and Z-averages are computed.

Tung [263] calculated two numerical solutions for general distributions. His procedure corrects for zone spreading on the column and allows to find the true distribution. Tung, Moore, and Knight [264] tested this method by refractionating cuts of a polymer sample, computing the true distribution of each, and determining the distribution of the original sample by summing up the distribution of the cuts. The result agreed very well with the distribution computed by Tung's method from a GPC run of the original sample. The correction is particularly important for narrow distributions, both single- and multi-peak. For broad distributions it is minor [265].

Hendrickson and Moore discussed in more detail the nature of separation in GPC [158] and the effects of molecular shape and of hydrogen bonding on small-to-medium sized molecules [266]. When bulkiness, branching, and polarity were taken into proper account, a large number of different compounds with only few exceptions correlated into a single calibration curve. The structural elements of small molecules were approximately additive in their effect on elution volume.

REFERENCES

1. P. Flodin, Dissertation, University of Uppsala, Sweden (1962) (available from Pharmacia, Uppsala, Sweden).
2. J. Porath and P. Flodin, *Protides Biol. Fluids, Proc. Colloq.* **10**, 290 (1963).
3. A. Tiselius, J. Porath, and P. A. Albertsson, *Science* **141**, 13 (1963).
4. C. J. O. R. Morris and P. Morris, "Separation Methods in Biochemistry," Wiley (Interscience), New York, 1963.
5. H. Determann, *Angew. Chem.* **76**, 635 (1964).
6. B. Gelotte, *in* "New Biochemical Separations" (A. T. James and L. J. Morris, eds.), p. 93. Van Nostrand, Princeton, New Jersey, 1963.
7. "Sephadex in Gel Filtration," Pharmacia, Uppsala, Sweden.
8. "Sephadex, A Unique Substance for Modern Chromatography," Pharmacia, Uppsala, Sweden.
9. "Sephadex Literature References." Pharmacia, Uppsala, Sweden.
10. J. Porath and P. Flodin, *Nature* **183**, 1657 (1959).
11. G. Ostling, *Acta Soc. Med. Upsalien.* **64**, 222 (1960).
12. J. Porath, *Clin. Chim. Acta* **4**, 776 (1959).
13. W. Björk and J. Porath, *Acta Chem. Scand.* **13**, 1256 (1959).
14. R. F. Hill and W. Konigsberg, *J. Biol. Chem.* **235**, PC21 (1960).
15. L. A. Hanson and B. G. Johansson, *Nature* **187**, 599 (1960).
16. F. A. Thoma and D. E. Koshland, *J. Biol. Chem.* **235**, 2511 (1960).
17. F. Miranda and S. Lissitzky, *Nature* **190**, 443 (1961).

18. H. Rasmussen and L. C. Craig, *J. Biol. Chem.* **236**, 759 (1961).
19. W. Björk, *Biochim. Biophys. Acta* **49**, 195 (1961).
20. G. Samuelsson, *Svensk Farm. Tidskr.* **65** (1961).
21. B. N. Ames, R. G. Martin, and B. G. Garry, *J. Biol. Chem.* **236**, 2019 (1961).
22. K. W. Daisley, *Nature* **191**, 868 (1961).
23. H. Bennich, *Biochim. Biophys. Acta* **51**, 265 (1961).
24. F. V. Pierce and M. E. Webster, *Biochem. Biophys. Res. Commun.* **5**, 353 (1961).
25. F. D. Gregory and L. Rodén, *Biochem. Biophys. Res. Commun.* **5**, 430 (1961).
26. F. Killander, F. Pontén, and L. Rodén, *Nature* **192**, 182 (1961).
27. W. J. Lipp, *J. Histochem. Cytochem.* **9**, 458 (1961).
28. F. E. Fothergill and R. C. Nairn, *Nature* **192**, 1073 (1961).
29. H. Rinderknecht, *Nature* **193**, 167 (1962).
30. L. Colobert and G. Dirheimer, *Biochim. Biophys. Acta* **54**, 455 (1961).
31. V. Stepanov, D. Handschuh, and F. A. Anderer, *Z. Naturforsch.* **16b**, 626 (1961).
32. I. Björk, *Exptl. Eye Res.* **1**, 145 (1961).
33. H. Determan and O. Zipp, *Ann. Chem.* **649**, 203 (1961).
34. L. Jacobsson, *Clin. Chim. Acta* **7**, 180 (1962).
35. S. Lissitzky, F. Bismuth, and M. Rolland, *Clin. Chim. Acta* **7**, 183 (1962).
36. H. Papkoff, C. H. Li, and W. K. Liu, *Arch. Biochem. Biophys.* **96**, 216 (1962).
37. H. Zuber and R. Jaques, *Angew. Chem.* **74**, 216 (1962).
38. E. M. Press and R. R. Porter, *Biochem. J.* **83**, 172 (1962).
39. W. M. Hunter and F. C. Greenwood, *Nature* **194**, 495 (1962).
40. S. E. Bresler, Kh. M. Rubina, R. A. Graevskaya, and N. N. Vasil'eva, *Biokhimiya* **26**, 745 (1961).
41. L. Jacobsson and G. Widström, *Scand. J. Clin. & Lab. Invest.* **14**, 285 (1962).
42. B. Gelotte, *J. Chromatog.* **3**, 330 (1960).
43. D. van Hoang, M. Rovery, and P. Desnuelle, *Biochim. Biophys. Acta* **58**, 613 (1962).
44. C. E. Cornelius and R. A. Freedland, *Cornell Vet.* **52**, 344 (1962).
45. R. Grasbeck and R. Karlsson, *Acta Chem. Scand.* **16**, 782 (1962).
46. W. V. Epstein and M. Tan, *J. Lab. Clin. Med.* **60**, 125 (1962).
47. T. L. Hardy and K. R. L. Mansford, *Biochem. J.* **83**, 34 (1962).
48. C. F. Högman and J. Killander, *Acta Pathol. Microbiol. Scand.* **55**, 357 (1962).
49. R. Guillemin, E. Yamazaki, M. Jutisz, and E. Sakiz, *Compt. Rend.* **255**, 1018 (1962).
50. L. Hána and B. Styk, *Acta Virol.* (*Prague*) **6**, 479 (1962).
51. C. Lapresle and T. Webb, *Biochem. J.* **84**, 455 (1962).
52. H. G. van Eijk, C. H. Monfoort, J. J. Witte, and H. G. K. Westenbrink, *Biochim. Biophys. Acta* **63**, 537 (1962).
53. B. Shapiro and J. L. Rabinowitz, *J. Nucl. Med.* **3**, 417 (1962).
54. M. Kuboyama, S. Takemori, and T. E. King, *Biochem. Biophys. Res. Commun.* **9**, 534 (1962).
55. W. Stumpf and E. H. Graul, *Med. Klin.* (*Munich*) **58**, 192 (1963).
56. E. E. Stinson and C. O. Willits, *J. Assoc. Offic. Agr. Chemists* **46**, 329 (1963).
57. E. Thureborn, *Nature* **197**, 1301 (1963).
58. E. B. Lindner, A. Elmquist, and J. Porath, *Nature* **184**, 1565 (1959).
59. B. Gelotte and A. B. Krantz, *Acta Chem. Scand.* **13**, 2127 (1959).
60. A. Bill, N. Marsden, and H. R. Ulfendahl, *Scand. J. Clin. & Lab. Invest.* **12**, 392 (1960).
61. Z. Pravda and E. Wisingerová, *J. Hyg., Epidemiol., Microbiol., Immunol.* (*Prague*) **4**, 509 (1960).
62. C. B. Anfinsen and E. Haber, *J. Biol. Chem.* **236**, 1361 (1961).
63. H. Palmstierna, *Sci. Tools* **7**, 39 (1961).

64. E. Bassett, S. M. Beiser, and S. W. Tanenbaum, *Science* **133**, 1475 (1961).
65. E. T. Bucovaz and J. W. Davis, *J. Biol. Chem.* **236**, 2015 (1961).
66. J. Zwan and A. F. van Dam, *Acta Histochem.* **11**, 306 (1961).
67. H. D. Matheka and G. Wittmann, *Zentr. Bakteriol., Parasitenk., Abt. I. Orig.* **182**, 169 (1961).
68. P. J. Knudsen and J. Koefoed, *Nature* **191**, 1306 (1961).
69. C. G. Beling, *Nature* **192**, 326 (1961).
70. L. Bosch, G. van der Wende, M. Sluyser, and H. Bloemendal, *Biochim. Biophys. Acta* **53**, 44 (1961).
71. H. Ishikura, *Biochim. Biophys. Acta* **51**, 189 (1961).
72. C. C. Curtain, *J. Histochem. Cytochem.* **9**, 484 (1961).
73. W. H. S. George and K. W. Walton, *Nature* **192**, 1188 (1961).
74. H. Rasmussen and L. C. Craig, *Biochim. Biophys. Acta* **56**, 332 (1962).
75. B. von Hofsten and J. Porath, *Acta Chem. Scand.* **15**, 1791 (1961).
76. S. Kuyama and D. Pramer, *Biochim. Biophys. Acta* **56**, 631 (1961).
77. M. A. Grodon, M. R. Edwards, and V. N. Tompkins, *Proc. Soc. Exptl. Biol. Med.* **109**, 96 (1962).
78. M. Murray and M. Chadwick, *Biochim. Biophys. Acta* **58**, 338 (1962).
79. U. Beiss and R. Marx, *Naturwissenschaften* **49**, 142 (1962).
80. M. Wagner, *Zentr. Bakteriol., Parasitenk., Abt. I. Orig.* **185**, 124 (1962).
81. Y. C. Lee and R. Montgomery, *Arch. Biochem. Biophys.* **97**, 9 (1962).
82. J. B. Woof, *Nature* **195**, 184 (1962).
83. J. Porath and A. V. Schally, *Endocrinology* **70**, 738 (1962).
84. R. C. Chandan and K. M. Shahani, *J. Diary Sci.* **45**, 645 (1962).
85. B. von Hofsten and J. Porath, *Biochim. Biophys. Acta* **64**, 1 (1962).
86. J. Porath, *Nature* **196**, 47 (1962).
87. E. Mammen and A. Ramien, *Thromb. Diath. Haemorrh.* **8**, 37 (1962).
88. G. Pettersson, E. B. Cowling, and J. Porath, *Biochim. Biophys. Acta* **67**, 1 (1963).
89. W. Appel, *Z. Physiol. Chem.* **330**, 193 (1963).
90. R. A. Reisfeld, B. G. Hallows, D. E. Williams, N. G. Brink, and S. L. Steelman, *Nature* **197**, 1206 (1963).
91. K. C. Robbins and L. Summaria, *J. Biol. Chem.* **238**, 952 (1963).
92. J. R. Sherman and J. Adler, *J. Biol. Chem.* **238**, 873 (1963).
92a. K. A. Granath, *in* "New Biochemical Separations" (A. T. James and L. J. Morris, eds.), p. 111, Van Nostrand, Princeton, New Jersey, 1963.
93. P. E. Wilcox and F. Lisowski, *Federation Proc.* **19**, 333 (1960).
94. R. L. Kisliuk, *Biochim. Biophys. Acta* **40**, 531 (1960).
95. K. O. Pedersen, *Arch. Biochem. Biophys.* Suppl. 1, 157 (1962).
96. G. Lundblad, *Acta Chem. Scand.* **15**, 212 (1961).
97. W. V. Epstein and M. Tan, *J. Chromatog.* **6**, 258 (1961).
98. J. Killander and P. Flodin, *Vox Sanguinis* [N.S.] **7**, 113 (1962).
99. B. Gelotte, P. Flodin, and J. Killander, *Arch. Biochem. Biophys.* Suppl. 1, 319 (1962).
100. P. Flodin and J. Killander, *Biochim. Biophys. Acta* **63**, 403 (1962).
101. P. I. Brewer, *Nature* **188**, 934 (1960).
102. J. Porath, *Biochim. Biophys. Acta* **39**, 193 (1960).
103. R. L. M. Synge and M. A. Youngson, *Biochem. J.* **78**, 31P (1961).
104. J. Porath and E. B. Lindner, *Nature* **191**, 69 (1961).
105. W. A. Klee, H. H. Richards, and G. L. Cantoni, *Biochim. Biophys. Acta* **54**, 157 (1961).
106. R. Djurtoft, *European Brewery Conv., Proc. 8th Congr., Vienna, 1961*, p. 298 (1961).
107. W. Nultsch, *Biochim. Biophys. Acta* **59**, 213 (1962).

108. S. Hjertén and R. Mosbach, *Anal. Biochem.* **3**, 109 (1962).
109. S. E. Bresler, Kh. M. Rubina, R. A. Graevskaya, and N. N. Vasil'eva, *Biokhimiya* **26**, 745 (1961).
110. W. H. S. George, *Nature* **195**, 155 (1962).
111. P. DeMoor, K. Heirwegh, J. F. Heremans, and M. Declerck-Raskin, *J. Clin. Invest.* **41**, 816 (1962).
112. B. Cortis-Jones, *Intern. Sugar J.* **64**, 133 and 165 (1962).
113. H. Determann, *Experimentia* **18**, 430 (1962).
114. C. F. Barlow, H. Firemark, and L. J. Roth, *J. Pharm. Pharmacol.* **14**, 550 (1962).
115. D. Givol, S. Fuchs, and M. Sela, *Biochim. Biophys. Acta* **63**, 222 (1962).
116. A. M. Crestfield, W. H. Stein, and S. Moore, *Arch. Biochem. Biophys.* Suppl. 1, 217 (1962).
117. H. Streuli, *Chimia (Aarau)* **16**, 371 (1962).
118. M. Kakei and G. B. J. Glass, *Proc. Soc. Exptl. Biol. Med.* **111**, 270 (1962).
119. M. Tan and W. V. Epstein, *Science* **139**, 53 (1963).
120. B. G. Johansson and L. Rymo, *Acta Chem. Scand.* **16**, 2067 (1962).
121. B. Lindquist, *Acta Chem. Scand.* **16**, 1794 (1962).
122. J. E. Coleman and B. L. Vallee, *J. Biol. Chem.* **237**, 3430 (1962).
123. H. Ishikura, *J. Biochem. (Tokyo)* **52**, 324 (1962).
124. J. H. Glick, Jr., *Arch. Biochem. Biophys.* **100**, 192 (1963).
125. G. R. Shepherd and D. F. Petersen, *J. Chromatog.* **9**, 445 (1962).
126. P. Cornillott, R. Bourrillon, J. Michon, and R. Got, *Biochim. Biophys. Acta* **71**, 89 (1963).
127. N. R. Ringertz and P. Reichard, *Acta Chem. Scand.* **14**, 303 (1960).
128. N. R. Ringertz, *Acta Chem. Scand.* **14**, 312 (1960).
129. H. Spitzy, H. Skrube, and K. Müller, *Mikrochim. Acta* p. 296 (1961).
130. G. E. Connell and R. W. Shaw, *Canad. J. Biochem. Physiol.* **39**, 1013 (1960).
131. P. Flodin and K. A. Granath, in "Symposium über Makromolekule," IIC6, Wiesbaden, Weinheim, 1959.
132. P. Flodin and K. Aspberg, in "Biological Structure and Function" (T. W. Goodwin and O. Lindberg, eds.), Vol. 1, p. 345, Academic Press, New York, 1961.
133. K. A. Granath and P. Flodin, *Makromol. Chem.* **48**, 160 (1961).
134. P. Nordin, *Arch. Biochem. Biophys.* **99**, 101 (1962).
135. B. Lindquist, *16th Intern. Dairy Congress, Proc.*, Copenhagen 1961, p. 673 (1962).
136. A. B. Edmunson, *Nature* **198**, 354 (1963).
137. G. Preaux and R. Lontie, *Arch. Intern. Physiol. Biochim.* **69**, 100 (1961).
138. A. Polson, *Biochim. Biophys. Acta* **50**, 565 (1961).
139. H. J. Cruft, *Biochim. Biophys. Acta* **54**, 611 (1961).
140. D. J. Lea and A. H. Sehon, *Can. J. Chem.* **40**, 159 (1962).
141. T. P. Kind and P. S. Norman, *Biochemistry* **1**, 709 (1962).
142. M. Nummi and T. M. Enari, *Brauwissenschaft* **15**, 203 (1962).
143. D. J. Millin and M. H. Smith, *Biochim. Biophys. Acta* **62**, 450 (1962).
144. E. Gross and B. Witkop, *J. Biol. Chem.* **237**, 1856 (1962).
145. G. Semanza and S. Auricchio, *Biochim. Biophys. Acta* **65**, 173 (1962).
146. B. Gelotte, *Naturwissenschaften* **48**, 554 (1961).
147. S. Zadražil, Z. Šormová, and F. Šorm, *Collection Czech. Chem. Commun.* **26**, 2643 (1961).
148. V. M. Ingram and J. G. Pierce, *Biochemistry* **1**, 580 (1962).
149. H. G. Boman and S. Hjertén, *Arch. Biochem. Biophys.* Suppl. **1**, 276 (1962).
150. A. W. Phillips and P. A. Gibbs, *Biochem. J.* **81**, 551 (1961).

151. C. Bengtsson, L. A. Hanson, and B. G. Johansson, *Acta Chem. Scand.* **16**, 127 (1962).
152. G. Guidotti. R. J. Hill, and W. Konigsberg, *J. Biol. Chem.* **237**, 2184 (1962).
153. G. S. Marks, R. D. Marshall, A. Neuberger, and H. Papkoff, *Biochim. Biophys. Acta* **63**, 340 (1962).
154. M. F. Vaughan, *Nature* **188**, 55 (1960).
155. M. F. Vaughan, *Nature* **195**, 801 (1962).
156. M. F. Vaughan and J. H. S. Green, "Techniques of Polymer Science," Soc. Chem. Ind. Monograph. Gordon & Breach, New York, 1963.
157. J. C. Moore, *J. Polymer Sci.* **A2**, 835 (1964).
157a. P. I. Brewer, *Polymer* **6**, 603 (1965).
158. J. C. Moore and J. G. Hendrickson, *J. Polymer Sci.* **C8**, 233 (1965).
159. L. E. Maley, *J. Polymer Sci.* **C8**, 253 (1965).
160. D. J. Harmon, *J. Polymer Sci.* **C8**, 243 (1965).
161. G. Lueben, Diplomarbeit, University Frankfurt (Main), Germany (1963).
162. H. Determann, G. Lueben, and T. Wieland, *Makromol. Chem.* **73**, 168 (1964).
163. K. H. Altgelt, *Makromol. Chem.* **88**, 75 (1965).
164. P. I. Brewer, *Nature* **190**, 625 (1961).
165. P. I. Brewer, *J. Inst. Petrol.* **48**, 277 (1962).
166. B. Cortis-Jones, *Nature* **191**, 272 (1961).
167. G. H. Lathe and C. R. J. Ruthven, *Biochem. J.* **62**, 665 (1956).
168. P. Andrews, *Nature* **196**, 36 (1962).
169. P. Andrews, *Biochem. J.* **91**, 222 (1964).
170. R. L. Steere and G. K. Ackers, *Nature* **194**, 114 (1962).
171. R. L. Steere and G. K. Ackers, *Nature* **196**, 475 (1962).
172. J. R. Whitaker, *Anal. Chem.* **35**, 1950 (1963).
173. D. J. Winzor and H. A. Scheraga, *Biochemistry* **2**, 1263 (1963).
174. D. J. Winzor and H. A. Scheraga, *J. Phys. Chem.* **68**, 338 (1964).
175. M. Iwatsubo and A. Curdel, *Compt. Rend.* **256**, 5224 (1963).
176. E. Ungerer, *Kolloid Z.* **36**, 228 (1925).
177. G. Wiegner, *J. Soc. Chem. Ind. (London)* **50**, 65T (1931).
178. N. Cernescu, Dissertation 661, Eidgenoess. Tech. Hochschule, Zuerich (1933).
179. J. Claesson and S. Claesson, *Arkiv. Kemi.* **19A**, No. 5, 1 (1944).
180. J. Claesson and S. Claesson, *Phys. Rev.* **73**, 1221 (1948).
181. S. Claesson, *Arkiv. Kemi.* **26A**, No. 24, 1 (1948).
182. H. Deuel, J. Solms, and L. Anyas-Weisz, *Helv. Chim. Acta* **33**, 2171 (1950).
183. H. Deuel and H. Neukom, *Advan. Chem. Ser.* **11**, 51 (1954).
184. R. M. Wheaton and W. C. Bauman, *Annal. N.Y. Acad. Sci.* **57**, 159 (1953).
185. B. Lindquist and T. Storgårds, *Nature* **175**, 511 (1955).
186. G. H. Lathe and C. R. J. Ruthven, *Biochem. J.* **60**, xxxiv (1955).
187. P. Flodin, *J. Chromatog.* **5**, 103 (1961).
188. J. P. Hummel and W. J. Dreyer, *Biochim. Biophys. Acta* **63**, 530 (1962).
189. S. Hjertén, *Arch. Biochem. Biophys.* Suppl. **1**, 147 (1962).
190. Folder of Bio-Rad Laboratories, 32nd and Griffin Ave., Richmond, California (1964).
190a. A. C. Allison and J. H. Humphrey, *Nature* **183**, 1590 (1959).
191. K. H. Altgelt, unpublished results (1962).
192. J. C. Moore, *18th Southwest Regional Am. Chem. Soc. Meeting, Dallas, Texas, 1962.*
193. J. Porath, *Pure Appl. Chem.* **6**, 233 (1963).
194. T. C. Laurent and J. Killander, *J. Chromatog.* **14**, 317 (1964).
194a. J. Porath, personal communication (1965).
195. A. G. Ogston, *Trans. Faraday Soc.* **54**, 1754 (1958).

196. T. Vermeulen, *Ind. Eng. Chem.* **45**, 1664 (1953).
197. H. Vink, *Acta Chem. Scand.* **18**, 409 (1964).
198. F. Helfferich, "Ionenaustauscher," Vol. 1. Weinheim, Bergstrasse, 1959; see "Ion Exchange," McGraw-Hill Series in Advanced Chemistry, 1962.
198a. T. C. Laurent and E. P. Laurent, *J. Chromatog.* **16**, 89 (1964).
198b. L. C. Craig, *Science* **144**, 1093 (1964).
199. G. K. Ackers and R. L. Steere, *Biochim. Biophys. Acta* **59**, 137 (1962).
200. E. M. Renkin, *J. Gen. Physiol.* **38**, 225 (1954).
201. G. K. Ackers, *Biochemistry* **3**, 723 (1964).
201a. W. F. Busse, *Phys. Today* **17**, 32 (1964).
202. H. L. Goldsmith and S. G. Mason, *J. Colloid Sci.* **17**, 448 (1962).
203. J. R. Millar, D. G. Smith, W. E. Marr, and T. R. E. Kressmann, *J. Chem. Soc.* p. 218 (1963).
204. K. W. Pepper, D. Reichenberger, and D. K. Hale, *J. Chem. Soc.* p. 3129 (1952).
205. E. Glueckauf, *in* "Ion Exchange and Its Applications," p. 34. Soc. Chem. Ind., London, 1955.
206. Report of Task Group One of Subcommittee 1 of ASTM Comm. E19 on Gas Chromatography (1962).
207. G. G. Purnell, "Gas Chromatography," p. 108. Wiley, New York, 1962.
208. L. S. Ettre, *J. Gas Chromatog.* **1**, 36 (1963).
209. J. L. Waters, "Resolving Power of GPC" (pamphlet). Waters Assoc., 1965, also *Abstr. Am. Chem. Soc. 149th Meeting, Detroit, 1965* (Div. *Polymer Chem. Preprints*) Vol. 6, No. 2, 1061.
210. H. Mark and G. Saito, *Monatsh. Chem.* **68**, 237 (1936).
211. A. Baum and E. Broda, *Trans. Faraday Soc.* **34**, 797 (1938).
212. G. R. Levi and A. Giera, *Gazz. Chim. Ital.* **67**, 719 (1937).
213. I. Landler, *Compt. Rend.* **225**, 234 (1947).
214. F. Miranda, H. Rochat, and S. Lissitzky, "Separation Methods in Biochemistry," Wiley (Interscience), New York, 1963.
215. A. Dobry and F. Boyer-Kawenoki, *J. Polymer Sci.* **2**, 90 (1947).
216. R. L. Scott, *J. Chem. Phys.* **17**, 297 (1949).
217. H. Tompa, *Trans. Faraday Soc.* **45**, 1142 (1949).
218. J. H. Hildebrand and R. L. Scott, "The Solubility of Nonelectrolytes," Reinhold, New York, 1950.
219. P. J. Flory, "Principles of Polymer Chemistry," Cornell Univ. Press, Ithaca, New York, 1953.
219a. G. Langhammer and L. Nestler, *Makromol. Chem.* **88**, 179 (1965).
219b. W. Haller, *Nature* **206**, 693 (1965).
219c. "Gel Filtration Materials," Bio-Rad Laboratories, Richmond, Calif., 1966.
220. J. A. Mikes, *J. Polymer Sci.* **30**, 615 (1958).
221. T. Alfrey, Jr. and W. G. Lloyd, *J. Polymer Sci.* **62**, 159 (1962).
222. W. G. Lloyd and T. Alfrey, Jr., *J. Polymer Sci.* **62**, 301 (1962).
223. K. Sollner, *Ann. N. Y. Acad. Sci.* **57**, 177 (1953).
224. D. W. Simpson and W. C. Bauman, U.S. Patent 2,771,163 (1956).
225. R. M. Wheaton, U.S. Patent, 2,911,362 (1959).
226. E. F. Meitzner and J. A. Oline (to Rohm and Haas), Union of South Africa Patent 592,393.
227. (Also to Rohm and Haas), Australia Patent 20-460-59563.
228. N. M. Bortnick (to Rohm and Haas), U.S. Patent 3,037,052 (1962).
229. T. R. E. Kressman and J. R. Millar (to Permutit), British Patent 889,304.
230. J. C. Moore, in preparation.

230a. *Natl. Bur. Std. (U.S.) Tech. News Bull.* (1966).
230b. E. Nystrom and J. Sjövall, *Anal. Biochem.* **12** (2), 235 (1965).
230c. "Sephadex LH 20 for Gel Filtration in Organic Solvents." Pharmacia, Uppsala, Sweden.
231. J. E. Salmon and D. K. Hale, "Ion Exchange—A Laboratory Manual." Academic Press, New York, 1959.
232. J. C. H. Hwa, U.S. Patent 2,689,832 (1954).
233. R. Kunin and F. X. McGarvey, U.S. Patent 2,692,244 (1954).
234. L. Tavani and M. Morini, U.S. Patent 2,885,371 (1959).
235. L. Thielen, U.S. Patent 2,960,480 (1960).
236. T. Kressman, U.S. Patent 3,030,318 (1962).
237. N. M. Bortnick, U.S. Patent 3,037,052 (1962).
238. P. B. Hamilton, *Anal. Chem.* **30**, 914 (1958); **32**, 1779 (1960).
239. W. W. Reynolds, *in* "Physical Chemistry of Petroleum Solvents," p. 42. Reinhold, New York, 1963.
240. R. Kunin, E. F. Meitzner, and N. M. Bortnick, *J. Am. Chem. Soc.* **84**, 305 (1962).
241. K. A. Kun and R. Kunin, *Polymer Letters* **2**, 587 (1964).
242. R. Kunin, E. F. Meitzner, J. A. Oline, S. Fischer, and N. Frisch, *Ind. Eng. Chem., Prod. Res. Develop.* **1**, 140 (1962).
243. G. K. Gordon, Ph.D. Thesis, Massachusetts Institute of Technology (1961).
244. E. Heftmann, "Chromatography." Reinhold, New York, 1961.
245. O. Samuelson, "Ion Exchange Separations in Analytical Chemistry." Stockholm, 1963.
246. J. Porath and H. Bennich, *Arch. Biochem. Biophys.* Suppl. 1, 152 (1962).
247. R. S. Porter and J. F. Johnson, *Nature* **183**, 391 (1959).
248. K. H. Altgelt, unpublished results (1965).
249. D. G. Lesnini, *J. Paint Technol.*, **38**, 498 (1966).
250. C. G. Beling, *Acta Endocrinol.* Suppl. 79, 9 (1963).
251. F. Rothstein, *J. Chromatog.* **18**, 36 (1965).
252. P. G. Condliffe and J. Porath, *Federation Proc.* **21**, A199b (1962).
253. A. T. James, J. R. Ravenhill, and R. P. W. Scott, *Chem. & Ind. (London)*, p. 746 (1964).
254. E. Haahti and T. Nikkari, *Acta Chem. Scand.* **17**, 2565 (1963).
255. J. E. Stouffer, T. E. Kerten, and P. M. Krueger, *Biochim. Biophys. Acta* **93**, 191–194.
256. "Sephadex Laboratory Columns." Pharmacia, Uppsala, Sweden.
257. L. Broman and K. Kjellin, *Biochim. Biophys. Acta* **82**, 101 (1964).
258. C. A. Baker and R. J. P. Williams, *J. Chem. Soc.* p. 2352 (1956).
259. B. G. Johansson and L. Rymo, *Acta Chem. Scand.* **18**, 217 (1964).
260. H. L. Berger and A. R. Shultz, *J. Polymer Sci.* **A3**, 3643 (1965).
261. F. Rodriguez and O. K. Clark, *Am. Chem. Soc. 150th Meeting, Atlantic City, 1965* (Div. Org. Coatings and Plastics Chem. Papers) **25**, (2) 220 (1965).
262. H. E. Pickett, M. J. R. Cantow, and J. F. Johnson, *J. Appl. Polymer Sci.*, in press.
263. L. H. Tung, *J. Appl. Polymer Sci.* **10**, 375 (1966).
264. L. H. Tung, J. C. Moore, and G. W. Knight, *J. Appl. Polymer Sci.*, **10**, 1261 (1966).
265. L. H. Tung, *J. Appl. Polymer Sci.*, **10**, 1271 (1966).
266. J. G. Hendrickson and J. C. Moore, *J. Polymer Sci.* **A1**, 167 (1966).

CHAPTER B.5

Thermal Diffusion

Alden H. Emery, Jr.
SCHOOL OF CHEMICAL ENGINEERING, PURDUE UNIVERSITY, LAFAYETTE, INDIANA

I. History...181
II. Basic Theory..182
   A. Convectionless Cells......................................182
   B. Thermal Diffusion Columns................................183
III. Studies of the Variables......................................183
   A. Standard Variables.......................................183
   B. Cell Studies of c and M..............................184
   C. Column Studies of c and M............................185
IV. Polymer Fractionations..186
V. Methods of Fractionation......................................188
   A. Cascade of Thermal Diffusion Columns.....................188
   B. Multiple Reservoirs......................................189
   C. Solvent Flow..189
   References..189

Thermal diffusion columns have been used successfully to fractionate solutions of high polymers, and in two cases have yielded molecular weight distributions in excellent agreement with those from fractional precipitation [1,2]. This chapter presents these encouraging results, a survey of polymer–solvent systems studied, and a review of the dismal state of the theory.

I. History

The first suggestion of the possibility of fractionating high polymers by means of thermal diffusion was made in an article in 1941 by Gralén and Svedberg [3], who built a small parallel-plate column with the intent of fractionating proteins. Another interesting effect occurred instead, an evaporative effect which merely concentrated without fractionating, caused by small air bubbles.

The first actual fractionation was conducted by Debye and Bueche [4] at Cornell, reported in 1948. They used concentric-tube columns with reservoirs and observed that in polystyrene in toluene, the molecular weight of the polymer in the bottom reservoir was higher than that in the top. At about the same time, Fritzemeier and Hermans [5] published a paper on the behavior of polymethyl methacrylate in a parallel-plate column. They found a very large concentrating effect, but reported that the molecular weight was the same in the two reservoirs.

Langhammer independently discovered the ability of the thermal diffusion column to fractionate high polymers, and in 1954 published the first of a series of papers on the subject [6]. An article of his in 1955 [7] reports the first use of a cascade of columns in high polymer work, in which the content of the top reservoir is fed to another column, and the content of the bottom to another. From the fractions so produced, one can construct a molecular weight distribution plot. In 1958, Guzmán and Fatou [8] independently conceived of using a cascade of columns, and fractionated polyvinyl chloride in cyclohexanone.

II. Basic Theory

This section will present information pertinent to the interpretation of results with columns, but not in itself practical. We have insufficient precise basic data (and possibly inadequate theory) to do a reasonable job of predicting column results, and must run column tests in any given system to see if thermal diffusion is worthwhile.

A. Convectionless Cells

The mass flux rate J_x is written as

$$J_x = -D\frac{\partial c}{\partial x} - D'c\frac{\partial T}{\partial x} \qquad (1)$$

in which c is concentration, x is distance, T is temperature, D is the diffusion coefficient, and D' is the thermal diffusion coefficient. The nomenclature of de Groot [9] is followed here, as in most of the literature on polymer solutions.

The moving-boundary cell used first by Hoffman and Zimm [10] uses this equation directly, taking advantage of the fact that for polymer solutions the first term is usually negligible in comparison with the second.

In the case of the open-column cell, Eq. (1) is rearranged for the steady state to give

$$\frac{d\ln c}{dx} = -\frac{D'}{D}\frac{dT}{dx} \qquad (2)$$

The ratio D'/D is also referred to as the Soret coefficient, s.

Solutions of high polymers have values of D' which are about the same as solutions of ordinary low molecular weight materials, but their diffusion coefficients are about one hundred times smaller. Thus one gets phenomenally high separations of polymer and solvent in a thermal diffusion cell.

B. Thermal Diffusion Columns

It is likely that most practical applications will be made with operation close to the steady state, so transient theory will be omitted. The steady state separation is given by [9]

$$\gamma = c_t/c_b = e^{-2\alpha} \tag{3}$$

$$\alpha = \frac{252\eta D'h}{\beta \rho g a^4} \tag{4}$$

in which η is viscosity, h is column height, ρ is density, $\beta = -\partial\rho/\partial T$, g is the acceleration of gravity, a is the slit width or plate spacing, and the subscripts t and b refer to the top and bottom reservoirs. All of these parameters have values in polymer solutions which are quite similar to those for low molecular weight materials, so the separations of polymer and solvent one obtains in columns are not extraordinary. We thus have the paradoxical result that one can get separations of polymer and solvent in a cell higher than experienced in most column studies.

The only parameters in Eq. (4) which are at all peculiar for polymer solutions are η and D'. The effect of the polymer solution itself on the operation of the thermal diffusion column ought to be analyzable in terms of the effect of concentration, c, and molecular weight, M, on the product $\eta D'$.

The literature can be split into articles in which the effects of the variables on the separation of polymer and solvent are considered, taken up in Section III, and those in which the separation of different molecular weights of polymer is measured, covered in Section IV. In the former, one characteristically uses a molecular weight fraction prepared by fractional precipitation and does not measure molecular weights in the reservoirs of the column; in the latter, one uses unfractionated polymer and measures the molecular weight of the polymer in the reservoirs. The latter study is the more important practically, but the former is potentially useful in interpreting results.

III. Studies of the Variables

A. Standard Variables

Certain variables of importance in any thermal diffusion column are considered in this section, and those two of peculiar importance to high polymers, namely c and M, are considered in Sections III,B and III,C.

The results of tests of the standard variables that have been made with polymer solutions coincide very well with the results from tests on low molecular weight materials. Debye and Bueche [4] verified the validity of the theoretical prediction that α should be proportional to some large power of the slit width, a. Their results, however, show the power to be three instead of the theoretical four. This agrees with observations on low molecular weight materials [11].

Langhammer et al. [1,12] verified the theoretical predictions on the effects of temperature level (no effect), temperature difference, height of column, and volume of reservoirs, and found that α varied inversely as the third power of a, as above, instead of the theoretical fourth. Thus the state of the theory in the case of the standard variables is the same for high polymers as for ordinary materials, and one can apply this part of the theory with some confidence.

B. Cell Studies of c and M

Since the only parameters in Eq. (4) that are unusual for polymer solutions are properties of the solutions, D' and η, one good way to study the peculiarities of polymer solutions would seem to be in convectionless cells. Most of these studies involved polystyrene in toluene, and just these are presented here.

Emery and Drickamer [13] used an open-column cell to determine essentially the Soret coefficient, as in Eq. (2), for polystyrene with M from 10,000 to 340,000 in toluene at c from 0.2% to 4%. They found a large effect of c, especially at low c, which correlated with thermodynamic data. Two sample correlation curves are shown in the form of D' in Fig. 1. D' values were calculated using the diffusion constants given by Meyerhoff [14]. They also found a fairly large effect of M on D'.

Hoffman and Zimm [10] used a moving-boundary cell to determine D' directly as a function of c for two molecular weights of polystyrene in toluene. Their results are also shown in Fig. 1. The effect of c noted above is borne out in their work, but they found less effect of M. Meyerhoff et al. [15], using an improvement of the technique, determined one point, also shown in Fig. 1.

Whitmore [16], using an open-column cell, obtained values for D' remarkably lower than anyone else. His curve for M of 310,000 for example, is at D' of about 0.1×10^{-7}, roughly independent of concentration.

Herren and Ham [17] measured D' as a function of M at a c of 1%, and found that D' above M of 300,000 was about constant at 1.5×10^{-7}.

In a convincing work, Meyerhoff and Nachtigall [18] used an optical method of analyzing the progress of separation in an open-column cell,

B.5. Thermal Diffusion

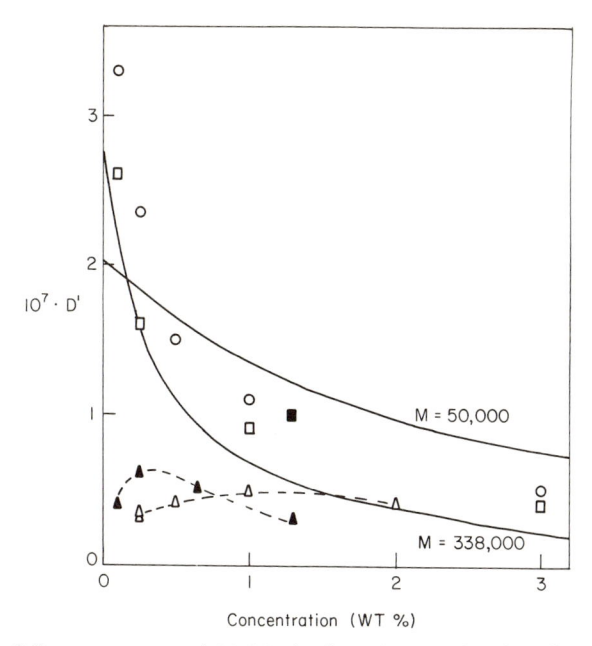

FIG. 1. Cell measurements of D'. Moving-boundary results of Hoffman and Zimm [10]: circles, $M = 60,000$; squares, $M = 1,100,000$. Black square, Meyerhoff et al. [15] moving-boundary measurement for $M = 2,000,000$. Solid lines, correlation of Emery and Drickamer [13] for data from open-column cell. Triangles, data of Debye and Bueche [4] from thermal diffusion column (not cell), for comparison.

and obtained D' of about 0.9×10^{-7}, independent of both c and M. More corroborating data were published later by Rauch and Meyerhoff [19].

C. Column Studies of c and M

Debye and Bueche [4] determined separation as a function of c for two molecular weights of polystyrene in toluene. Their results are shown in Figs. 1 and 2 in the form of D', as calculated by Hoffman and Zimm [10]. Clearly, the correspondence with cell measurements is bad, especially below c of 0.8%.

Langhammer et al. [1,12,20] determined s for polyvinyl pyrrolidone in water as a function of c and obtained straight lines, the slope of which was a linear function of M. They performed the same measurements on polystyrene in toluene, the results of which are plotted in Fig. 2, multiplied by 10 for convenience. About the only thing these data have in common with those of Debye and Bueche is an indication of a crossover of the curves for different molecular weights.

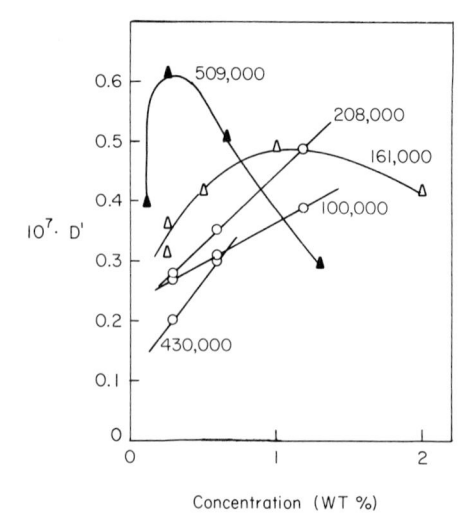

FIG. 2 Column measurements of D'. Curved lines, data of Debye and Bueche [4]. Straight lines, data of Langhammer et al. [1], multiplied by 10. Numbers on the lines indicate molecular weight.

IV. Polymer Fractionations

Studies in which the fractionation of high polymers was measured are summarized in Table I. The only safe generalization that can be drawn from these studies is that the polymer always becomes more concentrated in the bottom reservoir. Beyond that, one usually finds high molecular weight polymer concentrating in the bottom while the relative proportion of low molecular weight polymer increases in the top, but sometimes it is the other way around, and sometimes there is no separation of molecular weights. Kössler and Krejsa [21] observed what they call an "inversion," in which the proportion of high molecular weight material began to increase in the top reservoir, and then turned around and increased in the bottom. The total polymer concentration, meanwhile, increased steadily in the bottom as usual. All this is much like the situation in organic liquids, in which the generalization that aromatics go to the bottom and aliphatics to the top is generally valid, but there are exceptions, and even occasionally a system that exhibits "inversion."

It is somewhat disheartening to try to apply the results of the studies of c and M to the fractionations observed. In the case of polystyrene in toluene, the cell results suggest that the high molecular weight material should accumulate at the top. Langhammer, the second item in Table I, did find this, but Debye and Bueche, the first item, found the opposite.

TABLE I

Summary of Systems Studied

Polymer	Solvent	$10^{-5} M$	c (%)	High M in:	Distribution obtained	References
Polystyrene	Toluene	4.5	0.5	Bottom	—	Debye and Bueche [4]
Polystyrene	Toluene	0.8–6	1	Top	Good	Langhammer et al. [1]; Langhammer [12]
Polyvinyl pyrrolidone	Water	1–6	0.5–2	Bottom	—	Langhammer and Quitzsch [7]
Polyvinyl pyrrolidone	Water	0.3	—	No fractionation	—	Langhammer et al. [1]
Polyvinyl pyrrolidone	Water	0.3–6	—	Bottom	Poor	Langhammer et al. [1]
Polyvinyl alcohol	Water	—	0.4	Bottom	—	Langhammer [6]
Polacrylic acid	Water	—	0.4	Bottom	—	Langhammer [6]
Polyvinyl chloride	Cyclohexanone	—	0.8–5	Bottom	—	Guzmán and Fatou [8]
Polyacrylonitrile	Dimethyl formamide	—	0.4	Top	—	Langhammer [22]
Triacetyl cellulose	Ethyl chloride	—	0.4	Top	—	Langhammer [22]
Polymethyl methacrylate	Toluene and acetone	—	0.6	No fractionation	—	Fritzemeier and Hermans [5]
Polymethyl methacrylate	Benzene	—	1	Bottom	Good	Kössler and Krejsa [2]
Polybutyl methacrylate	Benzene	—	1	First top, then bottom	—	Kössler and Krejsa [21,23]
Polychloroprene	Benzene	—	0.9	Bottom	Poor	Krejsa [24]; Kössler and Stolka [25]

The result of Debye and Bueche on the study of the variables is consistent with their result, showing that at the concentration at which they performed the fractionation, 0.5%, the high molecular weight material had the higher D'. On the other hand, the study of the variables by Langhammer suggests that in all concentration ranges there will be a confusion of molecular weights in the column, whereas the complete fractionation they performed showed no such scrambling of molecular weights.

In the case of polyvinyl pyrrolidone, Langhammer et al., in a study of the variables, found a good linear relation between s and M which should indicate ease of fractionation at any M, but whole polymer of $M = 30,000$ did not fractionate, while high molecular weights did.

V. Methods of Fractionation

In addition to a simple single column, which yields two solutions, several other kinds of arrangements of apparatus have been used.

A. Cascade of Thermal Diffusion Columns

Langhammer et al. [1,7,12,20] first used a cascade of columns to get a larger number of solutions with varying M. By using three stages, as diagrammed below, one gets eight fractions.

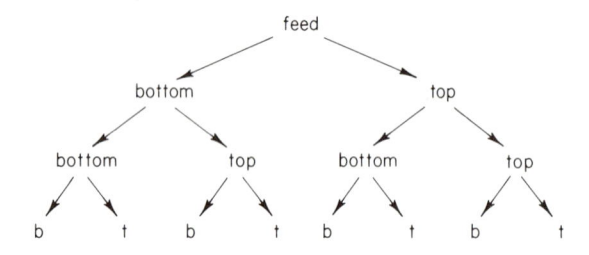

```
                          feed
                      ↙         ↘
                bottom             top
                ↙   ↘            ↙   ↘
          bottom    top     bottom    top
          ↙  ↘    ↙  ↘     ↙  ↘    ↙  ↘
          b   t   b   t    b   t   b   t
```

In the case of polyvinyl pyrrolidone, the integral distribution curve so obtained was much steeper than the curve produced from fractional precipitation; the thermal diffusion did not produce enough high or low molecular weights. In the case of polystyrene in toluene, however, the distribution curve obtained from three stages was in excellent agreement with that from fractional precipitation, and the curve obtained from two stages was a good approximation to this.

Guzmán and Fatou [8] used a cascade of columns with polyvinyl chloride in cyclohexanone. In their case, the content of the bottom reservoir of column 1 became the feed to column 2, the bottom of column 2 became the feed to column 3, and so on for eight stages in which only the bottom product was passed on.

B. Multiple Reservoirs

Kössler and Krejsa [23,24] devised a parallel-plate column with five reservoirs evenly spaced along the height. They obtained a reasonable distribution of M in the reservoirs, but they encountered a problem with this kind of apparatus. That is, while it is true that M is distributed in the reservoirs, so is c, and the quantity of polymer in the top reservoir is very small. To get a good M distribution curve, however, one wants roughly equal increments of polymer quantity. They partially alleviated this problem by reducing the size of the reservoirs in the bottom parts, and by diminishing the temperature difference at the bottom of the column.

C. Solvent Flow

Langhammer [26] used a suggestion of Korsching, in which the top reservoir is eliminated, solvent is fed slowly to the bottom reservoir, and the effluent from the top of the column yields a continuous distribution of M. Kössler and Krejsa [2] used this method to fractionate polymethyl methacrylate in benzene, stopping at a certain point and getting the remaining points by drawing a number of samples from various points in the column itself. Their distribution was in excellent agreement with that from fractional precipitation. This method appears to have great promise for obtaining M distributions, as it seems to do as good a job as a cascade, but is easier to manipulate. On the other hand, the effluent from a cascade can produce a larger quantity of polymer fractions for other work.

In conclusion, it may be noted that very promising results have been obtained in fractionating high polymers in solution by thermal diffusion, especially by cascades of columns and by solvent flow, but so far the theory does a poor job of interpreting the results.

REFERENCES

1. G. Langhammer, H. Pfennig, and K. Quitzsch, *Z. Electrochem.* **62**, 458 (1958).
2. I. Kössler and J. Krejsa, *J. Polymer Sci.* **57**, 509 (1962).
3. N. Gralén and T. Svedberg, *Naturwissenschaften* **29**, 270 (1941).
4. P. Debye and A. M. Bueche, *in* "High Polymer Physics" (H. A. Robinson, ed.), p. 497. Chem. Pub. Co., New York, 1948.
5. H. Fritzemeier and J. J. Hermans, *Bull. Soc. Chim. Belges* **57**, 136 (1948).
6. G. Langhammer, *Naturwissenschaften* **41**, 552 (1954).
7. G. Langhammer and K. Quitzsch, *Makromol. Chem.* **17**, 74 (1955).

8. G. M. Guzmán and J. M. Fatou, *Anales Real. Soc. Espan. Fis. Quim. (Madrid)* **B54**, 609 (1958).
9. S. R. de Groot, *Physica* **9**, 801 (1942).
10. J. D. Hoffman and B. H. Zimm, *J. Polymer Sci.* **15**, 405 (1955).
11. D. T. Hoffman, Jr. and A. H. Emery, Jr., *A.I.Ch.E.J.* **9**, 653 (1963).
12. G. Langhammer, *Svensk Kem. Tidskr.* **69**, 328 (1957).
13. A. H. Emery, Jr. and H. G. Drickamer, *J. Chem. Phys.* **23**, 2252 (1955).
14. G. Meyerhoff, *Z. Physik. Chem. (Frankfurt)* [N.S.] **4**, 334 (1955).
15. G. Meyerhoff, H. Lütje, and B. Rauch, *Makromol. Chem.* **44**, 489 (1961).
16. F. C. Whitmore, *J. Appl. Phys.* **31**, 1858 (1960).
17. C. L. Herren and J. S. Ham, *J. Chem. Phys.* **35**, 1479 (1961).
18. G. Meyerhoff and K. Nachtigall, *J. Polymer Sci.* **57**, 227 (1962).
19. B. Rauch and G. Meyerhoff, *J. Phys. Chem.* **67**, 946 (1963).
20. G. Langhammer, *J. Polymer. Sci.* **29**, 505 (1959).
21. I. Kössler and J. Krejsa, *J. Polymer Sci.* **29**, 69 (1958).
22. G. Langhammer, *Makromol. Chem.* **21**, 74 (1956).
23. I. Kössler and J. Krejsa, *J. Polymer Sci.* **35**, 308 (1959).
24. J. Krejsa, *Makromol. Chem.* **33**, 244 (1959).
25. I. Kössler and M. Stolka, *J. Polymer Sci.* **44**, 213 (1960).
26. G. Langhammer, *Z. Electrochem.* **65**, 706 (1961).

CHAPTER C.1

Turbidimetric Titration

Hanswalter Giesekus
INGENIEUR-ABTEILUNG ANGEWANDTE PHYSIK, FARBENFABRIKEN BAYER AG,
LEVERKUSEN, GERMANY

I. Introduction...191
II. Outline of the Method and Range of Application......................193
   A. Principle..193
   B. Efficiency and Limitation....................................193
III. Elaboration of the Method.......................................195
   A. Solvent–Precipitant System..................................195
   B. Polymer Concentration......................................197
   C. Rate of Addition of the Precipitant............................197
   D. Rate of Stirring..198
   E. Operating Temperature......................................198
IV. Apparatus...199
   A. Experimental Task..199
   B. Design of the Measuring Equipment..........................200
   C. Examples of Turbidimetric Titration Instruments................206
V. Evaluation...211
   A. The Time Dependence of Precipitation and Solution Titration......211
   B. Concentration of the Entire Polymer...........................213
   C. Concentration of the Precipitated Polymer......................214
   D. Estimation of the Molecular Weight Distribution................218
VI. Application..228
   A. Various Classes of Polymers Studied..........................228
   B. Application to Special Polymers...............................230
   References...246

I. Introduction

Preparative techniques of fractionation have the essential advantage of yielding substances that can be analyzed and, under certain circumstances, even further processed into test samples. However, a considerable disadvantage is that such techniques are normally very laborious and time consuming. As a rule, they can only be used for the clarification of fundamental questions and are unsuitable for investigating a larger number of samples (particularly for production control). Also, they require a relatively large quantity of material. Although this is of no importance for most technical products, it makes the application of the technique impossible for many physiological problems where sometimes only fractions of milligrams are available. In most cases, testing of the homogeneity of preparative fractions is equally impossible. Over a period of years, attempts have been made to find *analytical* methods which will allow statements on the molecular inhomogeneity of polymers to be made in a shorter time and with less material. It has thus been tried directly to evaluate gravimetrically or volumetrically the successive precipitation of such substances from a solution due to continuous addition of precipitant, see, e.g., Schulz [1]. However, the most successful technique to date has been turbidimetric titration in which the quantity of the precipitated polymer is measured optically by the turbidity produced.

The turbidimetric titration technique (turbidity titration; precipitation turbidimetry; German: Trübungstitration; French: titration turbidimetrique) as a method for determining the molecular weight distribution of high polymers was introduced by Morey and Tamblyn [2] in 1945. As a method for qualitative characterization of inhomogeneity, it had already been applied some years earlier by McNally [3], as well as Adams and Powers [4]. The method was considerably improved by Desreux *et al.* [5–8], both with respect to more favorable experimental conditions as well as with respect to the equipment needed. Wallenius *et al.* [9,10] recognized the importance of the method for investigating physiological materials and developed it into a microtechnique. Hengstenberg [11,12] introduced it as a method for controlling industrial operations. Melville *et al.* [13,14] described its suitability for the characterization of copolymers. Claesson [15] and Scholtan [16] contributed to the improvement and conceptual clarification of the method of evaluation. The process of particle formation and growth was first experimentally analyzed by Hastings *et al.* [17] and Allen *et al.* [18], but their conclusions were questioned by recent investigations of Ryabova *et al.* [18a], and Beattie [18b]. Giesekus [19] developed an automatic photometer which represented the prototype of a first commercial turbidimetric titrimeter [20]. More recently, Stearne and Urwin [21], Voronov and Volkova [21a], and Cantow [22] have introduced improvements of the equipment. The last-mentioned author for the first time applied turbidimetric solution titration systematically and found certain essential advantages for this modified technique. The temperature variation (used previously by Giesekus [23] instead of a precipitant for producing the turbidity) was applied by Taylor and Tung [24] to polyolefins. Gamble *et al.* [25], as well as Harrison and Peaker [25a] further developed this technique. Up to the present time, the turbidimetric titration is known to have been applied to more than 50 different homo- and copolymers; the actual number is certainly considerably larger.

Although the method is so widely applied, it has not yet found any detailed description in a monograph. Up to now, it has only been briefly dealt with in some review articles [1,12,26–28a].

II. Outline of the Method and Range of Application

A. Principle

The turbidimetric titration method is closely related to the method of fractional precipitation. In both methods, a nonsolvent (precipitant) is slowly added to a dilute polymer solution. The polymer, beginning with the highest molecular portions, is precipitated selectively. The time-consuming separation, working up, and analyzing of the fractions is here, in the words of Morey and Tamblyn, replaced by "optical weighing." The turbidity of the liquid produced by the precipitation is measured. In this manner, a conclusion is drawn as to the quantity of the substance precipitated at the precipitant concentration concerned. The precipitant concentration itself supplies a statement on the molecular weight of the respective portion.

A variant of the turbidimetric titration method which, in accordance with the procedure described above, might be termed more exactly a turbidimetric *precipitation* titration, is the turbidimetric *solution* titration. It is a counterpart to the method of fractional solution. A turbidity, which has been produced beforehand, is slowly dissolved by gradual addition of solvent, and the decrease of turbidity is measured. Although, in principle, both procedures are closely related, considerable differences may exist in practice owing to different speeds of equilibration and to the variation of the polymer concentration. These differences often result in making the one of these procedures more favorable than the other.

A further variant, to which the term *titration* no longer applies in the strictest sense of the word, is the production of or the redissolution of a turbidity by temperature variation. This method is the only alternative if no suitable solvent–precipitant system can be found. Unfortunately, in many cases, it requires covering a rather wide temperature range.

B. Efficiency and Limitation

The turbidimetric titration method primarily delivers information on the distribution of solubilities and is only a *relative* method for the determination of the molecular weight distribution. Further stipulations are a chemically uniform and unbranched polymeric material and calibration by means of preparative fractions of known molecular weight. Even if sufficiently narrow fractions are available, transformation of the turbidity curve into the molecular weight distribution curve is dependent on a number of conditions which sometimes are realized well and, in many other cases, poorly. This last fact seems to be the primary reason for the efficiency of the method being very differently assessed by the various authors.

In most practical cases, it is necessary to have a rapid, reproducible method which in some way characterizes the inhomogeneity of the samples. It is not necessary for the method to yield absolute results. In the opinion of most authors, this accounts for the primary importance of the turbidimetric titration method. The general opinion is that the simplicity, rapidity, and very good reproducibility of the turbidimetric method make it the technique of choice in a large polymer survey.

This is further confirmed by the increasing application of the method to polymer mixtures and copolymers. A quantitative conclusion as to the molecular weight distribution is very difficult in these cases owing to the chemical inhomogeneity which strongly influences the solubility behavior. A detailed discussion of this matter is presented in Chapter D. By means of turbidimetry, it is possible to obtain certain statements on the quantity ratio of homopolymer and copolymer even under the unfavorable condition of coprecipitation. Changes in the polymerization mechanism are reflected in a specific manner in the turbidity curve. Under certain conditions, the method may even supply information on the ratio of linear to branched polymer.

Whenever possible, a turbidity titration should not be used as the only method. Conclusions should not be drawn only on the basis of turbidity curves unless other standard techniques are applied in addition. A combination of turbidimetric titration and preparative fractionation methods is particularly useful. The results of the latter will not only check the turbidimetric titration, but the turbidimetric titration will also check the homogeneity of preparative fractions.

The above may be summarized by the remark of one of the author's colleagues who characterized the turbidimetric titration as follows: "It acts like a piano and not like a radio!" This means that the method cannot be properly applied by using it as a prescription, but its efficiency can only be really exploited when there is a critical head expertly judging the results. A certain degree of familiarity with the method and quite a lot of experience are necessary. This, of course, only refers to the application of the method to new polymer systems or to essential alterations in the conditions. When the equipment is suitably arranged, routine applications of the method will be very simple and, after a few instructions, may well be left to reliable laboratory assistants.

III. Elaboration of the Method

A. SOLVENT–PRECIPITANT SYSTEMS

Naturally, the same fundamental demand is made on the solvent–precipitant system as in the case of preparative fractionation, i.e., this system must have sufficient selectivity. In contrast to the usual precipitation technique, where coagulation and settling of the gel phase should occur as quickly as possible, turbidity measurement requires the precipitate to remain suspended in the sol phase. With increasing precipitation, new particles should always be formed, instead of the existing particles being enlarged by further aggregation. The particles should be shaped as regularly as possible and not possess too wide a size distribution. The particle mean size and its optical scattering power should not be too strongly dependent upon the molecular weight and the solvent–precipitant ratio. The equilibrium between the sol and the gel phases should be reached rapidly.

It is a general rule that these conditions may only be realized if the polymer concentration is sufficiently low. Unfortunately, no general statement can as yet be made about the conditions a solvent–precipitant system must have to fulfill the above requirements. The formation of stable turbidities is certainly dependent on the electrical charge of the particle surfaces, which may help or hinder further aggregation.

This assumption is supported by the well-known fact that the presence of an electrolyte, even in low concentration, decreases or completely eliminates the stability of the turbidity in most cases. There are, indeed, exceptions where such additions have a stabilizing effect. This last effect can be easily accounted for if the polymer particles, owing to secondary valence forces, can bind such ions and thus themselves become a kind of polyion. An example of this is offered by the results of Hoffmann [29]. He found that in the case of various polymers which are precipitated from methylene chloride solution, stable turbidities are formed only if HCl gas has been added beforehand to the solvent and/or precipitant.

We may illustrate the stabilizing effect of polyion formation with the case of polyamide in formic acid or *m*-cresol. In spite of all endeavors, it has, up to now, been impossible to obtain really stable turbidities of polyamide in *m*-cresol solution. However, Giesekus [30] was able to precipitate polyamide from formic acid solution as an extremely fine and perfectly stable precipitate using di-*n*-butyl ether. The particle size at the end of the precipitation was only about 20% larger than at the beginning. It is evident that polyamide forms polyions in formic acid but not in *m*-cresol from the concentration dependence of the viscosity.

Stabilization of a precipitate by addition of an emulsifier has been tried often. However, this approach has been effective in only a few cases, e.g., by

Taylor and Tung [24] and by Rabel and Ueberreiter [30a]. The addition of fine solid particles for modifying the process of nucleation was also ineffective.

Hengstenberg [12] has given the general rule: The combination of a poor solvent and weak precipitant yield the most stable turbidities. Rabel and Ueberreiter [30a] state, that in such combinations the speed of equilibration is maximized as well. As is the case for any rule, a number of exceptions are known.

In spite of certain recommendations of the type mentioned above, the best experimental method will generally be to look for a suitable pair of miscible liquids without following any specific rule. In many cases, better results will be obtained by using solvents or precipitants which themselves are compound systems. It is often easier to ensure that the refractive indexes of solvents and precipitants are approximately equal, but sufficiently different from that of the polymer if mixed systems are employed. While the latter of these conditions for obtaining a turbidity must necessarily be fulfilled, compliance with the first condition is at least desirable. The turbidity curves will acquire a simpler shape and can be interpreted and evaluated more easily if solvent and precipitant have nearly equal indexes of refraction.

Looking for a suitable solvent–precipitant system may be rather troublesome in individual cases. It is often possible to simply take over a system that has been used successfully for other polymers. Hoffmann [29], who investigated a relatively large number of diverse polymers, states that finding a system suitable for a new substance requires on an average about 100 preliminary tests. These tests may take about 2 weeks. His procedure is as follows:

"Dissolve a few milligrams of polymer in 20 ml of solvent. Place the solution in a beaker, equilibrate, and then add the precipitant slowly with a burette while stirring. After the beginning of the turbidity, interrupt stirring and the addition of the precipitant about once a minute, and observe if the precipitate flocculates. If the turbidity is stable for more than 15 minutes, tentative experiments in the turbidimetric titration apparatus may be carried out varying the addition and agitation speeds as well as the temperature and the concentration."

It is important only to use purified liquids for such preliminary experiments. Otherwise, there may be a danger of incorrectly rejecting a system which as such might be suitable. Concerning the pretreatment of the substances used, Hoffmann states: "It is advisable to dry the solvents and precipitants and then distill them. The substance to be measured should generally be free from inorganic salts, bases, and acids. These may be removed by reprecipitation or washing." In many cases, it may not be necessary to carry out such an operation; but it is safer to make sure first, instead of assuming from the beginning that such procedures will not be required. If analytical-grade reagents are used, pretreatment will, as a rule, be superfluous. When water is used as the solvent or precipitant, it is particularly important for the water to be free from any electrolytes. It is also advisable to deaerate the water, for oxygen and carbon dioxide, respectively, may influence the state of precipitate markedly. Under certain circumstances, specifically if one has to solve or titrate at higher temperatures, a nitrogen-protecting atmosphere may be necessary.

To get reproducible results it is also important that the solution has reached a state of equilibrium. Time of equilibration depends on a number of factors and varies over wide limits. As a rule, between dissolution and execution of the titration experiment a time interval of about 1–2 days should elapse. On the other hand, the solution should not be too old, for in many cases slight degradation occurs. This is particularly true if the solution has not been kept in the dark.

B. Polymer Concentration

As has been mentioned earlier, the polymer concentration should be low. Otherwise coagulation and flocculation may occur. Should these effects not be problems, complex precipitation-kinetic effects will often occur if the concentration is not sufficiently low. Such kinetic effects may simulate a complicated distribution of the molecular weight. Morey *et al.* [31] were the first to mention this. The separating effect will become better with decreasing concentration, but at the same time the variation of the precipitation with the concentration will become more and more pronounced. A lower limit of concentration will, in most cases, be given by the sensitivity of the measuring apparatus. Depending on the average molecular weight and the solvent–precipitant system selected, the most favorable initial concentrations will range from about 0.5 to 50 mg/100 ml in most cases.

C. Rate of Addition of the Precipitant

The slow rate of addition of the precipitant should be such that the distribution of the polymer between the sol and the gel phase at any time is not too far from equilibrium. The rate of addition should be fast enough that a slight instability of the turbidity has no influence. For practical reasons the test should not take too much time. If an initial volume of 50 ml of solution is taken as a basis, the range of the rates of addition to be applied lies between 0.05 and 2.0 ml/minute. Towards the end of the precipitation, the rate may be greater than at the beginning. It is advisable to accelerate the rate of addition in accordance with a fixed program to save time. In turbidimetric solution titration, preceded by precipitation, dissolution may take place at a faster rate than precipitation. Rates which are too fast will cause distortions in the turbidity curves, because of the greatly delayed establishment of the equilibrium, and of the unavoidable local overconcentration which causes coprecipitation. An important factor is the constancy or the precise reproducibility of the rate of addition of the precipitant. The size of the particles generated is dependent thereon. The rate of addition of the solute is not quite so critical.

Although, in general, continuous addition of a precipitant is more advisable, it is often carried out stepwise. This is recommended when the time for the establishment of the equilibrium is very long. This procedure is also advantageous in the case of an unstable turbidity. In this case the stirring time required after each addition will be short. If instability is more considerable, a preferable procedure is to prepare quite a number of solution batches and to add to each of them a different quantity of precipitant. The maximum turbidity, or the turbidity at a specified time after the beginning of the precipitation, is determined for each sample as a function of the concentration of the precipitant in each individual case. In the case of such test procedure, it will be important to standardize the precipitation conditions. For details of this technique see Beattie [18b].

D. Rate of Stirring

It is well known that rapid stirring increases the instability of the turbidity, so that the lowest possible rate of stirring should be chosen. This is particularly true if the turbidities are sensitive. It is important that the stirring rate is high enough to uniformly distribute the precipitant as quickly as possible and to prevent local overconcentration. Under no circumstances should air bubbles be stirred into the solution or produced by cavitation.

Both these conditions can be largely complied with by suitable agitator shapes (ring- or screw-type agitators). Time constancy and reproducibility are of major importance.

Under certain circumstances, it is an important advantage to produce the turbidity by temperature variation. In this case agitation can be dispensed with, provided that other means are applied to ensure the rapid equilibration of temperature.

E. Operating Temperature

When selecting the operating temperature, the boiling points of solvents and precipitants should be taken into account as a factor of primary importance; i.e., there should be no appreciable evaporation of one of the components during the experiment. On the other hand, the temperature should not be so low as to hinder the establishment of equilibrium. If precipitates are crystalline at room temperature or if they tend to postcrystallize (aging), it is an absolute must to choose a higher temperature. Otherwise, the fractionating effect will considerably deteriorate.

It will be necessary only in exceptional cases to resort to temperatures below room temperature. The customary experimental arrangements may then give rise to complications, e.g., moisture condensation on the optical system. There are a number of high polymers for which suitable solvent–precipitant systems are available only at elevated temperature. In particular, the polyolefins require temperatures up to about 150°C.

If the operating temperature is in the neighborhood of the room temperature and if the rate of precipitation or dissolution is low, thermostating of the liquid added can be dispensed with. At elevated temperatures thermostating is necessary to prevent chilling of the precipitant and solvent, which otherwise would cause coprecipitation of lower or higher molecular weight material. To minimize this disturbing effect, the added liquid should be maintained 5°–10°C higher than the liquid contained in the cell.

When producing turbidities by temperature variation, the cooling speeds used range from 10°–180°C/hour.

IV. Apparatus

A. EXPERIMENTAL TASK

In order to apply the turbidimetric titration method in the usual form of precipitation or solution titration, it is necessary to use equipment which will allow the turbidity of a liquid to be measured precisely over several hours. During this time the liquid is being mixed stepwise or continuously with a second liquid. Simultaneously, the temperature must be kept constant. When turbidity is produced by temperature variation, addition of a second liquid does not occur. Instead of this, the temperature of the measuring liquid must now be varied in accordance with a given program and must be measured together with the turbidity. The turbidity can be characterized both by measuring the extinction (optical density) as well as by measuring the scattered light at one or several angles. In order to obtain reliable and comparable results, it is of decisive importance to standardize all experimental conditions and to carry out the experimental procedure in a precisely reproducible manner. Automation is the most logical solution to the operational requirements.

B. Design of the Measuring Equipment

The measuring equipment for turbidimetric titration will normally contain the following units:

1. Light source and voltage stabilization unit
2. Optical equipment
3. Test cell with thermostating and agitating units
4. Injection device
5. Photocell with amplification, reading, or recording units.

The experimental task allows a variety of solutions to the problem. Thus, some authors use Beckmann [16], Pulfrich [32,33], or Brice-Phoenix photometers [34–36] or other equipment not specially designed for turbidimetric titration [10,24,37–40a], which has been modified for this purpose and equipped with additional units. Following Harris and Miller [41], many authors have used the Hilger-Spekker Photoelectric Absorptiometer [26,42–44] shown schematically in Fig. 1.

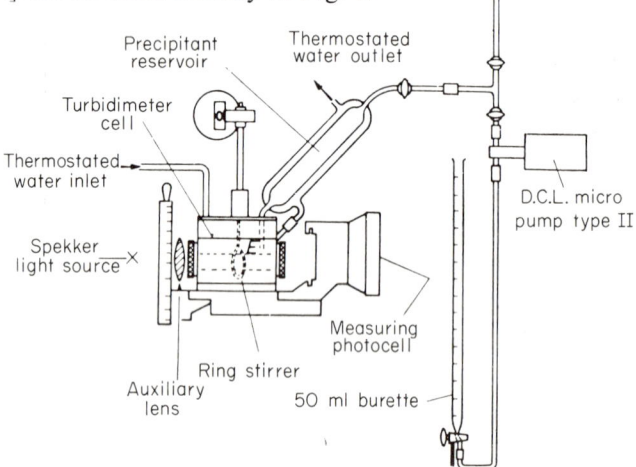

FIG. 1. Hilger-Spekker photometer adapted for turbidimetric titrations, according to Hall [26].

This apparatus measures the optical density by a comparison method using two different barrier layer photocells. As no stabilization or amplification units are required, the arrangement is very simple. Balancing is carried out manually so that it is neither possible to work continuously nor, in particular, automatically. In addition, owing to the use of two different barrier-layer photocells, the measuring accuracy is rather limited. The "Visomat-Spezialkolorimeter," built at the suggestion of Gordijenko et al. [45], is designed according to a similar principle but is equipped with vacuum phototubes.

Owing to the inconveniences and imperfections connected with the above-mentioned photometers, they should only be used for makeshift or less exacting measurements. For more rationalized and more precise operation, equipment specially designed for turbidimetric titration should be used. Naturally it will be advantageous to include commercial units, e.g., monochromators, dosimeters, stabilizers, amplifiers and recorders, whenever such are available. In particular, some units that have recently been designed for continuously operating industrial photometers, seem to be well suited for this purpose.

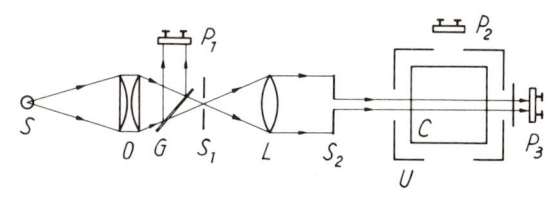

FIG. 2. Optical arrangement of a single-beam turbidimeter, according to Melville and Stead [14]. S, high-pressure mercury vapor lamp; O, condenser; L, lens; S_1 and S_2, apertures; G, light divider plate; C, measuring cell; P_1–P_3, photocells; U, thermostat.

Such turbidimetric titration photometers may either be designed using the balancing (double-beam) technique or the direct (single-beam) technique. It is possible to combine these two techniques in a way to be explained in greater detail later on. A typical single-beam arrangement—according to Melville and Stead [14]—is shown in Fig. 2; the photocell P_1 serves as a control device for keeping the light intensity constant. Figure 3 shows the diagram of a recording apparatus using the double-beam technique—according to Howard [46]. As contrasted to a setup functioning according to the principle of the Hilger-Spekker, this arrangement has, in addition to a balancing optical wedge servo mechanism, the fundamental advantage of using only one photocell. This is rendered possible by the application of the alternating light technique and allows a considerable increase in accuracy. A combination of the two techniques in the sense indicated above is realized, e.g., if the arrangement shown in Fig. 2 is modified so that the output of the reference photocell P_1 is fed directly into the measuring photocell amplifier. In this way, the output reduces the reading to constant intensity of the incident beam.

Details of all turbidimetric titrimeters so far designed [2,5–7,14,19–22,25,46,47] cannot be discussed here. However a brief survey will be given on the basis of our experience with various methods of optimizing equipment design.

1. Light Source and Voltage Stabilization

If the scattered light is to be measured at different angles in order to calculate the particle size, the use of monochromatic light is required. This is usually produced by a high-pressure mercury vapor lamp followed by a combination of filters. The use of water-cooled lamps is advisable.

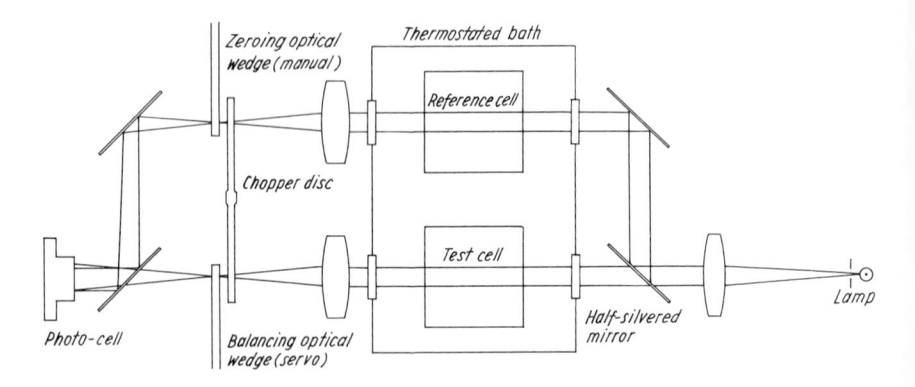

FIG. 3. Design of a double-beam, recording turbidity titration instrument, by Howard [46].

For measuring the optical density and/or the light scattered at 90°C, the use of white light is sufficient. For producing the white light, a low-voltage tungsten point light source, such as a microscopic lamp, is suitable.

The control of the light intensity depends on the light source used and the photometric technique. For the double-beam technique, no stabilization is required theoretically, especially if monochromatic light is used. In practice some stabilization is needed—compensation systems never work with perfect precision and without time delay.

On the other hand, the accuracy of the single-beam method is entirely dependent on the quality of light stabilization. For metal vapor lamps, current stabilization is not sufficient. The arcs are often subject to hot spotting, which must be corrected by a photocell monitor. A combined technique of current control and photocell monitoring is more satisfactory. When low-voltage tungsten lamps are used, it is possible to control the intensity of the light over prolonged periods of time with a precision greater than 1%, using two-stage current stabilization. For instance, Giesekus has used a magnetic stabilizer as the first stage and an iron-hydrogen ballast tube as the second stage in a current controller and obtained the current stability cited above [19]. Instead of the ballast tube, a simple transistor circuit controlled by a Zener diode may be applied [48]. It is also possible to use a reference photocell as the control device.

2. Optical Equipment

The demands made on the optical system are normally low. The only important limitation on the location of the diaphragms is that they should not be mounted in places where uncontrolled changes in temperature can occur. Thermal expansion may cause the diameters of the light beams to change during the experiment. This results in false readings.

When scattered light measurements are made, care should be taken to avoid stray light in the instrument. Reflection from cell surfaces must be avoided. Simple intermediate imaging of the incident and the scattered light prevents interference from stray light [22]. When multiangle measurements are used, the arrangement of the exit slits may be somewhat critical. This is particularly the case where forward-scattered light is being measured. Direct light may easily be scattered by means of edges, etc., into the beam being measured.

3. Test Cell with Thermostating and Stirring Arrangements

Although in a number of turbidimeters cylindrical test cells are used, it is certainly advantageous to make the surfaces through which the light enters and leaves the cell from plane optical glass or quartz. When measuring the optical density or light scattered at 90 degrees, a cell with a square or rectangular cross section is useful. Scattering measurements at several angles require more complicated cells, e.g., with octagonal cross section. Open cells should be designed so that the volume of the liquid can be increased at least fourfold.

Dilution of the solution to any desired extent without increasing the volume is possible by using a constant volume cell with overflow system as described by Desreux *et al.* [6,7]. The overflow design has the additional advantage of producing a more uniform concentration change when applied to constant speed titrations.

In a few cases, attempts have been made to use internal thermostating by controlled heating of the solution itself [6,47]. External thermostating is more satisfactory. External thermostating is usually carried out by means of a double-walled thermostating vessel provided with windows. The thermostat liquid flows through the double walls. Thermostating is also achieved by means of an open bath. Particularly elegant, although requiring more intricate technical means, is a system in which the test cell itself has two walls. The jacket thus formed can be thermostated directly [21]. In this case, these cells should be made from metal provided with glass windows. Care should be taken in the surface finishing of the metal so that no trace of metal compound can enter the test liquid. Metallic compounds will alter the precipitation conditions entirely.

For temperature precipitation, in which the temperature should be varied within wide limits (20°–180°C) in accordance with a given program, a metal cell may be used, thermostated by means of an electrically heated copper block provided with supply lines for the cooling liquid [25].

Flat magnetic stirrers are used for overflow constant volume cells. In the case of open cells, the contents of which steadily increase, it is advisable to use a screw-type agitator [19,20,47]. This kind of agitator ensures uniform mixing of the entire contents at low speed. Impeller-type stirrers are not well suited since, to have the same mixing effect, they require a higher local shear gradient. This high shear may coagulate sensitive precipitates. Stirring by hand is not advisable even if the precipitant is added stepwise. Air-driven agitators should be avoided. The safest driving means are synchronous motors equipped with suitable reduction gears. Speeds of agitation of 100–500 rpm are useful with screw-type agitators of about 1-cm diameter and the usual cell volumes.

Most cells are designed to require a minimum volume of 25–75 ml. A remarkable exception is the apparatus according to Wallenius et al. [9,10], designed for investigating physiological substances, which requires only 2 ml of solution.

4. Injection Device

Finely graduated burettes are often used when the precipitant is added stepwise. As various authors have reported, the intensity of the turbidity depends on the rate at which the precipitant is added. Therefore, equipment that allows the injection of the precipitant under precisely reproducible conditions should be given preference. In most cases, a syringe (capacity: 100–200 ml) is used for this purpose. The plunger is moved uniformly by means of a motor. Frequently, a variable speed drive is used; and the rate of addition is coupled with the speed of the recorder. Reproducibility is best when a synchronous motor is used in connection with a speed changing gear transmission [19,20]. An essential advantage of continuously variable speed control is the ability to alter at will the rate of precipitation in the course of an experiment by changing of the operating voltage or of the frequency of the driving motor [22]. By means of a programming control device it is possible to execute the stepwise addition of the precipitant in a fully automatic fashion [21a].

Following the example of Bischoff and Desreux [6], the syringe is often charged not directly with the precipitant, but with mercury. The precipitant is displaced by the mercury. This use of an intermediate liquid, which is inconvenient particularly when changing the precipitant, may be dispensed with by fitting the plunger so tightly that there is no need to use grease.

Teflon gaskets have been successfully used [19,20]. Preheating of the precipitant can be affected by thermostating either the entire liquid reservoir [14] or the injection line [6], (see Fig. 1).

It is important to introduce the precipitant into the test liquid as close to the stirrer as possible to ensure immediate mixing. For constant volume cells with an overflow system good mixing is mandatory for the calculation of the concentration change. When open cells are used, addition of the precipitant along the wall rather than in the center of the cell is advisable.

In turbidimetric solution titration, all remarks made here concerning the addition of the precipitant apply without any change to the addition of the solute.

5. *Photocell with Amplification and Reading or Recording Units*

Barrier-layer-type photocells were used in most of the older turbidimetric titration instruments [2,5–7,47]. It is simple to couple such a detector directly with a reading or recording galvanometer without using an intermediate amplifier. Owing to the well-known temperature dependence and aging properties of such cells, the precision attainable in this way is rather limited. In comparison, vacuum photocells [14,15] show considerably better stability. When the latter are used for direct indication of the light intensity, a stabilization of the cathode voltage will be required. This should be done simultaneously with the anode voltage stabilization of the amplifier tubes by means of a magnetic stabilizer and glow-discharge tubes.

Direct-current amplifiers are notoriously unstable over long operating periods. It is advisable to use alternating light and a tuned amplifier. The influence of surface leakage currents will be eliminated at the same time. Alternating current-operated mercury vapor lamps have a strong periodic intensity variation of double the main frequency. Additional modulation of the light intensity can then be dispensed with if this frequency is chosen as the resonance frequency of the amplifier [21]. However, the selecting of a resonance frequency which is not an integer multiple of the mains frequency has the advantage that any pickup from power supply or room illumination will not be amplified. Light modulated with such a frequency may be produced by means of a rotating chopper operated by a synchronous motor [19,20].

When measuring the scattered intensity of monochromatic light, a higher degree of amplification is normally required than in the case of white light. A photomultiplier should be used for this purpose. Except for voltage stabilization, the photomultiplier arrangement requires no sophisticated electronics. It is possible with photomultipliers to use the output of a reference photocell to cancel out intensity variations of the light source by controlling the dynode voltage. Compensation by means of a potentiometer recorder is also possible, as with other types of detectors [21]. A particular advantage of using photomultipliers is the ability to electronically correct for dilution effects (see Section V,C,3) by varying the dynode voltage [22].

C. EXAMPLES OF TURBIDIMETRIC TITRATION INSTRUMENTS

Two examples of turbidimetric titration photometers will now be mentioned which are typical of two different approaches to the development of such instruments. The first one, the automatic turbidimetric titration apparatus of Stearne and Urwin [21], is a device developed in a university institute. It was designed to meet the most exacting demands of research. The second apparatus, the automatic turbidimetric titration photometer of Giesekus [19], was developed in an industrial research laboratory. Primarily, it is intended to meet the requirements of practical applications and maintain the highest possible degree of accuracy and reproducibility of the results. We are unable to describe the interesting instrument recently developed by Cantow [22] due to lack of published information. It is reported to be capable of automatically carrying out precipitation and solution titrations as well as temperature precipitations. In several respects this instrument mediates between the two lines of development and shows a number of improvements. The improvements were referred to in detail in the preceding section. Published information on the automatic instrument of Voronov and Volkova [21a] is hardly sufficient for a detailed description. The most important advantages of this device are that it operates up to 200°C, and stepwise addition is executed automatically with a variable program. Another characteristic feature is the extensive use of commercial units (all of Russian provenance).

1. *Automatic Apparatus of Stearne and Urwin*

This device, a further development of the instrument of Oth and Desreux [7], is diagrammatically shown in Fig. 4. In common with its predecessor, simultaneous recording of the transmitted and of the scattered light at angles of 45 degrees, 90 degrees, and 135 degrees are possible. The replacement of the barrier-layer photocells by vacuum photocells and a multiplier and the use of alternating light are considerable improvements. Because of the high sensitivity thus achieved, it is possible to use relatively narrow slits, which avoids to a very large extent unwanted scattered radiation.

Voltage and current regulation are extremely good. With a recorder connected to the reference photocell, no variation in output current could be detected when the line voltage varied by 10%.

C.1. Turbidimetric Titration

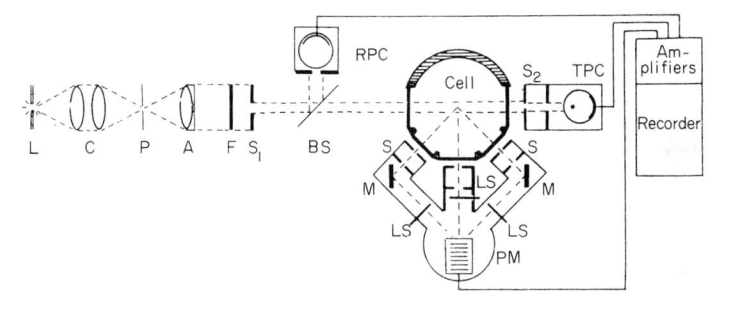

Fig. 4. Design of the turbidimeter by Stearne and Urwin [21]. L, light source (water-cooled mercury vapor lamp, compare ref. [50]); C, condenser; P, pinhole; A, achromat lens; F, Zeiss-Monochromat filter; S, S_1, S_2, slits; BS, beam splitter; RPC, reference photocell; TPC, transmission photocell; M, mirrors for deflecting the scattered light at 45° and 135° to photomultiplier; PM, photomultiplier (11 stages); LS, electrically operated light stops.

The test cell, shown in Fig. 5, is a closed cell with an overflow system. Unlike the original cell of Oth and Desreux, the newer version is made from black-mat, nickel-plated brass with double wall, so that it can be thermostated directly. The top, jacketed as well, is made of glass. The windows consist of optically plane quartz framed in Wood's metal. The effective volume is 55.9 ml. The temperature of the measuring liquid can be thermostated to ±0.02°C.

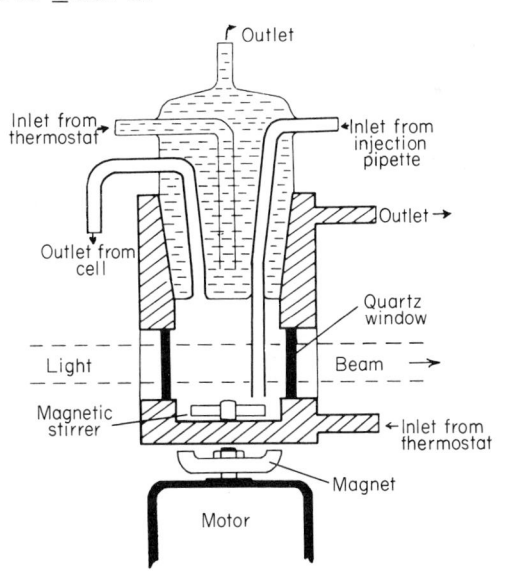

Fig. 5. Constant volume cell fitted with thermostated glass head, according to Stearne and Urwin [21].

The metering device, shown in Fig. 6, is exactly like the mercury displacement burette described by Bischoff and Desreux [6]. The packing rings of the polyethylene plunger are made of rubber. The driving motor is continuously variable and allows the adjustment of metering rates between 0.010 and 0.225 ml/minute. In addition, the number of rotations of the lead screw is recorded. A quantity of 0.651 ml corresponds to one revolution. The amplification and recording system is shown diagrammatically in Fig. 7. Only the alternating current component (100 cps) is amplified. A calibrating system indicates the linear operating range. The use of three identical amplifiers eliminates interferences due to individual amplifier characteristics. The amplifier system contains stabilizing stages for the supply current. Transmitted light and scattered light are recorded in periodic sequence by means of a multipoint recorder.

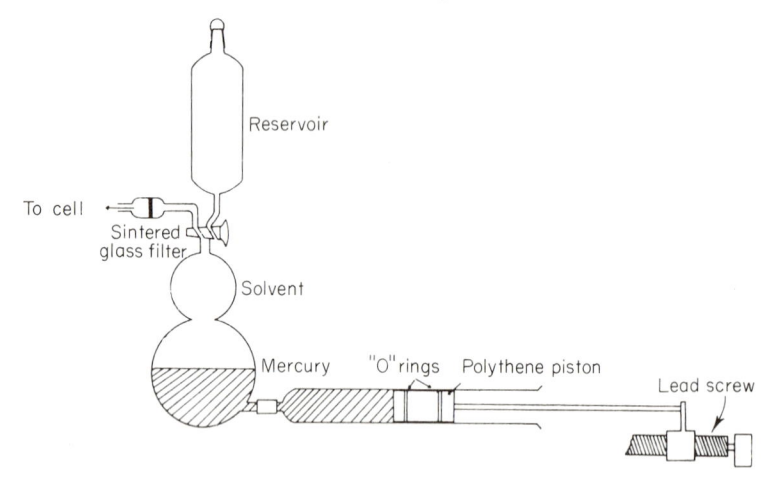

FIG. 6. Injection pipette of Stearne and Urwin [21].

2. *Automatic Instrument According to Giesekus*

This instrument records only the scattered light at 90 degrees to the direction of incidence. White light is used instead of monochromatic light. To avoid electrical interference and facilitate amplification, the photometer is operated with an alternating light. This frequency is a nonintegral multiple, four-thirds, of the mains frequency. Consequently, it can be operated in an illuminated laboratory under moderate cover. Its design is shown diagrammatically in Fig. 8.

The unit uses the single-beam principle, the lamp current being doubly stabilized as described in Section IV,B,1. The open rectangular cell

(40 × 50 × 180 mm) has a capacity of approximately 240 ml. It is kept at constant temperature by means of a double-walled vessel through which flows the thermostat liquid. The sample is agitated by means of a screw-shaped glass stirrer mounted in the cover. It is turned by means of a synchronous motor and a speed reduction gear (speed: 250 rpm and 60 rpm). The metering device consists of a glass cylinder (capacity: 200 ml) with a plunger sealed with Teflon rings. The plunger is driven by means of a synchronous motor and a speed reduction gear which allows the use of four different metering speeds between 0.12 and 0.6 ml/minute. A high-speed gear is used for filling the pump. The precipitant is fed directly into the cylinder and transferred by means of a capillary to the test cell where it runs down along the wall near the stirrer.

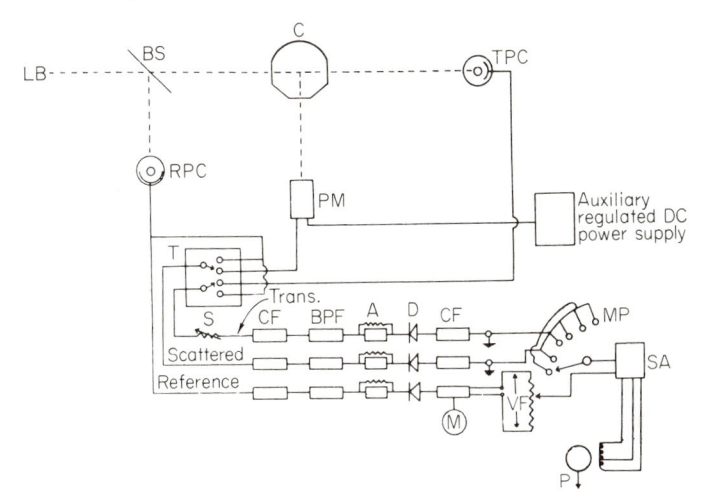

FIG. 7. Amplifier and recording system of the turbidimeter by Stearne and Urwin [21]. LB, light beam; BS, beam splitter; C, light-scattering cell; TPC, transmission photocell; RPC, reference photocell; PM, photomultiplier; T, test mechanism; S, gain switch for the transmission channel; CF, cathode follower; BPF, band pass filter; A, small amplifier circuit; D, rectifiers; M, metering system showing linear operating range of the equipment; VF, voltage from reference photocell fed to slide wire of recorder; MP, multipoint switch in recorder; SA, servoamplifier; P, multipoint recording pen.

The detector is a red-sensitive vacuum tube photocell. The output of the detector is amplified in a three-stage RC-coupled amplifier with a cathode follower final stage. There are five amplification ranges (1:3:10:30:100). The preferred recorder unit is a wide-band point recorder.

The light stability may be checked by interposing a set of mirrors in the light beam. This transfers the light, after reversing and weakening by gray filters, directly to the photocell. A glass turbidity standard inserted instead of the cell is used for calibration. The reading accuracy of this instrument over prolonged periods of time is about $\pm 0.5\%$.

The Agfa turbidity titration apparatus [20] is a direct further development of this instrument. In addition to small changes in the arrangement of the various components, the main difference lies in the design of the metering

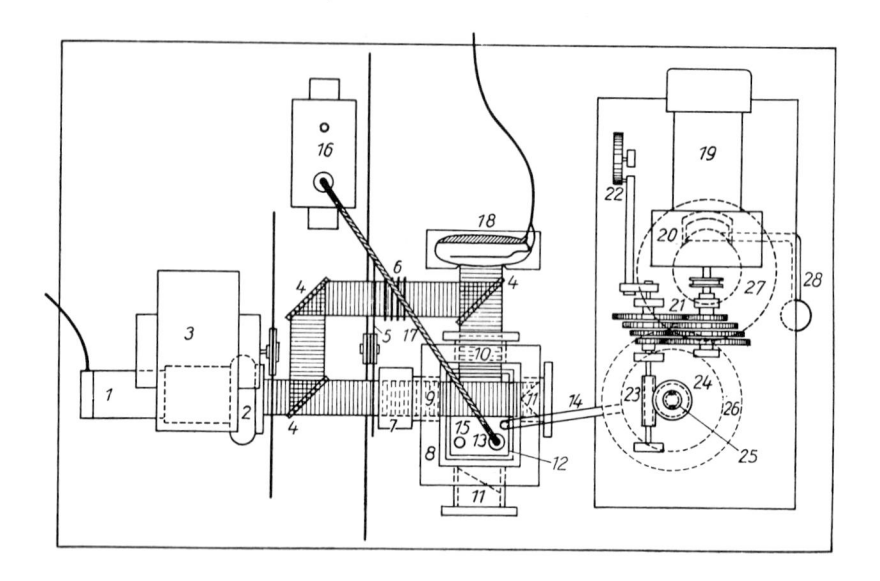

FIG. 8. Design of optical and mechanical systems of the Giesekus turbidimeter [19]. 1, light source (low-voltage incandescent lamp, 6 V, 15 W); 2, condenser with iris diaphragm and filter magazine; 3, synchronous motor with light-chopper; 4, plane-parallel deflecting plates; 5, shutter; 6, gray glass (removable); 7, diaphragm for scattered light; 8, thermostat; 9, entry slit; 10, exit slit; 11, light trap; 12, measuring cell; 13, stirrer; 14, supply line for precipitant; 15, thermometer; 16, synchronous motor for stirrer; 17, flexible shaft; 18, vacuum photocell with case; 19, synchronous motor to drive the metering device; 20, flanged reduction gear; 21, variable gear; 22, control wheel for gear box; 23, worm-gear drive; 24, worm gear; 25, piston spindle with guiding grooves; 26, metering device with heat insulation; 27, support for metering device; 28, lever for raising the metering device.

device. It is made of surface-treated steel with the plunger delivering the liquid in upward direction. The result is a more compact arrangement. The photometer unit of the Agfa turbidimetric titration apparatus (with the guard removed) is shown in Fig. 9.

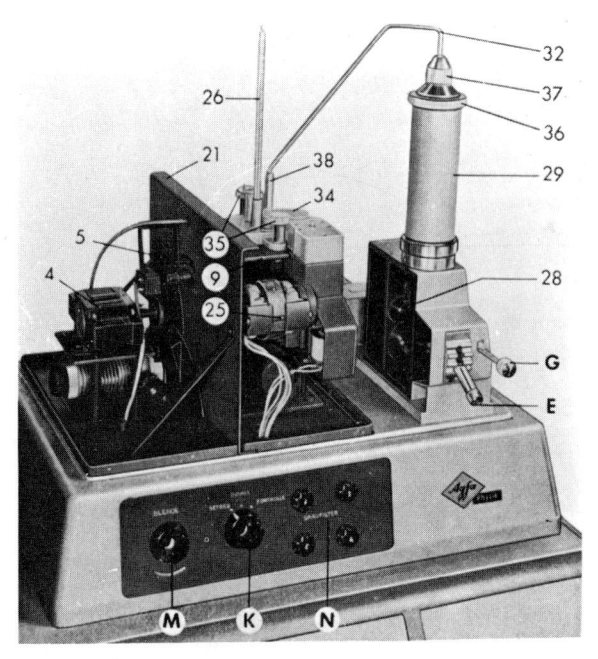

FIG. 9. View of the mechanical and optical parts of the Agfa turbidity titration apparatus [20]. 4, synchronous motor; 5, light chopper; 9, thermostat; 21, diaphragm for scattered light; 25, motor for stirrer; 26, thermometer; 28, variable gear for injection pipette; 29, piston of metering device; 32, supply line for precipitant; 34, cover for measuring cell with built-in stirrer; 35, knurled-head screw; 36, knurled knob; 37, screw cap; 38, Teflon socket, E, clutch lever for high gear; G, coupling rod for setting rate of injection; K, three-point switch (operation–dark current–control); M, adjustment for iris diaphragm; N, adjustment of gray filter in reference beam.

V. Evaluation

A. The Time Dependence of Precipitation and Solution Titration

If the solvent and precipitant volumes present in the test cell at the time t are termed $V^{(S)}(t)$ and $V^{(P)}(t)$ and the respective volumes at the beginning of the test $V_0^{(S)}$ and $V_0^{(P)}$, the precipitant-to-solvent ratio can be determined according to the equation

$$\chi(t) = V^{(P)}(t)/V^{(S)}(t), \qquad \chi_0 = V_0^{(P)}/V_0^{(S)} \tag{1}$$

The precipitant concentration is determined according to the equation

$$\gamma(t) = V^{(P)}(t)/[V^{(S)}(t) + V^{(P)}(t)] = \chi(t)/[1 + \chi(t)]$$
$$\gamma_0 = V_0^{(P)}/[V_0^{(S)} + V_0^{(P)}] = \chi_0/[1 + \chi_0] \tag{2}$$

1. Open Cell

If a precipitant is introduced into an open cell at the constant speed $v^{(P)}$, the following equations apply:

$$V^{(S)}(t) = V_0^{(S)} \qquad V^{(P)}(t) = V_0^{(P)} + v^{(P)}t \qquad (3)$$

$$\chi(t) = \chi_0 + u^{(P)}t \qquad u^{(P)} = v^{(P)}/V_0^{(S)} \qquad (4)$$

$$\gamma(t) = 1 - \frac{1-\gamma_0}{1+(1-\gamma_0)u^{(P)}t} \qquad (5)$$

If the precipitation is terminated at the time t_{max} and subsequent solution titration is carried out at the constant speed $v^{(S)}$, then, at the time t', calculated from the beginning of the dissolution process, the following conditions prevail:

$$V^{(S)}(t') = V_{max}^{(S)} + v^{(S)}t' \qquad V^{(P)}(t') = V_{max}^{(P)} \qquad (6)$$

$$1/\chi(t') = (1/\chi_{max}) + u^{(S)}t' \qquad u^{(S)} = v^{(S)}/V_{max}^{(P)} \qquad (7)$$

$$\gamma(t') = \gamma_{max}/(1 + \gamma_{max}u^{(S)}t') \qquad (8)$$

In the case of program-controlled precipitation or dissolution, in which $v^{(P)}$ and $v^{(S)}$ are functions of time, it is necessary to replace in Eqs. (3) to (8)

$$v^{(P)}t \quad \text{by} \quad \int_0^t v^{(P)}\,dt$$

$$v^{(S)}t' \quad \text{by} \quad \int_0^{t'} v^{(S)}\,dt' \quad \text{etc.}$$

2. Closed Cell with Overflow System

If a precipitant is introduced into a closed cell of the volume V_c at the speed $v^{(P)}$, the following equations apply:

$$V^{(S)}(t) + V^{(P)}(t) = V_0^{(S)} + V_0^{(P)}$$
$$= V_0^{(S)}(1+\chi_0)$$
$$= V_0^{(S)}/(1-\gamma_0) \qquad (9)$$
$$= V_c$$

$$dV^{(P)}(t) = [1 - \gamma(t)]v^{(P)}\,dt$$

Integration will easily supply the following result:

$$\chi(t) = (1 + \chi_0)\exp(u_c^{(P)}t) - 1 \qquad u_c^{(P)} = v^{(P)}/V_c \tag{10}$$

$$\gamma(t) = 1 - (1 - \gamma_0)\exp(-u_c^{(P)}t) \tag{11}$$

The increase in precipitant concentration, $\gamma(t)$, with increasing t is more rapid in this case than in an open cell. This can be seen from a comparison of Eq. (5) and (11).

The formulas for the dissolution process corresponding to the terms (7) and (8) are the following:

$$1/\chi(t') = (1 + 1/\chi_{max})\exp(u_c^{(S)}t) - 1 \qquad u_c^{(S)} = v^{(S)}/V_c \tag{12}$$

$$\gamma(t') = \gamma_{max}\exp(-u_c^{(S)}t') \tag{13}$$

The transition from constant to program-controlled precipitation and dissolution requires the same modifications as in the case of the open cell.

The formulas given here apply only to liquids miscible without volume contraction. If volume contraction occurs, the relations may be easily modified in the case of the open cell. For closed cells, considerable complications arise.

B. Concentration of the Entire Polymer

During precipitation and solution titration, the concentration of the entire polymer present, together with the precipitant concentration, is altered. This alteration is proportional to the solvent concentration $1 - \gamma$ in the case of precipitation. In the case of dissolution, the concentration change is proportional to the precipitant concentration γ. If c_i is the initial concentration referred to the solvent volume used, the following equation applies to the precipitation titration:

$$c(t) = c_i[1 - \gamma(t)] = c_i/[1 + \chi(t)] \tag{14}$$

whereas, the following equation applies to the solution titration

$$\begin{aligned} c(t') &= \frac{c_{max}}{\gamma_{max}}\gamma(t') = c_i\frac{1 - \gamma_{max}}{\gamma_{max}}\gamma(t') \\ &= c_{max}\frac{1 + \chi_{max}}{\chi_{max}}\frac{\chi(t')}{1 + \chi(t')} = \frac{c_i}{\chi_{max}}\frac{\chi(t')}{1 + \chi(t')} \end{aligned} \tag{15}$$

These relations apply in the same manner to titrations in both open and closed cells.

The above formulas assume no volume contraction. If contraction occurs, the following factor should be added to Eqs. (14) and (15):

$$1/\kappa = [V^{(S)} + V^{(P)}]/V^{(SP)} = \rho^{(SP)}/[(1-\gamma)\rho^{(S)} + \gamma\rho^{(P)}]$$
$$= \rho^{(SP)}(1+\chi)/[\rho^{(S)} + \chi\rho^{(P)}] \qquad (16)$$

where $V^{(SP)}$ represents the volume of the mixture and $\rho^{(S)}$, $\rho^{(P)}$, and $\rho^{(SP)}$ the densities of the solvent, the precipitant, and the mixture.

C. Concentration of the Precipitated Polymer

It is the aim of the turbidimetric titration method to determine, by measuring the turbidity, the concentration of the precipitated polymer at the precipitant concentration γ.

1. Determination of the Turbidity

The intensity of the light beam, I_t, the intensity of the reference beam, I_0, and the path length through the turbid solution, L, are related to the turbidity, τ, by the following relation:

$$\tau = (1/L)\ln(I_0/I_t) \qquad (17)$$

The turbidity of the solution can also be determined by measuring the intensity, i_θ, of the light which is scattered through a fixed angle, θ. Assuming the inherent turbidity of the unprecipitated solution has been subtracted, the following relationship is true:

$$\tau \sim i_\theta^*/I_0^* \qquad (18)$$

The * signifies values corrected for inherent turbidity. In the usual experimental arrangement, the scattering volume is located in the center of the cell and the light path of the scattered light in the turbid medium is $L/2$. Using the following relationships:

$$i_\theta = i_\theta^* \exp(-L\tau/2) \qquad I_t = I_0^* \exp(-L\tau/2) \qquad (19)$$

we may replace in Eq. (18) the two not directly observable intensities, i_θ^* and I_0^*, by the measured quantities, i_θ and I_t

$$\tau \sim i_\theta/I_t \qquad (20)$$

Normally, the intensity of the light, $i_{90°}$, scattered at right angles to the direction of incidence, is used for determining the turbidity.

2. Particle Size Correction

It is known from the theory of light scattering, cf. Stuart [49], that particles with a diameter small in comparison to the wavelength produce a turbidity proportional to the particle concentration. The turbidity is linearly related to the size of the particles. The particles occurring in turbidimetric titration normally have dimensions of the order of magnitude of the wavelength of the scattered light. In this case, the scattering behavior is more complicated. Owing to the internal interferences the scattered radiation is more or less weakened in different directions. This asymmetry can be characterized by means of the asymmetry factor

$$Z = i_{45°}/i_{135°} \qquad (21)$$

A knowledge of the asymmetry factor allows additional statements on the size and shape of the particles to be made. Provided the particles are approximately spherical and the difference of the refractive indexes between the particles and the suspending agent is not too great, it is possible to determine a function $f(Z)$ describing the influence of the particle size on the 90° scattering

$$f(Z)(i_{90°}/I_t) = Kc^* \qquad (22)$$

In Eq. (22) c^* means the concentration of the precipitated polymer; and K is a constant which depends on the difference of the refractive indexes, the wavelength, and the density of the particles. It is possible with some general assumptions to obtain, by measuring $i_{90°}$, $i_{45°}$, $i_{135°}$, and I_t, a quantity proportional to the concentration of the precipitated polymer, c^*.

Figure 10 shows the reciprocal of the function $f(Z)$. Instead of the asymmetry coefficient Z itself, a term, Z_{exp}, is plotted as abscissa, which is corrected according to the equation

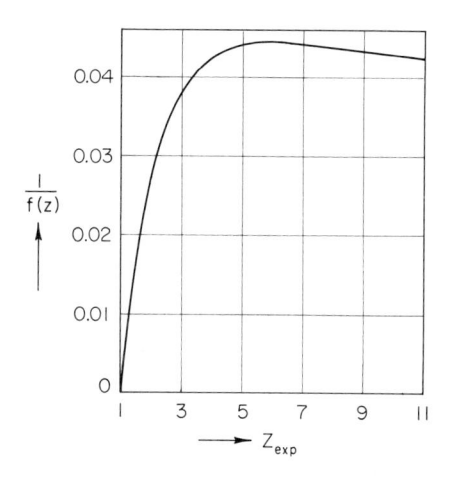

$$Z_{exp} = (Z + R)/(1 + RZ) \qquad (23)$$

FIG. 10. Reciprocal of the function $f(Z)$, according to Oth and Desreux [7].

for reflection losses at the outer cell walls. $R = [(n - 1)/(n + 1)]^2$ is the reflection coefficient at the glass/air interface; in Fig. 10, it is assumed to be 0.05. Urwin et al. [50] also take into account the reflection at the inner cell walls (with the reflection coefficient, r). In Eq. (23) they replace R by the term $(1 - r)^2 R + r$.

As can be seen from Fig. 10, the function $f(Z)$ passes through a minimum at $Z \approx 5$ and changes only very slowly above $Z > 5$. It is possible within a certain range of particle size to omit a measurement of the asymmetry without any appreciable loss of precision. When the polymer concentration is low, I_t is little different from I_0. Under such conditions, in spite of possible particle size variation, $i_{90°}$ alone is a satisfactory measure of the concentration, c^*, of the precipitated polymer.

According to Beattie [18b] an absolute determination of the concentration of precipitated polymer will be possible, if precipitation is executed discontinuously in a set of batches and the maximum of turbidity is observed with monochromatic light. Further conditions are that the refractive indexes of solvent and precipitant coincide and that the precipitant consists of spherical particles.

3. Dilution Correction

The concentration, c^*, is related to the overall concentration, $c(t)$ and $c(t')$, according to Eq. (14) and (15). The intention is to relate it to the overall quantity of the polymer. A function, $c_i^*(\gamma)$ is desired which is related to c_i as c^* is to c. Therefore, the measured turbidity, τ, the scattering intensity, $i_{90°}$, or the quantity, $f(Z)i_{90°}/I_t$, are corrected by adding a dilution factor. For a precipitation titration we obtain, using Eq. (14), e.g.,

$$\tau_{\text{corr}} = [1 + \chi(t)]\tau = \tau/[1 - \gamma(t)] \tag{24}$$

For a solution titration using Eq. (15) we obtain

$$\tau_{\text{corr}} = \chi_{\max} \frac{1 + \chi(t')}{\chi(t')} \tau = \frac{\gamma_{\max}}{1 - \gamma_{\max}} \gamma(t')\tau \tag{25}$$

4. Correction of the Refractive Index

The quantities derived above are proportional to the concentration of the precipitated polymer only if the constant K in Eq. (22) does not change during the titration. As this constant contains the refractive indexes of the solvent–precipitant mixture and of the particles, K will remain constant only if the refractive indexes of solvent and precipitant coincide.

If the refractive indexes of the solvent and the precipitant are not identical, the refractive index of the medium will change as the titration proceeds.

This results in the observation of a continued change in the turbidity of the solution even after the precipitate is completely formed.

In the case of small spherical particles, the constant K of the scattering formula (22) contains the factor

$$k = [(n_a^2 - n_l^2)/(n_a^2 + 2n_l^2)]^2$$
$$= (n_a - n_l)^2[(n_a + n_l)/(n_a^2 + 2n_l^2)]^2 \tag{26}$$

The terms n_a and n_l are the refractive indexes of the particles and of the surrounding liquid, respectively. For small differences [cf. Eq. (26)], the scattered light is approximately proportional to $(n_a - n_l)^2$. This was verified by Gooberman [43] in his investigations of polystyrene in the system benzene-methanol.

Precise correction of the refractive index requires a knowledge of the polymer concentration and the solvent-to-precipitant ratio in the precipitated particles. According to an investigation by Patat and Träxler [51] the solvent-to-precipitate ratio is independent of both the molecular weight of the precipitated particles and γ and corresponds to the γ value at which the polymer with the highest molecular weight starts precipitating. However, the polymer concentration in the aggregates usually depends considerably upon γ. It is true under favorable conditions, that the increase in polymer concentration is partly compensated by a corresponding decrease of the particle size, but this will not normally be the case quantitatively.

Owing to this complication, the theoretical determination of the refractive index correction will usually be omitted. Preference should be given to an empirical procedure. For this purpose, the turbidity, corrected for particle size and dilution, is plotted against χ or γ. Then, an attempt is made to represent the form of this function within the range of complete precipitation by a simple term, $\tau_u(\chi)$ or $\tau_u(\gamma)$. These terms supply a reasonable extrapolation for the range of smaller χ or γ values. The quotient $\tau_{\text{corr}}/\tau_u$ gives the relative measure of the concentration c_i^*.

Figure 11 shows an example of the empirical correction of the refractive index. The function $\tau_u(\gamma)$ is represented by a straight line with sufficient precision in this case.

Empirical procedures of the kind described above may serve, when the asymmetry coefficient Z cannot be measured, to eliminate partially the influence of particle size variation. The previous discussion applies only to homopolymers. No such conclusions can be drawn for copolymers. The different chemical structure of the monomer units causes additional changes in the refractive indexes. The swelling behavior of copolymers is a further complicating factor. At best, additional investigation of the pure components may furnish a basis for analysis.

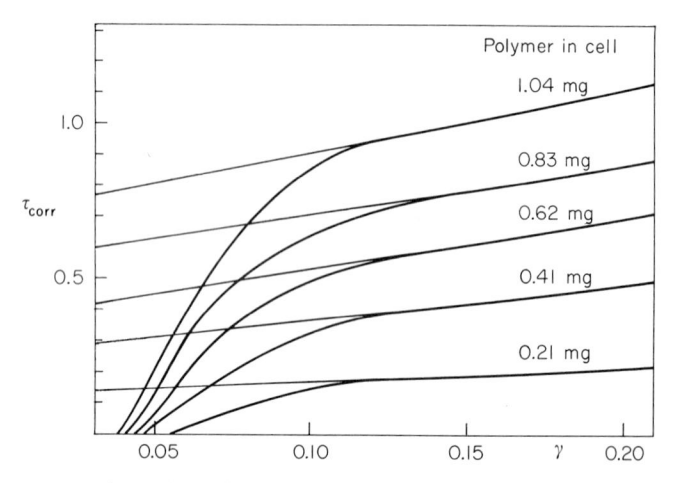

FIG. 11. Empirical determination of refractive index correction according to Allen and co-workers [18]. Turbidity titration of polystyrene in the system butyraldehyde-water-methanol (3:1) at 25°C; initial volume, 40 ml. Compare Section VI,B,1.

D. ESTIMATION OF THE MOLECULAR WEIGHT DISTRIBUTION

In most cases (see Section II,B) it is sufficient to obtain $c_i^*(\chi)$ or $c_i^*(\gamma)$ curves with the aid of the turbidimetric titration method and to compare these values for different samples. Frequently, the demands made are even lower: The previously discussed corrections are wholly or partly dispensed with. The turbidity curves may be determined under identical conditions from the extinction or the 90° scattering and used directly. On the other hand, there are problems for which the plotting of $c_i^*(\gamma)$ curves is not sufficient in itself, but the molecular weight distribution $C(M)$ is desired.

The transformation of the function $c_i^*(\gamma)$ into $C(M)$ requires calibration by means of molecular weight fractions obtained by preparative methods. Some authors try to gain from turbidity curves of these fractions, obtained under standard conditions, values of $\gamma_{1/2}$ (the turbidity at 50% of the final value), and to correlate these concentrations with molecular weights, M. The $c_i^*[\gamma(M)]$ curves obtained from such recalibration of the abscissa cannot be considered true molecular weight distribution curves. The concentration dependence of the precipitation is neglected completely, which results in systematic errors: The molecular weight obtained will always be too low, and the shape of the distribution curve will be considerably distorted. The distortion is greater for narrow than for broad distributions. For confident estimation, the equations describing the precipitation and dissolution process must be investigated more accurately and taken into account in the evaluation procedure.

1. Theoretical Basis

The precipitation of high polymer material from solution is a cooperative process. It should be possible to convert the molecular weight distribution $C(M)$ into the turbidity curve $c_i^*(\gamma)$ and *vice versa* by a functional transformation. That is to say, the quantity of the material $dc_i^*(\gamma)$ precipitated between γ and $\gamma + d\gamma$ depends, in varying degrees, upon the concentration of all molecular weights and particularly upon all molecules still in solution. No theory that might serve as a basis for the evaluation of turbidity curves has been developed on this general foundation. Morey [52] has suggested theoretical approaches. So far his work has been carried out to a limited extent and can furnish only first approximations as to the degree of interaction of two fractions with different molecular weights.

On the other hand, there is experimental evidence that interaction in the precipitation of different fractions is only to be expected, if the difference of molecular weights is rather small. It was shown by Klein *et al.* [52a–c] for samples of polystyrene, polyvinyl acetate and polyisobutylene in dilute solutions ($c < 1000$ mg/100 ml) that the addition of portions with lower molecular weights did not influence the precipitation point of the portion with higher molecular weight to any measurable degree, even though the percentage of the lower molecular weight portions was raised up to 600%. In these experiments the ratio of molecular weights of the different components was more than 2 : 1. At present there seems to be no experimental information on the increase of interaction with decreasing molecular weight difference. Obviously the difficulty in executing such experiments rests on the need of very sharp fractions.

The two methods of evaluation which have been described in the literature by Morey and Tamblyn [2] and by Claesson [15] are based on a common assumption. The assumption is that the concentration of the molecules already precipitated and the concentration of the molecules with a significantly different molecular weight do not appreciably affect the precipitation of any given molecular weight. However, this assumption is not valid for materials with molecular weights near that of the precipitated material. A proper evaluation should begin by determining a sequence of ΔM or $\Delta \gamma$ values which approximate the real continuous molecular weight distribution by a discrete sequence of fractions. For all of these fractions the precipitation may be considered to be independent of any other fraction.

In older publications dealing with the evaluation of turbidity curves, this fact has been implied but not clearly pointed out. Consequently, misunderstandings have occasionally occurred in the literature; and faulty conclusions have been drawn. The necessity of selecting correct $\Delta\gamma$ intervals was first clearly described by Scholtan [16] for the Morey-Tamblyn method. The Claesson method is clarified in this respect by Gooberman [43]. Therefore, our discussion does not precisely follow that of the original publications but includes, from the beginning, the above-mentioned improvements. Since the approximation of a continuous function by a step function will always be connected with a certain arbitrariness, we shall apply a moving average approximation.

2. Solubility Equation

The methods mentioned are based on the solubility equation. This equation establishes a relationship between the saturation concentration, $c^{(l)}$, of a uniform polymer of molecular weight, M, and the precipitant concentration, γ. This relationship can be represented to a good approximation by the formula

$$\gamma = -\Lambda(M) \ln c^{(l)} + \Gamma(M) \tag{27}$$

or in terms of $c^{(l)}$ by

$$c^{(l)}(M, \gamma) = \exp\{-[\gamma - \Gamma(M)]/\Lambda(M)\} \tag{28}$$

in which $\Lambda(M)$ and $\Gamma(M)$ are two functions of the molecular weight only which are often connected by a linear relation

$$\Lambda(M) = \lambda_0 + \lambda_1 \Gamma(M) \tag{29}$$

Frequently, $\Gamma(M)$ is represented in the form

$$\Gamma(M) = \Gamma_\infty + bM^{-a} \tag{30}$$

These relations were first derived and verified by Schulz et al. [1]. They started from a consideration of the partitioning of polymer between two liquid phases. However, Elias [53] was able to show that substantially the same equations can be obtained from a consideration of solution equilibria. When $c(\gamma) \leq c^{(l)}(\gamma)$ no precipitation occurs; if $c(\gamma) \geq c^{(l)}(\gamma)$, the precipitated proportion is given by the relationship $c^*(\gamma) = c(\gamma) - c^{(l)}(\gamma)$. Consequently, according to Eq. (28) the turbidity curve of a narrow fraction has the shape (for $c > c^{(l)}$)

$$c^*(\gamma) = c(\gamma) - \exp\{-[\gamma - \Gamma(M)]/\Lambda(M)\} \tag{31}$$

If a precipitation titration is being considered, $c(\gamma)$ should be substituted according to Eq. (14). For solution titrations, Eq. (15) should be applied. In the same way, $c^*(\gamma)$ may be replaced by $c_i^*(\gamma)$.

In a recent investigation Klein and Patat [52a,b] found that Eq. (27) does not apply to fractions of polystyrene in the system benzene-methanol (cf. Section VI,B,1), but that $\gamma = \gamma(\ln c^{(l)})$ is represented by a non-linear curve. However, the curvature is so small that it can only be detected by variation of $c^{(l)}$ over at least two decades. On the other hand, these authors verified with the above mentioned samples the fundamental formula $c^*(\gamma) = c(\gamma) - c^{(l)}(\gamma)$ to be valid quantitatively.

3. *Method of Morey–Tamblyn and Scholtan*

The basic assumption of this method is that it is possible to subdivide the experimentally determined turbidity curve $c^*(\gamma)$ into suitably selected $\Delta\gamma$ intervals. The associated intervals, $\Delta c_i^*(\gamma) = c_i^*(\gamma + \Delta\gamma) - c_i^*(\gamma)$, will then correspond to a uniform molecular weight, M, and the precipitates belonging to different $\Delta\gamma$ intervals will not influence one another.

For simplicity's sake, we shall assume that there is no dilution effect. Consequently, c and c_i, as well as c^* and c_i^*, are to be identified. This is the case for a sequence of experiments with the precipitant concentration, γ, increasing stepwise and the polymer concentration held constant. Also when precipitating by temperature variation, this condition is largely realized. It is necessary only to replace γ by T.

Under the above conditions, it is possible, when precipitation is complete, to describe the solubility equation for Δc^* in the following form:

$$\Delta c^*(\gamma) \approx (dc^*/d\gamma)\Delta\gamma = \exp\{-[\gamma - \Gamma(M)]/\Lambda(M)\} \tag{32}$$

The set of $\Delta\gamma$ should be determined empirically, as it depends on the interaction of the different molecular weights. No information can be obtained about these interactions from measurements on fractions. It is obvious to assume that this interaction is substantially limited to that γ range in which the solubility of a sharp fraction has dropped down to the $1/\alpha$ part. Thus, it follows from Eq. (27) that

$$\Delta\gamma = \Lambda(M)\ln\alpha \tag{33}$$

The only task remaining is to empirically establish the constant α. Its value will be presumed to lie roughly between 4 and 20 which corresponds to $1.5 < \ln\alpha < 3$.

Given $\Delta\gamma$ or α, Eq. (32) supplies a function $\gamma = \gamma(M)$, from which the differential molecular weight distribution

$$\frac{dC(M)}{dM} = \frac{dc^*(\gamma)}{d\gamma}\frac{d\gamma(M)}{dM} \tag{34}$$

and the integral molecular weight distribution

$$C(M) = \int_0^M \frac{dc^*(\gamma)}{d\gamma} d\gamma(M) \tag{35}$$

are determined at once. Equation (32) states that, as mentioned above, the fraction with the molecular weight $M(\gamma)$ is precipitated completely in the interval $\Delta\gamma$. Strictly speaking, this is not compatible with the solubility equation (28). However, this apparent inconsistency can be interpreted as follows: The residue not precipitated according to the solubility equation is compensated for by the material still precipitating from former fractions. Morey and Tamblyn [2] tried to avoid this assumption by inserting instead of $dc^*/d\gamma$ a term corresponding to the solubility equation

$$\frac{d\xi}{d\gamma} = \frac{dc^*/d\gamma}{1 - \exp(-\Delta\gamma/\Lambda)} \tag{36}$$

in Eq. (32). Since

$$\xi(1) = \int_0^1 d\xi(\gamma) > c^*(1) = c \tag{37}$$

Eq. (36) must still be corrected by a normalization introduced *ad hoc*

$$\left(\frac{d\xi}{d\gamma}\right)_{corr} = \frac{c}{\xi(1)} \left(\frac{d\xi}{d\gamma}\right) \tag{38}$$

If a constant exponent, $\Delta\gamma/\Lambda(M)$ is introduced into Eq. (36) [as it follows from Eq. (33)], Eqs. (32) and (38) will lead to identical results.

If the experiment takes place as a precipitation titration with continuous dilution according to Eq. (14), Eq. (32) will be replaced by

$$\frac{dc_i^*}{d\gamma} = \frac{\exp\{-[\gamma - \Gamma(M)]/\Lambda(M)\}}{(1 - \gamma)\Delta\gamma} \tag{39}$$

However, if a solution titration is involved in which the dilution is described by Eq. (15), the relation concerned will be

$$\frac{dc_i^*}{d\gamma} = \frac{\gamma_{max}}{(1 - \gamma_{max})} \frac{\exp\{-[\gamma - \Gamma(M)]/\Lambda(M)\}}{\gamma\Delta\gamma} \tag{40}$$

Further evaluation will be carried out in both cases in exactly the same manner as described above.

In the case of a precipitation titration, the Morey–Tamblyn Eq. (36)

acquires the form

$$\frac{d\xi_i}{d\gamma} = \left(\frac{dc_i^*}{d\gamma}\right) \bigg/ \left[1 - \frac{1-\gamma}{1-\gamma-\Delta\gamma}\exp\left(\frac{-\Delta\gamma}{\Lambda}\right)\right] \qquad (41)$$

and in the case of a solution titration Eq. (36) becomes

$$\frac{d\xi_i}{d\gamma} = \left(\frac{dc_i^*}{d\gamma}\right) \bigg/ \left[1 - \frac{\gamma}{\gamma+\Delta\gamma}\exp\left(\frac{-\Delta\gamma}{\Lambda}\right)\right] \qquad (42)$$

The factor in front of the exponential function is significant only if γ has values near either one or zero. At the same time, the significance of these equations for such values is particularly dubious. Ultimately, the second summand of the denominator will have to exceed the first one and, thus, give negative values for $d\xi_i/d\gamma$. Therefore, it appears more reasonable to apply the Eqs. (39) and (40) directly without the substitutions (41) and (42) and the subsequent normalization according to Eq. (38).

4. Modified Method of Claesson

The least satisfying assumption of the Morey–Tamblyn method is that the increase in turbidity in a $\Delta\gamma$ interval is attributed to the precipitation of particles of a single molecular weight. This difficulty is removed in the method of Claesson [15], for a simultaneous precipitation of particles with different molecular weights is presupposed.

This method is based on an assumed subdivision of the continuous molecular weight distribution, $C(M)$, into a number of discrete fractions, $\Delta C(M) \approx [dC(M)/dM]\Delta M$, which precipitate independently of one another. The solubility curves (28) are now considered to be functions of M, and γ is looked upon as parameter. Abbreviating, we write

$$s(M,\gamma) = \frac{c^{(l)}(M,\gamma)}{\Delta M} = \frac{\exp\{-[\gamma-\Gamma(M)]/\Lambda(M)\}}{\Delta M} \qquad (43)$$

Assuming as before that the precipitation takes place without dilution, the polymer precipitated at a certain precipitant concentration is given (see Fig. 12) by the area defined by the differential molecular weight distribution curve, dC/dM, and the curve $s(M,\gamma)$. The following equation applies to the point of intersection of the two curves:

$$dC(M)/dM = s(M,\gamma) \qquad (44)$$

This equation supplies a relationship between M and γ. It follows that the

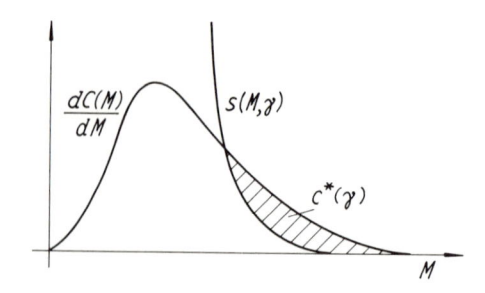

FIG. 12. Amount $c^*(\gamma)$ of a polymer with molecular weight distribution $C(M)$ precipitated at a concentration γ of the precipitant.

concentration of the precipitated polymer is given by

$$c^*(\gamma) = \int_M^\infty \left(\frac{dC}{dM'}\right) dM' - \int_M^\infty s[M', \gamma(M)] \, dM' \qquad (45)$$

By differentiation of this relation with respect to γ and inserting Eq. (44) we obtain

$$\frac{dc^*(\gamma)}{d\gamma} = -\int_M^\infty \frac{\partial s(M', \gamma(M))}{\partial \gamma} \, dM' \qquad (46)$$

The relation

$$\gamma = \gamma(M) \qquad (47)$$

is thus determined. If this is inserted into Eq. (44), the differential molecular weight distribution

$$dC(M)/dM = s[M, \gamma(M)] \qquad (48)$$

is obtained at once.

For a precipitation titration with successive dilution, it is only necessary to replace $s(M, \gamma)$ by $s_i = s(M, \gamma)/(1 - \gamma)$, so that instead of Eqs. (46) and (48) there are now the relations

$$\frac{dc_i^*(\gamma)}{d\gamma} = -\int_M^\infty \frac{\partial}{\partial \gamma} \left\{\frac{s[M', \gamma(M)]}{1 - \gamma(M)}\right\} dM' = -\int_M^\infty \frac{\partial s_i[M', \gamma(M)]}{\partial \gamma} \, dM' \qquad (49)$$

$$dC(M)/dM = s[M, \gamma(M)]/[1 - \gamma(M)] = s_i[M, \gamma(M)] \qquad (50)$$

For solution titration, $s(M, \gamma)$ is replaced correspondingly by

$$s_i(M, \gamma) = \frac{\gamma_{max}}{1 - \gamma_{max}} \frac{s(M, \gamma)}{\gamma} \qquad (51)$$

In the Morey–Tamblyn–Scholtan method, the intervals, $\Delta\gamma$, had to be determined empirically, and this is the case here for ΔM. This determination is somewhat more difficult, because the solubility equation (28) has a more complicated form with respect to M than with respect to γ. It may be the most convenient way to choose within the system of the solubility lines $c^{(l)}(M, \gamma)$, [considered to be functions of M] two suitable concentrations, $c_1^{(l)}$ and $c_2^{(l)} = \beta c_1^{(l)}$, and to determine $\Delta M(\gamma)$ by the condition

$$c^{(l)}(M, \gamma) = c_1^{(l)}$$

$$c^{(l)}(M + \Delta M, \gamma) = c_2^{(l)} = \beta c_1^{(l)} \tag{52}$$

The Claesson method will only lead to reasonable results, if the curve $dC(M)/dM$ has but one point of intersection with each of the curves $s_i(M, \gamma)$. A necessary, although by no means sufficient, condition in this respect is that (if M is selected large enough) the following formula applies:

$$\int_M^\infty s_i(M', \gamma) \, dM' < \infty \tag{53}$$

However, according to Eq. (43) this is not the case for $s_i(M, \gamma)$, if for $\Lambda(M)$ and $\Gamma(M)$ terms according to Eqs. (29) and (30) are inserted. Thus, $s_i(M, \gamma)$ converges towards the finite limit

$$s_i(\infty, \gamma) = \frac{\exp\{-(\gamma - \Gamma_\infty)/(\lambda_0 + \lambda_1 \Gamma_\infty)\}}{(1 - \gamma)\Delta M} \tag{54}$$

Therefore, these curves should be altered for small ordinate values in such a manner that (as indicated in Fig. 12) they reach the abscissa at a finite M value. If this does not agree with the experimental results, the integration should no longer be extended to $M' = \infty$. Instead it should be terminated at an $M_{\max}$ to be established empirically.

In practice, the Claesson method is not carried out analytically, but graphically, with the aid of a curvilinear coordinate system, the so-called Claesson grid. The first family of curves is thus determined by the curves $s_i(M, \gamma_\nu)$ with $\gamma_{\nu+1} = \gamma_\nu + \Delta\gamma$, while the second family is so designed that it subdivides the strips between $s_i(M, \gamma_\nu)$ and $s_i(M, \gamma_{\nu+1})$ into quadrangles of equal areas (see Fig. 13). The $\Delta c_i^*(\gamma)$ values belonging to the $\Delta\gamma$ differences are plotted in this grid, and the points thus obtained are interpolated. This curve describes the differential molecular weight distribution $dC(M)/dM$.

5. *Empirical Calibration*

Both the Morey–Tamblyn–Scholtan method as well as the modified Claesson method require quantities to be determined empirically, in particular the intervals $\Delta\gamma$ or ΔM. The form of the calculated molecular weight

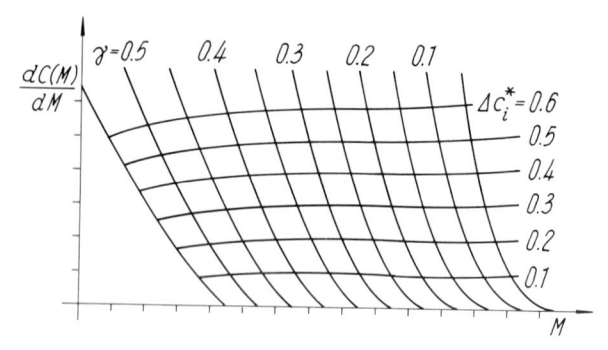

FIG. 13. Claesson grid.

distribution depends appreciably upon this choice. Reduction of the intervals usually causes a shifting of the curves to higher molecular weights. Menčík [54] emphasized this by way of an example. However, he did not recognize that this is an inherent feature of precipitation kinetics and by no means represents a specific shortcoming of the turbidimetric titration method.

As a first criterion, it should be demanded that the weight average molecular weight, $\overline{M}_w$, calculated from the molecular weight distribution determined by means of the method, coincides with the value determined with other methods. In addition, however, the method should, in general, be so adapted that the molecular weight distribution estimated from the turbidity curve $c_i^*(\gamma)$ largely coincides with that obtained from careful preparative fractionation.

Apart from the selection of $\Delta\gamma$ and ΔM, respectively, there is also some arbitrariness in the determination of the solubility curves $c^{(l)}(M, \gamma)$. They are obtained from the point of commencement of turbidity with narrow fractions or, according to Claesson [15], by relating the half-concentration, $\gamma_{1/2}$, according to $c_i^*(\gamma_{1/2}) = c_i^*(1)/2$ to $c^{(l)} = c^*(\gamma_{1/2})/2$. In other cases, the point of intersection of a tangent, applied to the turbidity curve under well-defined conditions, with the axis of abscissa may yield a reasonable value. A more complicated method making limited allowance for the inhomogeneity of the fractions was applied by Gooberman [43].

The finite width of the molecular weight distribution of even the best fractions introduces a noticeable uncertainty into the process of relating an M value, representative of the turbidity behavior, to the fraction in question. It can only be assumed that this value is greater than $\overline{M}_w$, but there is generally no possibility of estimating the difference. In view of this fact, Scholtan [16] has shifted the molecular weight curve determined from fractions to greater M values by a small amount. Thus, he has achieved close agreement between the molecular weight distribution obtained from the turbidity curve and those values gained from other methods.

6. *Comparison and Criticism of Evaluation Methods*

There will be a tendency towards giving the modified Claesson method preference over the Morey–Tamblyn–Scholtan method, because the assumptions made in the former case appear to agree more closely with the actual precipitation process. However, the system of solubility lines as used in the Claesson grid represents only a first approximation. This is illustrated in Fig. 14 which shows a molecular weight distribution, rather steeply declining towards greater molecular weights, and a single solubility line. In

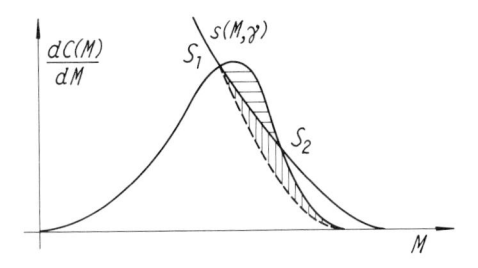

FIG. 14. Precipitation according to Claesson's assumptions and additional precipitation due to interaction of the high molecular weight components.

accordance with Claesson's assumption, the relatively low molecular weight part of the molecular weight distribution (characterized by horizontal shading) would have to be precipitated first, whereas the higher molecular weight part right of the point of intersection, S_2, should remain completely dissolved. This, of course, will not be the case in reality. As a consequence of the interaction of the higher and lower molecular weight portions, the solubility line will be steeper. This will cause the polymer with higher molecular weight to precipitate (as shown by the portion of the curve marked by vertical shading). Due to this interaction the real precipitation behavior lies halfway between the behavior of the Claesson and the behavior of the Morey–Tamblyn model. If the laws were known according to which the solubility lines are influenced by the distribution curve, it would be possible to improve, by some kind of successive approximation, the first approximation obtained by the Claesson method. Too much refinement would be unprofitable since even the corrected turbidity curve describes the real distribution, $c_i^*(\gamma)$, only more or less approximately.

Examining what has been published so far with respect to the application of the turbidimetric titration method for estimating the molecular weight distribution of polymers leads to the impression formulated by Howard [46], which is as follows: "It is fair to conclude that turbidimetric titration is not capable of accurately deducing the molecular weight distribution but that the rapidity of the method makes it suitable for less demanding work such as comparative studies of distribution changes." On the other hand, the singular judgement by Menčík [54]: "No confidence can be placed in the correctness of this method and, in spite of its seeming ease and speed, its application cannot be recommended," is neither sufficiently justified by the investigation of this author nor does it do justice to the results of scientists working in this field. A similar statement applies also to the "conclusions" of Rabel and Ueberreiter [30a] on the impossibility of a quantitative treatment of turbidimetric titrations. The investigations by these authors are based on only one material in connection with a few solvent–precipitant systems in which irregular aggregation takes place (see Section VI,B,1).

VI. Application

A. Various Classes of Polymers Studied

1. *Homopolymers*

According to what has been said before (Section V,D,6), the turbidimetric titration method is more suitable for supplying statements of a comparative nature than for determining the exact shape of the molecular weight distribution.

Using the turbidimetric titration method it is possible to follow various influences on the polymerization process, for example, discontinuous or continuous processing, polymerization in bulk, in solution, or in emulsion, variation of the temperature, the catalyst, or the rate of monomer addition. The change of the molecular weight distribution during the polymerization process can also be investigated. For polyacrylonitrile, Giesekus [55] found marked differences of the shape of the distribution curve at the beginning, in the middle, and at the end of the polymerization. In the same manner, it is also possible to characterize subsequent alterations of the polymer structure. Alterations occurring during the extrusion process may be easily detected, e.g., light degradation of polyamides [46], formation of less soluble, presumably cross-linked portions by oxidation of the melt. In the case of physiological substances, it is possible to study the degradation or the selective secretion in the respective organs (Wallenius *et al.* [9,10] and Scholtan *et al.* [56–58]). A particularly important application is the testing of preparative fractions for their homogeneity in comparison to that of the unfractionated product, the efficiency of extractions, and the like. Branching may also have an influence on the shape of the turbidity curve, see Melville and Stead [14].

2. Copolymers

When applying the turbidimetric titration method to copolymers, in particular block and graft copolymers, information is desired concerning the extent of copolymer formation and the amount of homopolymer present.

If quantitative statements are to be obtained, the different components should precipitate independently from one another. According to investigations by Allen *et al.* [18] this is not always the case. Frequently the less soluble component partly carries down with it a portion of the copolymer and, this in turn, the more soluble component. But even then, the turbidity curves can be looked upon as a function characterizing the homogeneity of the polymer, reflecting structural changes, and as useful for the relative characterization of the material. The applications then precisely correspond to those mentioned above for homopolymers.

Recently a modification of turbidimetric titration, the so-called *cloud point titration* (German: Fällpunkt-Titration), has successfully been applied to copolymer analysis. In this method, instead of the whole turbidity curve only its initial part is recorded, and from that the cloud point concentration $\gamma(c^{(l)})$ is determined. By variation of the polymer concentration $c^{(l)}$ the constant $\Gamma(M)$, cf. Eq. (27), may be extrapolated. For sufficiently large M this quantity depends only to a small extent on molecular weight and molecular weight distribution, see Eq. (30). However, it is dependant in a characteristic manner on the constitution of polymer molecules. Therefore, this method is well suited, i.e., for analyzing the composition of copolymers and for detecting small amounts of homopolymers in the presence of copolymers. For experimental details and application to different copolymers, see Gruber and Elias [58a–d].

3. Other Chemically Inhomogeneous Polymers

Very similar conditions to those prevailing in the case of copolymers also apply to other chemically inhomogeneous polymers. For instance, the solubility of cellulose esters depends on the degree of esterification, which may lead to errors in the statements concerning molecular weight distribution. Substitution of end groups also changes the solubility of the polymer. Although in many cases this dependence on chemical influences may render the interpretation of the measuring results more difficult, it may in other cases supply particularly important hints. Under certain circumstances, chemical reactions may be utilized for rendering possible the separation of various components by turbidimetric titration. (See discussion of the work of Hoffmann [59] in Section VI,B,11.) It is possible by means of turbidimetric titration to check slowly progressing chemical reactions, e.g., the degradation in certain solvents or the "maturing" of viscose solutions. Examples of extraordinary temperature dependence of precipitation are discussed in Sections VI,B,14–16.

B. APPLICATION TO SPECIAL POLYMERS

It is not possible to present here a critical investigation of all applications which have so far become known. The following data, which have been compiled from literature, should be looked upon as clues by a prospective user of the turbidimetric method. They are intended to make it easier for the user to elaborate the method for a concrete problem and not to render superfluous either the study of the original literature of the user's own critical investigations. For practical reasons, copolymers have always been treated after the homopolymers with which they have a main component in common.

1. *Polystyrene*

Of all polymers, polystyrene has been investigated by means of the turbidimetric titration method most frequently and most critically. A survey on the solvent–precipitant systems and the initial concentrations c_i as used by the various authors is given in Table I. In almost all cases the operating temperature was 25°C. As regards the solvent–precipitant system selected, it should be added that the combination of butyraldehyde and isopropanol has the advantage that the refractive indexes of the two components are practically equal. However, the system methyl ethyl ketone–acetone complies with the rule formulated by Hengstenberg (see Section III,A) and, consequently, yields a particularly high resolving power of the turbidity curve.

From investigations of the dissymmetry of angular scattering and electron micrographs, Hastings et al. [17] concluded that the shape of particles varies greatly in the course of titration. Pearl-string associates were found to an increasing degree. Steady enlargement of the aggregates has also been observed by Allen et al. [18] and by Rabel and Ueberreiter [30a]. In contrast to these results, Ryabova et al. [18a] did not find such variations by electron-microscopic means, if they removed the solvent from frozen samples by sublimation. They observed only spherical or oval particles of dimensions 0.08–0.4 μ. Even at the end of the titration no aggregation or coalescence was observed. Extensive coagulation occurred, if on preparing the samples

TABLE I
Turbidimetric Titration of Polystyrene

Author	Solvent	Precipitant	c_i (in mg/100 ml)	Remarks
Graham [60]	Methyl ethyl ketone	Methanol	2.5	Poor reproducibility
Hengstenberg [11,12]	Methyl ethyl ketone	Acetone (+1% water)	12.5–75	—
Melville and Stead [14]	Benzene	Methanol	1	Stepwise precipitation
Hastings et al. [17]	Benzene	Methanol	4	Pearl-string associates
Peaker and Robb [61]	Benzene	Methanol	4	Qualitative investigation
Hardy [62]	Methyl ethyl ketone	Water-methanol (3:1)	0.2–2	—
Gooberman [43]	Benzene	Methanol	2	Stepwise precipitation
Hoffmann [29]	Methylene chloride	Petrol ether	0.01–3	—
Allen et al. [18]	Benzene	Methanol	~1	—
	Butyraldehyde	Water-methanol (3:1)	~1	Large irregular aggregates
Mathieson [44]	Toluene	Methanol	0.3–12	Stepwise precipitation
Pravikova et al. [40]	Benzene	Methanol	2	—
Glöckner [62a]	Benzene	Methanol	<2	Stepwise precipitation (constancy after 30 min)
Urwin et al. [50]	Butyraldehyde	Isopropanol	0.4–2	Spherical particles
Cantow [22]	Methyl ethyl ketone	Acetone (+5% water)	3	Precipitation titration
Rabel and Ueberreiter [30a]	Benzene	Methanol	0.2–0.6 1.36	Solution titration Large growing particles
	Benzene	Ethanol	1.36	
	Dioxane	Methanol	1.36	
	Dioxane	Water	1.36	
	Cyclohexane	Methanol (?)	1.36	
Ryabova et al. [18a]	Benzene	Methanol	2–3	Nearly spherical particles
	Benzene	Ethanol	2–3	
Beattie [18b]	Methyl ethyl ketone	Isopropanol	5–20	Precipitation in a set of batches

the solvent was removed at temperatures somewhat above the freezing point. Therefore, it seems likely that in the previously cited investigations associates did not develop during the titration process but rather during sample preparation. This suggestion is further supported by an analysis of Beattie [18b], who showed that the course of dissymmetry, observed by Hastings et al. may be interpreted as well by assuming slightly growing spherical particles. In addition Mathieson [44] as well as Urwin et al. [50], and Gooberman [43] do not seem to have observed any irregularly shaped aggregates.

Rabel and Ueberreiter [30a] investigated the rate of equilibration in particular and found, that the rate increases, as the precipitant becomes weaker and the solvent poorer. In such systems the primary reason for this effect seems to be that smaller and fewer growing particles are generated.

Gooberman [43] and Mathieson [44] have carried out careful calibrations by means of fractionation. Particular attention has been devoted to the correction of the refractive index. Both authors have found very satisfactory agreement between the molecular weight distribution gained from turbidimetric titration with the aid of the Claesson method and that obtained by fractional precipitation.

Pravikova et al. [40] and earlier, Hengstenberg [12] have investigated not only polystyrene but also poly-α-methylstyrene and copolymers of both. Melville and Stead [14] have observed characteristic differences in the turbidity curves of linear and branched polystyrene (see Fig. 15). The branched polystyrene was prepared by bromination.

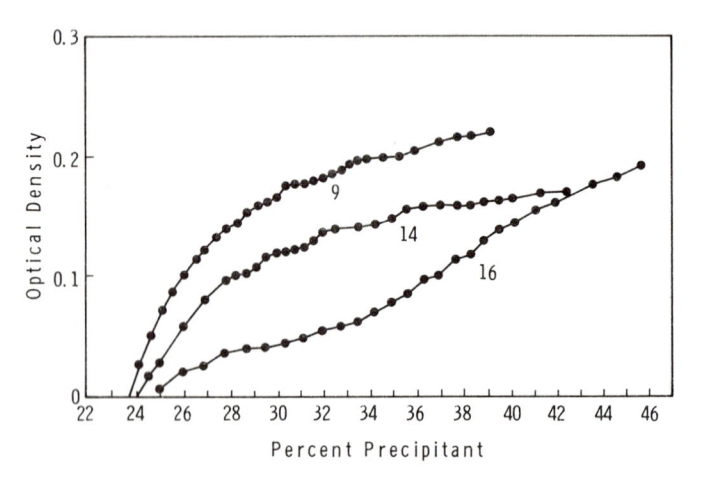

FIG. 15. Turbidity titration curves of a commercial polystyrene (Curve 9), of the brominated polymer prepared from it (Curve 14), and the branched polymer mixture prepared from the brominated product (Curve 16), according to Melville and Stead [14].

Finally Cantow [22] has found that in solution titration the establishment of equilibrium occurs much more quickly than in precipitation titration. In this case the concentration dependence is very low (see Fig. 16). Although mixtures of two polystyrenes of different molecular weight average do not behave entirely additively, the deviations are not considerable (see Fig. 17).

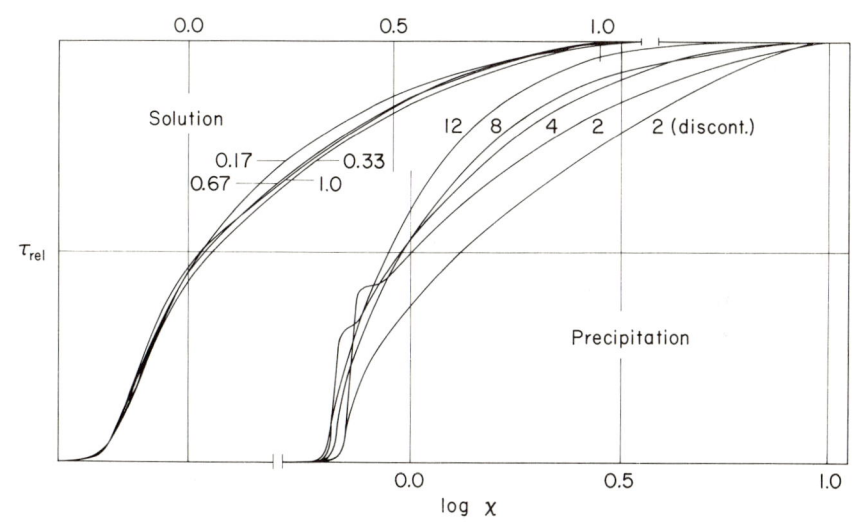

FIG. 16. Solution and precipitation titration of polystyrene ($\overline{M}_w = 444,000$) at several initial concentrations (in mg/100 ml), according to Cantow [22].

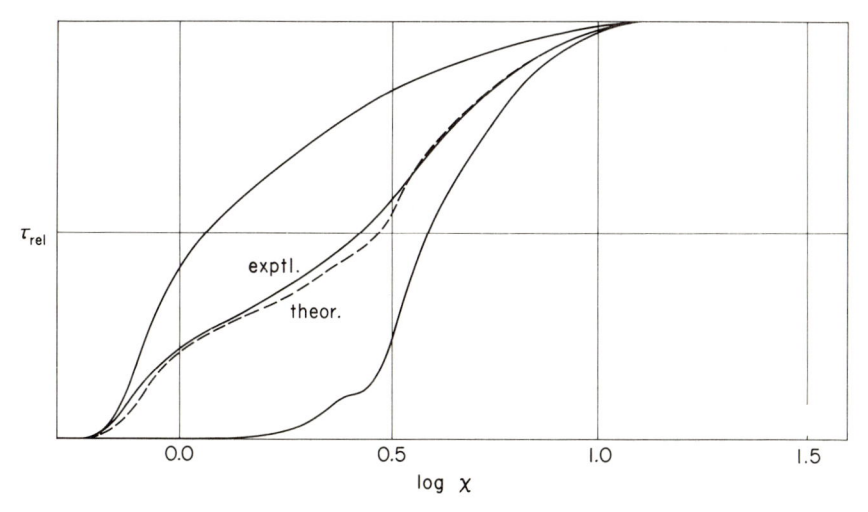

FIG. 17. Solution titration of a mixture of two polystyrenes ($\overline{M}_w = 43,000$ and $444,000$) at an initial concentration $c_i = 0.67$ mg/100 ml, according to Cantow [22].

As was shown by Harrison and Peaker [25a], precipitation and dissolution by temperature variation also give very satisfactory results. When methyl cyclohexane is chosen as the solvent ($c = 0.2$–0.65 mg/100 ml) and the temperature gradient is smaller than 10°C/hour, well-reproducible turbidity curves result. These are nearly identical for rising and lowering the temperature (20°–50°C). Dissymmetry observations prove, that particle dimensions are constant during precipitation or dissolution process but vary slightly from one experiment to another.

2. Polyvinyl Chloride

Polyvinyl chloride—as shown in Table II—has been investigated by several workers. Owing to the very limited number of solvents available, the versatility of the solvent–precipitant systems is lower. With one exception (Hengstenberg: 10°C), the operating temperature was 25°C.

Oth and Desreux [7] investigated with this polymer the variation of the particle size during titration and introduced the correction described in Section V,C,2. Grohn and Huu-Binh [33] evaluated the turbidity curves in accordance with the Morey–Tamblyn–Scholtan method and achieved very good agreement with the molecular weight distribution curves obtained by precipitation fractionation. Giesekus [63] used the method particularly for qualitative comparison of homopolymers with various graft copolymers based on polyvinyl chloride.

3. Polyvinyl Acetate

The earliest investigation of polyvinyl acetate was that by Morey et al. [31]. They used the solvent–precipitant system acetone–water and worked

TABLE II

TURBIDIMETRIC TITRATION OF POLYVINYL CHLORIDE

Author	Solvent	Precipitant	c_i (in mg/100 ml)
Mussa and Cernia [64]	Cyclohexanol	Ethyl alcohol or isobutyl alcohol	~40
Hengstenberg [11,12]	Tetrahydrofuran	Methanol	100
Oth and Desreux [7]	Cyclohexanone	Heptane–Carbon tetrachloride (9:1)	40
Hoffmann [29]	Cyclohexanone	Methanol	8
Grohn and Huu-Binh [33]	Cyclohexanone	Heptane–carbon tetrachloride (9:1)	30–70
Giesekus [63]	Tetrahydrofuran	Heptane–carbon tetrachloride (9:1)	5–10

at 25°C. In this investigation, a pronounced dependence of the turbidity curves on the initial concentration c_i was found for the first time. The effect was that reasonable curves could only be obtained for $c_i \leq 10$ mg/100 ml. But the influence of the change in the refractive index was overlooked. It was erroneously concluded that even at precipitant concentrations $\gamma > 0.9$ molecules are precipitated.

Various authors have investigated polyvinyl acetate in conjunction with vinyl acetate copolymers, e.g., Melville and Stead [14] and Allen et al. [18,65]. They too, have used the above-mentioned solvent–precipitant system and the same operating temperature. However, they chose initial concentrations $c_i = 1$ mg/100 ml and $c_i \leq 5$ mg/100 ml, respectively. Hoffmann [29] has used the system methylene chloride–methanol for copolymers of vinyl acetate and vinyl chloride or ethylene and worked with an initial concentration $c_i = 30–50$ mg/100 ml at 35°C.

Hartley [42] has investigated mixtures of polyvinyl alcohol and polyvinyl acetate, as well as graft copolymers of vinyl acetate and polyvinyl alcohol by means of a modified method. Instead of a solvent–precipitant system of the usual type, he has used two liquids, one of which—water—is a solvent for polyvinyl alcohol but a precipitant for polyvinyl acetate. The other liquid—acetone or ethyl alcohol—is, *vice versa*, a precipitant for polyvinyl alcohol, but a solvent for polyvinyl acetate. A discontinuous process in which the turbidity was determined on a number of samples of equal polymer concentration ($c = 50$ mg/100 ml) but different concentration of the two liquids produced interesting results. Curves were obtained which show turbidity which first decreases and later again increases with rising water content. The first turbidity can in this case be coordinated to the polyvinyl alcohol and to the copolymer rich in this component. The second turbidity can be coordinated to the polyvinyl acetate and to the copolymer rich in this component.

4. Polyvinyl Pyrrolidone

Polyvinyl pyrrolidone was first investigated by the continuous procedure of Hengstenberg [11,12]. In this procedure aqueous solutions ($c_i = 100$ mg/100 ml) are titrated with 16.8% Na_2SO_4 solution. Since such turbidities are not very stable, Campbell et al. [37] as well as Scholtan [16] modified the process. They used a set of Na_2SO_4 solutions of increasing concentration and added to each of them the same quantity of polyvinyl pyrrolidone solution. After a short agitation, the contents were left standing and the turbidity was measured at the same time after the addition of the solution or at the maximum turbidity.

Campbell et al. used Na_2SO_4 solutions of concentrations between 7 and 19 mg/100 ml and precipitated at 25°C. In order to precipitate shorter chains of polyvinyl pyrrolidone Scholtan increased the maximum concentration to 33.4 mg/100 ml and worked at 30°C.

Campbell *et al.* evaluated their results according to Morey and Tamblyn. Scholtan, when evaluating his results on this polymer, critically investigated the Morey–Tamblyn method and the Claesson method. He modified the former (see Section V,D,3). In a sequence of further investigations Scholtan *et al.* [56–58] applied the method to clarify the dependence of the permeability through the glomerular membranes of the kidney on the molecular weight and to investigate the difference between the excretion of the normal and the diseased kidney.

Giesekus [66] found that considerably more stable turbidities are obtained when working with the system methylene chloride–petroleum ether at initial concentrations of 2.5–5 mg/100 ml and at a temperature of 25°C. Intricate precipitation-kinetic effects leading to a further increase in turbidity and instability occur only at very high precipitant concentrations. However, the polymer is already completely precipitated at this point.

TABLE III

Turbidimetric Titration of Polymethyl Methacrylate

Author	Solvent	Precipitant	c_i (in mg/100 ml)	Remarks
Harris and Miller [41]	Acetone	Water	0.6	—
Mussa and Cernia [64]	Toluene	Ethanol	40	—
Melville *et al.* [13,14]	Acetone	Water	1–4	Block copolymers with styrene or vinyl acetate
Allen *et al.* [18]	Acetone	Water	~1	Also mixtures with polyvinyl acetate
	Butyraldehyde	Water-methanol (3:1)	~1	Also mixtures with polystyrene
Hoffmann [29]	Methylene chloride (containing HCl)	Petroleum ether (containing HCl)	10–40	Also copolymers with acrylates
Sorokin and Latov [66a]	Acetone	Water	10	Stepwise precipitation with 30 min intervals
Usmanov *et al.* [66b]	Tetrahydrofuran	Petroleum ether	5	Graft copolymers with vinyl chloride
Barnes *et al.* [66c]	Chloroform	Methanol	20	—

5. Polymethyl Methacrylate

A number of different polymethyl methacrylate types were first investigated by Harris and Miller [41]. The evaluation according to the Morey-Tamblyn method led to moderately good agreement with the molecular weight distribution obtained from a precipitation fractionation. As shown in Table III, several authors used different solvent–precipitant systems, partly with a view to the investigation of copolymers. The operating temperatures range from 20° to 25°C.

The critical investigation by Allen *et al.* [18] showed that polymethyl methacrylate when precipitated in the system acetone–water did not form any large aggregates but that the particle size did not exceed 1 μ. This is in contrast to their results with polystyrene.

6. Polyacrylonitrile

The turbidimetric titration method was elaborated for polyacrylonitrile at about the same time by Oth and Bisschops [8] and by Giesekus [55]. In both cases, dimethyl formamide was used as the solvent. The initial concentration c_i was 10–20 mg/100 ml and the operating temperature 25°C. While Oth and Bisschops chose the mixture heptane-dioxane (1 : 1) as the precipitant, Giesekus used di-*n*-butyl ether. Hoffmann [29] found that the separating effect at 50°C is better than at 25°C. He explained this by suggesting that the tendency of the precipitate to crystallize became greater with decreasing temperature.

No detailed information as to the properties of the system of Oth and Bisschops has been supplied. The stability of the turbidities as well as the separating effect of Giesekus' system is excellent. An example of this is given in Fig. 18, in which the differential turbidity curves of four fractions and their mixture are shown. The negative values of $d\tau_{rel}/d\chi$ observed with the fractions within a short interval are caused by the time delay in the establishment of the equilibrium and disappear when the precipitation velocity is lower.

In addition to homopolymers, Giesekus' system can also be applied to a number of copolymers of acrylonitrile, e.g., vinyl acetate, butyl acrylate [55], styrene, styrene sulfonic acid, butadiene, and chloroprene [29].

Climie and White [34] have used the system dimethyl formamide–benzene for a qualitative investigation of acrylonitrile copolymers with methyl methacrylate, methyl acrylate, styrene, and vinylidene chloride. They followed simultaneously the turbidity behavior and the variation of the viscosity with the concentration of the precipitant.

7. Polycarbonate

Krozer *et al.* [32] investigated polycarbonate with the system chloroform–

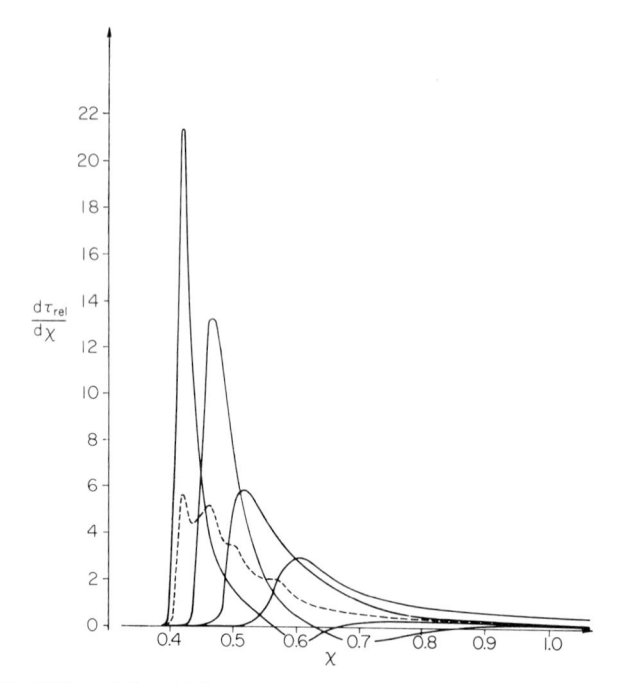

FIG. 18. Differential turbidity curves of four polyacrylonitrile fractions ($M \simeq 500{,}000$; 250,000; 75,000; and 20,000) and their mixture at a ratio 3:3:2:2. Solvent–nonsolvent system: dimethyl formamide–di-n-butyl ether; $c_i = 20$ mg/100 ml; $T = 50°C$; according to Agfa pamphlet [20].

methanol and the initial concentration $c_i = 2.5$ mg/100 ml at a temperature of 18°C, adding the precipitant stepwise. After applying a correction for the refractive index, they found no dependence between the size of the precipitated particles and the molecular weight of the particles. Using fractions, they determined the constants of the solubility equation and calculated the molecular weight distribution for two samples.

Satisfactorily stable turbidities and a good separating effect are also afforded by the system described by Müller [67]. This system uses the solvent dioxane-tetralin (tetrahydronaphthalene) (7:10) and the precipitant cyclohexane-decalin (decahydronaphthalene) (1:1), with the initial concentrations preferably ranging between 2 and 4 mg/100 ml and the temperature being 25°C. According to Hoffmann [29] the system methylene chloride–methanol ($c_i = 0.1$ mg/100 ml, 25°C) is also well suited.

Glöckner [67a] has used the last mentioned system ($c_i = 1.3$–2.6 mg/100 ml) for a determination of the molecular weight of polycarbonate fractions (Baker-Williams column) by cloud point titration (cf. Section

VI,A,2). In this way he was able to obtain the molecular weights of substances of which only a few milligrams were available.

8. *Polyamide and Polyurethane*

Gordijenko et al. [45] found that polyamide (Perlon, Nylon 6) can be investigated by means of a system consisting of the solvent *m*-cresol, the diluent isobutanol–petroleum hydrocarbons (1:9) and the precipitant petroleum hydrocarbons–acetylene tetrachloride (9:1). Solvent and diluent are used at a ratio of 5:8. The initial concentration is 6 mg/100 ml, and the operating temperature is 20°C. The petroleum hydrocarbons used should be aliphatic and free from olefins, have a boiling point between 92°C and 104°C and a refractive index of $n_D^{20} = 1.390$–1.395. The precipitates are not perfectly stable, so that the previously mentioned authors applied a stepwise procedure and little hand stirring. Giesekus [68] reduced the initial concentration to about 50%, replaced the poorly defined petroleum hydrocarbons by *n*-heptane, and worked continuously, stirring at a low speed with a screw-type stirrer. With the aid of this modified method, he investigated polyurethanes (Perlon U) in addition to polyamide (Perlon L).

Giesekus [30] also found that the solvent–precipitant system formic acid–di-*n*-butyl ether can be applied to polyamides and supplies extremely fine, stable turbidities (see Section III). However, a higher concentration $c_i \approx 20$ mg/100 ml must be chosen here. Correspondingly, the sharpness of separation seems to be somewhat lower than in the case of the above-mentioned system.

Howard [46] investigated polyamide (Nylon 66) by means of the system *m*-cresol-cyclohexane at concentrations between 1 and 2.7 mg/100 ml and an operating temperature of 25°C. He did not work continuously but used a sequence of samples with rising content of precipitant (similar to Campbell et al. [37] and Scholtan [16] in the case of polyvinyl pyrrolidone). A comparison of the distribution curve estimated from turbidimetric titrations with those obtained by repeated fractional precipitation shows that although a three-stage fractionation yields a better distribution than turbidimetric titration, turbidimetric titration is better than a single-stage fractionation.

Goodman and Scarso [35] investigated a graft copolymer of styrene on polyamide by means of the system toluene–hexane and an initial concentration $c_i = 1$ mg/100 ml and estimated the proportion of pure polystyrene.

9. *Cellulose Esters*

When the turbidimetric titration method was introduced and first applied, the emphasis was on cellulose esters. Thus, for instance, Morey and Tamblyn [2] investigated cellulose acetobutyrate with the solvent acetone and the precipitant ethanol–water (3:1), applying an initial concentration $c_i = 178$ mg/100 ml.

Dinitrocellulose was investigated by Oth [5] with the system acetone and methanol–water (9:1) at a concentration $c_i = 100$ mg/100 ml and an operating temperature of 25°C. Mussa and Cernia [64] used for this purpose the system acetone–water and a concentration $c_i = 40$ mg/100 ml. On the other hand, Claesson [69] suggests for this polymer a stepwise procedure in which various quantities of precipitant are added to a number of very dilute solutions ($c_i = 0.1$–1 mg/100 ml). The turbidities are then redissolved by a temperature increase and freshly formed by very slow cooling.

Bischoff and Desreux [6] investigated cellulose diacetate with a system consisting of the solvent butyraldehyde–ethyl alcohol (4:1) and the precipitant ethyl alcohol, choosing $c_i = 20$ mg/100 ml and the temperature 25°C. Giesekus [70] found that this system cannot be used for all cellulose diacetate types but that the system dimethyl formamide–di-n-butyl ether (which proved successful on polyacrylonitrile) can be applied at concentrations $c_i = 10$ mg/100 ml. This holds true also for the system introduced by Gordijenko et al. for polyamide (see Section VI,A,8). At a concentration of 2 mg/100 ml this supplies rather stable turbidities at 25°C. The Gordijenko system can even be applied to cellulose triacetate.

It should always be remembered in the case of cellulose esters that one of the decisive factors for the solubility is the degree of esterification. Thus, the turbidity curves contain only an integral statement about the molecular weight and the degree of esterification.

10. Dextran, Desoxyribonucleic Acid

Wallenius et al. [9,10] have utilized the turbidimetric titration method for the investigation of dextrans and developed the method into a micro technique. Only 1/100 mg of substance is needed per experiment. They were thus able to draw conclusions as to the molecular weight distribution of samples isolated from blood and urine. This led to the elucidation of the filtration processes in the glomerular membrane of the kidney in a manner similar to that applied later on by Scholtan et al. [56–58] to polyvinyl pyrrolidone. The above authors used the system water–ethanol with $c_i = 0.5$ mg/100 ml at 20°C and applied a stepwise procedure. They found that the precipitated particles were spherical and had a mean diameter of about 0.3–0.5 μ. For evaluation they used the Morey–Tamblyn method modified in some details for this purpose.

More stable turbidities were produced by Giesekus [70a] by substituting acetone for ethanol as precipitant ($c_i = 0.5$ mg/100 ml; 25°C). Also, fairly stable turbidities are obtained using dimethyl sulfoxide as solvent and dioxane or acetone as precipitants.

Shen and Eirich [71] investigated vinyl pyrrolidone graft copolymers on dextran by means of the system water–acetone in a qualitative manner.

The first-mentioned author [36] also studied graft copolymers of polyacrylic acid with desoxyribonucleic acid.

11. *Natural and Synthetic Rubber*

Polyisoprenes, polybutadienes, polychloroprenes, and copolymers of these substances were investigated in particular by Hoffmann [29]. The solvent–precipitant systems, the initial concentrations, and the operating temperatures used for this purpose are compiled in Table IV.

Giesekus [72] investigated polychloroprenes of medium molecular weight with the system dioxane–butanol and $c_i = 17$ mg/100 ml, $T = 25°C$. Very high molecular weight polychloroprenes, which are not soluble in dioxane, could be advantageously tested with the system toluene–methanol and $c_i = 3$ mg/100 ml, $T = 25°C$. Such polychloroprenes were degraded in methyl ethyl ketone in spite of certain precautionary measures [73]. Melkonyan and Bagdasaryan [73a] determined the molecular-weight distribution of polychloroprene (Standard- and π-Nairit), using the system benzene-methanol with $c_i = 2$–8 mg/100 ml.

Kovarskaya et al. [39] investigated block copolymers of rubber and polystyrene with the system acetone–methanol as well as block copolymers of rubber and Novolak resins with the system acetone–water ($c_i = 10$ mg/100 ml, $T = 30°C$). They compared these block copolymers with physical mixtures of homopolymers. Kargin et al. [38] tested block copolymers of butadiene–nitrile rubber and epoxy resin, prepared by mastication, with the aid of the system chloroform–methanol ($c_i = 20$ mg/100 ml).

In order to determine the content of polyisoprene of 3,4- or cyclized monomers, Hoffmann [59] combined turbidimetric titration with a chemical reaction. By reaction with perbenzoic acid, the solubility of the principal 1,4-components is strongly increased with respect to the admixtures mentioned. The admixture can be detected by the turbidimetric titration method with the system carbon tetrachloride–methanol—even when only fractions of a percent are present. This method can also be applied to polybutadiene. However, the solubility increase of 1,4-polybutadiene by reaction with perbenzoic acid is not so great as in the case of 1,4-polyisoprene.

12. *Resins*

Even before Morey and Tamblyn, the turbidimetric titration method for qualitative description of the distribution of polymers in resin was applied by McNally [3] as well as Adams and Powers [4]. The last-mentioned authors determined turbidity curves of a number of varnish resins (hydrocarbon resins, rosin-modified phenolic and maleic resins, and pure phenolic resins) using toluene as the solvent and methanol and hexane, respectively,

TABLE IV

TURBIDIMETRIC TITRATION OF RUBBERLIKE POLYMERS ACCORDING TO HOFFMANN[a]

Substance	Solvent	Precipitant	c_i (in mg/100 ml)	T (°C)
Polyisoprene (natural and synthetic), guttapercha, balata	Carbon tetrachloride	n-Butanol	0.5–20	30
Polybutadiene (cis-1,4, trans-1,4, and 1,2)	Carbon tetrachloride	n-Butanol	2	30
Chlorinated rubber	Carbon tetrachloride	n-Butanol	5	30
Cyclicized rubber	Carbon tetrachloride (containing HCl)	n-Butanol	0.5–20	30
Polychloroprene (Perbunan C)	Methyl ethyl ketone	n-Hexane	36	30
Styrene-butadiene copolymer (cold rubber)	Methylene chloride	Methanol	10	35
Rubber-methacrylate graft copolymer	Methylene chloride (containing HCl)	Methanol (containing HCl)	≤45	25
Acrylonitirile-butadiene copolymer (Perbunan N)	Dimethyl formamide	Di-n-butyl ether	4–24	50
Styrene-acrylonitrile-dichloroether-polychloroprene graft copolymer	Dimethyl formamide	Di-n-butyl ether	≤64	50

[a] Reference [29].

as the precipitants. Their concentrations were about 50 mg/100 ml. Their operating temperature was 25°C.

Čepelák [74] investigated phenol formaldehyde polycondensates (Resols, Novolaks) with the aid of the solvents acetone or methanol and the precipitant $1N\ H_2SO_4$ (i.e. "acidified water") at initial concentrations $c_i \geq 25$ mg/100 ml.

13. *Polyolefins*

Hoffmann [29] investigated high-pressure polyethylene (Lupolen H) at 100°C with the system decalin (decahydronaphthalene)–butanol and also ethylene propylene copolymers at 70°C with the system chlorobenzene–n-butanol. The results were not very satisfactory. Better stability was obtained by Tanaka *et al.* [40a] on polypropylene with tetralin (tetrahydronaphthalene) as solvent and "butyl cellosolve system" as precipitant ($c_i = 3$ mg/100 ml; 110°C). But, in general, the production of turbidity by temperature reduction, as reported by several authors, seems to be more favorable for this class of polymers.

Taylor and Tung [24] investigated linear polyethylenes in a solvent consisting of 70% α-chloronaphthalene and 30% dimethylphthalate with the addition of some ethyl cellulose for stabilization. The concentration was 2.5 mg/100 ml, and the temperature was varied within a range of about 160°–60°C. To prevent agglomeration of the turbidity, a special cell was used which was subdivided by a sequence of glass plates, so that convection was reduced.

With an improved experimental arrangement (see Section IV,B,3) Gamble et al. [25] have applied this method to ethylene-propylene copolymers using heptane–n-propanol (20:9) as the solvent and a concentration $c = 250$ mg/100 ml (related to heptane). The temperature is reduced at a speed of about 3°C/minute from 80° to 25°C. The results are certain and can be reproduced very easily.

14. *Polypropylene Oxide*

Linear and branched polypropylene oxides were investigated according to Giesekus [75] with the system methanol–water and an initial concentration $c_i = 40$ mg/100 ml at 25°C. This system possesses high resolution and yields stable turbidities. However, it is impossible to use it for the precipitation of low molecular weight polypropylene oxides (molecular weight lower than 1600). On the other hand, these can be covered too by using dimethyl formamide as solvent instead of methanol. However, it is necessary to apply a correction for the refractive index. The systems mentioned can also be applied to copolymers of ethylene oxide and propylene oxide. Turbidity curves are extremely sensitive to fluctuations in temperature.

Recent investigations by the same author [75a] have shown, that the method of temperature variation is more favorable than the above mentioned methods specifically for testing graft and block copolymers of propylene oxide and ethylene oxide. Solubility of this class of material in water is, in contrast to the usual behavior, reduced with increasing temperature, so that precipitation occurs above a critical point. At a concentration of 1.5 mg/100 ml turbidity curves range between about 10° and 90°C. Precipitates are absolutely stable and curves obtained by raising and lowering the temperature coincide very closely, if the temperature gradient is smaller than 1°C/min. The presence of ethylene oxide shifts the curves to higher temperatures. Not only the total content but also the position within the molecules is important. Ethylene oxide in the interior of the chains raises solubility more than when attached to the ends.

15. Ethoxylated Fatty Acid Derivatives

Ethoxylated fatty acid derivatives (sorbitol, triglyceride, oleic acid amide, coconut oil amine, etc.), used as viscose additives, were investigated by Giesekus [76] with the system methylene chloride–hexane. Depending on the kind of the product, the initial concentration was chosen between 10 and 250 mg/100 ml at a temperature of 25°C. The precipitates, only moderately stable, are sufficient for a qualitative test.

As was shown also by Giesekus [76a], ethoxylated castor oil should be investigated by the temperature variation method with water as solvent ($c = 15$ mg/100 ml). The turbidity is stable only below a critical temperature. However, the point of complete precipitation can be determined exactly by subsequent temperature reduction.

16. Alkylphenol Polyglycol Ethers

Giesekus [23] investigated nonylphenol hexaglycol ether (NP-6) and nonylphenol nonaglycol ether (NP-9) with the turbidimetric titration method. The turbidimetric precipitation titration could not be applied, and solution titration (i.e., the dissolution by dioxane at 65° or 75°C, respectively, of a turbidity obtained by temperature increase in aqueous solution) did not lead to fairly clear results. Therefore, the variation of the turbidity

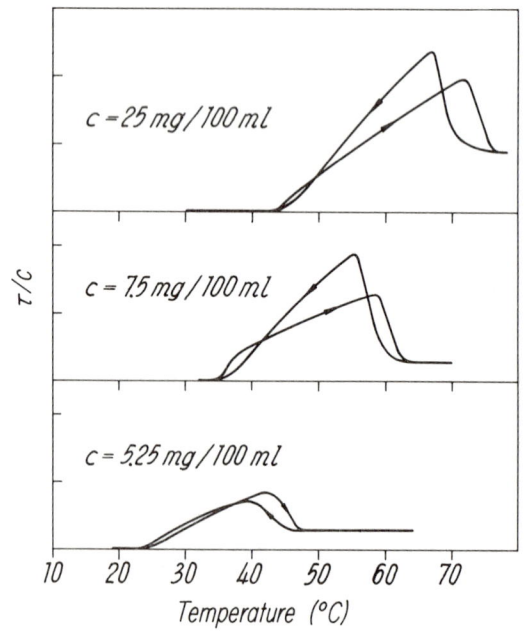

FIG. 19. Temperature dependence of the turbidity of aqueous NP-9 dispersions at different concentrations, according to Giesekus [23].

of aqueous solutions with the temperature was investigated within a larger range of concentrations. In the case of NP-9 the following course was observed (see Fig. 19): When the temperature was slowly increased, from a precisely defined temperature T_1, the originally clear aqueous solution began to become turbid. This turbidity became more and more pronounced until, at a temperature T_2, it reached a maximum. When the temperature was further increased, the turbidity decreased, and at T_3 it reached a low constant value. When, subsequently, the temperature was reduced, substantially the same process took place in reverse direction. Now, the turbidity maximum was at a somewhat lower temperature, T_4. This process could be repeated several times. To a first approximation, it was a reversible process.

However, the redissolution of the turbidity was observed only at low concentrations, $c < 30$ mg/100 ml. The concentration dependence on the four characteristic temperatures T_1 to T_4 shows an interesting trend, which is illustrated in a diagram in Fig. 20. With decreasing concentration, these

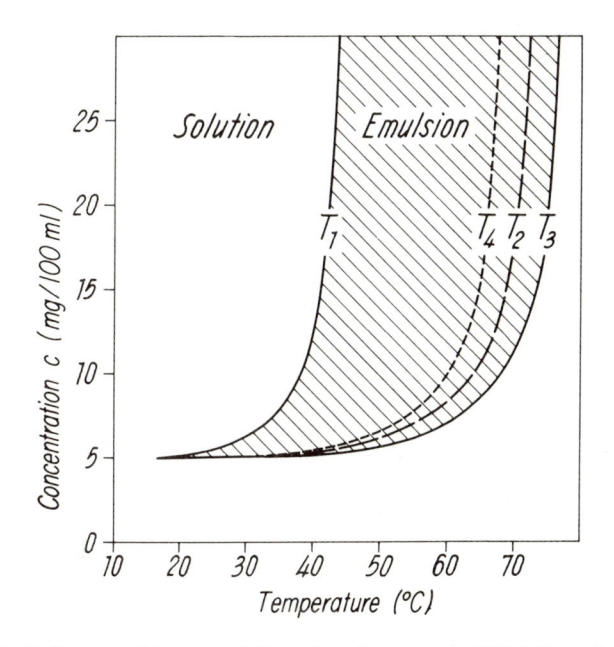

FIG. 20. (c, T) diagram of the state of dispersion of an aqueous NP-9 dispersion, according to Giesekus [23].

four characteristic values shift towards lower temperatures in such a manner that the temperature interval within which there is turbidity becomes smaller and smaller. At a critical concentration (in the above example:

$c_{min} \approx 5$ mg/100 ml) this interval equals zero. For $c < c_{min}$ turbidity will no longer appear at all.

Such diagrams will be different from product to product. In the case of mixtures the behavior observed is an intermediate one. Though the molecular weight distribution is not the decisive quantity here but turbidity behavior is determined substantially by the hydrophilic groups, this process can be considered a test specific to the substance concerned. The interpretation of the phenomenon is probably that at the temperature T_1 two or three of the four water molecules originally attached to an ethylene oxide group are lost. This greatly reduces the solubility (see Boehmke and Heusch [77]). The solubility of these complexes rises again with increasing temperature in such a manner that at the temperature T_2 complete redissolution is achieved.

The product NP-6 shows a somewhat more complicated behavior. But in this case also a turbidity point was observed which, with decreasing concentration, shifted towards the lower temperatures.

17. Siloxanes

Dimethylsiloxanes were investigated by Pravikova et al. [77a]. Best results were obtained with the solvent–precipitant system methyl ethyl ketone–methanol, in which no coagulation occurred.

ACKNOWLEDGMENTS

The author is indebted to Dr. E. A. Collins (B. F. Goodrich Chemical Company, Avon Lake, Ohio) for the use of his extensive and extremely useful list of references. Indebtedness must also be expressed to the authors colleagues, Dr. M. Hoffman, Dr. O. Müller and Dr. G. Gässler, for the use of their unpublished results. In addition, the author wishes to thank Dr. Hoffman for his helpful discussions.

The manuscript was prepared with the kind permission of the author's company, Farbenfabriken Bayer AG., Leverkusen. They also permitted the inclusion in the manuscript of many previously unpublished investigations. The author is obliged to Dr. E. M. Barrall II for revision of the English text.

REFERENCES

1. G. V. Schulz, in "Die Physik der Hochpolymeren" (H. A. Stuart, ed.), Vol. II, p. 748. Springer, Berlin, 1953.
2. D. R. Morey and J. W. Tamblyn, J. Appl. Phys. **16**, 419 (1945).
3. J. G. McNally, Lecture at the Gibson Island Conference on High Polymers 1942.
4. H. E. Adams and P. O. Powers, Ind. Eng. Chem., Anal. Ed. **15**, 711 (1943).
5. A. Oth, Bull. Soc. Chim. Belges **58**, 285 (1949).
6. J. Bischoff and V. Desreux, Bull. Soc. Chim. Belges **60**, 137 (1951).
7. A. Oth and V. Desreux, Bull. Soc. Chim. Belges **63**, 261 (1954).

8. A. Oth and J. Bisschops, *Compt. rend. 27th Congr. Intern. Chim. Ind., Brussels, 1954*, Vol. 3, p. 426.
9. A. Gronwall, H. Hint, B. Ingelman, G. Wallenius, and O. Wilander, *Scand. J. Chem. & Lab. Invest.* **4**, 363 (1952).
10. G. Wallenius, "Renal Clearance of Dextran as a Measure of Glomerular Permeability," p. 25. Almqvist & Wiksell, Uppsala, 1954.
11. J. Hengstenberg, *Angew. Chem.* **13**, 350 (1953).
12. J. Hengstenberg, *Z. Elektrochem.* **60**, 236 (1956).
13. A. S. Dunn, B. D. Stead, and H. W. Melville, *Trans. Faraday Soc.* **50**, 279 (1954).
14. H. W. Melville and B. D. Stead, *J. Polymer Sci.* **16**, 505 (1955).
15. S. Claesson, *J. Polymer Sci.* **16**, 193 (1955).
16. W. Scholtan, *Makromol. Chem.* **24**, 104 (1957).
17. G. W. Hastings, D. W. Ovenall, and F. W. Peaker, *Nature* **177**, 1091 (1956); G. W. Hastings and F. W. Peaker, *J. Polymer Sci.* **36**, 351 (1959).
18. P. E. M. Allen, R. Hardy, J. R. Majer, and P. Molyneux, *Makromol. Chem.* **39**, 52 (1960).
18a. L. G. Ryabova, Z. Ya. Berestneva, and N. A. Pravikova, *Visokomolekul. Soedin.* **7**, 1796 (1965).
18b. W. H. Beattie, *J. Polymer Sci.* **A3**, 527 (1965).
19. H. Giesekus, *Kolloid-Z.* **158**, 35 (1958).
20. Agfa A. G., Agfa-Trübungstitrationsmessgerät (pamphlet), Muenchen, 1959.
21. J. M. Stearne and J. R. Urwin, *Makromol. Chem.* **56**, 76 (1962).
21a. B. Ya. Voronov and G. I. Volkova, *Zavodsk. Lab.* **30**, 1411 (1964).
22. H.-J. Cantow, Trübungstitration von Hochpolymeren, Makromolekulares Kolloquium, Freiburg i. B., 6.3.1964.
23. H. Giesekus, unpublished report of Farbenfabriken Bayer (1960).
24. W. C. Taylor and L. H. Tung, *SPE (Soc. Plastics Engrs.) Trans.* **2**, 119 (1962).
25. L. W. Gamble, W. T. Wipke, and T. Lane, *J. Appl. Polymer Sci.* **9**, 1503 (1965).
25a. G. D. Harrison and F. W. Peaker, in "Techniques of Polymer Science", *Soc. Chem. Ind. (London) Monograph No. 17*, p. 96, Macmillan, New York, 1963.
26. R. W. Hall, in "Techniques of Polymer Characterization" (P. W. Allen, ed.), Chapter II, p. 53, Butterworth, London and Washington, D.C., 1959.
27. F. W. Peaker, *Analyst* **85**, 235 (1960).
28. S. G. Weissberg, S. Rothman, and M. Wales, in "Analytical Chemistry of Polymers" (G. M. Kline, ed.), p. 41. Wiley (Interscience), New York, 1962.
28a. N. S. Schneider, *J. Polymer Sci.* C, (8), 179 (1965).
29. M. Hoffmann, unpublished report of Farbenfabriken Bayer (1959).
30. H. Giesekus, unpublished results (1960).
30a. W. Rabel and K. Ueberreiter, *Kolloid-Z.* **198**, 1 (1964).
31. D. R. Morey, E. W. Taylor, and G. P. Waugh, *J. Colloid Sci.* **6**, 470 (1951).
32. S. Krozer, M. Vainryb, and L. Silina, *Vysokomolekul. Soedin.* **2**, 1876 (1960).
33. H. Grohn and H. Huu-Binh, *Plaste & Kautschuk* **8**, 63 (1961).
34. I. E. Climie and E. F. T. White, *J. Polymer Sci.* **47**, 149 (1960).
35. M. Goodman and L. Scarso, *Papers Am. Chem. Soc. 140 Meeting, Chicago, 1961 (Div. Polymer Chem. Preprints)* Vol. 2, p. 116.
36. K. P. Shen Kwei, *J. Polymer Sci.* **A1**, 2309 (1963).
37. H. Campbell, P. O. Kane, and I. G. Ottewill, *J. Polymer Sci.* **12**, 611 (1954).
38. V. A. Kargin, N. A. Plate, and A. S. Dobrynina, *Colloid J. (USSR) (English Transl.)* **20**, 315 (1958).
39. B. M. Kovarskaya, L. I. Golubenkova, M. S. Akutin, and I. I. Levantovskaya, *Vysokomolekul. Soedin.* **1**, 1042 (1959).

40. N. A. Pravikova, L. G. Ryabova, and Yu. Vyrskii, *Vysokomolekul. Soedin.* **5**, 1165 (1963).
40a. S. Tanaka, A. Nakamura, and H. Morikava, *Makromol. Chem.* **85**, 164 (1965).
41. I. Harris and R. G. J. Miller, *J. Polymer Sci.* **7**, 377 (1951).
42. F. D. Hartley, *J. Polymer Sci.* **34**, 397 (1959).
43. G. Gooberman, *J. Polymer Sci.* **40**, 469 (1959).
44. A. R. Mathieson, *J. Colloid Sci.* **15**, 387 (1960).
45. A. Gordienko, W. Griehl, and H. Seiber, *Faserforsch. Textiltech.* **6**, 105 (1955).
46. G. J. Howard, *J. Polymer Sci.* **A1**, 2667 (1963).
47. V. V. Guzeev, V. I. Morozov, B. P. Shtarkman, and E. E. Rylov, *Vysokomolekul. Soedin.* **1**, 1840 (1959).
48. G. Gässler, unpublished report of Farbenfabriken Bayer (1958).
49. H. A. Stuart, ed., "Die Physik der Hochpolymeren," Vol. I, p. 372. Springer, Berlin 1952.
50. J. R. Urwin, J. M. Stearne, D. O. Jordan, and R. A. Mills, *Makromol. Chem.* **72**, 53 (1964).
51. F. Patat and G. Träxler, *Makromol. Chem.* **33**, 113 (1959).
52. D. R. Morey, *J. Colloid Sci.* **6**, 407 (1951).
52a. F. Patat and J. Klein, *Polymer Letters* **3**, 615 (1965).
52b. J. Klein and F. Patat, *Makromol. Chemie* (1966) (in press).
52c. J. Klein and U. Wittenberger, *Kolloid-Z.* (in preparation).
53. H.-G. Elias, *Makromol. Chem.* **33**, 140 (1959).
54. Z. Menčík, *Collection Czech. Chem. Commun.* **24**, 3185 (1959).
55. H. Giesekus, unpublished report of Farbenfabriken Bayer (1954).
56. W. Scholtan, *Z. Ges. Exptl. Med.* **130**, 556 (1959).
57. G. Hecht and W. Scholtan, *Z. Ges. Exptl. Med.* **130**, 577 (1959).
58. K. Jahnke, W. Scholtan, and H. Lins, *Z. Ges. Exptl. Med.* **131**, 567 (1959).
58a. U. Gruber and H.-G. Elias, *Makromol. Chemie* **78**, 58 (1964).
58b. H.-G. Elias and U. Gruber, *Makromol. Chemie* **78**, 72 (1964).
58c. H.-G. Elias and U. Gruber, *J. Polymer Sci.* **B1**, 337 (1963).
58d. U. Gruber and H.-G. Elias, *Makromol. Chemie* **86**, 168 (1965).
59. M. Hoffmann, *Makromol. Chem.* **57**, 96 (1962).
60. J. P. Graham, in "Styrene" (R. H. Boundy and R. F. Boyer, eds.), p. 409. Reinhold, New York, 1952.
61. F. W. Peaker and J. D. Robb, *Nature* **182**, 1591 (1958).
62. R. Hardy, Ph.D. Thesis, Birmingham University (1958) (cited in Peaker [27], p. 239).
62a. G. Glöckner, *Abhandl. Deut. Akad. Wiss. Berlin, Kl. Chemie, Geol., Biol.* **1**, 89 (1963).
63. H. Giesekus, unpublished report of Farbenfabriken Bayer (1963).
64. C. Mussa and E. Cernia, *Ann. Chim. (Rome)* **41**, 130 (1951).
65. P. E. M. Allen, J. W. Downer, G. W. Hastings, H. W. Melville, P. Molyneux, and J. R. Urwin, *Nature* **177**, 910 (1956).
66. H. Giesekus, unpublished results (1958).
66a. M. F. Sorokin and V. K. Latov, *Zavodsk. Lab.* **31**, 547 (1965).
66b. Kh. U. Usmanov, A. A. Yul'chibaev, and T. Sirlibaev, *Plast. Massy.* **2**, 73 (1966).
66c. C. I. Barnes, J. K. Haken, and T. R. McKay, *Aust. J. Appl. Sci.* **15**, 84 (1964).
67. O. Müller, unpublished report of Farbenfabriken Bayer (1962).
67a. G. Glöckner, *Z. phys. Chem. (Leipzig)* **229**, 98 (1965).
68. H. Giesekus, unpublished report of Farbenfabriken Bayer (1955).
69. S. Claesson, discussion remark to Hengstenberg [12].
70. H. Giesekus, unpublished report of Farbenfabriken Bayer (1955).

70a. H. Giesekus, unpublished report of Farbenfabriken Bayer (1964).
71. K. P. Shen and F. R. Eirich, *J. Polymer Sci.* **53**, 81 (1961).
72. H. Giesekus, unpublished report of Farbenfabriken Bayer (1957).
73. H. Giesekus, unpublished report of Farbenfabriken Bayer (1961).
73a. L. G. Melkonyan and R. V. Bagdasaryan, *Izv. Akad. Nauk. Arm. SSR, Khim. Nauki* **18**, 333 (1965).
74. J. Čepelák, *Plaste & Kautschuk* **2**, 34 (1955).
75. H. Giesekus, unpublished report of Farbenfabriken Bayer (1964).
75a. H. Giesekus, unpublished report of Farbenfabriken Bayer (1965).
76. H. Giesekus, unpublished report of Farbenfabriken Bayer (1961).
76a. H. Giesekus, unpublished report of Farbenfabriken Bayer (1966).
77. G. Boehmke and R. Heusch, *Fette, Seifen, Anstrichmittel* **62**, 87 (1960).
77a. N. A. Pravıkova, V. P. Davydova, V. A. Kirichenko, and T. A. Yakushina, *Kauchuk i Rezina* **24**, 19 (1965).

References, marked by an additional letter, and all passages of the text, related to these, were added in proof.

CHAPTER C.2

Sedimentation

H. W. McCormick
PHYSICAL RESEARCH LABORATORY, THE DOW CHEMICAL COMPANY, MIDLAND, MICHIGAN

I. Introduction 251
II. Theory 252
III. Experimental 255
IV. Sedimentation Velocity 258
    A. Sedimentation Coefficient 258
    B. Polymer Heterogeneity 264
    C. Archibald Method 274
V. Sedimentation Equilibrium 276
    A. Average Molecular Weights 277
    B. Polymer Heterogeneity 278
    C. Time Required to Reach Equilibrium 279
    D. Density Gradient Sedimentation 280
    References 282

I. Introduction

Ultracentrifugation has long been considered a classical method for determining the molecular weights of high polymers since, in principle, it is capable of giving the entire molecular weight distribution curve as well as the different average molecular weights. Polymer chemists, however, have often become skeptical of the value of sedimentation measurements on account of experimental difficulties along with serious problems in the interpretation of data from nonideal solutions. Recent advances in the instrumentation, experimental technique, and theory have greatly increased interest in this field, with sedimentation methods now becoming reasonably successful in characterizing certain polymer systems in considerable detail. It will be the intent of this chapter to present those methods of sedimentation analysis which are useful for determining polymer heterogeneity.

In sedimentation analysis there are two types of experiments available. In sedimentation velocity experiments, the rate of sedimentation and diffusion are observed upon centrifugation at high centrifugal fields, while in sedimentation equilibrium experiments, the sedimentation and diffusion processes are allowed to come to a state of equilibrium upon centrifugation at lower centrifugal fields. In theory, molecular weight heterogeneity can be characterized by both types of experiment, sedimentation velocity giving a distribution of sedimentation constants and sedimentation equilibrium giving a distribution of molecular weights. Although a distribution of molecular weights is more readily interpreted by the average polymer chemist, it is found that a detailed distribution of sedimentation constants may be obtained by velocity experiments without making assumptions regarding the shape of the curve, as is generally necessary in the analysis of sedimentation equilibrium experiments. Velocity centrifugation has, therefore, received major emphasis in studying molecular weight heterogeneity, the data generally being combined with other measurements in converting to the more desirable molecular weight distribution. Sedimentation equilibrium is largely employed as a means of determining absolute average molecular weights; however, equilibrium studies in a mixed solvent system (density gradient centrifugation) have recently shown promise in the evaluation of polymer density heterogeneity.

Early analysis of sedimentation velocity experiments employed kinetic approaches for a description of the mass transport process. It is now recognized that an adequate description of this process requires a somewhat more involved theory based upon the thermodynamics of irreversible processes. On the other hand, classical thermodynamics is adequate for the description of a system at equilibrium. A brief account of the theory basic to analysis of sedimentation processes will be presented before describing the specific experiments employed in determining polymer heterogeneity. For a more detailed account of the basic theory the reader is directed to the excellent book by Fujita [1] on the "Mathematical Theory of Sedimentation Analysis."

II. Theory

Transport of molecules in a system under the influence of a centrifugal field is a rate process with movement in the direction of equilibrium. If the system is not greatly removed from equilibrium, as is the case for most sedimentation processes, it may be most rigorously treated by the thermodynamics of irreversible processes. In such a treatment the flow of J_i of component i is postulated to be linearly dependent upon the forces X_k which set up this flow. These relationships, known as phenomenological equations, take the form

$$J_i = \sum_k L_{ik} X_k \tag{1}$$

where L_{ik} are phenomenological coefficients which, like thermodynamic variables, are a function of temperature, pressure, and composition, but independent of the magnitude of X_k as long as the X_k remain small.

C.2. Sedimentation

The forces causing flow may be given by the negative gradient of the total potentials, which for isothermal sedimentation may be written

$$X_k = -\frac{\partial \Phi_k}{\partial r} - \frac{\partial \mu_k}{\partial r} \qquad (2)$$

where Φ_k is the centrifugal potential, μ_k is the chemical potential, and r is the radial distance from the axis of rotation. The gradient of centrifugal potential can be shown to be

$$\frac{\partial \Phi_k}{\partial r} = -\omega^2 r \qquad (3)$$

where ω is the angular velocity of rotation while the gradient of chemical potential may be expressed as

$$\frac{\partial \mu_k}{\partial r} = \sum_j \frac{\partial \mu_k}{\partial c_j} \frac{\partial c_j}{\partial r} + \frac{\partial \mu_k}{\partial P} \frac{\partial P}{\partial r} \qquad (4)$$

where c_j is the concentration of component j and P is the pressure. Under conditions of negligible compressibility the change in chemical potential with pressure is equal to the partial specific volume,

$$\partial \mu_k / \partial P = \bar{v}_k \qquad (5)$$

and the variation of pressure with position is given by

$$\partial P / \partial r = \rho \omega^2 r \qquad (6)$$

where ρ is the density of the solution. Substitution of Eqs. (3), (4), (5), and (6) into Eq. (2) gives the expression for the force

$$X_k = (1 - \bar{v}_k \rho)\omega^2 r - \sum_j \frac{\partial \mu_k}{\partial c_j} \frac{\partial c_j}{\partial r} \qquad (7)$$

The first term of this equation may be referred to as the sedimentation force while the second term is the diffusion force.

Description of the flow of component i under the condition of negligible compressibility can now be given in terms of measurable quantities by substitution of Eq. (7) into the phenomenological Eq. (1)

$$J_i = \sum_k L_{ik} \left[(1 - \bar{v}_k \rho)\omega^2 r - \sum_j \frac{\partial \mu_k}{\partial c_j} \frac{\partial c_j}{\partial r} \right] \qquad (8)$$

This may be put into the form

$$J_i = s_i c_i \omega^2 r - \sum_j D_{ij}(\partial c_j / \partial r) \qquad (9)$$

in which

$$s_i = \sum_k (L_{ik}/c_i)(1 - \bar{v}_k\rho) \tag{10}$$

$$D_{ij} = \sum_k L_{ik}(\partial \mu_k/\partial c_j) \tag{11}$$

are the sedimentation and diffusion coefficients, respectively. When the phenomenological coefficients are written as $M_i c_i/N_A f_i$ where M is the molecular weight, N_A is Avagadro's number, and f is the frictional coefficient per molecule and when the chemical potential is related to concentration by

$$\mu = \mu^\circ + (RT/M)\ln yc \tag{12}$$

where μ° is the reference chemical potential, y is the activity coefficient, R is the gas constant, and T is the temperature, Eqs. (10) and (11) can for a two-component system be put into the familiar form

$$s = M(1 - \bar{v}\rho)/Nf \tag{13}$$

$$D = (RT/N_A f)(1 + \ldots) \tag{14}$$

originally developed by Svedberg [2] using kinetic considerations.

Since the same phenomenological coefficients enter into the above definitions for sedimentation and diffusion coefficients, elimination of these coefficients allows evaluation of the molecular weight from measurements of partial specific volume, sedimentation coefficient, and diffusion coefficient. For a two-component system combination of Eqs. (10) and (11) along with the introduction of Eq. (12) for the chemical potential leads to

$$M = \frac{RTs[1 + c(\partial \ln y/\partial c)]}{D(1 - \bar{v}\rho)} \tag{15}$$

which at infinite dilution reduces to the familiar Svedberg relation

$$M = \frac{RTs_0}{D_0(1 - \bar{v}\rho)} \tag{16}$$

One of the major problems in the use of this equation for heterogeneous polymers is in the definition of the average molecular weight obtained. Corresponding equations have been derived by Baldwin [3] for the three-component system and by Peller [4] for the general multicomponent system; however, these complicated expressions have not yet been put into practical application.

The rate of change with time of the amount of a component in a section of volume between r and $r + \Delta r$ is equal to the amount of that component flowing into the section at r minus the amount flowing out at $r + \Delta r$. This is a statement of the conservation of mass for the transport process which in the limit of infinitesimally small Δr leads to the continuity equation

$$\frac{\partial c_i}{\partial t} = -\frac{1}{r}\frac{\partial (rJ_i)}{\partial r} \tag{17}$$

Substitution of Eq. (9) into Eq. (17) gives the differential equation for the ultracentrifuge

$$\frac{\partial c_i}{\partial t} = \frac{1}{r}\frac{\partial}{\partial r}\left(r\sum_j D_{ij}\frac{\partial c_j}{\partial r} - s_i\omega^2 r^2 c_i\right) \tag{18}$$

which for a two-component system reduces to the familiar Lamm [5] differential equation

$$\frac{\partial c}{\partial t} = \frac{1}{r}\frac{\partial}{\partial r}\left(rD\frac{\partial c}{\partial r} - s\omega^2 r^2 c\right) \tag{19}$$

This equation was first derived by Lamm using a kinetic approach; however, conditions required for its validity were not completely understood until derivation by the thermodynamics of irreversible processes.

The differential equation for the ultracentrifuge mathematically describes the isothermal sedimentation of a system being subjected to a centrifugal force provided partial specific volumes of the components are constant. Comparison of experimentally measured concentration distributions with the appropriate solutions to this equation makes it possible, in principle, to determine the sedimentation and diffusion coefficients which in turn may be substituted into the Svedberg equation to evaluate molecular weight. Sedimentation analysis, therefore, largely involves the solution of the Lamm differential equation subject to conditions encountered during experiments.

III. Experimental

Two general approaches may be taken in the investigation of polymer solutions by ultracentrifugation: sedimentation velocity carried out in a strong centrifugal field and sedimentation equilibrium carried out in a weak centrifugal field. In the early days of ultracentrifugation different types of apparatus were normally employed for the high- and low-speed centrifugation, with oil turbine or air-driven ultracentrifuges used for high speed and direct motor drive ultracentrifuges used for low speed. Continued developments in the field led to the electrically driven and magnetically suspended ultracentrifuges in more common usage today. The successful commercial production of the Spinco Model E Analytical Ultracentrifuge has been very instrumental in the recent progress in ultracentrifugation. This is an electrically driven ultracentrifuge which has been satisfactorily employed for both velocity and equilibrium centrifugation. The reader is directed to other sources [6–8] for a more detailed account of the various ultracentrifuges; however, a very brief, general description of the ultracentrifugation experiment will be given for those completely unfamiliar with this technique.

An ultracentrifuge consists of a metal rotor several inches in diameter which is spun in a vacuum at a constant temperature. The rotor is driven electrically, or otherwise, at a predetermined, constant speed between 1000 and 70,000 rpm. A small cylindrical cell containing the solution being centrifuged is held in the rotor near its periphery. Concentration gradients established in the cell are determined optically by using a beam of light traveling in a direction parallel to the axis of rotation, the beam being intercepted by the cell at each rotation. Refraction or absorption of rays of the beam passing through different portions of the cell is determined by use of the appropriate optics, which record the concentration or change in concentration along the direction of sedimentation within the cell. Progress of the sedimenting solute may be followed by photographically recording these optical patterns at increasing times of sedimentation.

The analytical cell consists of a short cylindrical centerpiece with an opening for the solution across its diameter, the open sides being sealed by two quartz or sapphire windows when assembled into a metal cell housing. This general cell design has not been altered over the years, while the usefulness of ultracentrifugation has been extended by the introduction of a variety of centerpieces. The conventional sector-shaped centerpiece comes in a choice of optical paths from 1.5 to 30 mm and sector angle from 2° to 4°. The smaller centerpieces are valuable when very small quantities of polymer are available, although greater optical sensitivity is obtained with the larger centerpieces. Use of both metal and plastic centerpieces allows a wide choice of solvents. The double-sector centerpiece developed for use in interference optics is now being routinely employed in schlieren optics. Greater accuracy is obtained by use of solvent in one sector to register a base line having the same window distortion as the solution curve. Synthetic boundary cells which permit the layering of one liquid over a more dense liquid during the operation of the ultracentrifuge have recently been introduced. Such cells have greatly extended the value of the ultracentrifuge by allowing formation of the initial sharp boundary in the center of the liquid column. Several versions of this cell are available, each being particularly suited for certain experimental applications. A multichannel cell has recently been developed for the simultaneous sedimentation of four short columns. Other cell designs have been reported for special purposes, although, as yet, most of these have seen little application by polymer chemists.

Optical methods commonly employed in the ultracentrifuge are based on the absorption or refraction of light passing through the polymer solution. The absorption method has found limited application for the polymer chemists since many polymers are not UV absorbers; however, if the polymer is a good absorber this method may be valuable in allowing accurate measurement at very low polymer concentration [9]. The refractive index methods commonly employed use either the schlieren optics or the interference optics. Schlieren optics give the refractive index gradient, dn/dr, as a function of the distance, r, from the center of rotation, as shown in Fig. 1 for the high-speed sedimentation of polystyrene in cyclohexane. Interference optics give the refractive index as a function of the distance r, as shown in Fig. 2 for the low-speed sedimentation of polystyrene in cyclohexane. Refractive index curves can be converted to concentration curves

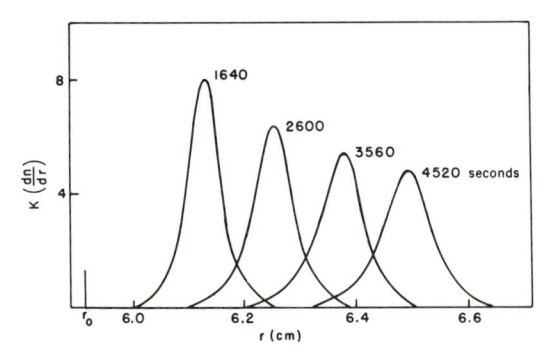

FIG. 1. Refractive index gradient curves obtained by the schlieren optical system for the sedimentation of a 0.30 gm/100 ml solution of Polystyrene S105 in cyclohexane at 59,780 rpm and 35°C.

with conversion constants determined from the refractive index change of standard solutions established by use of a synthetic boundary cell. Use of the refractive index methods requires a solvent having a refractive index greatly different from that of the polymer being examined.

Since temperature greatly affects the rate of sedimentation, it is necessary to carry out the sedimentation experiment under conditions that are as near isothermal as possible. In the early days of ultracentrifugation it was customary to average the rotor temperature determined before and after the sedimentation experiment. The newer ultracentrifuges are equipped with a temperature sensing device (thermistor) embedded in the base of the rotor which continually measures rotor temperature and controls heating elements in the rotor chamber. Combination of this temperature control

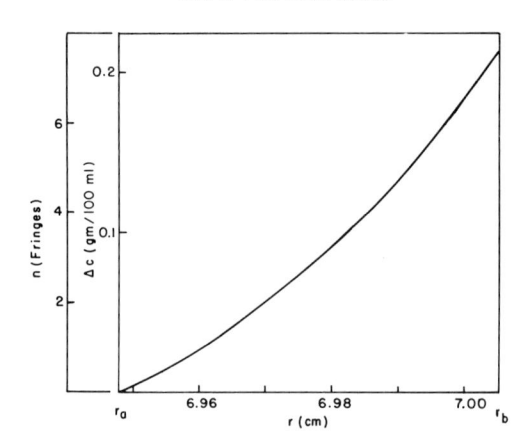

FIG. 2. Concentration curve obtained by the interference optical system for the sedimentation of a 0.21 gm/100 ml solution of polystyrene S105 in cyclohexane at 11,573 rpm and 35°C for 3 hours.

system with continuously operating refrigeration or high-temperature heaters allows ultracentrifugation at constant temperature between 0° and 150°C.

IV. Sedimentation Velocity

Sedimentation velocity experiments possess the valuable characteristic of pronounced sensitivity to polymer heterogeneity. This heterogeneity is recognized by the progressive spreading of the concentration gradient boundary as it traverses the ultracentrifuge cell. Although visual inspection of this boundary spreading can given qualitative information regarding polymer heterogeneity, such an examination may be misleading since the boundary spreading is not controlled entirely by polydispersity. Effects of diffusion, concentration, and hydrostatic pressure may also produce changes in the boundary shape. Despite substantial progress in the understanding of these effects, there is no simple rigorous treatment expressing all the factors affecting the shape of a sedimenting boundary. In the analysis of sedimentation velocity experiments these factors may, however, be evaluated separately in an effort to determine the boundary spreading caused only by polymer heterogeneity.

A. Sedimentation Coefficient

Examination of the simple diffusion-free sedimentation of a single solute shows that a sharp step boundary is formed between the solvent and solution as the solute sediments away from the air–liquid meniscus. This may be mathematically expressed by the Lamm equation with $D = 0$. Solution of this equation [1] shows that the boundary moves toward the bottom of the cell in accordance with the equation

$$r = r_0 \exp(s\omega^2 t) \qquad (20)$$

and that the concentration ahead of the boundary decreases exponentially with time in accordance with the equation

$$c = c_0 \exp(-2s\omega^2 t) \qquad (21)$$

where r_0 is the position of the boundary at zero time and c_0 is the concentration at zero time. Rearrangement of Eq. (20) into the form

$$\ln r = \ln r_0 + s\omega^2 t \qquad (22)$$

indicates that a plot of $\ln r$ versus t is a straight line with a slope of $s\omega^2$. Such a plot, shown in Fig. 3(a) for the sedimentation of a polystyrene

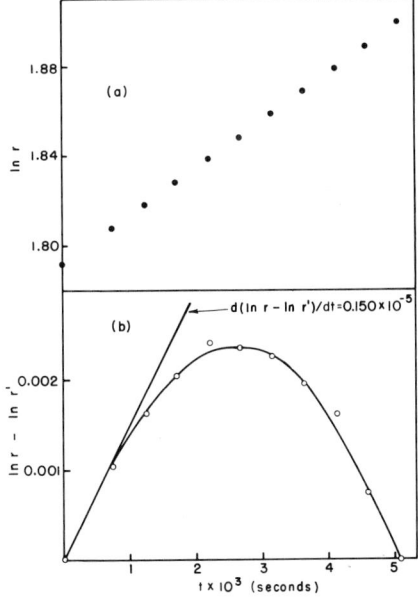

FIG. 3. (a) A plot of $\ln r$ versus time and (b) an expansion plot of $\ln r - \ln r'$ versus time for the sedimentation of a 0.25 gm/100 ml solution of polystyrene S105 in cyclohexane at 59,780 rpm and 35°C. $\ln r' = 1.79218 + 2.10927 \times 10^{-5} t$; $(d \ln r/dt)_{t=0} = 2.269 \times 10^{-5}$; $S_{t=0} = 5.79 \times 10^{-13}$.

fraction, is often employed for the evaluation of sedimentation coefficients. Rarely is a perfect straight line found in the practical application of this method due to pressure effects and dilution of the solute with time in accordance with Eq. (21). In order to avoid difficulties arising from these effects, data for the boundary movement may be treated in such a manner as to evaluate the sedimentation coefficient at zero time which corresponds to sedimentation at atmospheric pressure and the original solution concentration.

Evaluation of the sedimentation coefficient at zero time by the procedure of Trautman et al. [10] is illustrated in Fig. 3. The slight curvature in the plot of $\ln r$ versus t [Fig. 3(a)] is magnified by plotting against time the difference between $\ln r$ and the $\ln r'$ calculated from an equation for the straight line between the first and last points determined [Fig. 3(b)]. The initial slope of a smooth curve through these points is then added to the slope of the straight line to give a sedimentation coefficient at zero time.

1. Concentration Dependence

Sedimentation coefficients of nearly all materials have been found to be dependent upon the concentration of the solution being examined. Reasons for concentration dependence are not completely understood, although it is generally observed that at low concentrations the variation of sedimentation coefficient with concentration may be given by

$$s = s_0/(1 + kc) \tag{23}$$

where s_0 is the sedimentation coefficient at zero concentration, k is a constant, and s is the sedimentation coefficient at concentration c. At sufficiently low values of kc Eq. (23) can take the simple form

$$s = s_0(1 - kc) \tag{24}$$

Most interpretations of sedimentation velocity data presently require the coefficient at infinite dilution, which can be determined by the extrapolation of $1/s$ (or s) against c to $c = 0$. Such an extrapolation is illustrated in Fig. 4 for two polystyrene fractions centrifuged in a poor solvent (a theta solvent). This extrapolation indicates that at low concentrations the constant k may be satisfactorily represented by $k_1 s_0$ for a rather wide range in molecular weight. This is not surprising since it is reported [11,12] for many polymers that k is proportional to intrinsic viscosity, and under theta conditions the intrinsic viscosity is also proportional to s. Such a general relation

$$s = s_0(1 - k_1 s_0 c) \tag{25}$$

is useful in the analytical extrapolation of sedimentation data to infinite dilution. Concentration dependence is much greater when good solvents are employed.

2. Diffusion Effects

Treatment of the more general sedimentation velocity experiments which include diffusion effects requires a more complex solution to the Lamm

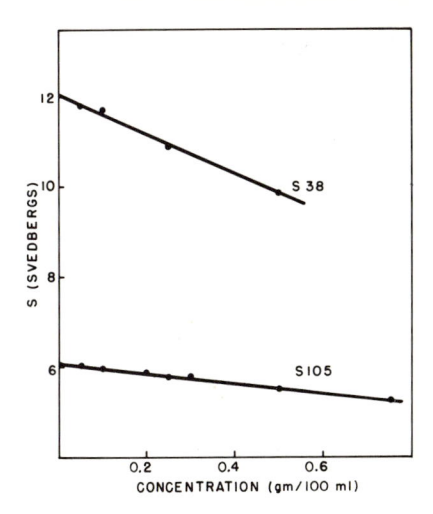

FIG. 4. Concentration dependence for the sedimentation of polystyrene S105 and S38 in cyclohexane at 59,780 rpm and 35°C, with the lines representing the best fit to the relation $s = s_0(1 - 0.030\, s_0 C)$.

equation. The first solution to the complete Lamm equation was provided by Faxen [13]. Even though his solution is only approximate with limited practical applicability, it has served as a basis for sedimentation analysis from the early days of the ultracentrifuge. This treatment, which assumes both sedimentation and diffusion to be concentration-independent, shows that a diffuse boundary of approximately Gaussian shape is formed by the sedimentation of a single solute in contrast to the sharp step boundary formed in diffusion-free sedimentation. This diffuse boundary formation does not alter the boundary position nor concentration ahead of the boundary.

Faxen's treatment indicates that a diffusion coefficient may be determined from sedimentation velocity experiments by examination of the shape of the diffuse sedimenting boundary. For a gradient curve represented by Faxen's solution, the area–height ratio (A/H) may be given by

$$\left(\frac{A}{H}\right)^2 = D\frac{2\pi}{s\omega^2}[\exp(2s\omega^2 t) - 1] \tag{26}$$

where

$$A = \int (\partial c/\partial r)\, dr \tag{27}$$

and

$$H = (\partial c/\partial r)_{\max} \tag{28}$$

Since Eq. (26) is based on the assumption that $2s\omega^2 t \ll 1$, it may be expanded in a power series, neglecting higher powers of $2s\omega^2 t$, to give

$$(A/H)^2 = 4\pi Dt \tag{29}$$

A plot of $(A/H)^2$ versus t should, therefore, give a straight line with a slope of $4\pi D$. Application of this method is illustrated in Fig. 5 for a polystyrene

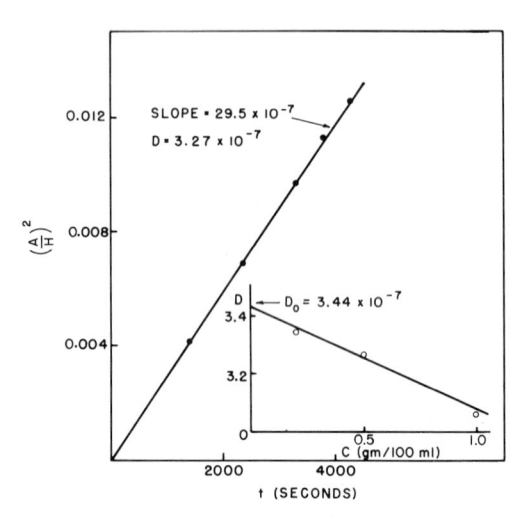

FIG. 5. A plot of the square of the area–height ratio against time for the boundary curve established by layering cyclohexane over a 0.5 gm/100 ml solution of polystyrene S105 at 4908 rpm and 35°C by means of a synthetic boundary cell. Insert shows an extrapolation to zero concentration of the diffusion coefficients obtained by this type of plot.

fraction centrifuged in a synthetic boundary cell at low centrifugal fields ($2s\omega^2 t = 0.0004$ to 0.0014). Although Faxen's treatment is actually valid only for concentration-independent sedimentation and diffusion, it holds fairly well for this experiment since the use of a synthetic boundary cell and a low rotational speed approaches free diffusion by minimization sedimentation effects.

In view of the concentration dependence of most sedimenting systems application of Faxen's approximate solution to the Lamm equation is quite limited. In fact, Baldwin [14] showed that even a slight dependence of s upon c causes appreciable inaccuracy in the determination of a diffusion coefficient when employing Faxen's method. Fujita [15] recently extended the solution of the Lamm equation to include a linear dependence of sedimentation on concentration. This decrease in s with c causes a marked sharpening of the boundary as a result of the molecules on the solvent side of the boundary having a higher sedimentation coefficient than do those on the solution side. Although the boundary position is still shown to be fairly well denoted by the position of the 'maximum gradient, Fujita's treatment provides an equation for determining the diffusion coefficient from the area–height ratio, which has been shown by Baldwin [14] and Fujita [16] to be far superior to Faxen's equation. These methods have found limited applicability in the determination of polymer diffusion coefficients since they are valid only for systems containing a single molecular weight solute.

3. *Molecular Weights from Sedimentation Coefficients*

Sedimentation coefficients are dependent upon the molecular weight of the sedimenting solute; however, the actual evaluation of the molecular weight requires a combination of the sedimentation coefficient with some other solution property for the same polymer. Classically this has been accomplished by application of the Svedberg equation (16) to sedimentation and diffusion measurements. Although the diffusion coefficient may, in principle, be calculated from the same data used for determining the sedimentation coefficient, a more static method is usually employed in order to give more accurate values. Application of the Svedberg equation may be illustrated by use of the sedimentation and diffusion data from Figs. 3 to 5, giving a molecular weight of 161,000 for the polystyrene fraction S105, values of 0.940 cc/gm and 0.7635 gm/cc being used for $\bar{v}$ and ρ. This molecular weight may later be compared with those obtained for the same fraction by other sedimentation methods.

The molecular weight of a polymer may also be calculated from the sedimentation coefficient by a combination with intrinsic viscosity data. Mandelkern and Flory [17] showed that the introduction of the relation between intrinsic viscosity and frictional coefficient into Eq. (13) leads to

$$s[\eta]^{1/3}/M^{2/3} = \Phi^{1/3}P^{-1}(1 - \bar{v}\rho)/\eta_0 N \qquad (30)$$

where $[\eta]$ is the intrinsic viscosity, η_0 is the viscosity of the solvent, and $\Phi^{1/3}P^{-1}$ is a constant supposed to be independent of the particular polymer–solvent system employed. Numerous experimental evaluations [17–19] indicate the approximate constancy of $\Phi^{1/3}P^{-1}$ with an average value of 2.5×10^6 for polymers having a random coil configuration. Although the value of $\Phi^{1/3}P^{-1}$ may differ with molecular configuration [20] and may not be entirely constant within a particular polymer–solvent system [21,22], critical use of Eq. (30) can be valuable for a simple calculation of approximate molecular weights. Application of Eq. (30) to the polystyrene fraction S105, which has an intrinsic viscosity of 0.310 dl/gm in cyclohexane at 35°C, gives a molecular weight of 135,000, using the value of 0.00748 poise for the solvent viscosity.

B. Polymer Heterogeneity

The study of polydispersity was one of the original topics undertaken in ultracentrifugal analysis [23,24]. Application of a centrifugal force to a non-uniform solute produces a separation of the solute components according to their rates of sedimentation. This provides a physical fractionation represented by a broadening of the sedimentation boundary. Polymer chemists have become involved in analyzing this boundary spreading to evaluate molecular weight distributions. Methods employed in most of the studies on polydispersity may be classified into two general approaches:

(1) measurement of certain parameters of the gradient curve which are used to calculate heterogeneity coefficients for assumed distribution functions;

(2) measurement of the distribution of sedimentation coefficients which is converted to a complete molecular weight distribution curve.

The first method usually involves measurement of the gradient curve width by the area–height ratio as introduced by Gralén [25] or by the standard deviation as introduced by Baldwin and Williams [26]. Although the aim of this method is to simplify computation, very little time is actually saved because of the time-consuming analysis of diffusion and concentration effects. Use of the first method is also limited to distributions which can be adequately described by simple distribution functions, while the second method can satisfactorily describe rather complex distributions.

1. Gralén Method

For a measure of boundary spreading Gralén [25] introduced the width, B, of the sedimentation curve defined by the ratio of the area, A, to the maximum height, H, as given by Eqs. (27) and (28). The derivative dB/dr, called the widening value, is obtained from a plot of B against the position of the boundary peak. Concentration effects are eliminated by extrapolation of dB/dr to zero concentration. If diffusion effects are negligible, as for materials studied by Gralén [25], Jullander [27], and Ranby [28], this widening value at infinite dilution, $(dB/dr)_0$, can be attributed entirely to polydispersity; however, if the diffusion contribution is large, as for many synthetic polymers, a further correction is necessary to exclude the spreading effects of diffusion. Eriksson [29] extended Gralén's treatment to include the contribution of diffusion by assuming that diffusion and sedimentation occur independently during centrifugation experiment. According to this treatment diffusion effects can be subtracted from the total width, B_{total}, by the relation

$$B_{\text{sed}} = (B_{\text{total}}^2 - 4\pi Dt)^{1/2} \tag{31}$$

to give the width, B_{sed}, due to sedimentation alone. Application of this method is illustrated in Fig. 6 for the polystyrene fraction S105, using the diffusion coefficients in Fig. 5 for the diffusion correction. Comparison of this corrected widening value of 0.14 with the uncorrected value of 0.26

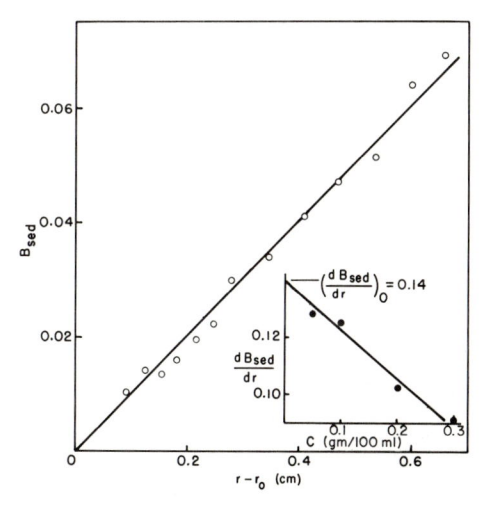

FIG. 6. A plot of the boundary gradient curve width against distance from the center of rotation for the sedimentation of a 0.2 gm/100 ml solution of polystyrene S105 in cyclohexane at 59,780 rpm and 35°C, diffusion effects being removed by $B_{\text{sed}} = (B_{\text{total}}^2 - 4\pi Dt)^{1/2}$ using $D = 3.44 \times 10^{-7}$. Insert shows an extrapolation of the widening value to zero concentration.

indicates the pronounced boundary spreading caused by diffusion. Eriksson reported diffusion corrections of this magnitude in his studies on polymethylmethacrylate, while these corrections were not considered in the earlier studies of Gralén and Lagermaln [30] and Kinell [31].

The dB/dr values representing polydispersity are sometimes used simply for distribution comparison. Ranby [28] introduced the application of dB/dr values to approximate fraction distributions for the construction of mass distribution curves from fractionation data. He assumed a symmetrical triangular distribution for each fraction having a base of $2sdB/dr$ and a height such that the area of the triangle is proportional to the weight of the fraction. Eriksson [29] extended this approach to approximate the fraction distribution by a Gaussian curve having an area proportional to the fraction weight and a standard deviation σ given by $s(dB/dr)/(2\pi)^{1/2}$. The study of Eriksson showed a very good agreement between the mass distributions calculated by the customary methods of Schulz [32], the triangle method of Ranby, and his Gaussian curve method. Agreement may not be as good for a less selective polymer fractionation.

Application of dB/dr values for calculating the coefficients of assumed distribution fractions is completely dependent upon the selection of a simple function to adequately represent the polymer distribution. Gralén [25] made the assumption that the logarithmic distribution function introduced by Lansing and Kraemer [33] for sedimentation equilibrium measurements could be applied to sedimentation constants by the equation

$$dc = K_s \exp(-y^2)\, ds \qquad (32)$$

where

$$y = (1/\gamma_s) \ln(s/s_m) \qquad (33)$$

s_m corresponds to the curve maximum, K_s is a constant expressing the maximum weight of the distribution curve, and γ_s is a distribution coefficient given by

$$(dB/dr)_0 = \pi^{1/2} \gamma_s \exp(-\gamma_s^2/4) \qquad (34)$$

The value γ_s is a direct measure of polymer heterogeneity. Although this distribution function is particularly useful for many polymers, Jullander [27] made the method more flexible by introducing a function with three parameters in order to permit independent variation of skewness and dispersion. Although the application of these assumed functions may be fairly rapid in certain cases and has received considerable attention [28,31,34] in the study of synthetic high polymers, it can never reveal details in a distribution and fails completely when the distribution contains more than one maximum.

2. Standard Deviation

Since second moments about the mean are additive in a combined distribution composed of independent distributions, Baldwin and Williams [26,35] introduced the use of moments as a measure of the position and width of the boundary gradient curve. The boundary position is given by the first moment, $\bar{r}$, of dc/dr about the center of rotation

$$\bar{r} = \frac{\int r(\partial c/\partial r)\, dr}{\int (\partial c/\partial r)\, dr} \qquad (35)$$

while the width of the boundary is given by the second moment, σ^2, of the curve about $\bar{r}$,

$$\sigma^2 = \frac{\int (r - \bar{r})^2 (\partial c/\partial r)\, dr}{\int (\partial c/\partial r)\, dr} \tag{36}$$

For a symmetrical boundary curve $\bar{r}$ coincides with the distance to the boundary maximum. Baldwin [36] showed for the case of a symmetrical distribution with constant s and D that the second moment σ^2 could be given by

$$\sigma^2 = (p\omega^2 \bar{r} t)^2 + 2\, Dt(1 + \bar{s}\omega^2 t) \tag{37}$$

where $\bar{s}$ and p^2 are the first and second moments about the mean of the distribution of sedimentation coefficients. The second moment, σ^2, of Eq. (37) consists of two components, $(p\omega^2 \bar{r} t)^2$ being due to heterogeneity in sedimentation coefficients and $2\, Dt(1 - \bar{s}\omega^2 t)$ being due to diffusion. These two effects may be separated since the boundary spreading due to sedimentation and diffusion depends upon different powers of time. Rearrangement of Eq. (37) after substitution of Eq. (20) leads to

$$\frac{\sigma^2(r_0/r)}{2t} = D + \left(\frac{p^2 \omega^4 r_0}{2}\right) \bar{r} t \tag{38}$$

showing that a plot of $\sigma^2(r_0/r)/2t$ against $\bar{r}t$ should give a straight line with D and p being obtained from the intercept and slope.

Even though this method is particularly useful for determining the boundary spreading free of diffusion effects, such a method, like that of Gralén, does not allow for measurement of details in polymer distribution.

3. Distribution of Sedimentation Coefficients

Boundary gradient curves observed in sedimentation velocity experiments reflect a rather high degree of sensitivity to polymer heterogeneity; however, full utilization of this resolving power requires a determination of the entire distribution of sedimentation coefficients. Normalized distributions of sedimentation coefficients are easily obtained from refractive index gradient curves (Fig. 1) by a simple transformation of coordinates [37,38], the ordinate

$$g^*(s) = \frac{(dn/dr) r^3 \omega^2 t / r_0^2}{\int (dn/dr)(r^2/r_0^2)\, dr} \tag{39}$$

representing the relative frequency of material having a sedimentation coefficient given by the abscissa

$$s = \frac{1}{\omega^2 t} \ln \frac{r}{r_0} \tag{40}$$

Apparent distributions obtained in this manner are true distributions of sedimentation coefficients only if the boundary spreading is not affected by pressure, diffusion, or concentration. Since this is not usually the case for the sedimentation of most polymers in organic solvents, it becomes necessary to consider these effects in order to obtain an accurate molecular weight distribution.

At the centrifugal fields normally employed in sedimentation velocity experiments, a large hydrostatic pressure is produced, varying from about 1 atm at the meniscus to several hundred atmospheres at the bottom of the cell. Since the solution density and viscosity as well as the partial specific volume of the solute vary with pressure, sedimentation occurring in such a pressure gradient varies as a function of distance from the meniscus. These pressure effects are most pronounced for the relatively compressible organic polymers and solvents of interest to polymer chemists. The problem of pressure-dependent sedimentation, first considered by Mosimann and Signer [39], has recently received considerable attention. Through a mathematical refinement of the empirical considerations of Oth and Desreux [40], Fujita [41] employed the Lamm equation to show that a linear dependence of sedimentation on pressure leads to

$$s^0 = \frac{1}{\omega^2 t} \ln \frac{r}{r_0} \left\{ 1 - \frac{m}{2} \left[\left(\frac{r}{r_0}\right)^2 - 1 \right] \right\}^{-1} \tag{41}$$

where

$$m = \tfrac{1}{2}\mu\omega^2 r_0^2 \rho^0 \tag{42}$$

the superscript zero denoting a value at 1 atm pressure and μ being a coefficient containing the dependence of the frictional coefficient, partial specific volume, and density on pressure. This gives a corresponding frequency function of

$$g^*(s^0) = \frac{(dn/dr)(r/r_0)^2 r\omega^2 t \{1 - m[(r/r_0)^2 - 1]\}^2}{\int (dn/dr)(r/r_0)^2 \{1 - m[(r/r_0)^2 - 1]\}\, dr} \tag{43}$$

neglecting the minor correction due to the pressure dependence of refractive index. Fujita extended this treatment to include concentration effects caused by the radial dilution accompanying boundary movement; however, Billick [42] showed that under θ conditions this error is generally within the experimental accuracy of sedimentation measurements. Such corrections may, however, be desirable for sedimentation in good solvents.

Application of Eqs. (41) and (43) requires the value of μ for the particular polymer–solvent system employed. If sufficient compressibility data is available μ may be calculated from its definition [41]

$$\mu = \lambda + \frac{\rho^0 \bar{v}^0 (\beta - 1/\kappa)}{1 - \rho^0 \bar{v}^0} \tag{44}$$

where β is the compressibility of the solvent, κ is the bulk modulus of the solute, and λ is the pressure dependence of the frictional coefficient which may be approximated from solvent viscosities by

$$\lambda = [(\eta/\eta^0) - 1]P \tag{45}$$

A value of $\mu = 2 \times 10^{-9}$ is calculated in this manner for the polystyrene-cyclohexane system at 35°C [42,43]. When compressibility data is not available the value of μ may be obtained directly from sedimentation data by a linear plot of $\ln(r/r_0)/\omega^2 t$ against $(r/r_0)^2 - 1$ or from data at different rotational speeds by a linear plot of $\ln(r/r_0)/\omega^2 t$ at fixed values of r/r_0 against ω^2, the slope of the first plot being $\frac{1}{4}\mu\omega^2 r_0^2 \rho^0 s^0$ and that of the second being $\frac{1}{4}\mu r_0^2 \rho^0 s^0 [(r/r_0)^2 - 1]$. By this technique Wales and Rehfeld [43] found a value of $\mu = 2.2 \times 10^{-9}$ for the polystyrene-cyclohexane system, while studies by Billick [42] and Blair and Williams [44] indicate a somewhat lower value. Using a value of $\mu = 2 \times 10^{-9}$ transformation of the refractive index gradient curves in Fig. 1 leads to the apparent distribution curves shown in Fig. 7.

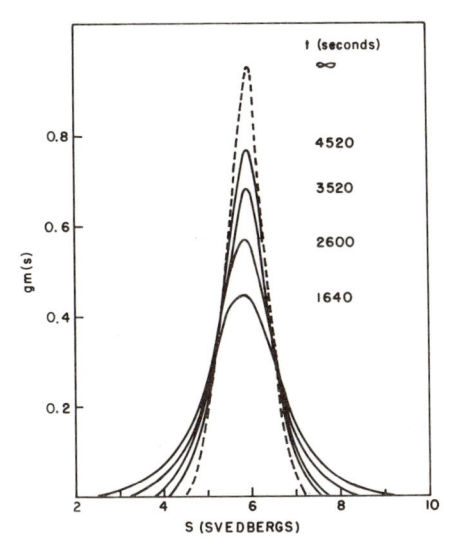

FIG. 7. Apparent distributions of sedimentation coefficients for increasing times of sedimentation of a 0.30 gm/100 ml solution of polystyrene S105 in cyclohexane at 59,780 rpm and 35°C, the curves being corrected to 1 atm pressure. Dashed line represents the curve corrected for diffusion.

The variation of apparent distribution curves with time is caused by the effects of diffusion upon boundary spreading. This diffusional contribution can be eliminated by extrapolation to infinite time, since spreading of the boundary due to difference in s is proportional to t while spreading due to diffusion is proportional to $t^{1/2}$. When diffusion effects are not too great this correction may be carried out by the method of Baldwin and Williams [26], where apparent distribution functions $g^*(s^0)$ at given values of s from several curves are plotted against $1/rt$ and extrapolated graphically to infinite time. Corrected frequency functions obtained at several sedimentation coefficients describe a distribution of sedimentation coefficients free of diffusional effects. Such a corrected curve is illustrated in Fig. 7. Although Gosting [45] showed that there is a range of time in which $g^*(s^0)$ is a linear function of $1/t$, this linear region is rarely reached in sedimentation velocity experiments due to the limited cell length. However, Baldwin [36] showed that an approximately linear region useful for this procedure is obtained when boundary spreading due to heterogeneity is at least three times that due to diffusion by the end of the experiment. Eriksson [46] and Baldwin [47] have presented other methods of extrapolation which may be used when diffusion is more pronounced, the most promising being the extrapolation of $(s - \bar{s})^2$ against $1/t \exp(\bar{s}\omega^2 t)$ at fixed values of $G(s)$, where $\bar{s}$ is the mean of the distribution of s and $G(s)$ is the integral distribution function

$$G(s) = \int_0^s g(s)\, ds \qquad (46)$$

Concentration dependence of sedimentation leads to the boundary sharpening effect illustrated in Fig. 8 which may be eliminated by an extrapolation to infinite dilution. Graphical extrapolations of distribution curves obtained at several concentrations have been accomplished by a plot of $g(s)$ against c at fixed values of s [48], a plot of s (or $1/s$) against c at fixed values of $g(s)/g(s)_{max}$ [49], or a plot of s (or $1/s$) against c at fixed values of $G(s)$ [50]. Extrapolation to zero concentration may also be accomplished by the analytical procedure of Baldwin [49] provided a relationship between sedimentation coefficient and concentration is known. This method essentially consists in determining the concentration and sedimentation coefficients of all species in the region ahead of the fastest boundary by means of a step-by-step correction for the appearance of each species, beginning at the solvent side of the boundary where only one species is present. Such a calculation can become quite tedious, but when a simple expression such as that given by Eq. (25) is available for the concentration dependence, this correction can be easily programmed for computer calculation. This concentration correction (Fig. 8) following previous correction for pressure and diffusion leads to a true distribution of sedimentation coefficients describing polymer heterogeneity.

a. *Molecular weight distribution.* For linear polymers the dependence of sedimentation coefficient upon molecular weight may be expressed by the relation

$$s = KM^a \tag{47}$$

where K and a are constants dependent upon the particular polymer–solvent system. Although this is valid only for linear polymers, deviation for

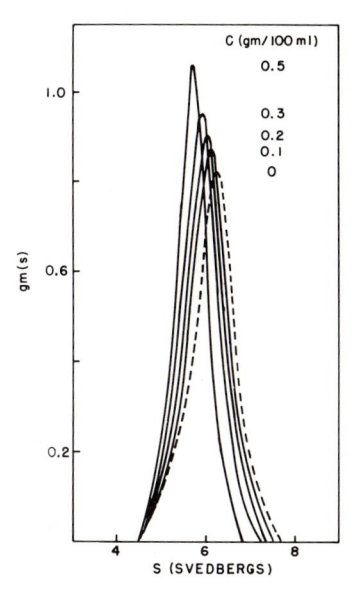

FIG. 8. Effect of concentration on the distribution of sedimentation coefficients for polystyrene S105 in cyclohexane at 35°C, the curves having been previously corrected for pressure and diffusion effects. Dashed line represents the true distribution of sedimentation coefficients obtained by extrapolation to zero concentration.

the relationship may be useful for studying the degree of branching [51]. Once the constants for Eq. (47) are known a complete molecular weight distribution can be obtained from the distribution of sedimentation coefficients merely by a change in variables. This relationship may be established by relating the sedimentation coefficients for several polymers to corresponding molecular weights obtained by an independent method. It is necessary in the application of this method to determine the sedimentation coefficient that corresponds to the average molecular weight employed, especially when the distribution of the calibration polymer is not sharp. The average sedimentation coefficient, $\bar{s}_a$, corresponding to a weight average molecular weight, may be calculated by the relation [52,53]

$$\bar{s}_a = \left[\int s^{1/a} g(s) \, ds \right]^a \tag{48}$$

where a is the constant in Eq. (47). Average sedimentation coefficients thus calculated for several polystyrene samples are listed in Table I along with

TABLE I

SEDIMENTATION COEFFICIENTS AND MOLECULAR WEIGHTS FOR POLYSTYRENE SAMPLES OF VARIOUS MOLECULAR WEIGHTS

Sample	$\bar{S}_a{}^a$	$\bar{M}_w{}^b$	$\bar{M}_w{}^c$	$\bar{M}_n{}^c$	$\bar{M}_w/\bar{M}_n{}^c$
S102	4.42	83,000	81,000	75,000	1.08
S103	5.52	127,000	127,000	118,000	1.07
S105	6.09	160,000	155,000	148,000	1.05
S109	6.71	185,000	187,000	176,000	1.06
S111	7.33	239,000	224,000	214,000	1.05
S108	7.90	267,000	260,000	242,000	1.07
S38	12.05	596,000	604,000	461,000	1.31
S13	15.05	945,000	968,000	642,000	1.51
S114	27.83	2,750,000	3,220,000	2,490,000	1.29
B8	—	—	279,000	113,000	2.47

[a] Determined from true distribution of sedimentation coefficients by Eq. (48) using $a = 0.5$.
[b] Determined by sedimentation equilibrium.
[c] Determined from molecular weight distribution curves obtained by sedimentation velocity analysis.

the weight average molecular weights obtained by sedimentation equilibrium measurements. The relation

$$s = 0.0155 \, M^{0.50} \tag{49}$$

is obtained by the best fit of this data to Eq. (47). The sedimentation coefficient will be somewhat less dependent upon molecular weight for good solvent systems ($a < \frac{1}{2}$). Molecular weights determined by the Archibald method [53,54], the Svedberg equation [55], and intrinsic viscosity calibration against light scattering [56] have been employed in establishing this relationship for several polymers, while Wales and Rehfeld [43] avoided this calibration procedure by assuming a value for K ($\frac{1}{2}$ for a θ solvent) and determining the molecular weight from intrinsic viscosity according to Eq. (30). Application of Eq. (49) leads to the molecular weight distributions in Fig. 9 for the anionically polymerized polystyrene S105 (the distribution of sedimentation constants given in Fig. 8) and for a thermally polymerized polystyrene. The different average molecular weights can be calculated from such complete molecular weight distributions by use of proper summation processes with the values so obtained for $\bar{M}_w$ and $\bar{M}_n$ listed in Table I. Molecular weight distributions and average molecular weights

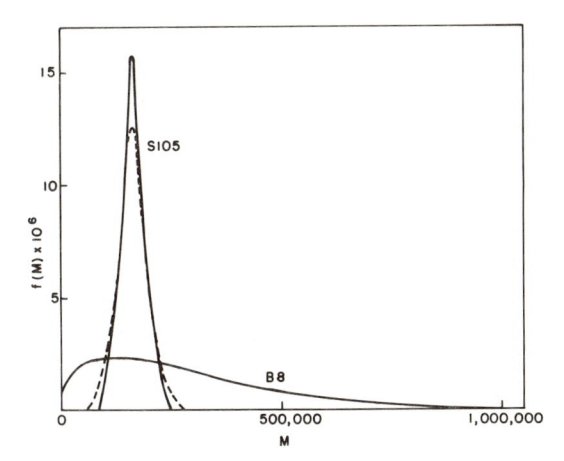

FIG. 9. Molecular weight distributions obtained by sedimentation velocity analysis for anionically polymerized polystyrene S105 and thermally polymerized polystyrene B8. Dashed line represents the apparent molecular weight distribution obtained by analytical correction for pressure and concentration effects (not diffusion effects) of a single curve after sedimentation of a 0.30 gm/100 ml solution of polystyrene S105 in cyclohexane at 59,780 rpm and 35°C for 4520 seconds.

obtained by this technique compare very favorably with those obtained from other methods by many workers throughout the world for several of these same polymers, although there are minor disagreements regarding the low or high molecular weight ends of the distribution curve.

Practical application of sedimentation velocity analysis for determining a molecular weight distribution appears to depend to a certain extent upon the solvent employed. Although this method should lead to a true molecular weight distribution irrespective of the solvent employed if apparent distributions obtained at finite concentrations are correctly extrapolated to infinite dilution [1], it is found that, at the concentrations required for most sedimentation measurements, a better molecular separation can be obtained by the use of a θ solvent. The degree of separation obtainable is illustrated in Fig. 10 by a comparison of the molecular weight distribution for a 50/50 mixture of two fractions with the distributions of the individual fractions. Satisfactory distributions have been obtained from sedimentation measurements carried out in better solvents [9,56]; however, reliable results are usually arrived at only for rather dilute solutions where optical measurements are generally less sensitive.

This type of analysis has become most useful to the polymer chemist due to the ability to depict the high degree of resolving power observed in sedimentation velocity experiments. Although the complete corrections necessary for obtaining a true distribution curve require considerable time

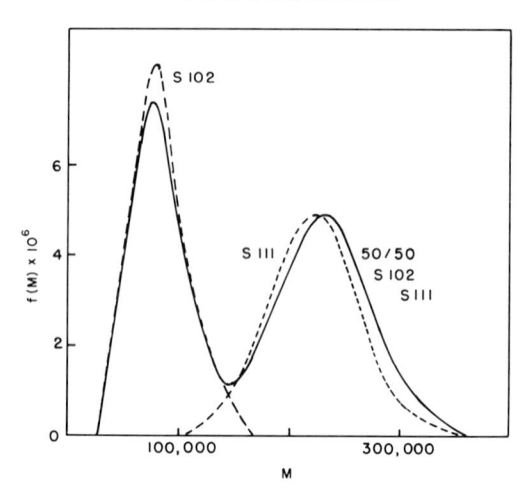

FIG. 10. Molecular weight distribution for a 50/50 mixture of two narrow-distribution polystyrenes with the dashed lines representing the distributions obtained on the separate samples. The $f(M)$ values for the separate samples are halved to permit better comparison.

and labor, it is often possible to obtain sufficiently accurate apparent distributions from a single schlieren pattern by the analytical computation (preferably on a computer) of many of these corrections. After correction constants are determined for a particular system, graphical extrapolation is required only for diffusion corrections, and the error involved in neglecting diffusion effects may usually be minimized by selecting a boundary curve obtained after long sedimentation time. Comparison of a true molecular weight distribution with an apparent distribution analytically corrected for only pressure and concentration effects is illustrated in Fig. 9, showing that such a rapidly determined distribution can be very useful in many applications.

C. Archibald Method

Archibald [57] proposed a method for determining the molecular weight of a solute from data obtained during the approach to sedimentation equilibrium. He pointed out that since there is no flow of solute across the meniscus and cell bottom ($J = 0$), Eq. (9) may be written for these two positions

$$sc_a\omega^2 r_a = D(dc/dr)_a, \qquad sc_b\omega^2 r_b = D(dc/dr)_b \qquad (50)$$

or

$$\frac{S}{D}\omega^2 = \frac{(dc/dr)_a}{c_a r_a} = \frac{(dc/dr)_b}{c_b r_b} \qquad (51)$$

where the subscripts a and b refer to values at the meniscus and cell bottom, respectively. Substitution into the Svedberg equation leads to the expression

$$M = \frac{RT(dc/dr)_a}{(1 - \bar{v}p)\omega^2 c_a r_a} = \frac{RT(dc/dr)_b}{(1 - \bar{v}p)\omega^2 c_b r_b} \tag{52}$$

showing that the molecular weight of a solute may be calculated from values of the concentrations and their gradients at the meniscus and cell bottom. For heterogeneous polymers sedimentation causes a concentration of the heavier solutes toward the cell bottom; consequently, the molecular weight increases with time at r_b and decreases at r_a. Extrapolation to zero time is, therefore, required for a determination of the average molecular weight of the original polymer.

Values of dc/dr at the ends of the cell are directly measurable from schlieren boundary curves while values of c must be calculated from the gradient curve. Klainer and Kegeles [58] have shown that, provided there remains a plateau region in the cell, the concentration at the meniscus can be expressed by

$$c_a = c_0 - \int_{r_a}^{r_p} \left(\frac{r}{r_0}\right)^2 \left(\frac{dc}{dr}\right) dr \tag{53}$$

where r_p is a reference position in the plateau region and c_0 is the concentration of the original solution which can be determined by integration

$$c_0 = \int_{r_a}^{r_b} \left(\frac{dc}{dr}\right) dr \tag{54}$$

of a separate schlieren curve obtained by a synthetic boundary cell prior to appreciable sedimentation. The concentration at the cell bottom is similarly determined by changing the minus sign in Eq. (53) to a plus sign. Best results are obtained from experiments performed under conditions that give moderately large values of $(dc/dr)_a$ and c_a. In general practice these conditions are empirically arrived at by varying the experimental speeds of rotation to yield desired values.

The molecular weight may be calculated directly from individual values of c_a, $(dc/dr)_a$, and r_a; however, Trautman [59] suggested that more precision could be obtained by determining the molecular weight from the slope of a plot of $(dc/dr)_a/\omega^2 r_a$ versus c_a for all possible data points. Erlander and Foster [60] have indicated that the molecular weight determined from the slope of this plot is probably a Z average, while it has previously been shown [61] that the molecular weight determined from the individual points is a weight average. Although the Trautman plot will yield a curved line for heterogeneous polymers, it can be usefully employed for determining the molecular weight at zero time, as illustrated by Fig. 11 for polystyrene S105 where a value of $\overline{M}_w = 160{,}000$ is obtained from the intercept at the original concentration ($t = 0$). For ideal polymer solutions (θ solvents) the molecular weight is concentration-independent. When better solvents are employed in the application of this method it is necessary to extrapolate the apparent molecular weights to infinite dilution [62]. Although the Archibald method has received considerable attention due to the short experimental times required, the method suffers from a loss in precision in extrapolating to the ends of the cell. Recent advances in the rapid attainment of sedimentation equilibrium have made the Archibald method somewhat less attractive.

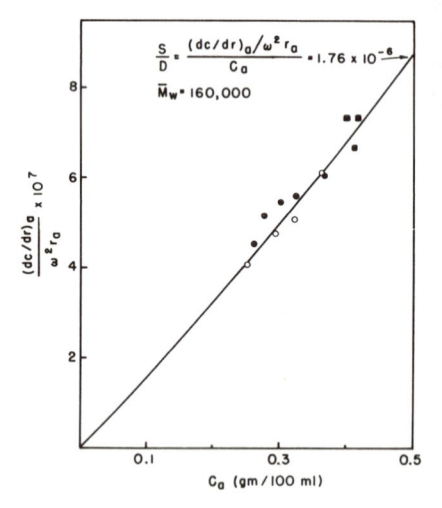

FIG. 11. A plot of $(dc/dr)_a/\omega^2 r_a$ versus concentration at the meniscus for the sedimentation of a 0.50 gm/100 ml solution of polystyrene S105 in cyclohexane at 35°C and rotational speeds of (○) 11,573 rpm, (●) 10,589 rpm, and (■) 7447 rpm.

V. Sedimentation Equilibrium

Sedimentation equilibrium has several advantages as an absolute method for measuring molecular weights of polymers. Large particles such as dust sediment out of the field, while low molecular weight impurities have little effect. Molecular weights varying from a few hundred to several million may be measured with only a small quantity of polymer, and in principle the heterogeneity can be determined. Disadvantages, which have included the length of the experiment and the difficulty in studying sufficiently low concentrations to enable reliable extrapolation to zero concentration, may be minimized by use of poor solvents and recent advances in the rapid attainment of equilibrium. Since sedimentation velocity gives greater sensitivity to details in molecular weight distribution, equilibrium measurements have been employed largely for determining only average molecular weights, although equilibrium studies in a mixed solvent (density gradient sedimentation) have recently shown promise in demonstrating other types of polymer heterogeneity.

A. Average Molecular Weights

The expression describing the concentration distribution in an ultracentrifuge cell at equilibrium can be derived either from thermodynamic or kinetic considerations, with the elementary expression for an ideal two-component system being derived from both approaches in the early work of Svedberg [63]. A more general expression may be obtained by putting the force on each component [given by X_k in Eq. (7)] equal to zero

$$(1 - \bar{v}_i p)\omega^2 r = \sum_j \frac{\partial \mu_i}{\partial c_j} \frac{dc_j}{dr} \tag{55}$$

Substitution of Eq. (12) into this expression leads to the differential equation for sedimentation equilibrium

$$M_i(1 - \bar{v}_i p)\omega^2 r = \frac{RT}{c_i} \frac{dc_i}{dr} + RT \sum \frac{\partial \ln y_i}{\partial c_j} \frac{dc_j}{dr} \tag{56}$$

Since the nonideality correction depends both on the total concentration at r and the distribution of species at that point, equations for such concentration-dependent systems are coupled in a very complex manner. For the thermodynamically ideal system the last term of Eq. (56) vanishes and we have

$$M_i = \frac{RT}{(1 - \bar{v}p)\omega^2} \frac{(dc_i/dr)}{c_i r} \tag{57}$$

The weight average molecular weight, $\bar{M}_{wr}$, of polymer molecules at point r in the cell is given by the summation

$$\bar{M}_{wr} = \frac{\sum c_i M_i}{\sum c_i} = \frac{RT}{(1 - \bar{v}p)\omega^2} \frac{dc/dr}{rc} \tag{58}$$

where c is the total concentration at point r. The weight average molecular weight of the whole polymer may be obtained by integration of this quantity over the entire cell

$$\bar{M}_w = \frac{\int_a^b \bar{M}_{wr} cr \, dr}{\int_a^b cr \, dr} = \frac{2RT(c_b - c_a)}{(1 - \bar{v}p)\omega^2 c(r_b^2 - r_a^2)} \tag{59}$$

Application of this analysis to the data in Fig. 2 gives a weight average molecular weight of 160,000 for polystyrene S105, which agrees very well with the molecular weights obtained by the Svedberg equation (Section IV,A,3) and the Archibald method (Section IV,C). Utilization of this data requires a determination of the concentration, c_a, at the meniscus which may be obtained, as done in this case, by a statement of the conservation of mass [33]

$$c_a = c_0 - \frac{r_b^2(c_b - c_a) - \int_{r_a}^{r_b} r^2 \, dc}{r_b^2 - r_a^2} \tag{60}$$

or approximated by assuming the concentration at $\frac{1}{2}(r_a - r_b)$ being equal to the original solution. Weight average molecular weights listed in Table I for several polystyrene samples were obtained by this method.

Lansing and Kraemer [33] and Wales [64] developed methods for calculating $\overline{M}_z$ and successively higher average molecular weights. In principle, it is possible to calculate as many average molecular weights as desired, while, in practice, experimental limitations rapidly decrease the reliability of averages higher than $\overline{M}_z$. Number average molecular weights cannot be calculated directly from sedimentation equilibrium data unless the solute concentration approaches zero at some point in the cell [65].

Concentration Dependence

Molecular weights obtained by Eq. (57) are concentration-independent only at the θ temperature where nonideality corrections are zero. Under non-θ conditions a complex concentration correction represented by the last term in Eq. (56) must be considered in determining the true molecular weight. Wales [64] proposed the approximation of $\partial \ln y_i / \partial c_k$ by use of virial coefficients

$$1/(\overline{M}_w)_{\text{app}} = (1/\overline{M}_w) + Bc \tag{61}$$

as done in osmotic pressure and light scattering; however, Mandelkern et al. [66] concluded that due to difficulties encountered by the variation in mass distribution along the length of the cell, the polymer molecular weight can be obtained in a straightforward manner only in ideal solutions. Fujita et al. [67] have shown that $(1/\overline{M}_w)_{\text{app}}$ has a reliable linear extrapolation with concentration for sufficiently low concentrations at temperatures near, but not necessarily at, the θ temperature.

B. POLYMER HETEROGENEITY

Sedimentation equilibrium is theoretically sensitive to the molecular weight distribution of a polymer solute. A relationship can be written between the molecular weight distribution and the concentration distribution at equilibrium. Several methods have been proposed by Rinde [68] and Wales [64] for the solution to this relation, but the results have generally been considered as unsatisfactory.

Another procedure for determining the molecular weight distribution is to find the distribution function yielding the average molecular weights calculatable from moments of the experimental concentration curve [65]. Since $\overline{M}_{z+1}$ is the highest average molecular weight that can be determined with any degree of accuracy, the number of moments of this distribution function that can be experimentally determined is limited to three or four. With such a limitation the fine details of a distribution cannot be reproduced. Determination of a molecular weight distribution from sedimentation equilibrium experiments has, therefore, usually evolved to the application of an assumed two-parameter function of reasonable form. Distribution functions most useful for this type of data fitting are discussed in Chapter E.

C. Time Required to Reach Equilibrium

Equilibrium measurement has often been looked upon with disfavor due to the length of time that has been considered necessary for attainment of equilibrium. This time has generally been estimated by Weaver's rule [69]

$$t = \frac{4(r_b - r_a)}{s\omega^2(r_a + r_b)} \tag{62}$$

although more recently Van Holde and Baldwin [70] have shown that under the experimental conditions suitable for most equilibrium measurements the time may be more satisfactorily given by

$$t_\delta = \frac{(r_b - r_a)^2 F(\alpha)}{D} \tag{63a}$$

where

$$\delta = \frac{\Delta C_{eq} - \Delta C_t}{\Delta C_{eq}} \tag{63b}$$

$$\Delta C = C_b - C_a \tag{63c}$$

$$F(\alpha) = -\frac{1}{\pi^2 U(\alpha)} \ln\left\{\frac{\pi^2 [U(\alpha)]^2 \delta}{4[1 + \cosh(\tfrac{1}{2}\alpha)]}\right\} \tag{63d}$$

$$U(\alpha) = 1 + \frac{1}{4\pi^2\alpha^2} \tag{63e}$$

$$\alpha = \frac{2RT}{\omega^2 M(1 - \bar{v}p)(r_b^2 - r_a^2)} \tag{63f}$$

δ being a measure of the displacement from equilibrium and t_δ the time required to come within δ of reaching equilibrium. For $\alpha > 0.6$ and $\delta = 10^{-3}$, $F(\alpha)$ is almost constant with a value of about 0.68. Using the 5-mm column often recommended for equilibrium experiments this would lead to a time of 137 hours for the attainment of equilibrium with polystyrene S105, while Weaver's rule would give a time three times as large. Van Holde and Baldwin pointed out that although Weaver's rule is adequate for estimating the approach to equilibrium when working at high rotational speeds ($\alpha < 0.6$), it greatly overestimates the time when working at the lower speeds normally employed in equilibrium measurements.

Since the time required to reach sedimentation equilibrium is proportional to the square of the depth of the solution, Van Holde and Baldwin suggested that the most practical method of reducing this time is to employ a short solution column. Reducing the column length to 0.67-mm, as was done in the experiment of Fig. 2 for polystyrene S105, allows the attainment of equilibrium in $2\frac{1}{2}$ hours instead of the 137 hours required in the 5-mm column. Yphantis [71] has extended this treatment by developing a multichannel cell which allows the simultaneous sedimentation of four short columns in one cell. Although the use of a very short column may reduce accuracy, it is possible to obtain fairly reliable weight average molecular weights in a very short time using this technique. For reducing the time required to attain equilibrium when using a larger column, Pasternak et al. [72] have suggested starting the centrifuge with a predetermined step distribution of concentration using a synthetic boundary cell, while Hexner et al. [73] have suggested the rapid establishment of this predetermined step distribution by initially running the centrifuge at a somewhat higher speed. These methods for the rapid attainment of sedimentation equilibrium have made the equilibrium method much more attractive for molecular weight determinations.

D. Density Gradient Sedimentation

Upon subjecting a mixture of two low molecular weight solvents of different densities to a centrifugal field, a density gradient is established. When a polymer is centrifuged in such a mixed solvent the polymer molecules concentrate at a position in this density gradient where the density is equal to the effective density of the polymer. This concentrating tendency is opposed by Brownian motion, which causes the polymer molecules to collect in a band of width inversely related to their molecular weight. Meselson et al. [74] showed that the concentration distribution at equilibrium of a single polymer species in a constant density gradient is Gaussian in shape with a standard deviation, σ, inversely proportional

C.2. Sedimentation

to the square root of the polymer molecular weight

$$\sigma^2 = \frac{RT}{M\bar{v}(d\rho/dr)_{r_0}\omega^2 r_0} \tag{64}$$

where $\bar{v}$ is the effective partial specific volume of the polymer molecules and $(d\rho/dr)_{r_0}$ is the density gradient at the position r_0 of the center of the Gaussian band, the density gradient being evaluated either by calculations from the relation for sedimentation equilibrium in a two-component system or by direct optical measurements in the ultracentrifuge. This treatment, developed on the assumption of constant partial specific volumes and unit polymer activity coefficients, has been extended by Fujita [75] to the more general case including thermodynamic interactions; however, the method does not appear to be extremely promising for systems with appreciable solute interaction.

If the polymer is heterogeneous with respect to molecular weight but homogeneous with respect to effective density, the polymer band observed at equilibrium will be a sum of many Gaussian curves with coincident means, each having a standard deviation related to the individual species molecular weight by Eq. (64). Moments of this composite curve may be employed to determine the various average molecular weights [74,76]. Due to the improbability of ever having θ conditions at all points in a density gradient established by a solvent mixture, it is always necessary to extrapolate the apparent molecular weights determined in this manner to infinite dilution. Hermans and Ende [76] have shown that a linear extrapolation of $(1/\overline{M}_n)_{app}$ and $(\overline{M}_w)_{app}$ against concentration to give satisfactory molecular weights can be obtained for the sedimentation of polystyrene in a θ solvent mixture of cyclohexanol and carbon tetrachloride. Under these conditions the apparent density of the polymer is strongly affected by preferential absorption of the good solvent by the polymer. Satisfactory molecular weights were not obtainable by this method in a good solvent mixture. Although the parameters for assumed distribution functions may be determined from average molecular weights obtained by this method, details in molecular weight distribution cannot be reproduced due to the limited number of moments available.

Density gradient sedimentation has found success not as a method for molecular weight determination but as a highly sensitive method for the detection of density differences in polymer solutes. If the effective densities of the polymeric species are sufficiently distinct, a distribution with more than one mode will be observed. Employing density gradient sedimentation techniques, Bresler et al. [77] have separated a block copolymer of styrene and isoprene from the corresponding homopolymers, while Buchdahl et al. [78] have split an acrylonitrile–vinyl acetate copolymer into three distinct fractions of different apparent partial specific volume with an observable difference of only 0.0005 ml/gm in the apparent partial specific volume of two of the fractions. Buchdahl et al. [79] have also separated atactic from stereoregular polystyrene, obtaining a difference in partial specific volume of 0.025 ml/gm. This is considerably larger than the 0.004 ml/gm difference found by Krigbaum et al. [80] using pycnometry, this discrepancy being attributed to differences in preferential adsorption by the two species. If the effective densities of the polymeric species are in the form of a continuous distribution rather than being distinct, a single model distribution may be observed which will still be Gaussian in shape if the polymer has a Gaussian distribution of effective densities [81]. Since this density distribution markedly increases the width of the polymer band in a density gradient, it is necessary to develop methods which distinguish between the effects of molecular weight and effective density. It appears that this cannot be accomplished exclusively by the experiment as described by Meselsen et al. [74]. In attempting to solve this problem Sueoka [82] suggested that the density gradient method be combined with sedimentation rate measurements, Baldwin and Shooter [83] suggested studying the sedimentation in a preformed density gradient, and Hermans [84] suggested the combination of equilibrium measurements in a density gradient with those in a single solvent. Although separation of molecular weight and density effects remains a problem in density gradient sedimentation, this method of ultracentrifugation possesses considerable promise in studying the heterogeneity in polymer composition.

REFERENCES

1. H. Fujita, "The Mathematical Theory of Sedimentation Analysis." Academic Press, New York, 1962.
2. T. Svedberg, *Kolloid-Z.* **36**, 53 (1925).
3. R. L. Baldwin, *J. Am. Chem. Soc.* **80**, 496 (1958).
4. L. Peller, *J. Chem. Phys.* **29**, 415 (1958).
5. O. Lamm, *Arkiv. Mat., Astron. Fysik.* **21B** (2) (1929).
6. T. Svedberg and K. O. Pedersen, "The Ultracentrifuge." Oxford Univ. Press, London and New York, 1940.
7. J. B. Nichols and E. D. Bailey, in "Physical Methods of Organic Chemistry" (A. Weissberger, ed.), 3rd ed., Vol. I, p. 1007. Wiley (Interscience), New York, 1960.
8. H. K. Schachman, "Ultracentrifugation in Biochemistry." Academic Press, New York, 1959.
9. J. A. V. Butler, A. B. Robins, and K. V. Shooter, *Proc. Roy. Soc.* **A241**, 299 (1957).
10. R. Trautman, V. N. Schumaker, W. F. Harrington, and H. K. Schachman, *J. Chem. Phys.* **22**, 555 (1954).
11. M. Wales and K. E. Van Holde, *J. Polymer Sci.* **14**, 81 (1954).
12. G. Meyerhoff, in "Ultracentrifugal Analysis in Theory and Experiment" (J. W. Williams, ed.), p. 47. Academic Press, New York, 1963.

13. H. Faxen, *Arkiv Mat., Astron. Fysik* **21B**, (3) (1929).
14. R. L. Baldwin, *Biochem. J.* **65**, 503 (1957).
15. H. Fujita, *J. Chem. Phys.* **24**, 1084 (1956).
16. H. Fujita, *J. Phys. Chem.* **63**, 1092 (1959).
17. L. Mandelkern and P. J. Flory, *J. Chem. Phys.* **20**, 212 (1952).
18. L. Mandelkern, W. R. Krigbaum, H. A. Scheraga, and P. J. Flory, *J. Chem. Phys.* **20**, 1392 (1952).
19. T. G. Fox and L. Mandelkern, *J. Chem. Phys.* **21**, 187 (1953).
20. H. A. Scheraga and L. Mandelkern, *J. Am. Chem. Soc.* **75**, 179 (1953).
21. M. Kurata and H. Yamakawa, *J. Chem. Phys.* **29**, 311 (1958).
22. W. H. Stockmayer and A. C. Albrecht, *J. Polymer Sci.* **32**, 215 (1958).
23. T. Svedberg and H. Rinde, *J. Am. Chem. Soc.* **45**, 943 (1923).
24. T. Svedberg and J. B. Nichols, *J. Am. Chem. Soc.* **45**, 2910 (1923).
25. N. Gralén, Dissertation, University of Uppsala (1944).
26. R. L. Baldwin and J. W. Williams, *J. Am. Chem. Soc.* **72**, 4325 (1950).
27. I. Jullander, *Arkiv Kemi, Mineral. Geol.* **21A**, No. 8 (1945).
28. B. G. Ranby, *in* "Advances in Colloid Science" (H. Mark and E. J. W. Verney, eds.), Vol. III, p. 198. Wiley (Interscience), New York, 1950.
29. A. F. V. Eriksson, *Acta Chem. Svand.* **7**, 623 (1953).
30. N. Gralén and G. Lagermaln, *Festskr. Tillagnad J. Arvid Hedval* p. 215 (1948).
31. P. O. Kinell, *Acta Chem. Scand.* **1**, 832 (1947).
32. G. V. Schulz, *Z. Physik Chem.* **B47**, 155 (1940).
33. W. D. Lansing and E. O. Kraemer, *J. Am. Chem. Soc.* **57**, 1369 (1935).
34. J. Oth and V. Desreux, *Intern. Symp. Macromol. Chem., Milan-Turin, 1954* p. 447.
35. J. W. Williams, R. L. Baldwin, W. M. Saunders, and P. G. Squire, *J. Am. Chem. Soc.* **74**, 1542 (1952).
36. R. L. Baldwin, *J. Phys. Chem.* **58**, 1081 (1954).
37. W. B. Bridgman, *J. Am. Chem. Soc.* **64**, 2349 (1942).
38. R. Signer and H. Gross, *Helv. Chim. Acta* **17**, 726 (1934).
39. H. Mosimann and R. Signer, *Helv. Chim. Acta* **27**, 1123 (1944).
40. J. Oth and V. Desreux, *Bull. Soc. Chim. Belges* **63**, 133 (1954).
41. H. Fujita, *J. Am. Chem. Soc.* **78**, 3598 (1958).
42. I. H. Billick, *J. Phys. Chem.* **66**, 1941 (1962).
43. M. Wales and S. J. Rehfeld, *J. Polymer Sci.* **62**, 179 (1962).
44. J. E. Blair and J. W. Williams, *J. Phys. Chem.* **68**, 161 (1964).
45. L. J. Gosting, *J. Am. Chem. Soc.* **74**, 1548 (1952).
46. A. F. V. Eriksson, *Acta Chem. Scand.* **10**, 360 (1956).
47. R. L. Baldwin, *J. Phys. Chem.* **63**, 1570 (1959).
48. J. W. Williams and W. M. Saunders, *J. Phys. Chem.* **58**, 854 (1954).
49. R. L. Baldwin, *J. Am. Chem. Soc.* **76**, 402 (1954).
50. N. Gralén and G. Lagermalm, *J. Phys. Chem.* **56**, 514 (1952).
51. L. D. Moore, Jr., G. R. Greear, and J. O. Sharp, *J. Polymer Sci.* **59**, 339 (1962).
52. R. L. Baldwin and K. E. Van Holde, *Fortschr. Hochpolymer-Forsch.* **1**, 451 (1960).
53. H. W. McCormick, *J. Polymer Sci.* **36**, 341 (1959).
54. H. W. McCormick, *J. Polymer Sci.* **41**, 327 (1959).
55. H. J. Cantow, *Makromol. Chem.* **30**, 169 (1959).
56. H. W. McCormick *J. Polymer Sci.* **A1**, 103 (1963).
57. W. J. Archibald, *J. Phys. & Colloid Chem.* **51**, 1204 (1947).
58. S. M. Klainer and G. Kegeles, *J. Phys. Chem.* **59**, 952 (1955).
59. R. Trautman, *J. Phys. Chem.* **60**, 1211 (1956).
60. S. R. Erlander and J. F. Foster, *J. Polymer Sci.* **37**, 103 (1959).

61. A. Ginsburg, P. Appel, and H. K. Schachman, *Arch. Biochem. Biophys.* **65**, 545 (1956).
62. N. E. Weston and F. W. Billmeyer, Jr., *J. Phys. Chem.* **67**, 2728 (1963).
63. T. Svedberg, *Z. Physik. Chem.* **121**, 65 (1926).
64. M. Wales, *J. Phys. & Colloid Chem.* **52**, 235 (1948).
65. M. Wales, F. T. Adler, and K. E. Van Holde, *J. Phys. & Colloid Chem.* **55**, 145 (1951).
66. L. Mandelkern, L. C. Williams, and S. G. Weissberg, *J. Phys. Chem.* **61**, 271 (1957).
67. H. Fujita, A. M. Linklater, and J. W. Williams, *J. Am. Chem. Soc.* **82**, 379 (1960).
68. H. Rinde, Dissertation, University of Uppsala (1928).
69. W. Weaver, *Phys. Rev.* **27**, 499 (1926).
70. K. E. Van Holde and R. L. Baldwin, *J. Phys. Chem.* **62**, 734 (1958).
71. D. A. Yphantis, *Ann. N.Y. Acad. Sci.* **88**, 586 (1960).
72. R. A. Pasternak, G. M. Nazarian, and J. R. Vinograd, *Nature* **179**, 92 (1957).
73. P. E. Hexner, L. E. Radford, and J. W. Beams, *Proc. Natl. Acad. Sci. U.S.* **47**, 1848 (1961).
74. M. Meselson, F. W. Stahl, and J. R. Vinograd, *Proc. Natl. Acad. Sci. U.S.* **43**, 581 (1957).
75. H. Fujita, *J. Phys. Soc. Japan* **15**, 336 (1960).
76. J. J. Hermans and H. A. Ende, *J. Polymer Sci.* **C1**, 161 (1963).
77. S. E. Bresler, L. M. Pyrkov, and S. Y. Frenkel, *Vysokomolekul. Soedin.* **2**, 216 (1960).
78. R. Buchdahl, H. A. Ende, and L. H. Peebles, *J. Polymer Sci.* **C1**, 143 (1963).
79. R. Buchdahl, H. A. Ende, and L. H. Peebles, *J. Phys. Chem.* **65**, 1468 (1961); *J. Polymer Sci.* **C1**, 153 (1963).
80. W. R. Krigbaum, D. K. Carpenter, and S. Newman, *J. Phys. Chem.* **62**, 1586 (1958).
81. R. L. Baldwin, *Proc. Natl. Acad. Sci. U.S.* **45**, 939 (1959).
82. N. Sueoka, *Proc. Natl. Acad. Sci. U.S.* **45**, 1480 (1959).
83. R. L. Baldwin and E. M. Shooter, *in* "Ultracentrifugal Analysis in Theory and Experiment" (J. W. Williams, ed.), p. 143. Academic Press, New York, 1963.
84. J. J. Hermans, *J. Colloid Sci.* **18**, 433 (1963).

CHAPTER C.3

Isothermal Diffusion

W. Burchard and H.-J. Cantow
LEHRSTUHL FÜR PHYSIKALISCHE CHEMIE DER MAKROMOLEKULAREN SUBSTANZEN
INSTITUT FÜR MAKROMOLEKULARE CHEMIE DER UNIVERSITÄT,
FREIBURG IM BREISGAU, GERMANY

I. Introduction... 285
II. Theory: The Differential Equation of Diffusion and Three Solutions...... 286
III. Experimental Methods for the Determination of Diffusion Constants.... 287
    A. Schlieren Optics, Ultraviolet Optics and Rayleigh Interference Optics. 287
    B. Jamin Interference Optics....................................... 287
    C. Diffusion Cells... 288
IV. Theory: Determination of Molecular Heterogeneity.................. 289
    A. Influence of Heterogeneity on Diffusion Curve Shape.............. 289
    B. Calculation of the Heterogeneity................................ 290
    C. Derivation of Molecular Heterogeneities from those of the Diffusion
       Constant.. 297
    D. Influence of Concentration on the Diffusion Coefficient............. 299
V. Methods for the Determination of the Diffusion Coefficient Distribution. 301
    A. Moment Method of Moacanin, Felicetta, and McCarthy............ 301
    B. Fourier Transformation.. 302
VI. Experimental Results.. 303
VII. Critical Remarks.. 304
    References... 305

I. Introduction

Recently, molecular heterogeneity has been determined frequently using the frictional properties exhibited by macromolecules in solution when subjected to ultracentrifugation.

Although the measurement of isothermal diffusion seems to be experimentally simple, it is a complex problem to obtain exact values. The essential difficulty is the generation of a sharp and unperturbed boundary.

The diffusion curve of polydisperse systems gives no direct indication of the molecular weight distribution. There is no spreading of different molecular weight molecules with increasing time as in sedimentation. The diffusions of the different molecules occurring with different velocities, are overlapping here with time.

II. Theory: The Differential Equation of Diffusion and Three Solutions

If a solution of concentration, c_0, is stratified at point $x = 0$ with a pure solvent, a diffusion process will be set up which will result in the equalization of concentration differences. A stream of particles or molecules is built up

$$j = -D \text{ grad } c \qquad (1)$$

(First law of Fick)

where D is the isothermal diffusion constant. Since the number of particles is constant

$$\text{div } j + (\partial c/\partial t) = 0 \qquad (2)$$

(Equation of continuity)

From the combination of Eqs. (1) and (2), the differential equation of diffusion (3) results

$$\partial c/\partial t = \text{div}(D \text{ grad } c) \qquad (3)$$

We shall now proceed to discuss those experimental arrangements by which diffusion is possible only in the x direction. This reduces Eq. (3) to the one-dimensional case

$$\frac{\partial c}{\partial t} = \frac{\partial}{\partial x}\left(D\frac{\partial c}{\partial x}\right) \qquad (4)$$

Given that in this case, D is independent of the concentration and an infinitely large diffusion cell with stratification beginning at point $x = 0$, we obtain two integrals as derived by Wiener [1] from Eq. (4)

$$\frac{dc}{dx} = -c_0 \frac{1}{2(\pi Dt)^{1/2}} \exp\left(-\frac{x^2}{4Dt}\right) \qquad (5)$$

and

$$c_x = \frac{c_0}{2}\left[1 - \frac{1}{(\pi Dt)^{1/2}}\right]\int_0^x \exp\left(-\frac{x^2}{4Dt}\right) dx \qquad (6)$$

Equation (5) is the symmetrical Gaussian curve, and Eq. (6) has the properties of an error integral. A straight line is obtained when Eq. (6) is plotted on probability paper in which the ordinate corresponds to the error integral.

If Eq. (6) is differentiated with respect to time, an additional and important result of the differential equation of diffusion is obtained. The change in concentration along the gradient is represented by

$$\Delta c_a = c(-a) - c(+a) \qquad (7)$$

The quantities $c(-a)$ and $c(+a)$ are the concentrations of the macromolecules at distance $x = -a$ and $x = +a$, respectively, measured from the starting point $x = 0$. Differentiation of Δc_a with respect to time gives

$$\frac{d}{dt}(\Delta c_a) = \frac{ac_0}{2(\pi D)^{1/2}} t^{-3/2} \exp\left(-\frac{a^2}{4Dt}\right) \tag{7a}$$

III. Experimental Methods for the Determination of Diffusion Constants

There are three principal methods by means of which the heterogeneity of a macromolecular mixture may be determined from isothermal diffusion.

(1) The entire concentration gradient for all values of x in the cell is made visible by some process. The change in concentration, dc/dx, is followed as a function of time. For the evaluation of the experimental data, Eq. (5) is used.

(2) Some quantity which is directly proportional to the concentration is measured, and its change with respect to time over the entire x domain is measured. Evaluation of these data can be carried out using Eq. (6).

(3) The variation of concentration with respect to time at two fixed points located symmetrically around the starting boundary is measured. These variations are evaluated using Eq. (7a).

A. Schlieren Optics, Ultraviolet Optics, and Rayleigh Interference Optics

The most common arrangement used for the measurement of heterogeneity is that given under (1) in the section above. The concentration gradient is made visible with schlieren optics. A convenient system has been designed by Philpot [2] and Svensson. The exact details of these optics will not be given here since they are available in several handbooks and manuals [3,4].

Concentration change with respect to time can be studied conveniently with either ultraviolet absorption or Rayleigh interference optics [2]. Since these arrangements are well known, we will omit discussion at this point.

B. Jamin Interference Optics

Figure 1 shows a schematic diagram of a Jamin system. This arrangement permits the measurement of concentration changes at two fixed points [3,5–7]. Essentially the apparatus consists of two partially mirror coated glass plates. These are arranged so that a monochromatic light beam, with a width, $2a$, equal to the distance between the two points of measurement, is passed through both reflecting plates and into the diffusion cell. Partial reflection will occur at the partitions. These reflections give rise, at the

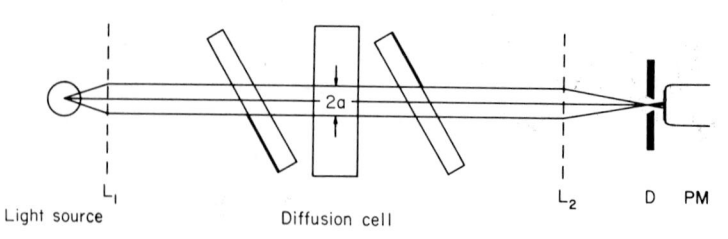

FIG. 1. Interferometric arrangement according to Jamin with photoelectric scanning system [5]. See text for explanation. PM, photomultiplier.

distance $2a$, to a second light beam which has a specific phase difference from the nonreflected beam. A lens, identified in the figure by L_2, focuses the two beams of light on the slit of the diaphragm D. With the boundary of the cell located in the optical axis, a concentration difference between the two measured points equal to the initial concentration, c_0, exists at the start of the diffusion. This difference in concentration at the two measured points results in a phase difference between the two light beams. At the plane of the diaphragm, N interference bands corresponding to the phase difference are manifested. For the purpose of demonstration suppose that the Nth band falls directly into the diaphragm opening. As the time after the beginning of the diffusion increases, the concentration differences in the cell progressively equalize. This causes the phase difference to become smaller. As the phase difference decreases, one interference band after another passes over the aperture of the diaphragm. Behind the diaphragm this movement of interference bands is seen as a continuous change from light to dark, etc. Daune et al. [5–7] have measured these moving bands using a photomultiplier tube and a strip chart recorder. The rate at which diffusion occurs is determined by measuring the time between two maxima of light intensity. The diffusion constant may be calculated using Eq. (7a). For the calculation see Section, IV,B,3.

C. Diffusion Cells

Diffusion cells used for the measurement of molecular heterogeneity are identical in most cases to those employed for the determination of molecular weight by Svedberg's equations [8]. Since these cells have been adequately described elsewhere, our consideration will be brief.

Boundary formation in these cells has been achieved by several unique methods: (a) slider, [9,10] (b) the two parts of the cell filled with solvent and solution, respectively, are moved towards each other, [11] (c) and suction at the boundary [5–7]. (d) Moacanin et al. [12,13] have described a very simple apparatus using microscope slides. They were able to generate a sharp and unperturbed boundary between a solution and a gelled solvent.

IV. Theory: Determination of Molecular Heterogeneity

A. Influence of Heterogeneity on Diffusion Curve Shape

Solutions to Eqs. (5) and (6) have been derived for monodisperse systems. The case of a heterogeneous mixture, characteristic of most polymers, is somewhat more complicated. In a heterogeneous mixture, the smaller molecules will diffuse more rapidly and the bigger molecules more slowly than a corresponding monodisperse substance with equal diffusion constant. This has profound consequences on the shape of the diffusion curve. In Figs. 2 and 3, a calculated example is shown graphically. The curves were calculated for a mixture of two components of identical concentration $c_0/2$, and diffusion constants $D_a = 1 \times 10^{-7}$ and $D_b = 4 \times 10^{-7}$ cm²/sec, respectively. The resulting curves are compared with that of a uniform substance, c, with a diffusion constant identical to that in the mixture

$$D_c = D_m = \tfrac{1}{2}(D_a + D_b) \tag{7b}$$

A symmetrical Gaussian curve describes the gradient of the monodisperse system. A straight line in the probability plot gives the concentration of the monodispersed material (solid line in Fig. 3). The curve describing the

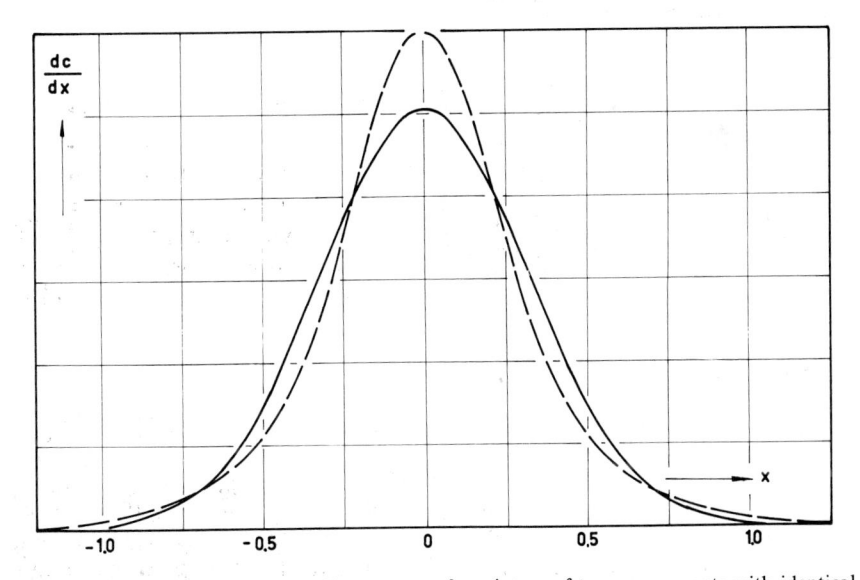

FIG. 2. Calculated diffusion gradient curves of a mixture of two components with identical concentrations $c_0/2$ with diffusion constants $D_a = 1 \times 10^{-7}$ and $D_b = 4 \times 10^{-7}$ cm² sec^{-1} (dotted line) and of a monodisperse substance with a diffusion constant identical with that of the mixture: $D_c = D_m = (D_a + D_b)/2$ (solid line).

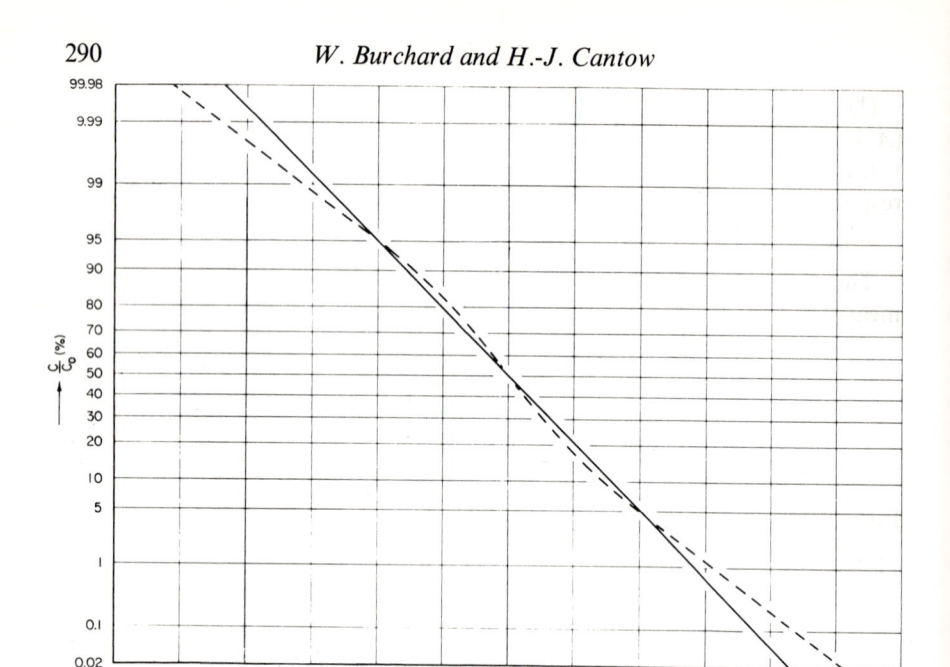

FIG. 3. Calculated concentration curves plotted on probability paper for the same mixture and a monodisperse sample used in Fig. 2.

gradient of the mixture (dotted line, Fig. 2), although symmetrical, is narrower at small abscissa values and tails more than a Gaussian curve (solid line, Fig. 2). The probability plot of the mixture concentration is a curve, not a straight line. These two figures demonstrate that the heterogeneity of a dissolved substance may be judged by comparing its gradient curve to that of a normal Gaussian distribution. This gives rise to a number of possibilities for the quantitative determination of heterogeneity by diffusion.

B. Calculation of the Heterogeneity

1. *From Time Dependence of the Concentration Gradient: Schlieren Optics*

It can be seen in Fig. 2 that both curves have the same total area but a different height at $x = 0$. According to Miller and Hamm [14], the ratio of the heights $(H/H_g)_{x=0}$ is a measure of the molecular heterogeneity for curves with normalized areas. H_g is the height of the Gaussian curve and H is the height of the measured curve at $x = 0$. The following steps are necessary for the evaluation of experimental results prior to the above comparison of height:

C.3. Isothermal Diffusion

(1) The area of the measured curve must be measured and normalized ($A = 1$).

(2) The average diffusion constant, D_m, must be determined. The corresponding normal Gaussian curve is calculated using D_m.

(3) The heights of both curves are compared.

The average diffusion constant is related to the second moment of the measured curve as follows:

$$D_m = \int_{-\infty}^{+\infty} x^2 f(x, D)\, dx \tag{8}$$

$$f(x, D) = \frac{1}{c_0} \sum_i \frac{c_i}{2(\pi D_i t)^{1/2}} \exp\left(-\frac{x^2}{4D_i t}\right) \tag{9}$$

For the previously given example, D_m is equal to $\tfrac{1}{2}(D_a + D_b)$. For the general case,

$$D_m = \sum_i \frac{c_i}{c_0} D_i = \sum_i w_i D_i \tag{10}$$

$$\sum c_i = c_0 \qquad \sum w_i = 1$$

The molecular weight of molecules with a diffusion constant, D_i, is M_i. These are present in the concentration c_i; c_0 is the initial concentration.

The area technique of Gralén [15] is similar to the analysis of Miller and Hamm. Gralén has characterized the heterogeneity by the ratio D_m/D_A with

$$D_A = \frac{1}{4\pi t} \frac{A^2}{H^2} \tag{11}$$

A is the area and H the maximum height of the diffusion curve.

For monodisperse samples, D_m is equal to D_A. This relationship can be verified by calculating the area and height for Eq. (5) and D_m from Eqs. (8) and (9). The calculated area and height according to Eq. (9) gives the following result for a polydisperse system [15]:

$$D_A = \frac{1}{[\sum (c_i/c_0) D^{-1/2}]^2} = \frac{1}{(\sum w_i D^{-1/2})^2} \tag{12}$$

From Eq. (10) it can be seen that D_m represents the weight average diffusion constant. D_A is similar to the number average constant. The following inequalities exist:

$$D_w = D_m \geq D_A \geq D_n \tag{13}$$

The subscripts w and n mean weight and number average, respectively. It follows from the above that

$$\frac{D_m}{D_A} = \left(\frac{H}{H_g}\right)^2_{x=0} \geq 1 \tag{14}$$

It is possible to derive the diffusion constant from the half-width of the diffusion curve. The width of a Gaussian curve for a monodisperse substance at its half-height is

$$B = 2(Dt)^{1/2} \log 2 \tag{15}$$

The diffusion constant is

$$D_B = \frac{B^2}{8(\log 2)t} \tag{16}$$

In the case of a polydisperse system, using Eq. (9), for $f(x, D)$ we obtain

$$D_B = [\sum (c_i/c_0)D^{1/2}]^2 = (\sum w_i D^{1/2})^2 \tag{17}$$

The half-width average lies between the weight and number average diffusion constants.

$$D_w = D_m \geq D_B \geq D_A \tag{18}$$

These ratios D_m/D_A, etc., give information concerning only the width of the distribution. To further characterize the distribution, it is necessary to calculate the higher moments of the diffusion curve.

$$m_{2n} = \int_{-\infty}^{+\infty} x^{2n} f(x, D)\, dx \tag{19}$$

If D is independent of concentration, as we have considered it to be in our previous arguments, all the odd moments of the above calculation are equal to zero. This means that all of the calculated curves are symmetrical.

More extensive information concerning molecular distribution may be obtained from diffusion curve widths measured at different fractions of the maximum height [16]. To obtain this information, Eq. (5) may be written as

$$-\log\left[\left(\frac{dc}{dx}\right)\bigg/\left(\frac{dc}{dx}\right)_{max}\right] = \frac{0.109}{Dt} x^2 \tag{5a}$$

$(dc/dx)_{max}$ is the concentration gradient at $x = 0$. This value lies at the maximum of the curve for ideal solutions. A straight line is obtained for monodispersed samples when the logarithmic expression on the left of Eq. (5a) is plotted as a function of x^2. The slope of this line is inversely proportional to the diffusion constant. A polydisperse mixture will give a curved line bending downwards.

In Fig. 4 the logarithmic expression is plotted versus $x^2/x_{1/2}^2$ ($x_{1/2}$ is the half-width). The curves are calculated for a molecular weight distribution according to Lansing and Kraemer [17]. The inhomogeneity term [18] $U = (M_w/M_n) - 1$, is chosen as the parameter. By comparing measured curve widths at different fractions of the maximum height of an experimental diffusion curve with the lines 0 to 20 in Fig. 4, it is possible to determine whether the distribution of the unknown sample follows the Lansing–Kraemer model or not [16].

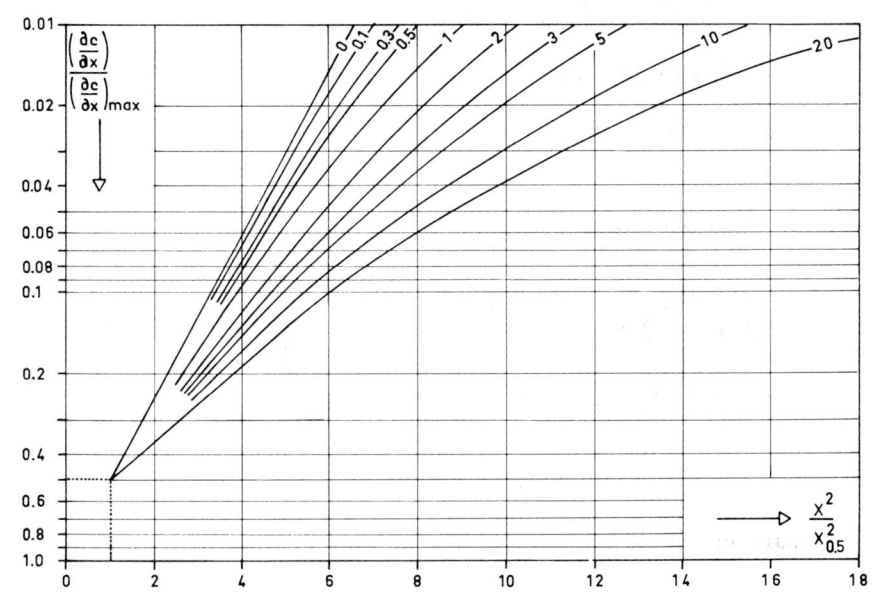

FIG. 4. $-\log[(dc/dx)/(dc/dx)_{max}]$ versus $x^2/x_{0.5}^2$ with inhomogeneity U as parameter [16]. Calculated for a Lansing–Kraemer distribution [17].

Even with agreement between the theoretical function and measured values, the existence of the postulated distribution function is not confirmed. Test calculations have shown that a mixture of three narrow molecular weight fractions gives a diffusion curve indistinguishable within experimental accuracy from that arising from a polymer with logarithmic normal distribution.

2. *From Time Dependence of Concentration: Ultraviolet and Rayleigh Interference Optics*

Using the usual ultraviolet absorption and interference devices, quantities are observed, which are directly proportional to the concentration of the dissolved material. To discuss heterogeneity we shall consider Eq. (6) and Fig. 3. Methods for the determination of inhomogeneity have been developed especially by Jacob, Freund, and Daune [5–7] and by Moacanin, Felicetta, and McCarthy [12,13].

If a mixture containing different molecular weights is under consideration, we must write for every component of the mixture a relation of the type of Eq. (6).

$$y_i = \frac{c_i}{c_0} = \frac{1}{2}\left[1 - \frac{1}{(\pi D_i t)^{1/2}} \int_0^x \exp\left(-\frac{x^2}{4D_i t}\right) dx\right] \quad (6a)$$

Being w_i the weight fraction of the ith component, we find for the mixture

$$y = \sum w_i y_i = \frac{1}{2}\sum w_i \left[1 - \frac{1}{(\pi D_i t)^{1/2}} \int_0^x \exp\left(-\frac{x^2}{4D_i t}\right) dx\right] \quad (6b)$$

Expanding the exponential function in terms of $x^2/4Dt$ and integrating gives

$$y = \sum w_i y_i = \frac{1}{2}\sum w_i \left\{1 - \frac{x}{(\pi D_i t)^{1/2}}\left[1 - \frac{x^2}{12 D_i t} + \frac{x^4}{160(D_i t)^2} - \cdots\right]\right\} \quad (20)$$

This can also be written in an abbreviated form

$$y = \frac{1}{2}\left\{1 - \frac{x}{(\pi D_1 t)^{1/2}}\left[1 - \frac{x^2}{12 D_2 t} + \frac{x^4}{160(D_3 t)^2} - \cdots\right]\right\} \quad (21)$$

Comparison of Eqs. (20) and (21) leads to the following averages:

$$D_1 = (1/\sum w_i D_i^{-1/2})^2$$

$$D_2 = \frac{\sum w_i D_i^{-1/2}}{\sum w_i D_i^{-3/2}} \quad (22)$$

$$D_3 = \frac{\sum w_i D_i^{-3/2}}{\sum w_i D_i^{-5/2}}$$

The last expression rests on the same considerations as Eq. (12), and therefore $D_1 = D_A$. For the other averages we have the inequalities $D_1 \geq D_2 \geq D_3$.

The averages can be determined experimentally. D_1 is obtained by plotting $y = c/c_0$ versus $x/t^{1/2}$. The initial slope of this curve is equal to $(\pi D_1)^{-1/2}$. The term D_2 can be derived by calculating at every point of the abscissa the difference between curve y and its initial tangent y'. Then $\Delta y = (y - y')$ is plotted versus $(x^2/t)^{3/2}$. The initial slope of this curve is equal to $12D_2(\pi D_1)^{-1/2}$. Plotting $(\Delta y - \Delta y')$ versus $(x^2/t)^{5/2}$ enables us to calculate D_3.

Ratios D_1/D_2 or D_2/D_3 can be used as a measure of inhomogeneity.

Moacanin et al. [12,13] have approached the problem in a different way. They first define the moments of a distribution of diffusion

$$\mu'_{n,D} = \sum w_i (D_i^{1/2})^n \tag{23}$$

The subscript n, D indicates the nth moment of diffusion distribution. It has to be distinguished from the moments of the experimental diffusion curve $y = c_x/c_0$, which are indicated with the subscripts n, y. This is defined by the following equation:

$$\mu'_{n,y} = \int_0^\infty x^n y \, dx = \int_0^\infty x^n \frac{c_x}{c_0} \, dx \tag{24}$$

Here, Eq. (6b) can be substituted for y. The moments $\mu_{n,y}$ can be determined from the measured diffusion curve without difficulty. Between these moments and those of the distribution of diffusion coefficients Moacanin and co-workers derived a relationship. Equation (6a) was multiplied by x^{n-1} and integrated over the entire x domain. By this procedure they obtain

$$\mu'_{n,D} = -\frac{n}{2^n t^{n/2} G} \mu'_{n-1,y} \tag{25}$$

$$G = \frac{1}{\sqrt{\pi}} \left(\frac{n-1}{2} \right)! \qquad \text{for odd values of } n \tag{26a}$$

$$G = \frac{(n-1)}{2} \frac{(n-2)}{2} \cdots \frac{1}{2} \qquad \text{for even values of } n \tag{26b}$$

The first four moments are

$$\mu'_{1,D} = \sum w_i D_i^{1/2} = \frac{0.886}{t^{1/2}} \mu'_{0,y} = \frac{0.866}{t^{1/2}} \int_0^\infty y \, dx \tag{27a}$$

$$\mu'_{2,D} = \sum w_i D_i = \frac{1}{t} \mu'_{1,y} = \frac{1}{t} \int_0^\infty xy \, dx \tag{27b}$$

$$\mu'_{3,D} = \sum w_i D_i^{3/2} = \frac{0.665}{t^{3/2}} \mu'_{2,y} = \frac{0.665}{t^{3/2}} \int_0^\infty x^2 y \, dx \tag{27c}$$

$$\mu'_{4,D} = \sum w_i D_i^2 = \frac{1}{3t^2} \mu'_{3,y} = \frac{1}{3t^2} \int_0^\infty x^3 y \, dx \tag{27d}$$

With these moments several of the averages defined in Eqs. (10) and (16) can be calculated. From Eq. (17) we obtain

$$(\mu'_{1,D})^2 = (\sum w_i D_i^{1/2})^2 = D_B$$

D_B is the diffusion constant, which can be derived from measurements of the width of the diffusion gradient curve. From Eq. (10) it follows that

$$\mu'_{2,D} = \sum w_i D_i = D_m = D_w$$

D_w is the weight average diffusion constant. The averages defined by Eq. (22) cannot be determined by this method. To estimate the molecular inhomogeneity the ratio D_m/D_B can be applied.

3. From Variation of Concentration Difference at Two Fixed Points with Time: Jamin Interference Optics

In the usual interferometric arrangement of Rayleigh we observe the pattern of concentration over the entire x domain. With the interferometric method of Jamin—as worked out by Daune, Freund, and Scheibling [5–7] —it is possible to see the variation with time of the concentration difference $c_{-a} - c_{+a}$ at the points $x = -a$ and $x = +a$. The point $x = 0$ is the stratifying boundary. This difference in concentration is proportional to the number of interference fringes in the plane of observation. During diffusion the number of interference fringes diminishes as the concentration between the points $x = -a$ and $x = +a$ becomes equal. Therefore, at any fixed point in the plane of observation, an alternation of light and dark fringes will be observed. Registration may be achieved by photomultiplier cell [5–7].

Given that dt in Eq. (7a) is the time between two light maxima at one point in the observation plane, and N is the number of interference fringes at the initial concentration c_0, we can write for Eq. (7a)

$$\frac{t^{3/2}}{\Delta t} = \frac{aN}{2(\pi D)^{1/2}} \exp\left(-\frac{a^2}{4Dt}\right) \tag{28}$$

or

$$\log y = \log \frac{t^{3/2}}{\Delta t} = \log \frac{aN}{2(\pi D)^{1/2}} - \frac{a^2}{4Dt} \tag{28a}$$

where t is the time since stratifying; N, the number of interference fringes at the time $t = 0$ (starting); a, the distance of the two rays from $x = 0$; and D, the diffusion constant for monodisperse substances.

Plotting $\log y$ versus $1/t$, produces a straight line with the slope $m_0 = -a^2/4D$ and the intercept $\log y_0 = \log[aN/2(\pi D)^{1/2}]$. In the case of a polydisperse sample, Eq. (28) is valid for every component with the diffusion constant D_i and the statistical weight $w_i = c_i/c$. We find for the entire system

$$y = \frac{t^{3/2}}{\Delta t} = \sum w_i \left[\frac{aN}{2(\pi D_i)^{1/2}} \exp\left(-\frac{a^2}{4D_i t}\right)\right] \tag{29}$$

$$\log y = \log \sum w_i \left[\frac{aN}{2(\pi D_i)^{1/2}} \exp\left(-\frac{a^2}{4D_i t}\right)\right] \tag{29a}$$

By plotting $\log y$ versus $1/t$ a curve with decreasing slope for increasing abscissa value is produced. For $a^2/4D_i t \ll 1$ we can expand Eq. (29) in terms of $a^2/4D_i t$

$$y = \frac{aN}{2\pi^{1/2}} \sum \frac{w_i}{D_i^{1/2}} \left[1 - \frac{a^2}{4D_i t} + \frac{1}{2}\frac{a^4}{(4D_i t)^2} - \cdots\right] \tag{30}$$

By stopping the series with the second term, we can approximate for small values of $a^2/4D_i t$

$$y \approx \frac{aN}{2(\pi D_1)^{1/2}} \exp\left(-\frac{a^2}{4D_2 t}\right) \tag{31}$$

$$\log y \approx \log \frac{aN}{2(\pi D_1)^{1/2}} - \frac{a^2}{4D_2 t} \tag{31a}$$

From the initial slope of the $\log y$ versus $1/t$ plot, the average diffusion constant can be obtained

$$D_2 = \frac{\sum w_i D_i^{-1/2}}{\sum w_i D_i^{-3/2}}$$

The ordinate intercept is the average value

$$D_1 = \left(\frac{1}{\sum w_i D_i^{-1/2}}\right)^2$$

These averages are identical to those obtained with the interferometric arrangement of Rayleigh.

C. Derivation of Molecular Heterogeneities from Those of the Diffusion Constants

In the preceding Sections (A and B) we have pointed out that different average diffusion constants may be obtained depending on the experimental arrangement. Therefore, we can obtain from the ratio of these averages a measure of the heterogeneity of the diffusion constants and of the breadth of the distribution curve producing these constants. The principal interest is not directed towards these distributions but to the molecular weight distributions. In particular, it is desirable to obtain the molecular inhomogeneity $\overline{M}_w/\overline{M}_n$ [18].

An interrelation exists between these two quantities. The diffusion constant, D, depends on the molecular weight, M. According to experimental evidence the following general empirical relation is valid:

$$D = KM^{-b} \qquad (32)$$

K and b are characteristic constants for an individual polymer–solvent system. The exponent b is >0.5 and <1 in a thermodynamically good solvent, but equal to 0.5 in a θ solvent [19]. In such a system there are no interaction forces between the segments of the dissolved macromolecules.

Substituting Eq. (32) into the expressions for D_m/D_A and for Eqs. (8) and (12), and solving Eq. (22) for D_1/D_2 and D_2/D_3, respectively, we obtain

$$\frac{D_m}{D_A} = \int f(M) M^{-b}\, dM \left(\int f(M) M^{b/2}\, dM\right)^2 \qquad (33)$$

$$\frac{D_1}{D_2} = \frac{\int f(M) M^{-3/2b}\, dM}{\left(\int f(M) M^{-1/2b}\, dM\right)^2} \qquad (34)$$

$$\frac{D_2}{D_3} = \frac{\left(\int f(M) M^{-1/2b}\, dM\right)\left(\int f(M) M^{-5/2b}\, dM\right)}{\left(\int f(M) M^{-3/2b}\, dM\right)^2} \qquad (35)$$

The continuous distribution $f(M)$ is used instead of the discontinuous distribution $w_i(M)$. In many cases molecular weight distributions can be approximated by functions of the Zimm–Schulz type [18,20]

$$f(M) = (\lambda^{y+1}/y!) M^y e^{-\lambda M} \qquad (36)$$

with

$$\frac{1}{y} = \frac{\overline{M}_w}{\overline{M}_n} - 1 \qquad \lambda = \frac{1}{\overline{M}_w - \overline{M}_n}$$

With these distributions we obtain

$$\frac{D_m}{D_A} = \frac{\Gamma(1 + y - b)}{\Gamma(1 + y)} \frac{\Gamma^2(1 + y + b/2)}{\Gamma^2(1 + y)}$$

$$\frac{D_1}{D_2} = \frac{\Gamma^2(1 + y)\Gamma(1 + y + 3/2b)}{\Gamma^3(1 + y + b/2)} \qquad (37)$$

$$\frac{D_2}{D_3} = \frac{\Gamma(1 + y + b/2)\Gamma(1 + y + 5/2b)}{\Gamma^2(1 + y + 3/2b)}$$

Γ is the well-known gamma function. With Eq. (37) the interrelationship between the molecular inhomogeneity $\overline{M}_w/\overline{M}_n = (y + 1)/y$ and the diffusion constant inhomogeneity has been established. The shape of $f(M)$ for different b values can be found in Fig. 5.

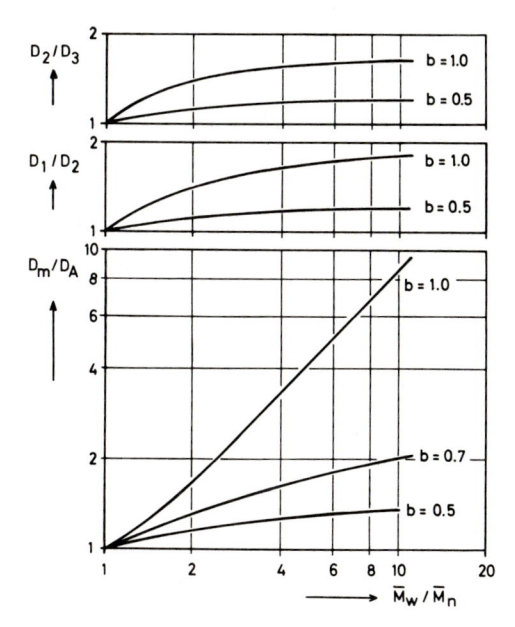

FIG. 5. Ratios of statistical moments of diffusion D_m/D_A, D_1/D_2 and D_2/D_3 with b from the equation $D = KM^{-b}$ as parameter. Calculation of ratios carried out with equations (37).

D. Influence of Concentration on the Diffusion Coefficient

For simplicity it was assumed hitherto that the diffusion constant should be independent of concentration. However, this is true only within certain limits in θ solvents [16]. In all other cases a more or less pronounced dependence of the diffusion on concentration is present. A diffusion determination is complicated by this dependence since the concentration gradient tends to become asymmetric with respect to the starting boundary. Consequently, it is impossible to derive the inhomogeneity from the diffusion curve at only one concentration. The ratios D_m/D_A, D_1/D_2, and D_2/D_3 have to be determined by measurements at several concentrations with subsequent extrapolation to infinite dilution. This is analogous to the determination of molecular weights by sedimentation and diffusion. Aside from a broader scattering of values, the relations given in the foregoing sections remain valid for the extrapolated average ratios.

Gillis and Kedem [21] have made an attempt to describe the trend of diffusion with increasing concentration.

The diffusion coefficient is a linear function of concentration for sufficiently low concentrations, c,

$$D = D_0(1 + kc) \tag{38}$$

According to Signer [22] this concentration dependence can be expressed in terms of the second virial coefficient of the osmotic pressure, B,

$$D = \frac{RT[1 + (2BM/RT)]c}{f_0(1 + k_s c)} \tag{39}$$

f_0 is the molar friction coefficient, and k_s is a hydrodynamic factor.

Gralén [15] has derived the expansion of Fick's equation for solutions in which D is dependent on c

$$\frac{\partial c}{\partial t} = \frac{\partial}{\partial x}\left(D\frac{\partial c}{\partial x}\right) = \frac{\partial}{\partial x}\left[D_0(1 + kc)\frac{\partial c}{\partial x}\right] \tag{40}$$

In principle, it is possible on the basis of this relationship to evaluate diffusion measurements made on polydisperse systems.

Gillis and Kedem [21] have given an approximate solution of the Fick equation.

Using schlieren optics c is related to Δn by

$$c = \Delta n/\beta \tag{41}$$

where Δn is the refractive index difference between solution and solvent and β is a constant. Introducing as new variables $\lambda = \tfrac{1}{2}x(D_0 t)^{-1/2}$, and $F = \Delta n$ we obtain $\Delta n(x, t) = F(\lambda)$ expanding $F(\lambda)$ in a power series of $h/\beta = \mu$ gives

$$F(\lambda) = F_0(\lambda) + \mu F_1(\lambda) + \mu^2 F_2(\lambda) + \ldots \tag{42}$$

The integrated expression has been found to be [21]

$$n(x, t) = n_0 + F_0(\lambda) + (h/\beta)F_1(\lambda) + (h^2/\beta^2)F_2(\lambda) \tag{43}$$

The function $F_0(\lambda)$, $F_1(\lambda)$ etc. has been determined by substitution of Eq. (43) in Eq. (40) and solving the resulting differential equation. The previously mentioned authors have given practical methods for the evaluation of concentration-dependent diffusion processes.

For a fraction of polymethacrylic acid ($P = 4600$) at concentrations ranging from 0.0064 to 0.064 unit mole/liter, Gillis and Kedem have given a plot of their calculations compared to experimentally determined values (Fig. 6). The solid line fits the experimental values quite well. Still better agreement is achieved by correcting for the initial boundary position (short dashed line).

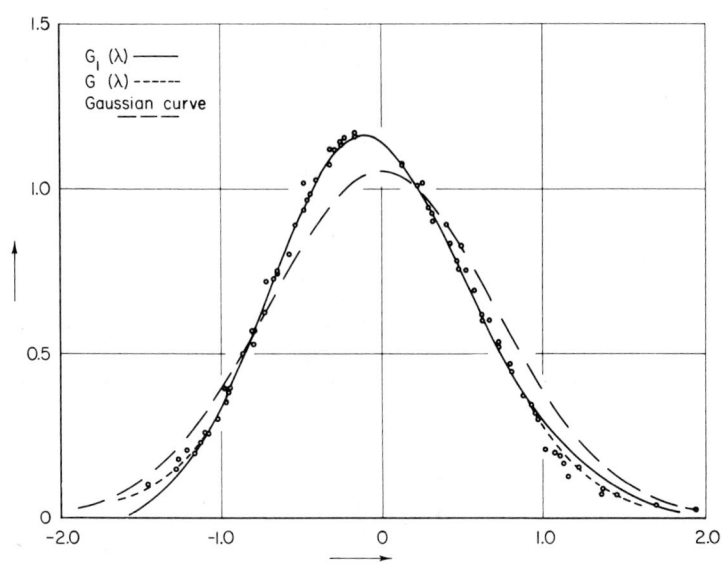

FIG. 6. Diffusion measurement in nonideal aqueous solutions of polymethacrylic acid [21]. See text for explanation.

Comparison with the symmetrical Gaussian curve for an ideal solution shows the shift of the curve maximum location and the asymmetry of the curve for nonideal solutions.

Although the authors have shown the possibility of the analysis of diffusion curves for monodisperse concentration-dependent systems, the analysis of polydisperse samples is possible only by the empirical extrapolation as made by Daune and Freund [5–7].

V. Methods for the Determination of the Diffusion Coefficient Distribution

A. Moment Method of Moacanin, Felicetta, and McCarthy

The gradient curve of sedimentation generally gives a qualitative picture of the molecular weight distribution. After a certain time of sedimentation every molecular species will correspond to a distinct point in the cell. However, a diffusion curve does not directly represent the molecular weight distribution, because there is no separation of molecular weight species with progressing time. Only a more or less rapid broadening of the Gaussian curve occurs. Therefore, the derivation of the molecular weight distribution from the diffusion curve can be carried out only after a detailed mathematical analysis of the curve has been made.

Moacanin and co-workers [12,13] approximate every distribution function by a Gaussian distribution and its derivatives (Bruns' series)

$$f(z) = \varphi(z) - \frac{a_3}{3!}\varphi'''(z) + \frac{a_4}{4!}\varphi^{IV}(z) - \ldots \quad (44)$$

with

$$\varphi(z) = \frac{1}{(2\pi)^{1/2}} \exp\left(-\frac{z^2}{2}\right) \qquad z = \frac{D^{1/2} - \mu'_{1,D}}{(\mu_{2,D})^{1/2}}$$

The terms $\mu'_{1,D}$ and $\mu_{2,D}$ are the first two moments of the diffusion constant distribution. These can be evaluated from the diffusion curves using Eqs. (27a) to (27d). The coefficients—often called "inclination" or "excess" of the distribution $f(z)$—are defined as

$$a_3 = \frac{\mu_{3,D}}{(\mu_{2,D})^{3/2}} \qquad a_4 = \frac{\mu_{4,D}}{(\mu_{2,D})^2} - 3 \quad (45)$$

The moments without primes are related to the average, while the primed moments relate to the origin of the ordinate. Between these two moments there exist the following relations (subscript D is omitted here)

$$\begin{aligned}\mu_2 &= \mu'_2 - \mu'^2_1 \\ \mu_3 &= \mu'_3 - 3\mu'_1\mu'_2 + 2\mu'^3_1 \\ \mu_4 &= \mu'_4 - 4\mu'_1\mu'_3 + 6\mu'^2_1\mu'_2 - 3\mu'^4_1\end{aligned} \quad (46)$$

The series (44) is stopped after the fourth derivative with the coefficient a_4. Higher moments than the fourth cannot be determined with sufficient accuracy. This is an important limitation of the Bruns' series. Equation (44) is a good approximation for curves with only one maximum. For curves with two maxima, Eq. (44) gives an inaccurate approximation. Frequently, it is not evident from the approximation that the distribution has two maxima.

B. Fourier Transformation

Benoit and co-workers [23,24] have illustrated another important property of the diffusion curve [Eq. (28)]. They proved with the aid of several transformations that this curve can be represented by a folding integral whose one component is completely determined by the distribution of the diffusion constants. The diffusion function

$$y(t) = \frac{a}{2(\pi)^{1/2}} \int_0^\infty \left[f(D)D^{-1/2} \exp\left(\frac{-a^2}{4Dt}\right)\right] dD$$

converts by two substitutions, $x = D_0/D$ and $f(D)dD = \varphi(x)dx$, into the expression

$$y(t) = \frac{1}{\pi^{1/2}} \int_0^\infty \left[\varphi(x) \frac{ax^{1/2}}{2(D_0 t)^{1/2}} \exp\left(-\frac{a^2}{4Dt}\right) \right] dx \qquad (47)$$

Here D_0 is a distinct, arbitrary diffusion constant, which can be regarded as a reference constant.

Introducing further the transformations $s = \log(a^2/4D_0 t)$ and $u = -\log x$ we may write

$$y(s) = \int_{-\infty}^{+\infty} h(u)[f(s-u)] \, du \qquad (48)$$

with

$$h(u) = \exp(-u)\varphi(\exp\{-u\})$$

$$f(s-u) = \frac{1}{\pi^{1/2}} \exp\left(\frac{s-u}{2}\right) \exp[-\exp(s-u)]$$

Equation (48) exhibits the type of folding integral, in which $h(u)$ determines unequivocally the distribution function of the diffusion constant. By Fourier transformation the folding integral is transformed into a simple product. However, Benoit and co-workers suggest no methods for the experimental determination of two of the three components. Furthermore, they have not cited an example to demonstrate the Fourier transformation.

VI. Experimental Results

The early measurements of Gralén were directed towards the determination of molecular heterogeneities of gluten proteins, cellulose, cellulose nitrate, xanthate, and glycolate from the ratios of the different moments [15]. He discussed the skew diffusion curve which arises from nonideal solution behavior. This diffusion curve was later treated in an exact mathematical approach by Gillis and Kedem [21]. The latter two authors tested their conclusions with polymethacrylic acid in aqueous solution (compare Section IV,D).

Miller and Hamm [14] have compared several methods for the estimation of molecular inhomogeneity. Using polyvinylpyrrolidone as a model system, they found isothermal diffusion inferior to sedimentation analysis for the determination of molecular heterogeneities.

Freund and Daune [25] have also compared the sedimentation and the diffusion methods. For polyvinyl alcohol in water they found good agreement between the two methods in the median and in the high molecular weight regions but large deviations at low molecular weights. Cantow [16] obtained good agreement between three distribution determination methods applied to styrene-butadiene copolymer. The methods were diffusion and sedimentation under θ conditions and fractional precipitation (Fig. 7). Daune et al. [26] have also examined polybutylmethacrylate and

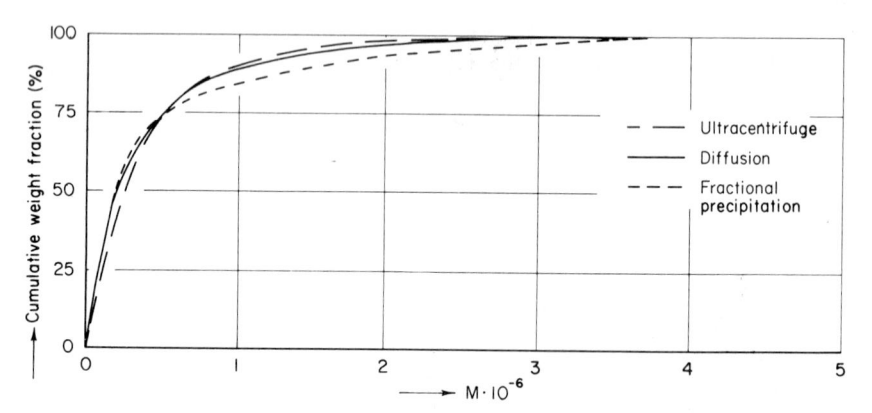

FIG. 7. Integral molecular weight distribution curves of styrene-butadiene copolymer (cold rubber) determined by various methods [16].

polystyrene in their diffusion apparatus [5–7] and in the ultracentrifuge under θ conditions. Although the sedimentation constant exhibited a strong concentration dependence, the diffusion constant was concentration-independent. Cantow, on the contrary, has found a slight concentration dependence of the diffusion process even under θ conditions for polystyrene [27] and styrene-butadiene copolymer [28].

At extremely low concentrations, Jacob et al. [29] were able to neglect the diffusion concentration dependence of desoxyribonucleic acid. Adsorption optics allowed the experiments to be made at very high dilutions. Daune et al. [30] have been able to carry out diffusion measurements at temperatures up to 150°C. They tested their apparatus with polystyrene.

Moacanin and co-workers [12] estimated the polydispersity of lignin sulfonate polymers using diffusion from solution into gel. A digital computer was used to calculate the various statistical moments.

VII. Critical Remarks

The interpretation of diffusion and sedimentation measurements was complicated and subject to question so long as only nonideal, concentration-dependent solutions had been investigated. The later use of pseudo-ideal

θ solutions [19] permitted a more exact derivation of molecular weight distributions from diffusion and ultracentrifugal measurements. Currently, the determination of molecular weight distributions is often carried out by sedimentation measurements. This is due to the development of the Archibald approach [31] by McCormick [32] and the velocity method in a θ solvent by Cantow [27]. Diffusion methods are rarely used for this purpose. However, Daune and co-workers [5–7,25,26,30] have greatly refined the theoretical and instrumental basis of the diffusion method. Given this equal instrumental and theoretical development, the diffusion method remains less popular since there is no spreading of the molecules of different molecular weight with time, such as is experienced in ultracentrifugation. The overlapping of the effects of molecules of high and medium weights limits the resolving power of the diffusion method.

On the other hand, diffusion measurements permit an exact determination of the lower end of the molecular weight distribution. Overlapping in this region is negligible. At low molecular weights sedimentation is inferior due to the dominating influence of isothermal diffusion. Combination of both methods should yield improved results.

Better results could be obtained from diffusion measurements if very sharp and unperturbed boundaries could be generated. Jamin interference equipment with photoelectric registration [5–7] is promising in this respect.

REFERENCES

1. O. Wiener, *Ann. Physik* [3] **49**, 105 (1893).
2. J. Philpot, *Nature* **141**, 283 (1938).
3. H. K. Schachman, "Ultracentrifugation in Biochemistry." Academic Press, New York 1959.
4. J. B. Nichols and E. D. Bailey, in "Physical Methods of Organic Chemistry" (A. Weissberger, ed.), 3rd ed., Vol. I, p. 1007. Wiley (Interscience), New York, 1960.
5. M. Daune, L. Freund, and G. Scheibling, *J. Chim. Phys.* **54**, 924 (1957).
6. M. Jacob, L. Freund, and M. Daune, *J. Chim. Phys.* **58**, 521 (1961).
7. R. Varoqui, M. Jacob, L. Freund, and M. Daune, *J. Chim. Phys.* **59**, 161 (1962).
8. T. Svedberg, *Kolloid-Z.* **36**, 53 (1925).
9. G. Meyerhoff, *Makromol. Chem.* **6**, 197 (1951).
10. G. Meyerhoff, *Makromol. Chem.* **15**, 68 (1955).
11. S. Claesson, *Nature* **158**, 834 (1946).
12. J. Moacanin, V. F. Felicetta, and J. L. McCarthy, *J. Am. Chem. Soc.* **81**, 2052 (1959).
13. J. Moacanin, H. Nelson, E. Back, V. F. Felicetta, and J. L. McCarthy, *J. Am. Chem. Soc.* **81**, 2054 (1959).
14. L. E. Miller and F. A. Hamm, *J. Phys. Chem.* **57**, 110 (1953).
15. N. Gralén, Dissertation, Uppsala (1944).
16. H.-J. Cantow, *Preprints Paper, IUPAC, Symp. Wiesbaden, 1959* Vol. II, Chapter 3. Verlag Chemie, Weinheim, 1959.
17. W. D. Lansing and E. O. Kraemer, *J. Am. Chem. Soc.* **57**, 1368 (1925).
18. G. V. Schulz, *Z. Physik. Chem.* **43B**, 25 (1939).

19. P. J. Flory, *J. Chem. Phys.* **17**, 1347 (1949).
20. B. H. Zimm, *J. Chem. Phys.* **16**, 1099 (1948).
21. J. Gillis and O. Kedem, *J. Polymer Sci.* **11**, 545 (1953).
22. R. Signer and H. Gross, *Helv. Chim. Acta* **17**, 726 (1934).
23. M. Daune and H. Benoit, *J. Chim. Phys.* **51**, 233 (1954).
24. M. Daune and L. Freund, *J. Polymer Sci.* **23**, 115 (1957).
25. L. Freund and M. Daune, *J. Polymer Sci.* **29**, 161 (1958).
26. M. Jacob, R. Varoqui, S. Klenine, and M. Daune, *J. Chim. Phys.* **59**, 865 (1962).
27. H.-J. Cantow, *Angew. Chem.* **70**, 318 (1958).
28. H.-J. Cantow, *Makromol. Chem.* **30**, 169 (1959).
29. M. Jacob, L. Freund, and M. Daune, *J. Chim. Phys.* **58**, 521 (1961).
30. R. Varoqui, M. Jacob, L. Freund, and M. Daune, *J. Chim. Phys.* **59**, 161 (1962).
31. W. J. Archibald, *J. Phys. & Colloid Chem.* **51**, 1204 (1947).
32. H. W. McCormick, *J. Polymer Sci.* **36**, 341 (1959).

CHAPTER C.4

Summative Fractionation

O. A. Battista
CENTRAL RESEARCH DEPARTMENT, FMC CORPORATION, PRINCETON, NEW JERSEY

I. The Summative Method... 307
II. Mathematical Interpretation of Summative Data..................... 309
III. Precipitating Power of Varying Compositions of Acetone-Water Precipitants... 311
References... 315

I. The Summative Method

In the fractionation of cellulose and other polymers it is important that the solute be exposed to the solvent for as short a time as possible. The summative method of fractional analysis allows for a minimum time of contact between solvent and solute.

The summative method consists of dissolving the polymer in a solvent and adding slowly a relatively large volume (one-third) of partial solvent in order that a small portion of the polymer will precipitate. The mixture is centrifuged and the precipitate is discarded. The polymer is quantitatively regenerated from an aliquot of the clear supernatant liquor. From the weight of the polymer obtained the weight percent in solution is calculated. This value is represented as $F_{(p)}$. A degree of polymerization measurement on this fraction is represented as $\bar{P}$, the weight average or viscosity value. The procedure is repeated with a fresh sample of polymer, but this time a stronger precipitant is used in order to precipitate more polymer. Thus, by varying the composition of the precipitant any number of values of $F_{(p)}$ with its companion $\bar{P}$ are obtained. The plot of $F_{(p)}$ versus $\bar{P}$ gives the summative distribution curve.

To facilitate handling, the summative fractionations should be carried out *at constant volume.*

The summative method has an unusual merit in that the conditions existing at any one cut are predetermined by the nature of the sample and not by the conditions existing after other fractions have been removed. This is of considerable importance in the reproducibility of experiments.

In the conventional stepwise precipitation procedures conditions of partial precipitation exist. Thus, material above a certain molecular weight may be almost totally precipitated whereas the components below this molecular weight are distributed between the two phases. No matter what the molecular weight of a particular species is, it will be more concentrated in the precipitate than in the supernatant liquid. Furthermore, the distribution coefficient will favor the precipitated phase to a greater extent as the molecular weight rises, and/or the amount of material precipitated increases. The amount of material precipitated up to a given point in a regular stepwise fractionation will, therefore, be a complicated function of the number and weight of the various fractions that have been obtained. Beall [1] has developed the mathematics of a satisfactory method for handling fractionation data subject to such limitations.

The resolving power of the summative process may not be as sharp as the resolving power of a stepwise fractionation because the low molecular weight components may be occluded with the precipitate. Nevertheless, fractional solution of the highly soluble short-chain components, from the precipitant into the supernatant liquor, will occur as conditions of equilibrium are approached during centrifuging to help reduce this limitation of the method. Also, as Jorgensen [2] has clearly shown, divergences between stepwise solution and stepwise precipitation procedures are not serious in the range of low molecular weights, but become extremely serious in the range of high molecular weights.

Coppick et al. [3] on the other hand preferred to apply the summative method, together with their mathematical equations, to the analysis of data obtained on summative fractions remaining in solution, without recovering the precipitates at all. By doing so, the procedure becomes even more rapid, and by working at constant volumes has additional advantages which they point out in their publication.

Boyer [4] proposed an optical method for measuring the polymolecularity of a heterogeneous polystyrene by observing the summative volumes of precipitates formed by the careful addition of varying amounts of nonsolvent.

Spencer [5] realized the need for a relatively rapid, inexpensive method of determining the molecular weight distribution of polymers. In essence he proposed the summative method of fractionation of polymers and provided a set of appropriate equations on theoretical grounds, but he did not demonstrate the applicability of the equations experimentally. He proposed working on the summative precipitates.

Billmeyer and Stockmayer [6] extended the theoretical consideration of Spencer's proposal, and published considerable experimental data utilizing information obtained by measuring the masses and the average molecular weights of the precipitates. They concluded that Spencer's method could not yield a detailed description of the polymolecularity, but that a single parameter indicating the breadth of the distribution could be obtained with fair precision. The procedure is a long and tedious one when low concentrations of solute are used, however.

Golub [7] pursued Spencer's [5] summative procedure based on cumulative precipitates, using GR-S as their polymer system. These data revealed that the summative precipitation procedure proposed by Spencer suffers from the relatively random nature of the precipitation from fraction to fraction, and the difficulty in obtaining true equilibrium conditions.

In an excellent series of papers Broda and co-workers [8–10] analyzed polymer fractionation procedures with special emphasis on the summative procedure, concluding that summative fractions can give comparative and reproducible results which would be very hard to obtain by the more conventional stepwise precipitation procedure. These advantages, coupled with the speed of the procedure, point up the inherent superiority of the summative method of fractionation.

The detailed summative method of fractional analysis selected as a typical example of how this procedure works was applied primarily to the fractionation of cellulose in a sodium hydroxide–solvent system. It is based on the summative solution fractionation procedure proposed by Coppick *et al.* [3]. However, the general procedures may be used for measurement of chain length distributions of other polymer systems where the solute may be dissolved in an inert organic solvent such as cellulose nitrate–acetone, or cellulose acetate–acetone systems. The application of the summative method of fractionation to the nitrate system has been applied by Tasman and Corey [11]. Rånby *et al.* [12] utilized the summative solution procedure [3] to fractionate with improved sharpness, pre-extracted cellulose nitrate, whereby interfering impurities are first reduced to a minimum prior to dissolving the primary polymer component.

II. Mathematical Interpretation of Summative Data

For the mathematical interpretation of summative fractionation data, it must be assumed that the average degree of polymerization of each summative fraction recovered from the supernatant liquid, after the precipitation and centrifuging steps, is the average of chains ranging from zero to n, where n is the degree of polymerization of the longest chains remaining in solution. In practice, it seems improbable that this assumption is fully met.

If P is the degree of polymerization of a homogeneous fraction of cellulose, let $F_{(p)}$ be some function of P such that $dF_{(p)}$ denotes the weight fraction having a degree of polymerization between P and $P + dP$; then

$$F_{(p)} = \int_{P=0}^{P=n} dF_{(p)} \tag{1}$$

Mark [13, 14] shows that the function $F_{(p)}$ is related to the number distribution curve in the following manner:

$$F_{(p)} = \int_{P=0}^{P=p} Pf_{(p)}\, dP \bigg/ \int_{P=0}^{P=\infty} Pf_{(p)}\, dP \qquad (2)$$

where $f_{(p)}$ is some function of P such that $f_{(p)}\, dP$ denotes the number of all the molecules having a degree of polymerization between P and $P + dP$.

If $\bar{P}_{m:n}$ is the weight average degree of polymerization for a heterogeneous sample of cellulose with varying degrees of polymerization such that P has intermediate values between m and n, then since P is an additive function [13,14]

$$\bar{P}_{m:n} = \sum_{P=m}^{P=n} \Delta F_{(p)} P \bigg/ \sum_{P=m}^{P=n} \Delta F_{(p)} \qquad (3)$$

where $F_{(p)}$ is a finite weight fraction of chain length equivalent to P. If $F_{(p)}$ is a continuous function of P, it follows that

$$\bar{P}_{m:n} = \int_{P=0}^{P=n} dF_{(p)} P \bigg/ \int_{P=0}^{P=n} dF_{(p)} \qquad (4)$$

in the summative analysis fractions of cellulose for which the degree of polymerization lies between zero and n are isolated.

Substituting Eq. (1),

$$\bar{P}_n F_n = \int_{P=0}^{P=n} dF_{(p)} P \qquad (5)$$

Now, on differentiating with respect to $F_{(p)}$ and converting to the general rather than the particular values of the functions involved,

$$\frac{d\bar{P}[F_{(p)}]}{dF_{(p)}} = P$$

so that if a finite value of $\Delta F_{(p)}$ is taken as small

$$P = \frac{\Delta(\bar{P}F)}{\Delta F} = \frac{\bar{P}_1 F_1 - \bar{P}_2 F_2}{F_1 - F_2} \qquad (6)$$

By means of Eq. (6), the experimentally determined summative curves are readily converted to the integral expression of $F_{(p)}$ against P by calculating the value of P for each value of $F_{(p)}$.

Further conversion to the differential distribution curve, $g_{(p)}$ against P, is accomplished by graphical differentiation where $g_{(p)} = d[F_{(p)}]/dP$.

A bar graph makes the interpretation of the distribution data easier to visualize. It is obtained by a plot of ΔF versus P. Such graphs have no mathematical significance but they are widely used because they give a picture of polymer heterogeneity.

The assumptions used in the above derivation probably lead to an oversimplified picture of the chain length distribution. As indicated mathematically by Beall [1], they may lead to an underestimation of the amount of high degree of polymerization material in the sample. Nevertheless, the analysis of summative fractionation data depends on the application of the simple expression given by Eq. (6). Table I shows that in the summative process the solute and precipitate fractions are reasonably additive. In other words, Eq. (6) is, for all practical purposes, valid.

TABLE I

COMPARISON OF SUMMATIVE SOLUTE AND PRECIPITATE DATA FOR WOOD PULP[a]

Summative fractions (from solution)			Summative fractions (from precipitate)			Arithmetic average[b] D.P. of combined fractions
Fraction	% by wt	D.P.	Fraction	% by wt (difference)	D.P.	
1-S	96.0	991	1-P	4.0	1892	1020
2-S	89.5	953	2-P	10.5	1658	1025
3-S	73.5	930	3-P	26.5	1468	1072
4-S	62.0	759	4-P	38.0	1397	1010
5-S	52.0	620	5-P	48.0	1297	942
6-S	30.3	348	6-P	69.7	1352	1050
7-S	26.5	289	7-P	73.5	1313	1043
8-S	21.5	211	8-P	78.5	1216	1000
9-S	14.4	187	9-P	85.6	1210	1055
10-S	7.4	120	10-P	92.6	1095	1017
11-S	4.9	120	11-P	95.1	1065	1012

[a] Average D.P. = 1015.
[b] Arithmetic average = [(% by wt solute × D.P.) + (% by wt precipitate × D.P.)].

To provide a practical example of the application of the summative method of fractionation to the measurement of the polymolecularity of cellulose, we have selected representative data obtained for a rayon. The reader is referred to the original publication by Coppick, Battista, and Lytton [3] for the experimental details.

III. Precipitating Power of Varying Compositions of Acetone-Water Precipitants

A curve showing the precipitating power of a constant volume of acetone-water mixtures for a sample of rayon wood pulp which was regenerated from cupriethylenediamine solution and dissolved in 2 N sodium hydroxide is illustrated in Fig. 1. By varying the concentrations of acetone from 20% to 50% by volume a satisfactory number of fractions may be obtained. The resolution obtained by this fractionation method is good, as indicated by the relatively gentle slope of the curve.

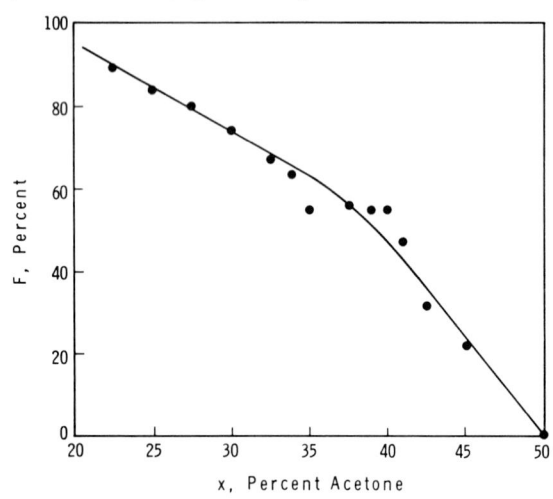

Fig. 1. Solubility characteristics in fractionation system. Fraction of pulp remaining in solution versus composition of acetone-water precipitant.

A typical viscose rayon yarn manufactured from wood pulp was fractionated by the summative method into 16 fractions; the results of this fractionation are given in Table II. The corresponding summative curve is shown in Fig. 2.

TABLE II

EXPERIMENTAL DATA FOR SUMMATIVE CURVE[a] FOR VISCOSE RAYON

$F_{(p)}$ (%)	$\bar{P}$	$F_{(p)}$ (%)	$\bar{P}$
100.0	347	22.8	177
94.0	330	20.7	173
88.0	327	13.2	95
83.2	306	12.3	93
58.0	257	11.6	116
38.3	211	9.35	59
34.6	227	7.78	46
30.4	209	7.40	46
24.1	161		

[a] See Fig. 2 for summative curve.

C.4. Summative Fractionation

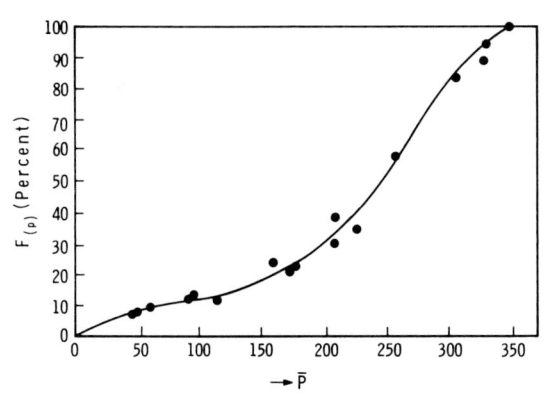

FIG. 2. Summative distribution curve for typical rayon prepared from wood pulp.

By proceeding along the smooth summative curve of Fig. 2, starting at $F_{(p)}$ equal to 100% and taking intervals of $\Delta F_{(p)}$ of 5%, the data in Table III were obtained by means of Eq. (6).

TABLE III

Calculated Data for Converting Summative Curve[a] to Integral Weight Distribution Curve[b] for Viscose Rayon

F (%)	$\bar{P}$	$F\bar{P}$	$P = \Delta F \bar{P}/\Delta F$
100	347	34,700	640
95	332	31,500	540
90	320	28,800	500
85	310	26,300	460
80	300	24,000	440
75	290	21,800	440
70	280	19,600	380
65	273	17,700	360
60	265	15,900	380
55	255	14,000	350
50	245	12,250	310
45	235	10,600	344
40	222	8,880	306
35	210	7,350	288
30	197	5,910	272
25	182	4,550	258
20	162	3,240	252
15	132	1,980	266
10	65	650	105
5	25	125	25

[a] Figure 2. [b] Figure 3.

These results give the stepwise graph of Fig. 3; a smooth curve through these steps provides the integral weight distribution curve.

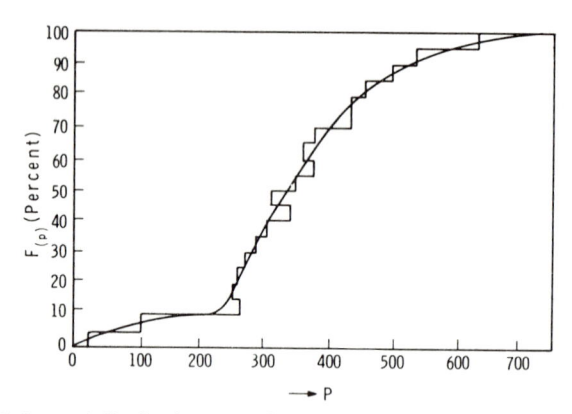

FIG. 3. Integral distribution curve for typical rayon prepared from wood pulp.

Graphical determination of the slope of this curve leads to the differential weight distribution curve of Fig. 4.

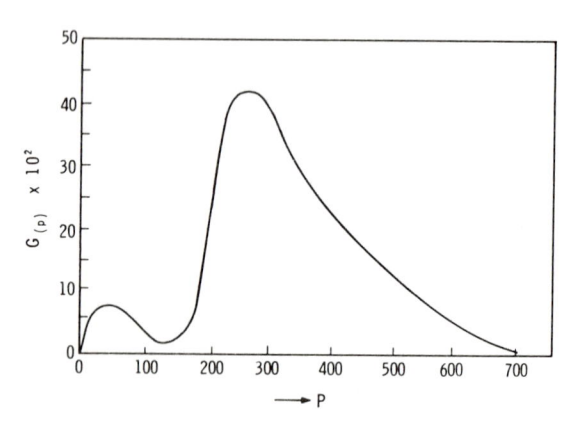

FIG. 4. Differential distribution curve for typical rayon prepared from wood pulp.

Using the integral distribution curve in Fig. 3, the data in Table IV are obtained, from which the bar graph or pictorial representation of Fig. 5 is derived.

TABLE IV

GRAPHICAL DIFFERENTIATION DATA[a] FOR VISCOSE RAYON

P	i (%)	ΔF_i (%)
0–50	5	5
50–100	7.5	2.5
100–150	9	1.5
150–200	10	1
200–250	15	5
250–300	37	22
300–350	55	18
350–400	69	14
400–450	80	11
450–500	88	8
500–550	94	6
550–600	97	3
600–650	99	2
650–700	100	1

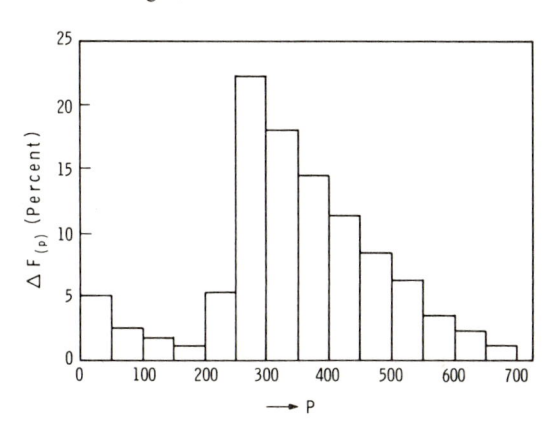

[a] Figure 5.

FIG. 5. Pictorial or bar graph representation of distribution for typical rayon prepared from wood pulp.

REFERENCES

1. G. Beall, *J. Polymer Sci.* **4**, 483–513 (1949).
2. L. Jorgensen, "Studies on Partial Hydrolysis of Cellulose." Trykt Hos Emil Moestue A.S., Oslo, 1950.
3. S. Coppick, O. A. Battista, and M. R. Lytton, *Ind. Eng. Chem.* **42**, 2533–2538 (1950).
4. R. F. Boyer, *J. Polymer Sci.* **9**, 197–218 (1952).
5. R. S. Spencer, *J. Polymer Sci.* **3**, 606–607 (1948).

6. F. W. Billmeyer, Jr. and W. H. Stockmayer, *J. Polymer Sci.* **5**, 121–137 (1950).
7. M. A. Golub, *J. Polymer Sci.* **11**, 281–285 (1953).
8. A. Broda, *J. Polymer Sci.* **25**, 117–118 (1957).
9. A. Broda, T. Niwinska, and S. Polowinski, *J. Polymer Sci.* **29**, 183–189 (1958).
10. A. Broda, T. Niwinska, and S. Polowinski, *J. Polymer Sci.* **32**, 343–355 (1958).
11. J. E. Tasman and A. J. Corey, *Pulp Paper Mag. Can.* **48**, No. 3, 166–170 (1947).
12. B. G. Rånby, O. W. Woltersdorf, and O. A. Battista, *Svensk Papperstid.* **60**, 373–378 (1957).
13. H. Mark, *Paper Trade J.* **113**, 34–40 (1941).
14. H. Mark and R. Simha, *Trans. Faraday Soc.* **36**, 611–618 (1940).

CHAPTER C.5

Rheological Methods

J. Schurz
INSTITUT FÜR PHYSIKALISCHE CHEMIE, UNIVERSITÄT GRAZ, GRAZ, AUSTRIA

I. Introduction to the Problem.. 317
II. Parameter Methods... 318
   A. Viscosity Measurements in Two Solvents......................... 319
   B. Parameters from Non-Newtonian Flow........................... 319
   C. Miscellaneous.. 320
III. Evaluation of Flow Curves... 321
   A. General Remarks... 321
   B. Polydispersity Parameters from Linearized Flow Curves............ 325
   C. Evaluation of Total Flow Curves................................ 326
IV. Evaluation of Relaxation Measurements............................. 332
V. Miscellaneous Methods... 335
   References.. 336

I. Introduction to the Problem

The first attempts to use rheological methods for the estimation or determination of polydispersity and/or molecular weight distribution (MWD) were made a few decades ago. Originally, the reason for this effort was the hope of escaping the time-consuming, cumbersome classical methods, such as fractional precipitation or solution. Later, an additional aspect proved attractive, i.e., to eventually devise methods for the determination of molecular weight and polydispersity in the solid state.

In general, there are two ways of obtaining the desired information. First, we may try to derive from rheological measurements two or more characteristic parameters, which depend on polydispersity in such a way, that their combination furnishes a polydispersity index. We may call these the "parameter methods." Obviously, these methods will have all the well-known disadvantages of classifying polydispersity by but one parameter. Second, we may make use of the entire flow curve. A flow curve is a plot of the apparent viscosity or shear stress versus the average shear rate, from the initial (first) range of Newtonian viscosity up to the final (second) Newtonian range. Such a flow curve represents the response of a solution or melt to variable shearing forces and contains a wealth of information, among which is the distribution curve of polydispersity. The problem is to isolate the desired piece of knowledge. One may avoid this intricate task by plotting flow curves in such a way that straight lines are obtained. Using the slope of these lines, a polydispersity parameter may be obtained. But, then, little is gained over the parameter methods mentioned above; and the result hardly merits the quite difficult experimental work. The most profound method, of course, would be a detailed analysis of the flow curve with the aim of deriving from it the explicit distribution curve. Although, in the authors opinion, this should be basically possible, the practical solution has not yet advanced very far. Attempts have been made from both the theoretical and the practical side, but the gap between the two has so far not yet been bridged. Also, this is true for the attempt to evaluate relaxation curves for this purpose. At present, the theoretical concepts are not yet ripe for the hands of the experimental worker. The practical attempts remain more or less empirical. The approximations and limitations inherent in the empirical approach are not always made sufficiently clear. It is true that some success has been obtained. Rheological means can already provide some information on the polydispersity of a sample and its variation in the course of certain reactions.

In this chapter, the present state of the outlined problem is presented and the pertinent literature made available. Limitation of space forbids a detailed discussion of the basic rheological concepts and experimental methods. However, sufficient references are given from which this knowledge may be ascertained [1].

II. Parameter Methods

Most of the parameter methods bear some resemblance to the inhomogeneity number U (from "Uneinheitlichkeit") as introduced by Schulz [2]: $U = [\overline{M}_w/\overline{M}_n] - 1$. In cases where the exponent a of the viscosity equation is 1, and thus the viscosity average of molecular weight, $\overline{M}_v$, equals the weight average, $\overline{M}_w$, the latter may be determined from the limiting viscosity number $[\eta]$. Recently, it was proposed to use $\sqrt{U}$ rather than U as a polydispersity index, particularly for narrow distributions [3]. The common trait of the U value and similar polydispersity indices is that the comparison of two (or more) moments of the distribution function $F(M)$ is used to determine the spread in this function. Usually, different moments, e.g., different averages of molecular weight, must be determined by different methods. Here we shall concern ourselves only with procedures in which rheological measurements are used.

A. Viscosity Measurements in Two Solvents

The constants in the viscosity equation $[\eta] = KM^a$ are sensitive to polydispersity [4]. With increasing homogeneity, K becomes smaller, while a remains unchanged [5] or increases. In particular, in different solvents with different exponents a, the viscosity average of molecular weight will represent different moments of the molecular weight distribution $F(M)$, since it depends on the value of a

$$\overline{M}_v = \left\{ \frac{\sum N_i M_i^{a+1}}{\sum N_i M_i} \right\}^{1/a}$$

Therefore, the comparison of $\overline{M}_v$ values obtained in different solvents should provide an index of polydispersity. This method has been proposed by Onyon [6] and Frisch [7]. The last author has worked out a method which introduces the polydispersity parameter $\delta(r,s)$, determined from measurements of the limiting viscosity numbers in two solvents for which the exponents are r and s

$$\delta(r, s) = [\overline{M}_v(r)/\overline{M}_v(s)] - 1$$

where $[\eta]_r = K_r \overline{M}_v^r(r), [\eta]_s = K_s \overline{M}_v^s(s)$, and $r > s$. This polydispersity parameter is related to the U value of Schulz according to

$$U = [2/(r - s)]\delta(r, s)$$

In later papers, Frisch et al. [8] have presented experiments to substantiate the method. They claimed good results if the difference between r and s was sufficiently large, and if this difference was accounted for in the molecular weight sums. Mussa [9] has suggested that this polydispersity parameter may also be determined from whole samples if only r and s are known for highly homogeneous materials. He proposed to vary both solvent and temperature in order to obtain large differences. Recently, however, Breitenbach [10] has published a critical account of these "viscosity methods." On the basis of experiments both from the literature and of his own he arrived at the following conclusion: The general accuracy of viscosity measurements is not sufficient to make possible a reliable estimation of polydispersity by viscosimetric methods.

B. Parameters from Non-Newtonian Flow

The deviations from Newtonian flow, as found for polymer solutions even at rather high dilution are partially dependent on the polydispersity [11]. Generally speaking, the decrease in viscosity with increasing shear rate will be less sharp and more spread out if the MWD is broader. Therefore, appropriate measurements may be used to estimate the polydispersity [12].

There are some indications that the constants in the viscosity–concentration equation as extended for non-Newtonian flow [13] are sensitive to polydispersity. Extensive evidence is still lacking. Reichmann [14] used plots of $\eta_D/\eta_{D=0}$ versus α (where η_D and $\eta_{D=0}$ are the viscosity at shear rate D and $D = 0$, respectively, and α is the ratio of shear rate and rotary diffusion constant). He found that for monodisperse systems at low values of α a sharper decrease was observed than for polydisperse systems. Sabia [15] used flow relations valid for melts or concentrated solutions to derive a "dispersity factor"

$$KM_r/\overline{M}_w = \tau/\eta_0^n$$

where η_0 is the zero shear viscosity, n is a constant, τ is the characteristic relaxation time, and M_r the related molecular weight according to

$$\tau = K'M_r\eta_0$$

M_r depends on the number of molecules per unit volume. For a given η_0, the non-Newtonian flow should be the more pronounced, the broader the distribution. This dispersity factor compares well with the expression $\overline{M}_z\overline{M}_{z+1}/\overline{M}_w$ for polyisobutylene but poorly with $\overline{M}_w/\overline{M}_n$ for polystyrene and polyethylene. In another investigation, indications were found that with blends of polystyrene fractions, studied as melts, the apparent viscosity varies with $\overline{M}_w$ at low shear rate, but with $\overline{M}_n$ at high shear rate [16].

C. Miscellaneous

Obviously, measurements at different states of solution, indicative of polydispersity, may be obtained in different ways. It has been proposed that viscosity measurements of both diluted and concentrated solutions should be compared [17]. In this way, polydispersity figures have been obtained for viscose. These figures included not only the actual polymolecularity, that is, the MWD, but also the polydispersity due to the formation of supermolecular aggregates. Measurements of viscosity, non-Newtonian flow, and heats of activation of viscous flow both in different solvents and at different temperatures have been used to obtain information about the polydispersity (mainly with respect to supermolecular aggregates) of polystyrene solutions [18]. In some experiments, the irregularities in the viscosity–concentration function at extremely low concentrations have been used to speculate about the effect of polydispersity upon the shape of the obtained maxima [19]. As these effects in themselves are by no means clearly understood at present, extreme care in interpretation is advisable. Ward and Whitmore [20] have shown that at high concentrations the viscosity of a suspension of polydisperse spheres increases less rapidly than for

III. Evaluation of Flow Curves

A. General Remarks

The term "flow curve" refers to a plot of a figure which characterizes the force imposing flow, versus a figure describing the corresponding response. In capillary viscosimeters, the shear stress at the capillary wall σ_m is plotted versus the so-called reduced flow velocity D (also called mean shear rate). The quantities are given by

$$\sigma_m = Rp/2l$$
$$D = 4Q/\pi R^3$$

where R is the radius of the capillary; l, the length of the capillary; p, the driving pressure; and Q, the outflow volume per second. An apparent viscosity $\eta' = \sigma_m/D$ may be calculated. The flow curve can be given as a plot of D versus σ_m, or η' versus D, or η' versus σ_m. As it is necessary to plot several orders of magnitude in both D and σ_m to ascertain the initial as well as the final Newtonian range η_0 and η_∞, in most cases plots of $\log D$ versus $\log \sigma_m$ and $\log \eta'$ versus $\log D$ are used, as shown in Fig. 1. As these flow curves are made up directly from experimentally determined data, they depend on the instrument used. However, by means of integral equations they may be converted to consistent values of shear rate and viscosity. For the capillary, the well-known equation by Weissenberg

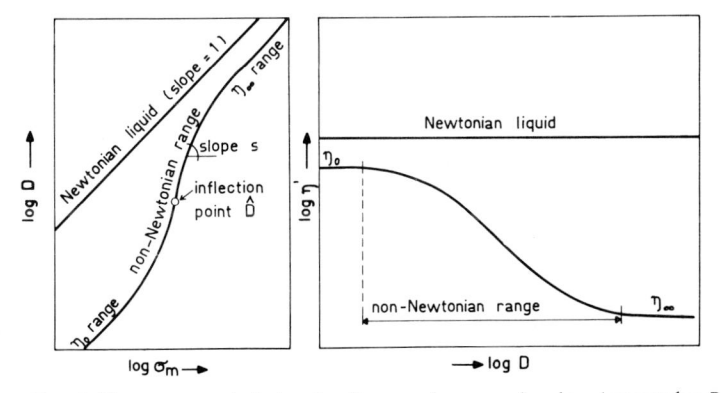

Fig. 1. Flow curves, plotted as $\log D$ versus $\log \sigma_m$ and as $\log \eta'$ versus $\log D$.

represents a solution; for flow curves obtained in Couette-type viscosimeters, appropriate formulas have also been given. These conversions make use of the experimental flow curve and its first differential quotient; details are found in the volumes listed in [1]. Such calculations must be carried out if comparisons are to be made with theories, or if absolute determinations are to be attempted. In empirical methods, they may, in many cases, be dispensed with.

Dynamic measurements should be equivalent to the steady state methods. A flow curve, obtained by plotting dynamic viscosity versus frequency, should be able to serve the same purpose as a plot of η' versus D. A number of experiments appear to substantiate this. Therefore, all considerations which we are going to make hold equally well for flow curves from capillary or Couette experiments and for dynamic measurements. Flow curves are obtained mostly for dilute solutions. The evaluation of concentrated solutions is more intricate, although the effect of non-Newtonian flow as such is larger. Complete flow curves are experimentally difficult to measure. The unequivocal determination of the η_0 range requires in many cases extremely small shear stresses, while for η_∞ values of the order of magnitude of 10^4 to 10^6 dynes/cm^2 may be required. At high rates viscous heating poses a serious problem. Most of the experimental difficulties of accurate flow curve measurement have been amply discussed in the literature [1,22,23]. It is obvious that the reliability of any polydispersity measure derived from flow curves depends in the first place on the accuracy with which the flow curve has been measured. A further improvement of the present state of experimental technique at this level is certainly desirable.

Let us now consider the influence of polydispersity upon the shape of flow curves. The flow curve of a whole polymer is made up from several contributions: (a) The flow curve of the individual particle, be it molecule, aggregate, or any "flow unit." For chain molecules, this term also contains the statistics of the coiled chains. These can be accounted for by choosing appropriate averages without introduction of a second term. (b) The contribution of mutual interactions of the particles or flow units. This term may be kept small by working with highly diluted solutions or eliminated by extrapolation to zero concentration. In some cases, a better picture of reality is obtained if we do regard solutions with a certain degree of interaction, or even use the interaction as a basis for theory. Such has been done by Bueche in his infinite network concept. (c) The distribution of size and/or shape of the particles or flow units. This is the polydispersity or polymolecularity (MWD) of a system. In most practical cases, the shape of the particles will be sufficiently uniform.

The problem is now to extract the polydispersity curve from the flow curve. This should be possible immediately if the last-mentioned term, i.e., that due to polydispersity, plays a role considerably larger than the other ones. It appears that this may actually be the case if certain conditions are met (e.g., appropriate materials, solvents, and dilute solutions). However, if the polydispersity term is small and affects the flow curve in only a minor way, the matter becomes rather hopeless. For this problem a knowledge of the flow curve of single, uniform particles is of utmost importance, as then the deviations from this flow curve would describe the polydispersity effect. Unfortunately, there is neither a generally accepted theory, nor do we possess experimental flow curves measured with very uniform material. However, such measurements are under way at present.

It is known that, in general, polydispersity acts in such a way as to broaden the flow curves and to render them more flat, as shown experimentally long ago by Philippoff [24] (cf. schematic drawing, Fig. 2). This behavior is easily understood on the basis of the additivity of the viscosity.

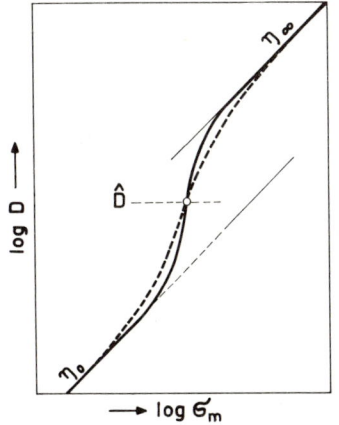

FIG. 2. Schematic drawing of flow curves: ———, uniform sample; – – – – –, polydisperse sample.

On account of it we may assume also additivity for the contributions of the individual particles to the flow curve, i.e., to the deviations from Newtonian flow. Thus, the onset of an individual molecule's contribution will be proportional to some power of its molecular weight M^x, and the amount of its contribution proportional to M^x times the number of such molecules present in the solution. Therefore, the low molecular weight portions will generally contribute less than the high ones. Furthermore, non-Newtonian flow will be observed earlier in the flow curve for a polymolecular than for a uniform substance of the same average molecular weight. In the polymolecular case, chains of greater length are also present. They are present only in a rather small number, so that the amount of the deviation from

Newtonian flow will be small at first. For uniform particles the deviation will start later, but the amount will be larger from the beginning. Therefore, the flow curve of the polydisperse mixture will be flatter, less pronounced, but broadened. It will extend over a larger range of shear rate in comparison to the flow curve of a uniform material of the same average molecular weight. This is shown schematically in Fig. 2.

We may attribute to each molecule M_i (or to each particle) a shear rate D_i, at which it begins to deviate from Newtonian flow. For any given D_i, all molecules equal to or larger than M_i will contribute to non-Newtonian flow. Thus, the flow curve as a whole should correspond to an integral distribution curve, but modified by the other contributions mentioned before.

For a given molecular weight, on the one hand, the shear dependence of the viscosity (i.e., the deviations from Newtonian flow) will be largest for the monodisperse substance, as it is caused by and confined to only one type of particle. On the other hand, the polydisperse mixture will always exhibit shear dependence of viscosity at smaller shear rates than a uniform sample. We may expect the polydispersity curve to manifest itself in the flow curve in such a way that the amount of shear dependence at its maximum rate should be a measure of the height of the distribution curve. The shear rate, for which shear dependence is first observed, will be indicative of the largest molecule present. If we assume a symmetrical distribution, this will give the half-width. This simple concept is somewhat obscured by the fact that the amount of the deviation from Newtonian flow due to a molecule of type M_i will depend on both the number of such molecules, and on M^x. Unfortunately, little is known about the power of x of this dependence. No shear rate limit can be found for the smallest molecule present. The point at which the shear dependence ceases, i.e., the transition of the flow curve to the second Newtonian range η_∞, merely indicates the smallest molecule contributing to the shear dependence of viscosity. This need not be (and generally will not be) the smallest molecule present. The low molecular weight tail of the distribution curve will then escape detection by flow curve analysis.

We may put this argument in another way. In dynamic measurements the molecules or particles are caused to vibrate. In shear flow too, according to the Bueche theory, the particles perform a compression–extension vibration while rotating. These vibrations are related to the deviations from Newtonian flow. Obviously, each particle will have a frequency or shear rate at which it responds most easily to the applied force. This can be compared with a resonance phenomenon; the exciting force is maximally absorbed, the rate of the deviation from Newtonian flow (the variation of viscosity with varying shear rate) is largest. This characteristic frequency or shear rate is a function of the particle mass, probably in the form of a "natural relaxation time," as proposed by Bueche. Therefore, from the frequency or shear rate of this resonance position, the mass of the respective molecule can be determined. In other words, its molecular weight. The amount of non-Newtonian flow, that is, the energy absorbed, will depend on the number of molecules resonating at this frequency. In general, the frequency of these resonance vibrations indicates the mass of the resonating particles, and the amplitude their number. This simple concept should be applicable to both dynamic and steady state measurements. Again, the complicated makeup of flow curves from many more contributions than the polydispersity effect will obscure the picture and render the extraction of polydispersity curves actually a rather complex task.

B. Polydispersity Parameters from Linearized Flow Curves

It is possible to devise approximations which represent flow curves over certain ranges as straight lines. In the light of the previous discussion, it is clear that the slope of such a plot must be a measure of the polydispersity of the sample—among many other things, of course. It has been shown, that for moderately concentrated solutions, a plot of $\log \eta'$ versus σ_m yields straight lines over a range of several orders of magnitude. Due to the non-logarithmic plot of σ_m, the range of very small shear stresses is not properly accounted for. Thus, the straight lines are obtained at the best for σ_m values between 10^2 and 10^4 dynes/cm^2. Such plots have been presented for cellulose dissolved in Cuoxam by Meskat and Linsert [25]. With the help of crude fractionations performed on the same samples, a quantitative relation between the slope of the straight line flow curves and the polydispersity (pulps and cottons) was established. Higher slopes indicated less disperse samples. The relation was rather poor, the scatter of the points wide, and, above all, solute concentration and average molecular weight appeared to play a preponderant role. Solute concentration and average molecular weight had to be kept constant before a comparison of polydispersity was possible. These difficulties not withstanding, the feasibility of the method was clearly established. The same plot was used by Schurz [26] for viscose. He too derived polydispersity relationships, among other figures. The influence of average molecular weight and solute concentration were again disturbing effects.

A method roughly equivalent to that previously given is to plot $\log \eta'$ versus $\log \sigma_m$ on the assumption that a power law governs the flow. Here the exponent of the power law, n, may serve as an index of polydispersity. It is subject to the same restrictions as the slope of $\log \eta'$ versus σ_m plots. It is greatly influenced by other factors, such as solute concentration and average molecular weight. Thus, the polydispersity, a minor effect in itself, is obscured by major contributions from other variables.

Some compensation for solute concentration and, in some cases, the average molecular weight, is furnished if the same flow curve as that under investigation is used to determine the scale of the plot. This has been done in an approximating method put forward by Umstätter [27]. Here the quantity

$$\frac{\ln(\eta'/\eta_\infty)}{\ln(\eta_0/\eta_\infty)}$$

is plotted versus $\ln D$ on probability paper. Thus, η_0 and η_∞ determine the scale used. It was assumed that the above expression obeys a Gaussian sum curve (error function). Umstätter maintained that this approximation could be explained on theoretical grounds. He met serious objections. This method should be regarded as an empirical approximation presupposing symmetrical flow curves resembling error functions. The plot is independent of solute concentration (but not of average molecular weight). The slope of the straight lines, which is the half-width of the Gaussian curve, may be used as a measure of polydispersity. The larger the slope, the more uniform is the sample. This method has been applied by Edelmann for solutions of various high polymers, such as cellulose nitrate [28], polyacrylonitrile [29], and natural rubber (effect of mastification) [30]. Comparisons have been made with the U value by Schulz, and fair agreement was found. Mašura [31] has studied the aging of alkali cellulose of this method (cellulose nitrate solutions). His results are not clearly in favor of the method, as the MWD of his samples deviated from Gaussian curves.

C. Evaluation of Total Flow Curves

The experimental observations concerning the flattening effect of polydispersity on flow curves are numerous. Philippoff's experiments [24] were good examples, as are those by Meskat and Linsert [25]. Rouse and Sittel [32] measured the real part of the dynamic intrinsic viscosity as a function of MWD. The curve for the narrow fraction was significantly steeper than that of a "natural distribution." The experiments by Mills [33] on polyethylene melts also substantiate this behavior. A discussion on the influence of MWD on the shape of flow curves, plotted as $\log \eta_r / \log \eta_{r,0}$ versus $\log(\tau D)$ has been given by Rodriguez and Goettler [33a]; here η_r and $\eta_{r,0}$ are the relative viscosity at the shear rate measured, respectively at zero shear rate. The broadening effect of MWD was demonstrated graphically. More examples will be given later. Thus, the existence of a polydispersity effect on flow curves appears experimentally established. For more explicit evaluations, the general relationships discussed in Section III,A must be cast into equations.

1. *Empirical Flow Curve Analysis (PD-Kennkurven Method)*

With certain simplifications, the non-Newtonian viscosity (expressed as apparent viscosity η') at a given shear stress σ_m may be written as follows [34]:

$$\eta' = ck \int F(M) M^n f(\sigma, M)\, dM \tag{1}$$

where c is the concentration of the solution, k a constant, $F(M)$ the MWD curve, and $f(\sigma, M)$ the flow curve of the single, isolated particle, here the molecule. Actually, $f(\sigma, M)$ should also contain a term taking into account the particle shape or stiffness of molecular chains. This very important figure is certainly not adequately covered by the term M^n, where n is usually between 0.5 and 3.5. A general difficulty here is that the behavior of single particles depends on the total shear stress, which in turn is dependent on all particles present. The shear stress as contained above in the function $f(\sigma, M)$ is, therefore, a sum over all σ_i, and the integral must be taken over all $f_i(\sigma)$ rather than $f(\sigma_i)$. Instead of the shear stress, σ, we may use the shear rate q. Now, the problem is to derive an explicit expression for $F(M)$ from measured values of η' or $d\eta'/dq$ as functions of q (or from equivalent figures, e.g., D as a function of σ_m). A general solution of this problem meets considerable difficulties, which so far have not yet been overcome. However, both theoretical and empirical approaches have been proposed.

An example of the latter type is the empirical flow curve analysis as proposed by the author [35]. It is based on the observation that the inflection point of a $\log D$ versus $\log \sigma_m$ flow curve, which is characterized by its value of the mean shear rate, $\hat{D}$, is related to the molecular weight of the solute by the equation

$$\hat{D} = aM^{-b} \tag{2}$$

where a and b are empirical constants, which have already been determined for a great number of solute–solvent systems [36]. $\hat{D}$ is dependent on concentration. It is small at high concentrations, increases as the concentration is lowered until, eventually, a constant $\hat{D}$ is reached. This may reflect a decrease in the size of the flow units as the concentration becomes smaller. A constant $\hat{D}$ would be an indication of invariable flow units, presumably molecules. Further, it has been shown that the slope s of $\log D$ versus $\log \sigma_m$ flow curves is related to the solute concentration c by the equation

$$s - 1 = \beta(M)c \tag{3}$$

where $\beta(M)$ is some function of M. In general $\beta(M)$ is not known but can be determined experimentally. We shall assume that for a given solute concentration (generally for $c = 1\%$) Eq. (3) can be applied to the various molecular weight species present in the polydisperse mixture. Then, at a certain shear rate value D_i, the corresponding slope s_i should give the "concentration" of the molecular species M_i. M_i is calculated from D_i with the help of Eq. (2), as soon as the empirical constants a and b are known. Thus, a plot of $s - 1/\beta(M)$ versus M [calculated according to Eq. (2) from D] should give a curve resembling the differential mass distribution curve of the sample. It is blurred, of course, by various other effects. For this reason, we have called it "PD-Kennkurve" (PD curve) in order to indicate its crude approximative nature. A major obstacle is the effect of average molecular weight as contained in $\beta(M)$. Attempts have been made to eliminate this effect by an experimental determination of the function $\beta(M)$. For certain cases it has been found to be a simple proportionality, namely kM. Here, a correction was possible using the expression $s - 1/M$. This procedure was never completely satisfactory. To make matters worse, it has been found not to hold in many cases. In general, a correction for $\beta(M)$ is better dispensed with, and the PD curves are plotted simply as $s - 1$ versus M. At this point a previously discussed difficulty becomes apparent: for small values of M (generally corresponding to a degree of polymerization of about 100–200), all PD curves show an abrupt downturn. This has nothing to do with the polydispersity curve but indicates that such small molecules no longer contribute to the shear dependence of viscosity. Here the very effect on which the method is based ceases, and the method becomes meaningless as a result.

With due care, this empirical method may be used to obtain certain information on the polydispersity. It has been applied to several polymers. Comparison with fractional precipitation showed fair agreement [35]. PD curves were obtained for viscose and cellulose nitrate [36], carboxymethyl cellulose [37], cellulose [38], polyisobutylene and polyvinylpyrrolidone [39], natural rubber [40], polystyrene [41], and polymethylmethacrylate [42]. The method has been used to study polydispersity during the aging of alkali cellulose [43]. It has been tested with mixtures, in which case PD curves with two peaks could be found [44]. Recently, PD curves of polystyrene of usual polydispersity and of a "monodisperse" sample made by the anionic technique were obtained [45]. They are reproduced, together with the corresponding flow curves, in Fig. 3. The difference in polydispersity is clearly seen, though the concentration used is too high. A recent study using the PD curves method with acrylonitrile–methyl methacrylate copolymers by Chinai [45a] reported satisfactory results.

It must be added that in some cases PD curves tell little about the polydispersity, as they contain many disturbing effects. This is especially true if PD curves are constructed from flow curves obtained with solutions of high concentration. The method works best for polymers which give

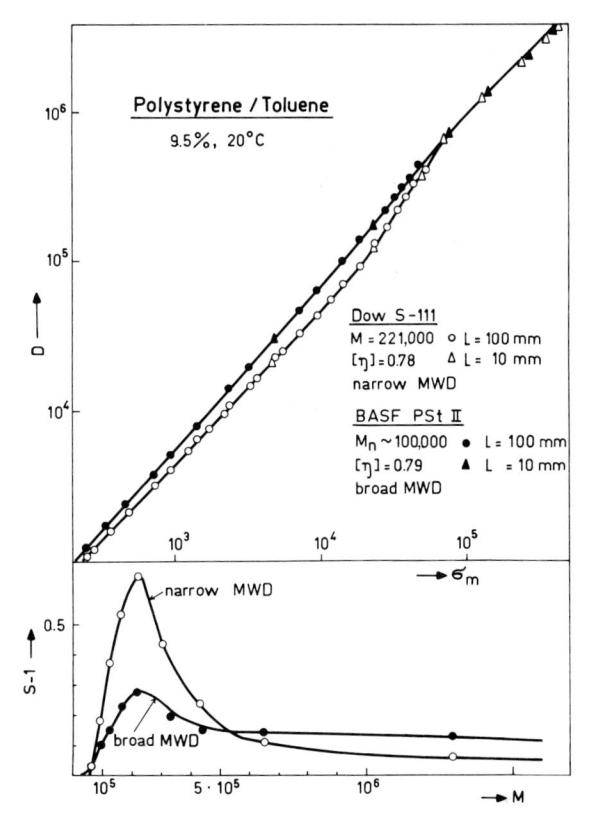

FIG. 3. Flow curves (top) and PD curves (bottom) of narrow and broad molecular weight distribution polystyrene in toluene. Despite the excessively high solute concentration, 9.5%, the difference is clearly seen.

well-defined flow curves at low concentration. In solution such polymers consist of large, anisotropic, and stiff molecules (chain molecules). It is important that the flow curve be determined with high accuracy. Care has to be taken to avoid spurious effects, such as viscous heating at high shear rates, dissolution of the driving gas into the solution, poor thermostating, Hagenbach-Couette, and relaxation effects. Such influences can distort the shape of the flow curve to such a degree that any interpretation is rendered completely meaningless.

2. *Theoretical Approaches to the Flow Curve Evaluation*

In some cases, an exact solution of the problem outlined in Section III,C,1 is possible, when explicit functions can be selected for the general terms in the integral given in Eq. (1).

Peticolas et al. [46] have worked out a specific path on the basis of the normal coordinate theory of viscosity [47]. As this theory sets up the flow curve as a sum over the various modes of relaxation of the flow unit, this summation appears in the equation instead of a closed function for the flow behavior. The necessary inversions for the explicit calculation of the MWD function $F(M)$ are made possible by theorems from the theory of numbers. The method starts out from an experimentally determined curve giving the frequency dependence of some viscoelastic property $\Phi(\omega)$. This is preferably complex viscosity or stress relaxation. The function $\Phi(\omega)$ consists of a sum over the terms λ_p characterizing the relaxation of the various normal modes of the motion and a function $f(\omega, \lambda_1^{-m})$. The term $f(\omega, \lambda_1^{-m})$ is related to $F(M)$ by an integral equation. For certain cases this equation can be converted to give a direct expression for $F(M)$. The treatment is exact within the limits of the underlying theory. For instance, using certain approximations and simplification, the following relation has been derived for the dynamic viscosity:

$$F(M) \simeq -\frac{4\omega^2}{\pi cRT}e^2\left[\frac{dS(\omega)}{d(\omega^2)} + \frac{\omega^2}{2}\frac{d^2S(\omega)}{d(\omega^2)^2}\right] \qquad (4)$$

where

$$S(\omega) = \sum_{n=1}^{\infty}\left[\frac{\mu(n)}{n^2}\right]\Phi\left(\frac{\omega}{n^2}\right)$$

and μ is the Moebius function. In the function $S(\omega)$, the term $\Phi(\omega/n^2)$ stands for the flow curve. The relationship with what has been said before is seen if we consider that the function $S(\omega)$ in Eq. (4) approximately represents the flow curve. Under the same conditions, the frequency ω represents the shear rate. If these assumptions are made in Eq. (4), the polydispersity $F(M)$ is again some function of $d\eta'/d(q^2)$. The authors have used their derivations to test the normal coordinate theory of viscosity with respect to both the calculation of average molecular weights and the evaluation of polydispersity. They point out that a test of these theories is possible only if due care is taken of the polydispersity. Comparison with experiments is scanty, and no attempt has yet been made to derive the MWD from rheological measurements for any practical case. However, the method appears well founded and is certainly worth testing with experiments on various high polymer solutions. The method is, as it stands, applicable to dynamic measurements of complex viscosity and stress relaxation. An extension into steady state viscosity measurements should be possible.

The flow theory of Eyring-Ree [48] may be used too as a basis for polydispersity determination. Faucher [49] has pointed out that, according to this theory, the flow curve $f(q)$ may be expressed by an integral

$$f(q) = \int_0^\infty G(t) \operatorname{arsinh}(qt)\, dt$$

If $f(q)$ is known, the integral may be solved to yield the distribution of flow units $G(t)$, which should be connected with the MWD.

3. *Other Observations*

In this section, we shall try to point out briefly other observations pertinent to our problem. Mainly, we shall refer to papers in which rheological measurements have been used to gain information about polydispersity by methods more complicated than those concerned with the determination of a single polydispersity parameter.

Rudd [50] measured melt viscosities for both narrow and broad MWD polystyrene samples. Plots of η' versus σ_m showed clear differences between narrow and broad MWD. Narrow MWD material had higher viscosities at high shear rate. It was found that zero shear viscosity and non-Newtonian flow depended on $\overline{M}_w$, while high shear viscosity depended on some average between $\overline{M}_w$ and $\overline{M}_n$. Schreiber et al. [51], in work on polyethylene melt viscosity, observed the initial shear stress for non-Newtonian flow, $\sigma_{m,1}$, which is characteristic for every molecular species present, because at $\sigma_{m,1}$ it begins to contribute to the shear dependence of viscosity. Their argument is similar to what has been said in Section III,A. They worked out the relation $1/\sigma_{m,1} \propto (1 - M_c/M)^n$, where M_c is the critical molecular weight, according to Bueche, above which entanglement takes place. Plots of $\log(1/\sigma_{m,1})$ versus $\log(M/M_c)$ give straight lines, which intersect the ordinate at larger values of $\log[1/\sigma_{m,1}]$ and have larger slopes, the broader the MWD. The same authors reported work on other polymers [52]. Mills [53] studied the melt viscosity of polyethylene and found that an increase in MWD for a given average molecular weight increased the sensitivity to pressure, i.e., the shear rate. It was also found that polyethylene with wide MWD was less sensitive to temperature changes than the same polymer with a narrower MWD. Similar results were reported by Martinovich et al. [54]. Thus, it appears that a broader MWD increases the "shear sensitivity," i.e., the viscosity will show a larger decrease on increasing shear stress. Leaderman et al. [55] found with polyisobutylene that a blend containing a small amount of high molecular weight material possessed an elastic recovery much greater than for any fraction and showed much more pronounced non-Newtonian flow. On the other hand, small amounts of low molecular weight fractions in high molecular weight material had no effect. Tschoegl [56] has recently discussed the theories of Zimm, and of Rouse, and suggested that polydispersity may cause an apparent change from

Zimm model to Rouse model (first includes interaction effects). However, he also expressed the opinion that, in this case, the MWD is not the decisive factor. The deviations from theory in the measurements by Smith [57] on polyethylene may also be explained by the MWD effect. Mills *et al.* [58] studied melt fracture and found it occurred at a higher shear stress for a broader MWD. The MWD also effects stress cracking in solid high polymers. Stress cracking is enhanced by increased amounts of low molecular weight material [59]. References to older work may be found in a review on polydispersity determination given by this author some time ago [60].

A great number of papers report related observations concerning the influence of polydispersity on certain rheological characteristics, e.g., the critical viscosity for entanglement and the slope of $\log \eta'$ versus $\log M$ plots at constant shear rate. This slope is higher for broader MWD [60a]. The flow activation energy at constant shear stress decreases more rapidly with shear rate for narrow MWD than for broad distributions [60b]. Polydispersity was found to decrease the flow index and lower the activation energy [60c]. A narrow MWD decreases the flow rate at high shear stress [60d]. Polydispersity causes deviations from Bueche's viscosity curve and affects the activation energy [60e]. Connections with viscoelastic properties of solutions [60f] and branching [60g] have been explored. The steady-state melt flow behavior of polyethylene blends, with consideration of the effect of a gel-fraction, has been studied by Nakajima and Wong [60h]. They found, that a gel fraction manifests itself as a "tailing up" effect of the flow curve at small shear rates. Some more papers reporting on the relation between melt flow behavior and MWD are compiled in [60i].

IV. Evaluation of Relaxation Measurements

It is known that certain rheological measurements of viscoelastic properties of solutions, melts, or solids may be used to calculate the spectrum of relaxation or retardation times, the so-called relaxation (or retardation) time distribution (RTD) [61]. Beside dynamic measurements, stress relaxation studies have been principally used for this purpose. Approximative methods are available to transform the stress relaxation curve $E(t)$, which is thought as consisting of the contributions of a set of relaxation times τ_i

$$E(t) \simeq \sum E_{i,0} \, e^{-t/\tau_i} = \int_0^\infty \frac{dE_{i,0}}{d\tau} e^{-t/\tau} \, d\tau$$

where $dE_{i,0}/d\tau = W(\tau)$, into the corresponding distribution curve of relaxation times $W(\tau)$. It is a reasonable assumption, supported by many experimental results, that the relaxation time distribution, RTD, should reflect the polydispersity. This can be understood easily if we assume that each particle species present is manifested by its natural relaxation time. Thus a relation must hold between the relaxation time of a particle and its mass or molecular weight. Several reasons for this relation have been put forward, based on existing theories of viscoelastic behavior. Consequently, a number of papers have been devoted to attempts to use these general relations for the calculation of the MWD from the RTD curve. Experimentally the RTD curve has been obtained mainly from stress relaxation measurements.

The basic difficulty here is of a similar nature to that encountered with flow curves. That is to say, any particle has numerous ways to relax, resulting in a spectrum of relaxation times. In addition, the MWD gives rise to a further spectrum of values due to the different molecules present. Each molecule may be represented by an average relaxation time. Again, experimental evidence suggests that the evaluation of the MWD may still be possible. In fact, if we compare experimentally measured relaxation or creep functions with theoretical functions, there is a striking similarity between them and flow curves modified by polydispersity. In both cases, the experimental curves appear flattened, increasing more gradually and extending over a longer time interval than a theoretical curve would predict (cf. the measurements of Leaderman on polyisobutylene) [62]. Although this behavior is not necessarily caused by polydispersity, the effect of this property would certainly be expected to bring about a flattening. Ferry [63] has recently discussed the bearing of the MWD on relaxation curves with the help of several experimental studies. He states that experimental curves are, in general, more gradual in slope and lie at lower frequencies than calculated curves. This deviation is to be expected due to polydispersity, since components larger than the average molecular weight will contribute much longer relaxation times. These discrepancies will be greater for broader distributions.

It has been shown by work on solids [64] and concentrated solutions [65] that the distribution of mechanical relaxation times in the rubbery region of a linear amorphous high polymer is closely correlated with the shape and width of the MWD. Fujita and Ninomiya [66] correlated the mechanical relaxation distribution of linear amorphous polymers with the molecular weight distribution by means of an integral equation. For this equation they were able to present a solution on the basis of certain special assumptions. One of these assumptions was that both relaxation time and viscosity vary with the 3.4th power of the molecular weight. With this method, they could predict the MWD from experimental relaxation data obtained over the entire region of rubbery time scale. Stress relaxation experiments with unfractioned samples of polystyrene and polyvinyl acetate were analyzed in this way with satisfactory results, except in the range of very low molecular weights. The authors pointed out that the RTD may also be obtained from dynamic mechanical measurements in the audiofrequency range as performed by Ferry et al. [67]. They tested the method using data on polyisobutylene taken from the literature [68], and found that some of these assumptions were not valid.

Tobolsky [69] has shown that the stress relaxation properties of linear amorphous polymers, in the region of rubbery flow, depend on both molecular weight and polydispersity. He tested these assumptions on two polystyrene samples, one a "monodisperse" material obtained by the anionic technique, the other an ordinary one [70]. It was found that the monodisperse sample had a clearly narrower distribution of relaxation times. Furthermore, its RTD was of the "box type." Stress relaxation could definitely and easily distinguish between the two samples investigated, although the heterogeneity index $\overline{M}_w/\overline{M}_n$ of 1.5 for the polydisperse material was by no means very high. Thus, it appears that the RTD in the rubbery region of linear amorphous polymers is connected with polydispersity: the height of the relaxation spectrum increases, and the RTD approaches a "box," with increasing sharpness of the molecular weight distribution. Sobue and Murakami [71] have shown that these two criteria are identical. They also discuss the method of Tobolsky and Murakami [72] which uses as an index of polydispersity the following expression:

$$\alpha_B = \tau_{\max} E_{\max}/\eta$$

where $\tau_{\max}$ and $E_{\max}$ are the maximum relaxation time and the associated partial modulus and η is the flow viscosity. The last-mentioned authors studied the effect of polydispersity of polystyrene on the stress relaxation. They measured stress relaxation and flow viscosity at 115°C for samples whose heterogeneity (expressed as $\overline{M}_w/\overline{M}_n$) varied between 1 and 15. They observed a correlation between the $\overline{M}_w/\overline{M}_n$ ratio and the α_B value. Peticolas [73] extended the calculation of the relaxation times of whole polymers to include branching and presented equations by which both the effect of branching and of MWD on rheological measurements may be calculated. He could correlate the MWD with stress relaxation measurements in polymer melts (polyethylene and polystyrene) using the theory of Menefee and Peticolas [46]. Watkins et al. [74] measured stress relaxation in concentrated solutions of polymethyl methacrylate and were able to calculate the MWD, assuming the friction factor to increase with the 1.5th power of the molecular weight. For a critical discussion see ref. [75]. Bueche [76] discussed the effect of polydispersity on steady state elastic compliance. The compliance is proportional to $\overline{M}_{z+1}\overline{M}_z/M_w$ [77] and, thus, depends greatly on high molecular weight material. A polymer having a wide MWD will show more elastic effects than a homogeneous polymer with the same weight average molecular weight. Mussa et al. [77a] found that a higher polydispersity causes a shift of the entire normalized stress relaxation curve towards higher values.

V. Miscellaneous Methods

An important, although at present little used, method is the measurement of optical birefringence. It appears that it holds some promise for polydispersity determination. Little has been published on the subject. As early as 1938, Sadron treated the polydispersity effect and presented measurements of flow birefringence on a mixture of molecular and micellar cellulose acetate in cyclohexanone [78]. A discussion of these results has recently been given by Peterlin [79]. Signer tried to use flow birefringence to obtain information on the molecular weight distribution from the extinction angle, but his results were dubious [80]. The modifications brought about by the heterogeneity in flow birefringence measurements are caused by the most asymmetric molecules retaining their orientation to the lowest values of the shear rate. Therefore, by applying the usual formula, values for the rotary diffusion coefficient obtained from the limiting slopes of the extinction angle versus shear rate curves are too small. A discussion of this topic has been given by Cerf and Scheraga [81]. For some special distribution functions of rodlike particles, birefringence and extinction angle have been calculated [82]. A rigorous theory of flow birefringence is available for monodisperse solutions of particles of ellipsoidal shape. This theory has been extended to cover polydispersity [82]. For chain molecules, such a theory is still lacking. At present, the polydispersity problem can be treated only experimentally, as has been done in several cases [70,80]. It appears that significant effects are only observed if mixtures of two very different fractions are studied. According to a private communication by Janeschitz-Kriegl, the formation of supermolecular aggregates can be detected. If even a very small amount of such aggregates has formed, the extinction angle decreases rapidly at small shear rates. With higher values of shear rate the extinction angle assumes the value expected from molecular solution. This method is claimed to be very sensitive.

A very sensitive parameter can be derived from flow birefringence by plotting E_n versus β [where $E_n = (1/3C)(\Delta n \cos 2x/vkT)$ and $\beta = q(\eta - \eta_0) \times M/RTc)$]. E_n is higher, the broader the MWD. By putting $E_n = pE$

(with E the theoretical value for monodisperse material), the parameter p is obtained, which can become as large as 100 (e.g., for technical polyethylene) and is independent of the molecular weight [82a]

Rudd and Gurnee [83] measured stress birefringence during stress relaxation on solid compression-molded samples of polystyrene. In their plots of birefringence over strain versus log time, and birefringence versus log time, they could clearly distinguish between narrow and broad distributions. The narrow distribution showed a sharper maximum in the first-mentioned plot and a steeper decrease in the second plot. As with previous cases, polydispersity appears to have a broadening or flattening effect on the curve expected from a monodisperse sample.

Dielectric measurements have also been used for the evaluation of polydispersity. This appears reasonable, as charged particles and those having dipole moments can easily be excited into vibration by an electromagnetic wave. These vibrations will be characterized by a relaxation time connected with the respective molecular or particle weight. In general, what has been said for mechanical relaxation or for resonance phenomena will be, with some reservation, applicable to the dielectric method. Scherer and Testerman [84] have measured the dielectric dispersion of cellulose nitrate. From the dispersion curves they could construct integral distribution curves, for which they claimed good agreement with MWD curves as obtained from fractional precipitation. Mark has reported a method devised by Debye [85], in which a solution is placed into an inhomogeneous electrical field. The polarized molecules (in the example described, polystyrene) migrate to sites of higher field strength. The resulting concentration gradient enforces a back-diffusion. The time function of the dielectric constant as measured for the solution should yield information on the molecular weight distribution. For rodlike particles (polypeptides), the shape of the dielectric absorption spectrum has been found to be extremely sensitive to MWD [86].

REFERENCES

1. F. R. Eirich, ed., "Rheology," Vols. 1–3. Academic Press, New York, 1956, 1958, 1960; A. J. Staverman and F. Schwarzl, *in* "Die Physik der Hochpolymeren" (H. A. Stuart, ed.), Vol. IV, pp. 1 and 126. Springer, Berlin, 1956; W. Meskat, *in* "Messen und Regeln in der chemischen Technik" (J. Hengstenberg, B. Sturm, and O. Winkler, eds.), p. 698. Springer, Berlin, 1957; M. Reiner, "Deformation and Flow." Lewis, London, 1949.
2. G. V. Schulz, *Z. Physik. Chem.* **B43**, 25 (1939).
3. G. G. Lowry, *Polymer Letters* **1**, 489 (1963).
4. H. P. Frank, *J. Polymer Sci.* **7**, 567 (1951).
5. H. P. Frank and J. W. Breitenbach, *J. Polymer Sci.* **6**, 609 (1951).
6. P. F. Onyon, *Nature* **183**, 1670 (1959).
7. H. L. Frisch and J. L. Lundberg, *J. Polymer Sci.* **37**, 123 (1959).
8. J. L. Lundberg, M. Y. Hellman, and H. L. Frisch, *J. Polymer Sci.* **46**, 3 (1960).

9. C. Mussa, *J. Polymer Sci.* **41**, 541 (1959).
10. J. W. Breitenbach, *Makromol. Chem.* **60**, 18 (1963).
11. J. Schurz, *Makromol. Chem.* **10**, 194 (1953); **12**, 127 (1954); cf. J. Schurz, *Oesterr. Chemiker-Z.* **62**, 139 (1961).
12. F. Breazeale, *J. Polymer Sci.* **3**, 141 (1948).
13. J. Schurz, *Monatsh. Chem.* **86**, 454 (1955); **94**, 859 (1963).
14. M. E. Reichmann, *J. Phys. Chem.* **63**, 638 (1959).
15. R. Sabia, *J. Appl. Polymer Sci.* **7**, 347 (1963).
16. W. P. Cox and R. L. Ballman, *J. Appl. Polymer Sci.* **4**, 121 (1960).
17. E. A. Pakshver and G. V. Vinogradov, *Khim. Volokna* No. 2, p. 25 (1963) [cf. *Chem. Abstr.* **59**, 3001c (1963)].
18. V. E. Dreval, A. A. Tager, and A. S. Fomina, *Vysokomolekul. Soedin.* **5**, 1404 (1963).
19. K. Edelmann, *Kolloid-Z.* **145**, 92 (1956).
20. S. G. Ward and R. L. Whitmore, *Brit. J. Appl. Phys.* **1**, 286 (1952).
21. K. A. Wolf, ed., "Struktur und physikalisches Verhalten von Kunststoffen," Table 5, pp. 61 and 76. Springer, Berlin, 1962.
22. J. Schurz, *Kolloid.-Z* **148**, 76 (1956); *Rheol. Acta* **2**, 143 (1962); H. Umstätter, *Kolloid-Z.* **145**, 102 (1956); H. Umstätter and R. Schwaben, "Einführung in die Viskosimetrie und Rheometrie." Springer, Berlin, 1952 (cf. Meskat [1]).
23. W. Philippoff, "Viskosität der Kolloide." Steinkopff, Darmstadt, 1942.
24. W. Philippoff, *Chem. Ber.* **70**, 827 (1937); W. Philippoff and K. Hess, *A. Physikal. Chem.* **B31**, 237 (1936 [Philippoff [23], p. 330]).
25. W. Meskat, *Chem.-Ing.-Tech.* **24**, 333 (1952); *Dechema Monograph.* **25**, 9 (1955); F. Linsert, Ph.D. Thesis, University of Köln (1950).
26. J. Schurz, *Papier* **9**, 45 (1955); *Kolloid-Z.* **138**, 149 (1954).
27. H. Umstätter, *Kolloid-Z.* **139** 120 (1954); H. Umstätter and R. Schwaben, "Einführung in die Viskosimetrie und Rheometrie." Springer, Berlin, 1952.
28. K. Edelmann, *Faserforsch. Textiltech.* **5**, 59 (1954); **8**, 184 (1957).
29. K. Edelmann, *Faserforsch. Textiltech.* **5**, 325 (1954); **6**, 269 (1955); *Kolloid-Z.* **145**, 92 (1956).
30. K. Edelmann, *Rheol. Acta* **1**, 53 (1958); *Gummi Asbest* **11**, 251 (1958); **12**, 66 (1959); *Rubber Chem. Technol.* **30**, 470 (1957).
31. V. Mašura, *Chem. Zvesti* **16**, 232 (1962).
32. P. E. Rouse and K. Sittel, *J. Appl. Phys.* **24**, 690 (1953).
33. D. R. Mills, G. E. Moore, and D. W. Pugh, *SPE (Soc. Plastics Engrs.) Tech. Papers* **6**, 10 (1960).
33a. F. Rodriguez and L. A. Goettler, *Trans. Soc. Rheol.* **8**, 3 (1964).
34. J. Schurz, *Rheol. Acta* **1**, 58 (1958).
35. J. Schurz, *Kolloid-Z.* **154**, 97 (1957); **155**, 45 and 55 (1957).
36. J. Schurz, *J. Colloid Sci.* **14**, 492 (1959); *Proc. Intern. Symp. on Second Order Effects in Elasticity, Plasticity and Fluid Dynamics, Haifa 1962*, 427.
37. J. Schurz and H. Streitzig, *Monatsh. Chem.* **87**, 632 (1956); **88**, 325 (1957).
38. K. H. Schäfer, Ph.D. Thesis, Graz University (1959); cf. J. Schurz and K. H. Schäfer, *Papier* **14**, 139 (1960).
39. J. Schurz, K. H. Schäfer, T. Steiner, and M. Hermann, *Gummi, Asbest, Kunststoffe* **14**, 1122 (1961).
40. K. Windisch, Ph.D. Thesis, Graz University (1960); cf. J. Schurz and K. Windisch, *Kolloid-Z.* **177**, 149 (1961).
41. K. H. Schäfer, Ph.D. Thesis, Graz University (1959).
42. H. Gröblinghoff, Ph.D. Thesis, Graz University (1961).

43. J. Schurz, G. Gröblinghoff, and K. Windisch, *Holzforschung* **15**, 8 (1961).
44. J. Schurz and H. Streitzig, *Monatsh. Chem.* **89**, 229 (1958).
45. M. Hermann, Ph.D. Thesis, Graz University (1962).
45a. S. N. Chinai and W. C. Schneider, *Rheol. Acta* **3**, 148 (1964).
46. E. Menefee and W. L. Peticolas, *Nature* **189**, 745 (1961); *J. Chem. Phys.* **35**, 946 (1961); W. L. Peticolas and E. Menefee, *ibid.* p. 951; W. L. Peticolas, *ibid.* p. 2128.
47. B. H. Zimm, in "Rheology" (F. R. Eirich, ed.), Vol. 3, p. 1. Academic Press, New York, 1960.
48. T. Ree and H. Eyring, *J. Appl. Phys.* **26**, 793 and 800 (1955); W. K. Kim, N. Hirai, T. Ree, and H. Eyring, *ibid.* **31**, 385 (1960).
49. J. A. Faucher, *J. Appl. Phys.* **32**, 2336 (1961).
50. J. F. Rudd, *J. Polymer Sci.* **44**, 459 (1960); **60**, S9 (1962).
51. H. P. Schreiber, E. B. Bagley, and D. C. West, *Polymer* **4**, 365 (1963).
52. E. B. Bagley and D. C. West, *J. Appl. Phys.* **29**, 1511 (1958).
53. D. R. Mills, G. E. Moore, and D. W. Pugh, *SPE* (*Soc. Plastics Engrs.*) *Tech. Papers* **6**, 10 (1960) [cf. *Chem. Abstr.* **54**, 16908c (1960)].
54. R. J. Martinovich, P. J. Boeke, and R. A. McCord, *SPE* (*Soc. Plastics Engrs.*) *Tech. Papers* **6**, 30 (1960).
55. H. Leaderman, R. G. Smith, and L. C. Williams, *J. Polymer Sci.* **36**, 233 (1959).
56. N. W. Tschoegl, *J. Chem. Phys.* **39**, 149 (1963); N. W. Tschoegl and J. D. Ferry, *Kolloid-Z.* **189**, 37 (1963).
57. T. G. Smith, Ph.D. Thesis, Washington University, St. Louis (1960); cf. J. M. McKelvey, "Polymer Processing," p. 33. Wiley, New York, 1962.
58. D. R. Mills, G. E. Moore, and D. W. Pugh, *SPE* (*Soc. Plastics Engrs.*) *Tech. Papers* **6**, 4-1 (1960).
59. E. T. Severy, "Rheology of Polymers," p. 167. Reinhold, New York, 1962.
60. J. Schurz, *Oesterr. Chemiker-Z.* **56**, 312 (1955).
60a. R. Sabia, *J. Appl. Polymer Sci.* **8**, 1053 (1964).
60b. R. S. Porter and J. F. Johnson, *Polymer* **5**, 201 (1964).
60c. D. R. Mills, G. E. Moore, and D. W. Pugh, *SPE* (*Soc. Plastics Engrs.*) *Tech. Papers* **6**, (1960).
60d. J. Ferguson, B. Wright, and R. N. Haward, *J. Appl. Chem.* (*London*) **14**, 53 (1964).
60e. R. L. Ballman and R. H. M. Simon, *J. Polymer Sci.* **A2**, 3557 (1964).
60f. J. Lamp and A. J. Matheson, *Proc. Roy. Soc.* **A281**, 207 (1964).
60g. D. P. Wyman, L. J. Elyash, and W. J. Frazer, *J. Polymer Sci.* **A3**, 681 (1965).
60h. N. Nakajima and P. S. L. Wong, *Trans. Soc. Rheol.* **9**, 3 (1965).
60i. B. J. Cottam, *J. Appl. Polymer Sci.* **9**, 1853 (1965). H. P. Schreiber, *J. Appl. Polymer Sci.* **9**, 2101 (1965). L. F. Shalaeva, I. A. Marakhonov, L. N. Veselovskaya, N. M. Domareva, P. A. Ilchenko, A. S. Semenova, and I. I. Nikolaeva, *Plastich. Massy* 1965, No. 4, p. 5. H. W. McCormick, F. M. Brower, and L. Kin, *J. Polymer Sci.* **39**, 87 (1959). K. Kamide, Y. Inamoto, and K. Ohuo, *Chem. High Polymers* (*Tokyo*) **22**, 505, 529 (1965). A. Coen and G. Petraglia, *Mat. Plast. Elast.* **31**, 1057 (1965).
61. A. J. Staverman and F. Schwarzl, in "Die Physik der Hochpolymeren" (H. A. Stuart, ed.), Vol. IV, p. 1. Springer, Berlin, 1965; J. D. Ferry, *ibid.* p. 373; J. D. Ferry, "Viscoelastic Properties of Polymers." Wiley, New York, 1961.
62. H. Leaderman, R. G. Smith, and R. W. Jones, *J. Polymer Sci.* **14**, 47 (1954); H. Leaderman, *Proc. 2nd Intern. Congr. Rheol., Oxford, 1953* p. 203. Academic Press, New York, 1957.
63. J. D. Ferry, "Viscoelastic Properties of Polymers," pp. 169, 172, 176, 177, 178, 287, 290, 383, and 384. Wiley, New York, 1961.

64. R. D. Andrews and A. V. Tobolsky, *J. Polymer Sci.* **7**, 221 (1951); J. R. McLoughlin and A. V. Tobolsky, *J. Colloid Sci.* **7**, 555 (1952).
65. J. D. Ferry, I. Jordan, W. W. Evans, and M. F. Johnson, *J. Polymer Sci.* **14**, 261 (1954); J. M. Watkins, *J. Appl. Phys.* **27**, 419 (1956).
66. H. Fujita and K. Ninomiya, *J. Polymer Sci.* **24**, 233 (1957).
67. L. D. Grandine and J. D. Ferry, *J. Appl. Phys.* **24**, 679 (1953); J. D. Ferry, L. D. Grandine, and E. R. Fitzgerald, *ibid.* p. 911; J. D. Ferry and M. L. Williams, *J. Colloid Sci.* **7**, 347 (1952); G. W. Becker, *Kolloid-Z.* **140**, 1 (1955).
68. H. Fujita and K. Ninomiya, *J. Phys. Chem.* **61**, 814 (1957).
69. A. V. Tobolsky, *J. Appl. Phys.* **27**, 673 (1956); *J. Polymvr Sci.* **40**, 443 (1959).
70. A. V. Tobolsky, A. Mercurio, and K. Murakami, *J. Colloid Sci.* **13**, 196 (1958); A. V. Tobolsky, R. Schaffhauser, and R. Böhme, *Polymer Letters* **2**, 103 (1964).
71. H. Sobue and K. Murakami, *J. Polymer Sci.* **51**, S29 (1961).
72. A. V. Tobolsky and K. Murakami, *J. Polymer Sci.* **47**, 55 (1960).
73. W. L. Peticolas, *J. Polymer Sci.* **58**, 1397 (1962); *J. Chem. Phys.* **39**, 3392 (1963).
74. J. M. Watkins, R. D. Spangler, and E. C. McKannan, *J. Appl. Phys.* **27**, 685 (1956).
75. P. U. A. Grossman, *J. Polymer Sci.* **46**, 257 (1960).
76. F. Bueche, "Physical Properties of Polymers," p. 223. Wiley (Interscience), New York, 1962.
77. F. Bueche, *J. Appl. Phys.* **26**, 738 (1955); **24**, 423 (1953); J. D. Ferry, I. Jordan, W. W. Evans, and M. F. Johnson, *J. Polymer Sci.* **14**, 261 (1954).
77a. C. Mussa, P. Sacerdote, P. Guglielmino, and V. Tablino, *J. Appl. Polymer Sci.* **8**, 385 (1964).
78. C. Sadron and H. Mosimann, *J. Phys. Radium* **9**, 384 (1938).
79. A. Peterlin, in "Rheology" (F. R. Eirich, ed.), Vol. 1, p. 618. Academic Press, New York, 1956.
80. R. Signer, *Makromol. Chem.* **95**, 188 (1941).
81. R. Cerf and H. A. Scheraga, *Chem. Rev.* **51**, 185 (1952).
82. H. A. Scheraga, *J. Chem. Phys.* **19**, 983 (1951); M. Goldstein, *J. Chem. Phys.* **20**, 677 (1952).
82a. H. Janeschitz-Kriegl, paper given at the Makromolek. Koll. Freiburg/Br. 13.3. 1965 [cf. U. Daum and H. Janeschitz-Kreigl, Preprint P4, *Intern. Macromol. Symp. Prague 1965*].
83. J. F. Rudd and E. F. Gurnee, *J. Polymer Sci.* **A1**, 2857 (1963).
84. P. C. Scherer and M. K. Testerman, *J. Polymer Sci.* **7**, 549 (1951).
85. H. Mark, *Angew Chem.* **67**, 79 (1955); *Kunststoffe* **44**, 541 and 577 (1954).
86. J. Marchal and E. Marchal, *Vysokomolekul. Soedin.* **6**, 561 (1964) [cf. *Chem. Abstr.* **61**, 1954g (1964)].

CHAPTER D

Chemical Inhomogeneity and its Determination

O. Fuchs
and
W. Schmieder
FARBWERKE HOECHST AG., VORM. MEISTER LUCIUS & BRÜNING, FRANKFURT (M)-HOECHST, GERMANY

   I. Causes for the Chemical Inhomogeneity of Macromolecular Compounds.. 341
  II. Determination of Chemical Inhomogeneity........................... 344
     A. Fractionation... 344
     B. Additional Methods of Fractionation............................ 352
     C. Fractionation of Mixtures...................................... 353
     D. Characterization of Fractions.................................. 354
     E. Indirect Methods for Detection of Chemical Inhomogeneity.......... 355
 III. Quantitative Description of Chemical Inhomogeneity.................. 356
     A. Tabular and Graphical Representation........................... 356
     B. Average Chemical Inhomogeneity and Chemical Partition Ratio...... 357
     C. Average Value of Chemical Composition......................... 360
     D. Comparison of the Two Methods................................ 361
 IV. Literature Survey.. 362
    References... 371

I. Causes for the Chemical Inhomogeneity of Macromolecular Compounds

During the fractionation of macromolecular compounds, the assumption is generally made that the macromolecules differ only with respect to molecular weight and that they otherwise have identical chemical composition. However, this assumption is not fulfilled for many copolymers, for compounds which have been partially modified after synthesis, for several natural products, and for many other polymers. Here the individual molecules can be chemically different despite, for example, identical molecular weights. An extreme case is represented by mixtures of two or more homopolymers. In these cases there is an inhomogeneity with respect to chemical composition in addition to the inhomogeneity with respect to molecular weight. It is to be expected that the physical properties of a product depend upon the extent of molecular weight inhomogeneity as well as upon the degree of the chemical heterogeneity of the individual molecules. This assumption is fully justified by experience [1–4]. Consequently, for the recognition of these relations and for the evaluation of kinetic data, it is also necessary to consider chemical inhomogeneity and its determination.

The investigation of these phenomena in macromolecular systems is still in the preliminary phase. The extent and significance of these phenomena on bulk properties is poorly understood. Therefore, for convenience of discussion, the possible kinds of chemical inhomogeneity are listed in Table I. An adequate understanding of these categories of chemical inhomogeneity is necessary prior to the discussion of methods of fractionation. In the second column, this tabulation contains references to products for which chemical inhomogeneity can be expected. Details of the examples given in Col. 3 can be found in the literature in Section IV of this chapter.

For the purpose of classifying macromolecular compounds, it is convenient to use the kind and arrangement of the basic units in the macromolecules. For instance, consider the case of the partially saponified polyvinyl acetate. The polymer contains two basic chemical units, vinyl alcohol and vinyl acetate. Products which contain homopolymers are not mentioned separately in Table I since they are contained in the scheme. As an example, a copolymer of the two compounds X and Y can contain true copolymers and, at the same time, either homopolymers of the kind X and Y or a graft copolymer. The system styrene on cellulose can contain a true graft copolymer, pure polystyrene, and unaltered cellulose.

A chemical inhomogeneity can be expected most frequently in the categories A and B in Table I. The origin of chemical inhomogeneity and its extent depend upon the conditions of manufacture—temperature, solvent, the way of adding the individual components, the conditions of stirring, reaction in a continuous or discontinuous manner. In the case of partially modified products, the appearance of a pronounced chemical inhomogeneity is expected if a heterogeneous phase appears during the chemical reaction. The partial hydrolysis of polyvinyl acetate is an example. In the case of cocondensates and coaddition polymers, a chemical inhomogeneity can only occur if at least three different basic molecules are present during manufacture. The cocondensate of a polyester from a diol and a mixture of terephthalic acid and isophthalic acid belongs here. The coordination polymers which have been described in Table I are metal chelates such as nickel chelates from bis(1,2-dioximes) or zinc chelates from tetraacetylene ethane [5].

In the case of Category B, the specification of an average sequence length is not sufficient to characterize the compound. It is to be expected that during the synthesis of the product for some kinetic reasons the sequence length at the start of chain growth is different from that at the end (precipitation of the polymer above a certain molecular weight, etc.). Thus, the formation of these various sequence lengths changes during the course of

TABLE I

An Outline of Chemical Inhomogeneities Possible in Polymer Systems

1 Nature of chemical inhomogeneity	2 Products which may contain such chemical inhomogeneities	3 Examples of products for which such chemical inhomogeneities have been proven	4 Method of characterization of chemical inhomogeneity
A. The basic units forming the macromolecule are chemically *different* and are present in unequal amounts; however, they are distributed *randomly* within each molecule	Copolymers, cocondensates, co-addition polymers, coordination polymers; partially modified polymers; natural products	Copolymers of vinyl chloride and vinyl acetate; chlorinated polyolefins; partially esterified polysiloxanes; proteins; polycobalto oxanorgano siloxane	Chemical composition of the fractions
B. The basic units forming the macromolecule are chemically *different* and are present in equal or unequal amounts; however, they are distributed *nonrandomly* in different ways within each molecule	Possible for all compounds named above, especially for block copolymers and graft copolymers	Block copolymer of α-methyl styrene and methyl methacrylate; graft copolymer of methyl methacrylate on poly(vinyl chloride) or polystyrene	Sequence length of identical basic units and distribution of sequence length; length of the grafted branches and their distribution; distribution of points of grafting along the backbone chain
C. The basic units forming the macromolecule are chemically *identical*; however, they differ in the manner of their *nonrandom* steric arrangement	Valid for all products which can form the following structural isomers: head-to-head and head-to-tail structures; *cis-trans* structures; structural isomers of optically active polymers; formation of noncyclic and cyclic structures from identical monomers; stereoblocks with respect to tactic placement of the monomers within the polymer	Polypropylene, polypropylene oxide, poly(alkyl acrylate), polyacetaldehyde, optically active poly-4-methylhexene-1	Nature and number of structural isomers and their distribution within the macromolecule
D. The basic units forming the macromolecule are chemically *different* and differ in the manner of their steric *nonrandom* arrangement	The same materials as in Row C and including the compounds named under A and B	No example known	As under C

343

the reaction. It is necessary to specify the distribution of the sequence length in the macromolecules for the various fractions.

Categories C and D are concerned with the appearance of structural isomers. These can be present for identical basic molecules (Category C) as well as for different basic molecules (Category D). In general, such isomers are distributed randomly so that one cannot speak of a chemical inhomogeneity. However, there are cases known for which a nonstatistical distribution is present. (See the examples in Col. 3 of Table I.) The simultaneous formation of noncyclic and cyclic structures during the polymerization of a monomer (for instance, diallyl esters) could be considered as a copolymerization. Since the starting material is only one monomer, this case was classified there.

The fact that all kinds of macromolecules can be branched was not considered in the scheme of Table I. A partial selection occurs in the course of the fractionation, according to the degree of branching if appreciable branching is present and if the degree of branching differs between individual molecules. In the fractionations of either polyvinyl acetate or polysaccharides, it is possible to obtain fractions of differing degrees of branching, because solubility of the molecules depends upon branching. (See details in Section IV.) For these reasons, branching has to be considered as a kind of chemical inhomogeneity. In order to characterize a product, we have to define the following factors: the number of branching points (for instance, per 100 basic units), the nature of the branches (length, distribution of length, and chemical structure), and the distribution of the branches along the chain backbone. For a polymer containing many branches, the chain backbone is not well defined, since it is impossible to distinguish between the main chain and its branches.

It should be noted that the method of characterization of the chemical inhomogeneity, Col. 4, Table I, is sometimes very difficult and for some cases beyond our present means. The latter statement is especially true for the determination of the distribution of the individual compounds along the main chains and for evaluating the properties of sequences and branches.

II. Determination of Chemical Inhomogeneity

A. FRACTIONATION

For the purpose of fractionation of a chemically inhomogeneous macromolecular compound, any of the well-known solubility methods described in other chapters can be used. For a chemically homogeneous material, solubility depends on molecular weight only. In the case of a chemically inhomogeneous material, however, solubility is determined by molecular

D. Chemical Inhomogeneity and its Determination 345

weight as well as by the chemical structure of the molecules. During the fractionation of a chemically inhomogeneous compound by means of the solubility behavior of the individual molecules, a superposition of the contributions of molecular weight and chemical structure occurs. In cases of pronounced chemical inhomogeneity, the separation of the molecules during fractionation does not take place according to molecular weight. The varying chemical composition overcompensates the influence of molecular weight upon solubility. In this case both properties change discontinuously during the course of the fractionation. (Examples will be given later.)

Solubility is also a function of the nature of the solvent–nonsolvent used for fractionation, and of temperature. In general, solvent–nonsolvent, and temperature will have a different influence upon the dependence of solubility upon molecular weight and upon chemical composition. From this it follows that for the case of chemically inhomogeneous macromolecular compounds the results of fractionation depend strongly upon the conditions of fractionation. This phenomenon also appears in special cases for chemically homogeneous products, for instance in the case of association.

There is a fundamental difficulty in evaluating the results of fractionation where chemically inhomogeneous compounds are concerned. The question arises as to which results among several obtained under different conditions are the correct ones. In principle an answer may be obtained by varying the conditions of fractionation and taking the data which show the widest separation according to both the properties of molecular weight and chemical nature. This result still does not have to be correct. Further fractionation experiments may separate the product even further. Due to limitations of time and effort, the search for better fractionation conditions must be held to a few preliminary experiments. The conditions, once formulated, must be observed rigorously and specified with each fractionation result. In any case, these conditions are usually determined separately for each polymer. This approach is not free from arbitrariness, but for practical reasons this is unavoidable.

The phenomena appearing during the fractionation of chemically inhomogeneous compounds may be explained using two actual examples of very different polarity: a copolymer from vinyl acetate and vinyl chloride with an average chlorine content of 31.3 wt % and a copolymer of 64 mole % ethylene and 36 mole % propylene. The vinyl acetate–vinyl chloride copolymer was fractionated using acetone as solvent and petroleum ether as nonsolvent at room temperature [6]. The chlorine content of fractions was determined. It varied between the limits of 29% and 33%. From this it could be concluded that the copolymer approaches chemical uniformity.

That this conclusion is completely wrong may be demonstrated if the same product is fractionated using again acetone as a solvent but now methanol as nonsolvent. The chlorine content of the fractions is now between 19.6% and 38.3%. In reality, the product is thus chemically very inhomogeneous. Since the same result was obtained with six other solvent–nonsolvent systems [6], it can be assumed that the second experiment describes, to a first approximation, the actual chemical composition.

The fractionation of the ethylene-propylene copolymer according to chemical composition is more difficult than the previously mentioned vinyl acetate–vinyl chloride copolymer fractionation. This is due to the chemical similarity of the monomers, ethylene and propylene. The solubility of the ethylene-propylene copolymers in all cases is more dependent on molecular weight. This is shown in Tables II and III. Table III shows wide fluctuations in propylene content with fraction number; whereas, Table II shows a most regular increase in chlorine content with fraction number. The

TABLE II

FRACTIONATION OF A COPOLYMER OF VINYL ACETATE AND VINYL CHLORIDE BY ELUTION IN ACETONE-METHANOL AT ROOM TEMPERATURE

Fraction Number	Weight (mg)	% Cl	E_{1_i}[a]
1	56.0	19.6	0.346
2	41.0	20.6	0.363
3	43.5	23.5	0.414
4	61.5	28.9	0.510
5	78.5	23.4	0.412
6	64.5	32.7	0.577
7	72.5	35.4	0.625
8	65.5	36.2	0.638
9	51.0	36.0	0.636
10	32.0	36.4	0.642
11	38.0	33.7	0.595
12	48.0	38.3	0.676
13	38.0	38.2	0.673
14	56.0	37.7	0.665
15	26.5	33.3	0.587
		31.3[b]	

[a] The meaning of E_{1_i} is defined in Section III,B.
[b] Average value.

D. Chemical Inhomogeneity and its Determination

TABLE III

Fractionation of a Copolymer of Ethylene and Propylene by Elution in p-Xylene–Dimethylformamide at 85°C

Fraction number	Weight (mg)	Mole % propylene
1	16.0	44.9
2	16.0	35.3
3	10.5	32.5
4	8.0	36.2
5	14.0	41.2
6	12.5	50.0
7	16.0	41.2
8	17.0	39.1
9	17.5	42.2
10	26.5	37.6
11	27.0	34.6
12	110.0	33.0
13	51.0	37.1
14	14.5	28.2
15	7.5	25.1
16	4.0	—
		36.1[a]

[a] Average value.

molecular weight of the ethylene-propylene copolymer (expressed as $\eta_{sp/C}$, with $C = 0.1\%$, measured in decahydronaphthalene at 135°C) increases steadily. See also Table IV* [8]. In general, when the two monomers in a copolymer are very similar in structure to one another, the solubility is more dependent on molecular weight than composition.

The varying separation efficiency with respect to molecular weight and chemical composition for the same copolymer of the different fractionation systems can be seen from Table V [8]. System 1 separates the least and System 2 the most with respect to molecular weight. On the other hand, System 1 is more sensitive to propylene content than System 2. These data clearly show the problems which appear during the fractionation of macromolecular compounds which have an inhomogeneity in chemical structure in addition to differences in molecular weight. The course of fractionation in System 1 for propylene content and in System 2 for molecular weight is summarized in Tables III and IV. Table VI shows a fractionation using

* The molecular weight of each fraction may be calculated according to Moraglio [7] if $\eta_{sp/C}$ as well as the propylene content of each fraction are determined.

TABLE IV

FRACTIONATION OF A COPOLYMER OF ETHYLENE AND PROPYLENE BY ELUTION IN CARBON TETRACHLORIDE–ETHYL ACETATE AT 50°C

Fraction number	Weight (mg)	η_{sp}/C
1	55.0	0.03
2	40.5	0.17
3	35.0	0.43
4	23.0	0.61
5	26.0	1.10
6	44.0	2.27
7	38.5	2.73
8	48.0	3.29
9	62.5	4.83
10	65.5	4.92
11	64.0	6.54
12	60.5	7.43
13	26.0	10.72
14	26.5	20.2
15	4.5	—

TABLE V

FRACTIONATION OF A COPOLYMER OF ETHYLENE AND PROPYLENE UNDER DIFFERENT CONDITIONS

System Number	Fractionation conditions	Separation effect[a]			
		Molecular weight		Mole % propylene	
		Minimum	Maximum	Minimum	Maximum
1	p-Xylene–Dimethylformamide at 85°C	0.02	7.2	25	51
2	Carbon tetrachloride–Ethyl acetate at 50°C	0.02	20.2	25	39

[a] Separation effect is expressed by the minimum and maximum values for η_{sp}/C and propylene content obtained by fractionation.

D. CHEMICAL INHOMOGENEITY AND ITS DETERMINATION

System 1 of Table V. Here the reduced specific viscosity of the fractions was measured. The specific viscosity increases to a maximum value of 7.16 and then decreases slightly during further fractionation.

TABLE VI

FRACTIONATION OF A COPOLYMER OF ETHYLENE AND PROPYLENE BY ELUTION IN p-XYLENE–DIMETHYLFORMAMIDE AT 85°C

Fraction number	Weight (mg)	η_{sp}/C
1	38.0	0.02
2	23.5	0.07
3	24.5	0.17
4	39.0	0.38
5	44.5	0.65
6	39.5	1.11
7	40.0	3.57
8	152.5	6.53
9	93.5	7.16
10	59.5	6.44
11	25.0	6.32
12	3.5	—

As mentioned before, fractionation of chemically inhomogeneous compounds can be carried out by means of an elution as well as a precipitation fractionation. The methods differ with respect to the sequence of the fractions obtained; in the first case the most easily soluble molecules are isolated first, in the second case the least soluble. The question arises here as to how the superposition of molecular weight and chemical composition influences the fractionation according to the two methods, all other conditions being kept constant. Since we could not find a literature example to answer this question, we fractionated by both methods another copolymer of vinyl acetate and vinyl chloride with an average chlorine content of 46.5%. Acetone was solvent and methanol was nonsolvent [9]. The results are compared in Tables VII and VIII. Both methods gave approximately identical results. A comparative fractionation according to the precipitation and the elution method was also carried out for a polyvinyl alcohol which contained 10% acetyl groups using water as solvent and n-propanol as nonsolvent. The acetyl content of the fractions differed, which means the product is chemically heterogeneous. However, both methods yielded corresponding results [10].

TABLE VII

FRACTIONATION OF A COPOLYMER OF VINYL ACETATE AND VINYL CHLORIDE BY ELUTION IN ACETONE-METHANOL AT ROOM TEMPERATURE

Fraction number	Weight (mg)	Cl content (%)
1	100.7	40.0
2	192.5	42.7
3	235.5	44.5
4	367.0	45.5
5	236.4	47.0
6	167.4	46.8
7	189.2	48.2
8	536.2	48.3
9	129.1	48.3
10	69.8	45.9
11	88.7	45.0
12	28.8	43.8
		46.2[a]

[a] Average value.

TABLE VIII

FRACTIONATION OF A COPOLYMER OF VINYL ACETATE AND VINYL CHLORIDE BY PRECIPITATION IN ACETONE-METHANOL AT ROOM TEMPERATURE

Fraction number	Weight (mg)	Cl content (%)
1	253.0	49.4
2	257.2	48.8
3	275.5	48.3
4	679.7	48.6
5	489.9	46.3
6	150.2	46.8
7	105.8	46.3
8	233.3	45.0
9	124.1	43.5
10	58.1	41.7
11	55.0	41.4
12	89.6	29.5
		46.6[a]

[a] Average value.

D. CHEMICAL INHOMOGENEITY AND ITS DETERMINATION

To illustrate the dependence of fractionation results upon the nature of solvent and nonsolvent, the following examples are given:

1. With graft copolymer of styrene on polymethyl methacrylate or of methyl methacrylate on polystyrene separation of the products into fractions of different composition is possible using benzene and petroleum ether. Composition fractionation is not possible using benzene and methanol [11].

2. Cellulose acetate is fractionated with acetone and heptane. The fraction of highest molecular weight has the smallest acetyl content. If water is used instead of heptane, then a high acetyl content is connected with a high molecular weight [12].

3. In fractionation of diethyl acetamide–cellulose xanthate with dimethyl sulfoxide and water the degree of substitution of the first fractions is higher than that of the following fractions. If the system dimethyl sulfoxide and acetone is used, the components with the lowest degree of substitution appear first [13].

4. It has been shown for nitrocellulose that the result of fractionation depends upon the temperature of fractionation [14].

A separation of macromolecules into fractions with varying composition or varying structure can be obtained by variation of temperature alone. For instance, by cooling a 60°C solution of polypropylene oxide in isooctane, a separation according to tacticity and to molecular weight was obtained [15]. Fractionation of agar in water at different temperatures produced fractions of varying degree of esterification and calcium content [16].

In addition to the previously mentioned difficulties in the fractionation of a macromolecular, chemically inhomogeneous compound, there is yet another, i.e., the interpretation of the analytical data on the chemical composition of the fractions. For the examples we have cited, we were satisfied with the analysis of a chemical property (for instance, the chlorine or propylene content) for each individual fraction. However, there are macromolecules which, despite identical molecular weights and identical overall composition, differ in the arrangement of monomer groups along the chain. The components X and Y of the copolymer may be distributed randomly along the polymer chain or part of the molecule may preferentially contain the component X and the other contain component Y. This nonidentical, nonstatistical distribution of the components along the chains will have an effect upon the solubility and, consequently, upon the results of the fractionation. The same phenomenon can appear for all other chemically inhomogeneous polymers mentioned in Table I. For example, it is possible to have a nonuniform distribution of the ether groups in a partially etherified cellulose or of the chlorine atoms in a partially chlorinated polyolefin.

The determination of such a nonstatistical chain structure is still very difficult. Possibly some of the methods mentioned below in Section II,E might help to clarify the situation.

B. Additional Methods of Fractionation

We mention here column chromatography (see Chapter B.3), electrophoresis, and gel permeation (see Chapter B.4). In the first method a separation of the macromolecules under investigation takes place on the basis of the dependency of the solubility upon molecular weight and chemical composition. Since solubility decreases with increasing molecular weight, a separation according to molecular weight can easily be understood. However, solubility also depends upon the presence of polar groups within the polymer so that in the case of the presence of a chemical inhomogeneity there will also be a fractionation according to composition. If the polarity of the individual groups along the main chain of a macromolecule is approximately equal, the fractionation is determined mainly by the magnitude of the molecular weight. For larger differences in polarity of the various groups, this influence may outweigh that of the molecular weight and separation can occur preferentially according to chemical structure. In this manner a horse serum albumin and egg albumin were separated in an aqueous phosphate solution into fractions with varying phosphoric acid group content [17]. Sulfur-rich proteins were separated by chromatography into fractions with different sulfur and nitrogen content [18]. The influence of the affinity between the compound to be fractionated and the carrier material has been investigated for the racemic mixture of poly-4-methyl-1-hexene [19]. During the chromatographic fractionation of human fibrinogen on cellulose, fractions of varying immunological properties were obtained [20].

Electrophoresis is restricted to macromolecules which contain ionizable groups. The possibility of separating the abovementioned sulfur-rich proteins into compounds with varying sulfur and nitrogen content by means of electrophoresis was demonstrated by Gillespie [18].

In the case of gel permeation chromatography, the macromolecules are separated according to molecular size. As long as one is dealing with molecules which differ only with respect to molecular weight, the separation effect depends on molecular weight only. Even this sample situation may be complicated if, in a chemically inhomogeneous polymer, some groups are more subject to solvation than others. Then, despite identical molecular weights some chains may have larger molecular volumes. Naturally, this disturbs the sample molecular volume–molecular weight ratio for which

D. Chemical Inhomogeneity and its Determination

the gel was calibrated. Therefore, superposition phenomena are possible in gel permeation chromatography.

C. Fractionation of Mixtures

A mixture of two homopolymers can be considered as a material with an extremely high degree of chemical inhomogeneity. Fractionation can be carried out according to one of the above-mentioned methods. For the case of precipitation fractionation, the following are sources of errors:

1. The components of the mixture to be separated possess very different solubility properties. It would be expected that by using a solvent which is a good solvent for one component and a precipitant for the other, a separation of both components in this solvent can be obtained easily. This high separation efficiency is valid only if the concentration of the common solution is small. For example, if a 5% solution of polystyrene and polyvinyl acetate in benzene is introduced into great excess of methanol, the polystyrene will be precipitated quantitatively; but the precipitate will also contain a great amount of polyvinyl acetate. This is the case even though the latter is easily soluble in the benzene-methanol mixture [21]. The coprecipitated amount of polyvinyl acetate is smaller the lower the initial concentration. Under otherwise equivalent conditions better separation is obtained if the nonsolvent is poured into the solution of both polymers.

2. Frequently, mixtures are analyzed simply by an intensive extraction using a suitable solvent for one component of the mixture. The usefulness of this method is very limited according to our experience. For example, if an intimate mixture of 75% polystyrene and 25% polyvinyl acetate is prepared by precipitating a common solution in benzene with hexane, and extracted for 24 hours with methanol, the extracted product still contains 3% polyvinyl acetate [21]. A continued extraction lowers the polyvinyl acetate content very little. Obviously, a portion of the polyvinyl acetate molecules is enveloped so strongly by the polystyrene molecules that diffusion of the former out of the latter takes place very slowly. This effect is even more pronounced if an intimate mixture of a cross-linked (and consequently insoluble) product and a soluble product is present. Such a model mixture was prepared as follows [21]: 2.5 gm of partially saponified polyvinyl acetate with 36% acetyl content was dissolved in dimethyl formamide together with 0.5 gm of polyvinyl chloride. To this solution 1.5 gm of hexamethylene diisocyanate was added. A cross-linking takes place along the OH groups of the partially saponified polyvinyl acetate after heating to 80°C. According to control experiments, the polyvinyl chloride does not react with the isocyanate. This cross-linking reaction produced a highly swollen gel which was contracted by the addition of

methanol and pulverized after drying. Although this powder was extracted for 24 hours with tetrahydrofuran, it still contained 14% polyvinyl chloride. The same phenomenon was found for several other systems. From these experiences it follows that extraction of such a mixture can possess very limited value.

Let us now consider the frequently described extraction of crystalline polymers for the determination of the "soluble portion." It has been shown [22] that this method can yield misleading results for the content of crystalline products. These errors can be circumvented by dissolving the whole polymer first in a suitable solvent at elevated temperature and then slowly cooling this solution without stirring. Thus, a prefractionation of the precipitating product takes place by means of which the soluble portions become accessible for subsequent extraction. In addition to this, the filtrate of the prefractionated precipitate also contains soluble portions which should be added to those obtained by extraction.

D. Characterization of Fractions

The fractions obtained by one of the methods in Sections II,A and II,B can be characterized in different ways. Since we are primarily interested in the chemical constitution of the macromolecules, chemical analysis, UV, or IR spectrophotometry are preferred. In some cases, relative quantities are all that are required. These can be easily defined by determination of density, refractive index, dielectric and mechanical losses, melting point, transition temperature, surface tension, solubility, and upper and lower cloud points during heating and cooling of the solutions.

In addition to this, it is possible to estimate the amount of optically active component in the polymer by measurement of the specific optical rotation. For crystalline material one can measure the degree of crystallization; and for polyelectrolytes, the electric conductivity and the degree of dissociation. Nuclear magnetic resonance is useful for the determination of the degree of stereoregularity and sequence length in the main chain. Sequence length can be determined by IR spectrophotometry, differential thermal analysis, and pyrolysis gas chromatography. Examples for the latter methods are cited in Sections II,E and IV.

In addition to these, it is desirable to develop new methods of measurement for the characterization of the entire product and of the fractions. This is especially important for graft and block copolymers and for polymers which contain components of different stereoregularity. Today, it is impossible to see where these new methods will be found. Every measurable quantity which makes a statement concerning the composition of the macromolecules is important. Until the goal of an exact characterization of chemically inhomogeneous macromolecular compounds has been reached, we must be satisfied with the measurement of several properties whose exact interpretation is not yet possible.

E. Indirect Methods for Detection of Chemical Inhomogeneity

The presence of chemical inhomogeneity can sometimes be detected by investigation of the original product according to specific methods without previous fractionation. The results obtained in this manner have only qualitative value. On the other hand, they can be obtained more rapidly than by complete fractionation and subsequent characterization of the fractions.

From the temperature dependence of the elastic modulus and the mechanical loss factor, d, of a copolymer of vinyl chloride and 2-ethyl hexyl acrylate, it is possible to detect the presence of an acrylate-deficient product (d_{max} is at 80°C) and an acrylate-enriched product (d_{max} is at 0°C) [23]. The chemical polydispersity of copolymers may be determined by light scattering measurements in solvents of different refractive index, n. The molecular weights, $\overline{M}_w$, obtained in this manner depend upon the magnitude of n; should the $\overline{M}_w$ values differ for the various solvents a high degree of chemical inhomogeneity is indicated. Measurements of this kind have been described for a copolymer of styrene and methyl methacrylate [24,25].

The infrared spectra of copolymers of ethylene and propylene show an absorption at 13.7 μ if sequences of three methylene groups are present in the backbone and also an absorption at 13.9 μ if sequences of at least five methylene groups are present. If absorption at these frequencies occurs, it is possible to make certain conclusions concerning the distribution of methylene groups within the chain [26]. For the copolymer of vinyl chloride and vinylidene chloride, the sequence length has been determined by infrared measurements [27].

Centrifugation analyses can also be helpful in characterization. When a solution of a copolymer of acrylonitrile and vinyl acetate in dimethyl formamide plus bromoform is centrifuged for 150 hours at 33,500 rpm, three separate zones appear in the centrifuge cell due to the differing degrees of branching in the molecules. Similarly, a separation of atactic and stereoregular polystyrene is possible by using benzene plus bromoform [28].

An investigation of graft copolymers is possible in certain cases after previous chemical modification. For instance, when the graft copolymer of methyl methacrylate on natural rubber is ozonized, only the polyisoprene chain is decomposed while the polymethyl methacrylate branches remain [29]. If the latter are then fractionated according to molecular weight, information regarding the chain length distribution of the branches in the original product is obtained.

It is possible to determine the degree of grafting of a graft copolymer of styrene on cellulose from the number average molecular weight of the initial cellulose, that of the copolymer, and that of the polystyrene after saponification [30].

The presence of a chemical inhomogeneity for the graft copolymer of styrene on polyethylene could be demonstrated by heating a sample above the melting point and observing it under a polarizing microscope. Macromolecules with great differences in their chemical structure are usually noncompatible in melt and consequently form two phases [31].

The sequence length of the propylene oxide in the block copolymers from acetaldehyde and propylene oxide can be determined, after hydrolysis with hydrochloric acid, from the degree of polymerization of the unchanged propylene oxide blocks [32].

The separation of linear and branched polysaccharides is possible by precipitation of the linear component from an aqueous $CaCl_2$ solution with iodine (formation of an iodine complex). The complex disintegrates again in water. The soluble portion is highly branched. The two fractions also differ in their ratios of galactose:arabinose:xylose [33].

The sequence length of the syndiotactic blocks for several samples of polymethyl methacrylate synthesized under different conditions has been determined by nuclear magnetic resonance [34]. Similar investigations (determination of isotactic, syndiotactic, and stereoblock components) were carried out for other samples of polymethyl methacrylate [35]. The amounts of isotactic and syndiotactic structures in polymethacrylic acid anhydride have been determined from nuclear resonance spectra [36]. Pyrolysis of copolymers at elevated temperatures and chromatographic investigation of the products of pyrolysis can give information regarding the sequence length of the comonomers [37]. By means of differential thermal analysis it was determined that a copolymer of propylene and styrene represents a mixture of true copolymer and polystyrene [38]. By means of the same methods, information regarding the sequence length can also be obtained [39].

Further examples for these and other methods are compiled in Section IV. A survey on methods for determining sequence length distributions is given by Harwood [39a, b].

III. Quantitative Description of Chemical Inhomogeneity

A. Tabular and Graphical Representation

The presence of a chemical inhomogeneity can be demonstrated best by a tabular compilation of experimental data as in Tables II, III, VII,

D. Chemical Inhomogeneity and its Determination

and VIII. In comparisons of different products, this form of representation of results is not always clear. In these cases, it is more useful to plot the data in a way similar to that used to express an inhomogeneity in molecular weight, using as abscissa the analytical value and as ordinate the sum of the fractions in percent. As mentioned in Section II,B, the effects of molecular weight and chemical composition overlap during fractionation. For that reason, the analytical results on the fractions cannot be expected to change continuously in the order of the fraction number. It is necessary to order the fractions according to increasing analytical values. The curve in Fig. 1 was obtained in this manner from the data in Table II. As is customary for molecular weight distributions, each fraction was cut in half numerically and added to the sum of the previous values.

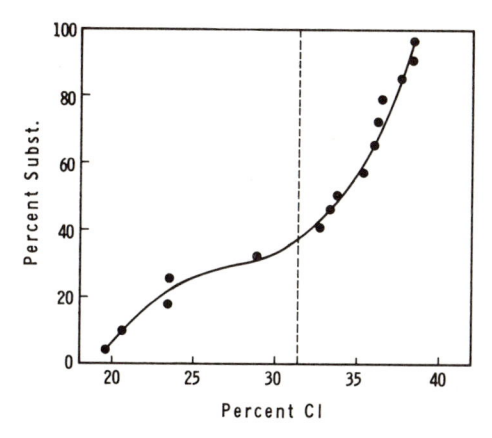

FIG. 1. Chemical inhomogeneity of a copolymer of vinyl acetate and vinyl chloride.

B. Average Chemical Inhomogeneity and Chemical Partition Ratio

In addition, it is often desirable to represent the degree of chemical inhomogeneity by one or two quantities as is done for molecular weight distributions by stating molecular weight averages and a degree of inhomogeneity, $U = (\overline{M}_w/\overline{M}_n) - 1$ [40]. In general, such a simple representation is not possible for chemical inhomogeneity. Cantow and Fuchs [41] have made two suggestions for the numerical determination of chemical inhomogeneity. They are the average chemical inhomogeneity and the chemical partition ratio as defined in this section and the calculation of average values of chemical composition as explained in Section III,C. Both cases will be explained briefly using data from Table II and Fig. 1.

In the case of a copolymer obtained from the two monomers 1 and 2, the weight averages $\overline{E}_{1_w}$ and $\overline{E}_{2_w}$ of the composition are defined by the following equations:

$$\bar{E}_{1w} = \sum_i c_i E_{1_i} \Big/ \sum_i c_i \quad \text{and} \quad \bar{E}_{2w} = \sum_i c_i E_{2_i} \Big/ \sum_i c_i \tag{1}$$

where c_i represents the amounts of the fractions, and E_{1_i} and E_{2_i} represent the amounts of the components 1 and 2, respectively, in the ith fraction. If the E values are expressed as fractional amounts of 1, then $\bar{E}_{1w} + \bar{E}_{2w} = 1$.

From this it follows that the mean Cl content of the copolymer described in Table II is 31.3%. From this we obtain an average vinyl chloride content of 55.2% in the copolymer, i.e., $\bar{E}_{1w} = 0.552$. The average vinyl acetate content is therefore 44.8%; thus, $\bar{E}_{2w} = 0.448$. The E_{1_i} values calculated for the individual fractions are given in the last column of Table II.

To describe the average chemical inhomogeneity U_1 with respect to component 1, we introduce the following symbols: $c_{1_i}^+$ shall denote the amounts of those fractions which contain a larger amount of component E_1 than that corresponding to the average value $\bar{E}_{1w}$; correspondingly, $c_{1_i}^-$ represents the amounts containing less of component E_1 than the average. In the first case, the difference $E_{1_i} - \bar{E}_{1w}$ is positive and is denoted by $\Delta E_{1_i}^+$; analogously, the difference $E_{1_i} - \bar{E}_{1w}$ in the second case is negative and is therefore denoted by $\Delta E_{1_i}^-$. From this, we obtain

$$U_1 = U_1^+ + U_1^-$$

where the partial inhomogeneities U_1^+ and U_1^- are defined by

$$U_1^+ = \sum_i c_{1_i}^+ \Delta E_{1_i}^+ \Big/ \sum_i c_{1_i}^+$$

and

$$U_1^- = \sum_i c_{1_i}^- \Delta E_{1_i}^- \Big/ \sum_i c_{1_i}^-$$

Since the numerators in the expressions for U_1^+ and U_1^- are of the same magnitude, the following may be substituted for U_1^-:

$$U_1^- = \sum_i c_{1_i}^+ \Delta E_{1_i}^+ \Big/ \sum_i c_{1_i}^-$$

Corresponding expressions result for component 2.

In addition to the average chemical inhomogeneity U_1, the asymmetry of the inhomogeneity of each component is important for the characterization of the product. The chemical partition ratio serves as a measure for this; for component 1, this ratio is given by

$$V_1 = U_1^+/U_1^- = \sum_i c_{1_i}^- \Big/ \sum_i c_{1_i}^+ = 1/V_2$$

D. Chemical Inhomogeneity and its Determination

These equations should also be applied to the copolymer in Table II. If the fitted curve shown in Fig. 1 is divided into sections of 5% by weight each, the data shown in Table IX are obtained for component 1 (vinyl chloride).

TABLE IX

CALCULATION OF THE AVERAGE CHEMICAL INHOMOGENEITY AND OF THE CHEMICAL PARTITION RATIO

c_{1_i}	$\Delta E_{1_i}^+$	$\Delta E_{1_i}^-$	E_{1_i}
0.025	—	0.208	0.344
0.075	—	0.196	0.356
0.125	—	0.180	0.372
0.175	—	0.161	0.391
0.225	—	0.138	0.414
0.275	—	0.093	0.459
0.325	—	0.032	0.520
0.375	—	—	0.552
0.425	0.021	—	0.573
0.475	0.039	—	0.591
0.525	0.053	—	0.605
0.575	0.065	—	0.617
0.625	0.076	—	0.628
0.675	0.086	—	0.638
0.725	0.095	—	0.647
0.775	0.102	—	0.654
0.825	0.109	—	0.661
0.875	0.115	—	0.667
0.925	0.120	—	0.672
0.975	0.125	—	0.677

From these data, by means of the above-mentioned equations, the inhomogeneity values are found to be[1]

$$U_1^+ = 0.05 \times 1.006/0.625 = 0.080 \qquad U_1^- = 0.05 \times 1.008/0.375 = 0.134$$

$$U_1 = U_1^+ + U_1^- = 0.214 \qquad V_1 = U_1^+/U_1^- = 0.597$$

$V_1 < 1$ means that the deviations to the negative are, on an average, greater than those to the positive.

The copolymer can therefore be characterized by the quantities U_1 and V_1. Instead of U_1 and V_1 it is also possible to use the values referring to

[1] For the same example some incorrect values have been reported by Cantow and Fuchs [41].

component 2 (vinyl acetate), i.e., $U_2 = U_1 = 0.214$ and $V_2 = 1/V_1 = 1.675$. These considerations can be generalized to include the case of the copolymer containing more than two components [41]. If there are j components, the description of the chemical inhomogeneity requires jU and jV values. The application of these general equations to calculate the U and V values from the fractionating data is illustrated in ref. [41] by means of a tercopolymer from ethylene, propylene, and a diene.

C. Average Value of Chemical Composition

During the fractionation of the original product, one obtains i fractions. (In the example in Table II, $i = 15$.) From the weights and the values of the corresponding measured chemical property of the fractions one can define [as in Eq. (1) in Section III,B] average z values of this measured property comparable to the averaged z values found for molecular weights

$$\bar{E}_{1_z} = \sum_i c_i E_{1_i}^2 \bigg/ \sum_i c_i E_{1_i} \qquad (2)$$

From the Eqs. (1) and (2) the chemical inhomogeneity with regard to the component 1 can be calculated

$$U_{1_{zw}} = (\bar{E}_{1_z}/\bar{E}_{1_w}) - 1$$

For the calculation of $\bar{E}_{1_w}$, $\bar{E}_{1_z}$, and $U_{1_{zw}}$, it is really not necessary to arrange the fractions according to increasing E_{1_i} or to draw the average curve. One can start directly from the individual measured values of the fractions.

The values for $\bar{E}_{1_w}$, $\bar{E}_{1_z}$, and $U_{1_{zw}}$ calculated from Tables II, VII, and VIII are compiled in Table X together with the values for U_1 and V_1. The indication of decimals for $\bar{E}_{1_w}$ and $\bar{E}_{1_z}$ is not caused by an optimistic estimation of the accuracy of the measurements of c_i and E_{1_i}. It is caused by purely mathematical reasons. Since the difference between $\bar{E}_{1_w}$ and $\bar{E}_{1_z}$ is generally very small and, consequently, $\bar{E}_{1_z}$ and $\bar{E}_{1_w}$ is only slightly

TABLE X

Values for U and E Calculated from the Data in Tables II, VII, and VIII

Table	U_1	V_1	E_w	E_z	U_{zw}
II	0.214	0.597	31.3146	32.6339	0.042
VII	0.065	0.733	46.1629	46.2670	0.002
VIII	0.078	0.648	46.6334	46.9247	0.006

D. Chemical Inhomogeneity and its Determination

larger than 1, additional errors which can result from rounding off the products $c_i E_{1_i}$ and $c_i E_{1_i}^2$ have to be avoided. This is only possible if the mathematical accuracy is better than the measured accuracy. The experimental errors of c_i and E_{1_i} which amount to about 0.5–1% of the values in Table II are contained in a similar manner in $\bar{E}_{1_w}$ and $\bar{E}_{1_z}$.

It is possible to define $\bar{E}_{1_n}$ and $U_{1_{wn}}$ values using the same technique as above. These are

$$\bar{E}_{1_n} = \sum_i c_i \bigg/ \left(\sum_i c_i/E_{1_i}\right) \quad \text{and} \quad U_{1_{wn}} = (\bar{E}_{1_w}/\bar{E}_{1_n}) - 1$$

However, in case there is only little homopolymer contained in a copolymer (i.e., for $E_{1_i} = 0$ for fraction i) we obtain $\bar{E}_{1_n} = 0$ and, consequently, $U_{1_{wn}}$ becomes infinite. For this reason, we refrain from using these quantities.

D. Comparison of the Two Methods

Both of the methods described in Section III,B and C for numerical representation of chemical inhomogeneity have advantages and disadvantages. The U_1 and V_1 values have the advantage of being more easily understandable. They are especially suited for a comparison of products of similar composition. On the other hand, in some special cases a differentiation according to U_1 and V_1 is less pronounced than according to E_{zw}. As an illustration, Table XI contains the values for U_1, V_1, and U_{zw}

TABLE XI

Values for U and V Calculated from Curves 1–4 in Fig. 2

Curve	U_1	V_1	U_{zw}
1	0.0050	1.00	0.037
2	0.0070	1.00	0.099
3	1.000	0.11	0.11
4	1.000	1.00	1.00

calculated for the four hypothetical distribution curves shown in Fig. 2. In Curve 1, E_i increases linearly with w_i, while Curve 2 contains an additional amount of 10% each of $E_i < 1\%$ and $E_i > 2\%$. Curves 3 and 4 represent mixtures of two homopolymers. Further examples and discussions can be found in the paper by Cantow and Fuchs [41].

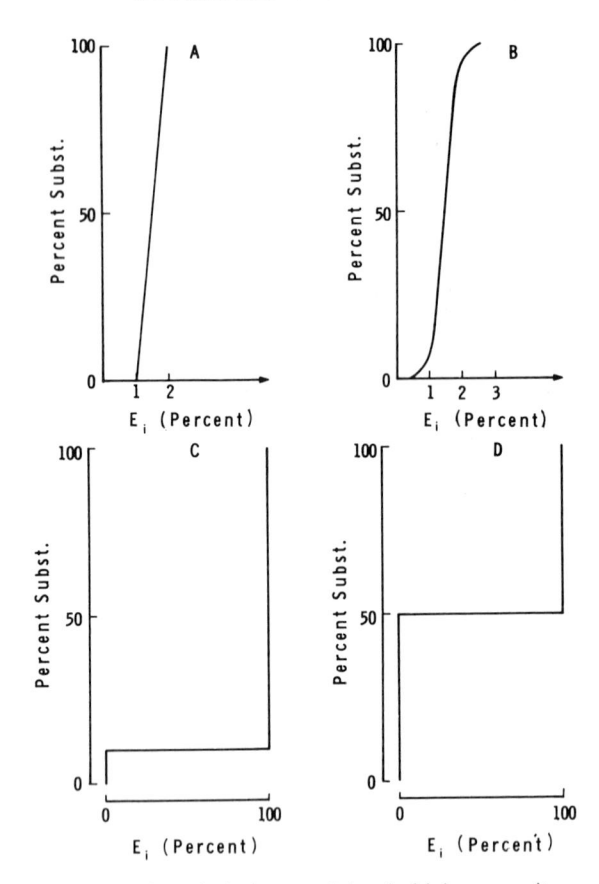

FIG. 2. Four hypothetical cases of chemical inhomogeneity.

IV. Literature Survey

Table XII represents a compilation of results from the literature on polymers which have been investigated with respect to their chemical inhomogeneity. The first column indicates the material under investigation, the second column mentions the method employed, and the third column refers to the literature citation. References to any peculiarities are given in parentheses in Column 1.

Earlier results on the chemical inhomogeneity of cellulose derivatives can be found in the book by Ott et al. (ref. [3], pp. 428, 679, 690, 896, 913, 1100, and 1183). Theoretical considerations on chemical inhomogeneity can be found in Harwood et al. [see 39a, 41a–l].

D. Chemical Inhomogeneity and its Determination

TABLE XII
Literature Survey

Product under investigation	Method	Reference
Copolymers of		
Styrene and methyl methacrylate	Light scattering	[24,25,42,43]
Styrene and methyl methacrylate	Fractionation	[43a]
Styrene and methyl methacrylate	Fractionation	[43,44,44a]
Styrene and butadiene	Light scattering	[45]
Styrene and butadiene	Fractionation	[46–48,48a]
Styrene and butadiene	Sedimentation and diffusion	[49]
Styrene and butadiene	Centrifugation of latex	[49a]
Styrene and butadiene	Extraction	[49b]
Styrene and divinylbenzene	Light scattering	[50]
Styrene and vinyltoluene (product homogeneous)	Fractionation	[51]
Styrene and propylene	Differential thermal analysis	[38]
Styrene and propylene	Extraction	[51a]
Poly-d-glucose (structure)	Extraction	[51b]
Styrene and vinylnaphthalene	Fractionation	[43a]
Styrene and 1,1-diphenylethylene	Extraction	[51c]
Styrene and *trans*-stilbene	Extraction	[51c]
Styrene and *o*-methylstyrene	Fractionation	[52]
Styrene and α-methylstyrene	Fractionation, Infrared analysis	[53]
Styrene and *p*-fluorostyrene	Fractionation	[52]
Styrene and *p*-bromstyrene	Fractionation	[43a]
Styrene and iodostyrene	Density gradient centrifugation	[53a]
Styrene and 2,6-dimethylstyrene	Fractionation	[52]
Styrene and *p-tert*-butylstyrene	Fractionation	[52]
Styrene and 4-methyl-1-pentene	Fractionation	[54]
Styrene and 3-methyl-l-butene	Fractionation	[54]
Styrene and 4-vinyl-1-cyclohexene	Fractionation, Light scattering	[55]
Styrene and acrylonitrile	Fractionation, Infrared analysis	[53]
Styrene and acrylonitrile	Light scattering	[55a]
Styrene and acrylonitrile	Fractionation	[55b]
Styrene, acrylonitrile and vinyl acetate	Kinetics	[55c]
Styrene and maleic acid anhydride	Fractionation, Infrared analysis	[53,55d]
Styrene and styrene *p*-laurylamide	Fractionation	[56]
Vinyl chloride and vinyl acetate	Fractionation	[6,57]
Vinyl chloride and vinyl acetate (product homogeneous)	Method of polymerization	[58]
Vinyl chloride and acrylonitrile	Fractionation	[59–61]
Vinyl chloride and vinylidene cyanide	Infrared analysis	[27]
Vinyl chloride and 2-ethyl hexylacrylate	Dynamic mechanical testing	[23,62]

TABLE XII—continued

LITERATURE SURVEY

Product under investigation	Method	Reference
Copolymers of		
Vinyl chloride and acrylic ester	Fractionation, Mechanical properties	[63]
Vinyl chloride and ethylene-1,2-dicarboxylic acid ester	Mechanical properties	[64,65]
Vinyl acetate and vinylidene cyanide	Fractionation	[66]
Vinyl acetate, acrylonitrile, and methylvinylpyridine	Fractionation	[67]
Vinyl acetate, acrylonitrile, and 2-methyl-5-vinylpyridine	Fractionation	[68]
Vinyl acetate and acetaldehyde	Saponification and fractionation	[68a]
Methyl methacrylate and acrylonitrile	Fractionation	[67]
Methyl methacrylate and methacrylic acid	Fractionation	[68b]
Methyl methacrylate, methacrylic acid and Li methacrylate	Fractionation	[68c]
Ethylene and propylene	Fractionation	[8]
Ethylene and propylene	Turbidity titration	[69]
Ethylene and butadiene	Fractionation	[69a]
Ethylene and cyclopentene	Extraction	[69b]
Ethylene and butene-2	Fractionation	[69c]
Butadiene and methacrylic acid (product branched)	Dynamic mechanical testing	[70]
Butadiene and vinyl alcyl ether	Fractionation	[71]
Butadiene and isoprene	Fractionation	[71a]
Isoprene and pentadiene-(1,3)	Fractionation	[71b]
Acrylonitrile and vinylidene chloride	Solubility	[72]
Acrylonitrile and α-methylstyrene	Fractionation, Infrared analysis	[53]
Acrylonitrile and β-propiolactone	Turbidity titration	[73]
Acrylonitrile and butadiene	Oxidation	[74]
Acrylonitrile and hexene-1	Fractionation	[74a]
Acrylonitrile and diisobutylene	Fractionation	[74a]
Acrylonitrile and 2-methylbutene-2	Fractionation	[74a]
Acrylonitrile and 2-methylpentene-1	Fractionation	[74a]
Acetaldehyde and propylene oxide	Extraction	[32]
β-Propiolactone and propylene oxide	Fractionation, Infrared analysis	[75]
β-Propiolactone and epichlorhydrin	Fractionation, Infrared analysis	[75]
β-Propiolactone and tetrahydrofuran	Fractionation, Infrared analysis	[75]
β-Propiolactone and 3,3-bis(chloromethyl)oxycyclobutane (determination of sequence length)	Extraction and hydrolysis	[76]
3-Methylbutene and butene-1	Solubility	[76a]

D. Chemical Inhomogeneity and its Determination

TABLE XII—continued

LITERATURE SURVEY

Product under investigation	Method	Reference
Copolymers of		
α-Vinylpyrrolidone and ε-caprolactam	Infrared analysis	[77]
Ethyl acrylate and 2-vinylpyridine	Fractionation	[78]
Butadiene, acrylonitrile, and epoxy resin	Extraction	[79]
Formaldehyde and 1,3-dioxacyclo-heptane	Preparation, Degradation	[79a]
Trioxane and ethylene oxide	Preparation	[79b]
Trioxane and dioxolane	Preparation	[79b]
Epichlorhydrine and tetrahydrofurane	Solubility	[79c]
Adenosine and uracil (determination of sequence length)	Kinetics	[80]
Cocondensates from		
Castor oil and phthalic anhydride	Fractionation	[81]
Adipic acid-hexamethylene diamine and ε-caprolactam	Fractionation, Osmotic measurements	[82]
Coadducts from		
Phenol formaldehyde	Fractionation	[83]
Phenol formaldehyde and phenyl isocyanate	Fractionation	[84]
Coordination polymers		
Poly(organoaluminocobalto siloxane) (inhomogenous with respect to Co content)	Fractionation	[85]
Chemically modified polymers		
Acetylcellulose	Fractionation	[12,86–88]
Acetylcellulose	Electrical conductivity	[88a]
Nitrocellulose	Fractionation	[14,89–92]
Methylcellulose	Fractionation	[93]
Ethylcellulose	Fractionation	[94,95]
Carboxymethylcellulose	Fractionation	[96–98]
Alkali cellulose, benzylated	Microscopic investigation	[99]
Lignin sulfonic acid	Fractionation	[99a]
Polyethylene, chlorinated	Fractionation	[6]
Polyethylene, chlorinated	Softening temperature	[100]
Poly(vinyl acetate), partially saponified	Fractionation	[10,101,102]
Poly(vinyl acetate), partially saponified	Light scattering	[102]
Polycaprolactam, ethoxylated	Extraction	[103]
Poly(methyl acrylate), partially saponified	Viscosity measurements	[103a]
Poly(vinyl alcohol), partially acetalized	Preparation	[103a]

TABLE XII—continued

LITERATURE SURVEY

Product under investigation	Method	Reference
Natural products		
Yeast ferment	Electrophoresis	[104]
Hemicellulose	Fractionation	[105]
Casein	Fractionation	[106]
Pepsin	Fractionation	[107]
Agar	Fractionation	[16,108]
Wool protein	Fractionation, Electrophoresis, Ultracentrifuge, Chromatography	[18]
4-O-Methyl-d-glucouronoxylane	Fractionation	[109]
Bovine serum albumin	Chromatography	[110]
Hemoglobin	Chromatography	[111]
Human fibrinogen	Chromatography	[20]
Horse serum albumin	Chromatography	[17]
Egg albumin	Chromatography	[17]
β-Lactoglobulin	Electrophoresis	[112]
Human acidglobin	Fractionation	[112a]
Human serumalbumin	Chromatography	[112b]
Barley albumin	Fractionation	[112c]
Thiolignin	Fractionation	[112d]
Chloritholocellulose	Extraction	[112e]
Block copolymers from		
Styrene and methyl methacrylate	Fractionation	[113–115,115a,b]
Styrene and methyl methacrylate	Fractionation, Extraction	[115c]
Styrene and methyl methacrylate	Extraction	[116]
Styrene and methyl methacrylate	Turbidity titration	[117,117a]
Styrene and acrylonitrile	Extraction	[116]
Styrene and methacrylonitrile	Fractionation	[118]
Styrene and isopropyl acrylate	Fractionation	[118]
Styrene and isoprene	Fractionation	[119]
Styrene and butadiene	Extraction	[120]
Styrene and α-methylstyrene	Fractionation, Extraction	[115c]
Styrene and polyoxymethylene	Fractionation	[121]
Styrene and ethylene oxide	Fractionation, Extraction	[115c]
Methyl methacrylate and vinyl acetate	Turbidity titration	[117]
Methyl methacrylate and acrylonitrile	Fractionation, Extraction	[116,118,122]
Acrylonitrile and ethylene oxide	Fractionation, Infrared analysis	[123]
Acrylonitrile and isopropyl acrylate	Fractionation	[118]
Ethylene and propylene	Infrared analysis	[26]
Ethylene and propylene (determination of sequence length)	Mechanical properties	[124,125]
Graft copolymers from		
Styrene on poly(methyl methacrylate)	Fractionation, Chromatography	[11]
Styrene on poly(methyl methacrylate)	Fractionation	[126–128]

D. Chemical Inhomogeneity and its Determination

TABLE XII—continued
Literature Survey

Product under investigation	Method	Reference
Graft copolymers from		
Styrene on poly(methyl methacrylate)	Fractionation, Light scattering	[129]
Styrene on poly(methyl methacrylate)	Radiochemical measurements	[130]
Styrene on polyethylene	Nuclear magnetic resonance, Measurements with polarized light	[31]
Styrene on polyethylene	Fractionation	[130a]
Styrene on polyethylene	Extraction	[130b]
Styrene on butadiene-styrene copolymer	Fractionation	[131]
Styrene on polyisobutylene	Fractionation	[132,133]
Styrene on polyisobutylene	Fractionation	[133a]
Styrene on polystyrene	Fractionation	[134]
Styrene on oxidized polystyrene	Fractionation, Light scattering	[55]
Styrene on poly(vinyl chloride)	Fractionation	[126]
Styrene on poly(vinyl acetate)	Fractionation	[127]
Styrene on poly(butyl methacrylate)	Fractionation	[127]
Styrene on rubber	Mechanical properties	[135]
Styrene on starch	Extraction	[136]
Styrene on cellulose	Extraction	[137]
Styrene on cellulose	Determination of $\overline{M}_n$	[30]
Styrene on cellulose treated with Ce salt (determination of branching)	Extraction, Acetolysis	[138]
Styrene on cellulose acetate	Extraction, Fractionation	[139,140]
Styrene on cellulose acetate	Density gradient centrifuge	[140a]
Styrene on desoxythioethyl cellulose	Fractionation	[141]
Styrene on poly-ω-hydroxyenantic acid methacrylate	Fractionation	[141a]
Styrene on fumaric acid polyester	Hydrolysis and Extraction	[141b]
Methyl methacrylate on polystyrene	Fractionation	[11,128,142]
Methyl methacrylate on rubber	Fractionation	[114,143–145]
Methyl methacrylate on rubber	Extraction	[29,146,147]
Methyl methacrylate on poly(vinyl acetate)	Fractionation	[127]
Methyl methacrylate on poly(vinyl alcohol)	Fractionation	[148]
Methyl methacrylate on poly(vinyl alcohol)	Extraction	[148a]
Methyl methacrylate on poly(vinyl benzoate)	Fractionation, Infrared analysis	[149]
Methyl methacrylate on amylose	Determination of $\overline{M}_n$	[30]
Methyl methacrylate on cellulose	Extraction	[148a]
Methyl methacrylate on epoxy resin	Extraction	[136]

TABLE XII—continued

LITERATURE SURVEY

Product under investigation	Method	Reference
Graft copolymers from		
Vinyl acetate on poly(methyl methacrylate)	Fractionation	[127,128,142]
Vinyl acetate on poly(methyl methacrylate)	Radiochemical measurements	[130]
Vinyl acetate on poly(vinyl alcohol)	Fractionation, Turbidity titration	[150]
Vinyl acetate on poly(butyl methacrylate)	Fractionation	[127]
Vinyl acetate on polystyrene	Fractionation	[128]
Vinyl acetate on polyvinylbenzoate	Fractionation, Infrared analysis	[149]
Vinyl acetate on poly(ethyl α-chloroacrylate)	Fractionation	[142]
Acrylonitrile on poly(methyl methacrylate)	Fractionation	[151]
Acrylonitrile on poly(methyl methacrylate)	Extraction	[136]
Acrylonitrile on poly(ethyl acrylate)	Fractionation	[152]
Acrylonitrile on polystyrene	Fractionation	[153]
Acrylonitrile on polystyrene	Extraction	[153a]
Acrylonitrile on cellulose	Extraction	[148a,153b]
Acrylonitrile on cellulose acetate	Extraction	[153b]
Acrylonitrile on poly(vinyl alcohol)	Extraction	[148a]
Acrylonitrile on cellulose triacetate	Fractionation	[141]
Acrylonitrile on desoxydi-n-propylaminoethyl cellulose	Fractionation	[142]
Methacrylonitrile on cellulose	Extraction	[148a]
Methacrylonitrile on poly(vinyl alcohol)	Extraction	[148a]
Vinyl chloride on poly(methyl methacrylate)	Fractionation	[128]
Butyl methacrylate on poly(vinyl acetate)	Fractionation	[127]
Acenaphthylene on natural rubber	Extraction	[136]
Isobutylene on poly(chloromethyl styrene)	Fractionation	[154]
Isoprene on poly(methyl methacrylate)	Density gradient centrifuge	[154a]
Acrylic acid on ethylene-propylene copolymer	Extraction	[154b]
Hydroxyenentic acid on poly(methyl acrylate)	Fractionation	[154c]
Glycidyl methacrylate on poly(vinyl chloride)	Fractionation, Hydrolysis	[155]

D. Chemical Inhomogeneity and its Determination

TABLE XII—continued

Literature Survey

Product under investigation	Method	Reference
Homopolymers		
Polystyrene (tacticity)	Fractionation	[28,156]
Polystyrene (tacticity)	Solubility, Infrared analysis	[157]
Polystyrene (branching)	Turbidity titration	[117]
Polystyrene (branching)	Fractionation	[158]
Poly-p-methylstyrene (crystallinity)	Fractionation, Infrared analysis	[159]
Polymethylmethacrylate (tacticity)	Nuclear magnetic resonance	[34,35]
Polymethacrylic acid anhydride (tacticity)	Nuclear magnetic resonance	[36]
Poly(vinyl chloride) (branching)	Fractionation	[160,161,161a]
Poly(vinyl chloride) (tacticity)	Fractionation	[162]
Poly(vinyl acetate) (branching)	Fractionation	[8,163]
Poly(vinyl acetate) (branching)	Fractionation, Light scattering	[164]
Poly(vinyl acetate) (head-to-head structure)	—	[164a]
Polyethylene (branching)	Fractionation	[165,166,166a]
Polypropylene (crystallinity)	Fractionation	[167]
Polypropylene (crystallinity)	Fractionation, Solubility, Infrared analysis	[168]
Polypropylene (tacticity)	Extraction	[22]
Polypropylene (tacticity)	Fractionation	[168a,168b]
Polybutadiene (branching)	Fractionation	[147]
Polybutadiene (1,2 and 1,4 structure)	Fractionation	[169]
Poly(propylene oxide) (optical activity)	Extraction	[170]
Poly(propylene oxide) (tacticity)	Fractionation	[15,170a]
Poly-4-methyl-1-hexene (optical activity)	Fractionation	[19]
Polyglucose (branching)	Fractionation	[171,172]
Poly(vinyl alcohol) (head-to-head structure)	Oxidation	[173]
Poly(vinyl alcohol) (head-to-head structure)	Fractionation	[174]
Poly(vinyl alcohol) (tacticity)	Fractionation	[175]
Polyacetaldehyde (crystallinity)	Extraction	[176]
Polybenzofuran (optical activity)	Fractionation	[177]
Hemicellulose (branching)	Fractionation	[33]
Polycyclobutene (tacticity)	Nuclear magnetic resonance, Infrared analysis	[178]
Poly(vinyl acetal) (branching)	Fractionation	[179]
Polyester (branching)	Fractionation	[179a]
Polymer of $(CH_3)_3-C-(CH_2)_3-CH=CH_2$	Extraction	[179b]

TABLE XII—continued

LITERATURE SURVEY

Product under investigation	Method	Reference
Homopolymers		
Polymer of $(CH_3)_3-Si-(CH_2)_3-CH=CH_2$	Extraction	[179b]
Mixtures		
Polystyrene and ethyl cellulose	Solubility	[180]
Polystyrenes and polybutadiene	Solubility	[181]
Polystyrene and poly(methyl methacrylate)	Solubility	[182]
Poly(vinyl acetate) and poly(vinyl chloride)	Chromatography	[183]
Poly(vinyl acetate) and polystyrene	Precipitation	[21]
Poly(vinyl acetate) and polystyrene	Extraction	[21]
Partially saponified cross-linked poly(vinyl acetate) and poly(vinyl chloride)	Extraction	[21]
ADDENDA		
Copolymers of		
Styrene and methyl methacrylate	Fractionation, Light scattering	[184]
Styrene and methyl methacrylate	Fractionation	[185]
Styrene and methyl methacrylate	Extraction	[186]
Styrene and methyl methacrylate	Light scattering	[187]
Styrene and butadiene	Fractionation	[188]
Styrene, butadiene, and acrylonitrile	Extraction	[189]
Styrene and acrylonitrile	Fractionation	[190]
Methyl methacrylate and vinyl carbazole	Fractionation	[191]
Ethylene and propylene	Fractionation	[192]
Ethylene and propylene	Infrared analysis	[193]
Acrolein and vinyl phenyl ether	Fractionation	[194]
Chemically modified polymers		
Acetylcellulose	Fractionation	[195]
Natural products		
Histone	Precipitation	[196]
Human lactoserum proteins	Fractionation, Gelpermeation	[197]
Desoxyribonucleic acid	Chromatography	[198]
Block copolymers from		
Styrene and 2-vinylpyridine	Extraction	[199]
Acetaldehyde and olefins	Extraction	[200]
Graft copolymers from		
Styrene on butadiene-styrene copolymer	Fractionation	[201]
Methyl methacrylate on polystyrene	Light scattering	[202]
Methyl methacrylate on poly(vinyl acetate)	Fractionation	[203]

D. Chemical Inhomogeneity and its Determination

TABLE XII—continued
LITERATURE SURVEY

Product under investigation	Method	Reference
Methyl methacrylate on ethylene propylene copolymer	Extraction	[204]
Homopolymers		
Polystyrene (structure)	Fractionation	[205]
Polystyrene (tacticity)	Extraction	[206]
Polybutadiene, hydrogenated (double bond)	Fractionation	[207]
Poly(propylene oxide) (crystallinity)	Fractionation	[208]
Poly(vinyl chloride) (structure)	Fractionation	[209]
Polychloroprene (branching)	Fractionation	[210]
Polyethyleneimine (branching)	Electrophoresis	[211]
Poly [(R)(S)-4-methyl-1-hexene] (optical activity)	Chromatography	[212]
Poly [(R)(S)-3, 7-dimethyl-1-octene] (optical activity)	Chromatography	[212]
Poly [(R)(S)-3-methyl-1-pentene] (optical activity)	Chromatography	[212]
Polydimethylketene (crystallinity)	Extraction	[213]
Poly-d-glucose (structure)	Gelpermeation	[214]
Poly-2, 3-dihydropyrane (structure)	Fractionation	[215]
Mixtures		
Polystyrene and styrene–isopropyl-styrene copolymer	Precipitation	[216]

REFERENCES

1. O. Fuchs and H. J. Leugering, in "Kunststoffe" (R. Nitsche and K. A. Wolf, eds.), Vol. 1, p. 118. Springer, Berlin, 1962; O. Fuchs and H. Hellfritz, in "Wandel in der chemischen Tecknik," p. 356. Frankfurt, Main, 1963.
2. J. F. Mahoney and C. B. Purves, J. Am. Chem. Soc. **64**, 9 (1942).
3. E. Ott, H. M. Spurlin, and M. W. Grafflin, in "Cellulose and Cellulose Derivatives," Part II. Wiley (Interscience), New York, 1954.
4. J. C. Robb and F. W. Peaker, in "Progress in High Polymers," p. 172. Heywood, London, 1961.
5. H. Kiehne, Gummi, Asbest, Kunststoffe **15**, 969 (1962).
6. O. Fuchs, Verhandl ber. Kolloid-Ges **18**, 75 (1958).
7. G. Moraglio, Chim. Ind. (Milan) **41**, 984 (1959).
8. O. Fuchs, Makromol. Chem. **58**, 65 (1962).
9. W. Schmieder, unpublished results (1961).
10. M. Matsumoto and G. Takayama, Chem. High Polymers (Tokyo) **18**, 169 (1961).
11. G. J. K. Acres and F. L. Dalton, J. Polymer Sci. **A1**, 2419 (1963).
12. A. J. Rosenthal and B. B. White, Ind. Eng. Chem. **44**, 2693 (1952).
13. K. Yamada and S. Mukoyama, Kolloid-Z. **163**, 98 (1956).
14. H. Sihtola and K. Aejmelaeus, Suomen Kemistilehti **B28**, No. 1, 79 (1955).
15. C. Booth, M. N. Jones, and E. Powell, Nature **196**, 772 (1962).

16. S. A. Glikman and I. G. Shubtsowa, *Vestn. Sloven. Kem. Drustva* **3**, 19 (1956).
17. A. S. Cyperovic and I. P. Galic, *Ukr. Biokhim. Zh.* **34**, 666 (1962).
18. J. M. Gillespie, *Australian J. Biol. Sci.* **16**, 259 (1963).
19. P. Pino, F. Ciardelli, G. P. Lorenzi, and G. Natta, *J. Am. Chem. Soc.* **84**, 1487 (1962).
20. J. S. Finlayson and M. W. Wosesson, *Biochemistry* **2**, 42 (1963).
21. H. Dexheimer and O. Fuchs, *Makromol. Chem.* (1966) (in press).
22. O. Fuchs, *Makromol. Chem.* **58**, 247 (1962).
23. L. Bohn, *Kunststoffe* **53**, 93 (1963).
24. W. Bushuk and H. Benoit, *Can. J. Chem.* **36**, 1616 (1958).
25. H. Benoit and M. Lang, *Ind. Plastiques Mod.* (*Paris*) **12**, No. 9, 25 (1960).
26. T. A. Veerkamp and A. Veermans, *Makromol. Chem.* **50**, 147 (1961).
27. S. Enomoto, *J. Polymer Sci.* **55**, 95 (1961).
28. R. Buchdahl, H. A. Ende, and L. H. Peebles, *J. Phys. Chem.* **65**, 1468 (1961).
29. Y. Ogata, N. Yasumoto, T. Fujine, Y. Minoura, and M. Imoto, *J. Chem. Soc. Japan, Ind. Chem. Sect.* **65**, 1136 (1962).
30. C. P. J. Glaudemans and E. Passaglia, *J. Polymer Sci.* **C2**, 189 (1963).
31. D. Ballantine, D. J. Metz, J. Gard, and G. Adler, *J. Appl. Polymer Sci.* **1**, 371 (1959).
32. H. Fujii, T. Fujii, T. Saegusa, and J. Furukawa, *Makromol. Chem.* **63**, 147 (1963).
33. B. D. E. Gaillard, *Nature* **191**, 1295 (1961).
34. T. G. Fox and H. W. Schnecko, *Polymer* **3**, 575 (1962).
35. A. Nishioka, H. Watanabe, I. Yamaguchi, and H. Schimiza, *J. Polymer Sci.* **45**, 232 (1960).
36. W. L. Miller, W. S. Brey and G. B. Butler, *J. Polymer Sci.* **54**, 329 (1961).
37. L. A. Wall, in "Analytical Chemistry of Polymers" (G. M. Kline, ed.), II, p. 5. Wiley (Interscience), New York, 1962.
38. I. Hayashi and R. I. Ichikawa, *J. Chem. Soc. Japan, Ind. Chem. Sect.* **66**, 1350 (1963).
39. B. Ke, in "New Methods of Polymer Characterization" (H. Mark, ed.), pp. 368 and 374. Wiley (Interscience), New York, 1964.
39a. H. J. Harwood, *Angew. Chem.* **77**, 405 (1965).
39b. H. J. Harwood, *Angew. Chem.* (1966) (in press).
40. G. V. Schulz, *Z. Physik. Chem.* **B43**, 25 (1939).
41. H.-J. Cantow and O. Fuchs, *Makromol. Chem.* **83**, 244 (1965).
41a. J. J. Hermans, *J. Colloid Sci.* **18**, 433 (1963).
41b. R. S. Kenn, *Chem. High Polymers* (*Tokyo*) **15**, 18 (1958).
41c. R. W. Kilb and A. M. Bueche, *J. Polymer Sci.* **28**, 285 (1958).
41d. S. Krause, *J. Polymer Sci.* **35**, 558 (1959).
41e. D. Prevorsek, *J. Polymer Sci.* **B1**, 229 (1963).
41f. B. D. Coleman and T. G. Fox, *J. Am. Chem. Soc.* **85**, 1241 (1963).
41g. I. M. Spinner, C. Y. Lu, and W. F. Graydon, *J. Am. Chem. Soc.* **77**, 2198 (1955).
41h. A. D. Litmanovich and A. V. Topchiev, *Neftekhimiya* **3**, 336 (1963).
41i. D. Prevorsek, *J. Polymer Sci.* **B1**, 229 (1963).
41j. H. J. Harwood and W. M. Ritchey, *Am. Chem. Soc. Polymer Preprints Meeting, Philadelphia, 1964* Vol. 5, p. 299.
41k. A. D. Litmanovich and V. Y. Shtern, *IUPAC, Symp. Prague, 1965* p. 130.
41l. S. A. Zlatina and A. Levin, *Plasticheskie Massy* p. 3 (1963).
42. W. Bushuk and H. Benoit, *Compt. Rend.* **246**, 3167 (1958).
43. W. H. Stockmayer, L. D. Moore, Jr., and M. Fixman, *J. Polymer Sci.* **16**, 517 (1955).
43a. J. Herz, D. Decker-Freyss, and P. Rempp, *IUPAC, Symp. Prague, 1965* p. 530.
44. A. D. Litmanovich, V. Y. Shtern, and A. V. Topchiev, *Neftekhimiya* **3**, 217 (1963).
44a. L. G. Kudryavtseva, A. D. Litmanovich, A. V. Topchiev, and V. Y. Shtern, *Neftekhimiya* **3**, 343 (1963).

D. Chemical Inhomogeneity and its Determination

45. R. Tremblay, M. Rinfret, and R. Rivest, *J. Chem. Phys.* **20**, 523 (1952).
46. M. A. Golub, *J. Polymer Sci.* **10**, 591 (1953); **11**, 281 and 583 (1953).
47. J. A. Yanko, *J. Polymer Sci.* **3**, 576 (1948).
48. L. S. Rosik and B. Krabal, *Chem. Prumysl.* **9**, 377 (1959).
48a. J. Blackford and R. F. Robertson, *J. Polymer Sci.* **A3**, 1289 (1965).
49. I. Y. Poddubnyi, V. A. Grechanovskii, and M. I. Mosevitskii, *Vysekomolekul. Soedin.* **5**, 1042 (1963).
49a. J. B. Yannas, *J. Polymer Sci.* **A2**, 1633 (1964); J. B. Yannas and I. E. Isgur, *ibid.* 4719.
49b. S. Wada, *Chem. High Polymers (Tokyo)* **18**, 733 (1961).
50. C. D. Thurmond and B. H. Zimm, *J. Polymer Sci.* **8**, 477 (1952).
51. R. A. Mock, C. A. Marshall, and V. D. Floria, *J. Polymer Sci.* **11**, 447 (1953).
51a. I. Hayashi, *J. Chem. Soc. Japan, Ind. Chem. Sect.* **67**, 258 (1964); 1126 (1965).
51b. K. Yanagisawa, *Chem. High Polymers (Tokyo)* **22**, 58 (1965).
51c. H. Yuki, K. Kosai, S. Murahashi, and J. Hotta, *J. Polymer Sci.* **B2**, 1121 (1964).
52. C. G. Overberger and S. Nozakura, *J. Polymer Sci.* **A1**, 1439 (1963).
53. A. W. Hanson and R. L. Zimmerman, *Ind. Eng. Chem.* **49**, 1803 (1957).
53a. J. J. Hermans and H. A. Ende, *J. Polymer Sci.* **C4**, 519 (1963).
54. C. G. Overberger and K. Miyamichi, *J. Polymer Sci.* **A1**, 2021 (1963).
55. J. A. Manson and L. H. Cragg, *J. Polymer Sci.* **33**, 193 (1958).
55a. Y. Shimura, I. Mita, and H. Kambe, *J. Polymer Sci.* **B2**, 403 (1964).
55b. G. Mino, *J. Polymer Sci.* **22**, 369 (1956).
55c. O. Solomon, M. Tomescu, N. Demian, and M. Dimonie, *Bull. Inst. Politiekn. Bucuresti* **22**, 97 (1960).
55d. T. L. Aug and H. J. Harwood, *Am. Chem. Soc. Polymer Preprints Meeting, Philadelphia, 1964*, Vol. 5, p. 306.
56. N. T. Notley, *J. Polymer Sci.* **A1**, 227 (1963).
57. F. Krasovec, *Vestn. Sloven. Kem. Drustva* **4**, 97 (1957).
58. R. J. Hanna, *Ind. Eng. Chem.* **49**, 208 (1957).
59. G. Centola and G. Prati, *Ric. Sci. Suppl.* 23, 1975 (1953).
60. S. G. Selikman and N. W. Michailow, *Kolloidn. Zh.* **19**, 35 (1957).
61. J. Schurz, T. Steiner, and H. Streitzig, *Makromol. Chem.* **23**, 141 (1957).
62. W. Albert, *Kunststoffe* **53**, 86 (1963).
63. L. E. Nielsen, *J. Am. Chem. Soc.* **75**, 1435 (1953).
64. H. W. Ebersbach and K. H. Michl, *Kunststoffe* **49**, 513 (1959).
65. Wacker-Chemie G.m.b.H., München, Belgian Patent 579,155 (1959).
66. J. A. Yanko, *J. Polymer Sci.* **22**, 153 (1956).
67. G. R. Cotten and W. C. Schneider, *J. Appl. Polymer Sci.* **7**, 1243 (1963).
68. M. Wishman, F. E. Detoro, M. C. Botty, C. Felton, and R. E. Anderson, *J. Appl. Polymer Sci.* **7**, 833 (1963).
68a. G. Takayama, *J. Chem. Soc. Japan, Ind. Chem. Sect.* **59**, 1432 (1956).
68b. V. G. Aldoshin and S. Y. Frenkel, *Vysokomolekul. Soedin.* **4**, 116 (1962).
68c. V. A. Myagchenkov, E. V. Kuznetsov, O. A. Iskhakov, and V. M. Luchkina, *Vysokomolekul. Soedin.* **5**, 724 (1963).
69. L. W. Gamble, W. T. Wipke, and T. Lane, *Am. Chem. Soc., Div. Polymer Chem., Preprints* **4**, 162 (1963).
69a. G. Natta, A. Zambelli, I. Pasquon, and F. Ciampelli, *Makromol. Chem.* **79**, 161 (1964).
69b. G. Natta, G. Dall'Asta, G. Mazzanti, I. Pasquon, A. Valvassori, and A. Zambelli, *Makromol. Chem.* **54**, 95 (1962).
69c. G. Natta, G. Dall'Asta, G. Mazzanti, I. Pasquon, A. Valvassori, and A. Zambelli, *J. Am. Chem. Soc.* **83**, 3343 (1961).
70. W. Cooper, *J. Polymer Sci.* **28**, 195 (1958).

71. S. N. Ushakov, S. P. Mitsengendler, N. V. Krasulina, *Bull. Acad. Sci. USSR, Div. Chem. Sci.* (*English Transl.*), **3**, 366 (1957).
71a. T. Suminoe, K. Sasaki, N. Yacmazaki, and S. Kambara, *Chem. High Polymers* (*Tokyo*) **21**, 9 (1964).
71b. G. Natta, L. Porri, and A. Carbonaro, *Makromol. Chem.* **77**, 126 (1964).
72. E. H. Hill and J. R. Caldwell, *J. Polymer Sci.* **47**, 397 (1960).
73. Y. Shimosaka, T. Tsuruta, and J. Furukawa, *J. Chem. Soc. Japan, Ind. Chem. Sect.* **66**, 1498 (1963).
74. A. I. Jakubtschik and A. I. Spasskowa, *Zh. Obshch. Khim.* **30**, 2172 (1960).
74a. C. S. Y. Kim, E. O. Hook, F. Veatch, and E. C. Hughes, *Kunststoff-Rundschau* **12**, 65 (1965).
75. K. Tada, T. Saegusa, and J. Furukawa, *J. Chem. Soc. Japan, Ind. Chem. Sect.* **66**, 1501 (1963).
76. K. Tada, T. Saegusa, and J. Furukawa, *Makromol. Chem.* **71**, 71 (1964).
76a. A. D. Ketley, *J. Polymer Sci.* **B1**, 121 (1963).
77. F. Kobayashi and K. Matsuya, *J. Polymer Sci.* **A1**, 111 (1963).
78. A. G. Schen Van and G. Smets, *Bull. Soc. Chim. Belges* **64**, 173 (1955).
79. V. A. Kargin, N. A. Plate, and A. S. Dobrynina, *Kolloidn. Zh.* **20**, 332 (1958).
79a. H.-D. Hermann and K. Weissermel, *Makromol. Chem.* (1966) (in press).
79b. K. Weissermel, E. Fischer, K. Gutweiler, and H.-D. Hermann, *Kunststoffe* **54**, 410 (1964).
79c. T. Saegusa, T. Neshima, H. Imai, and J. Furukawa, *Makromol. Chem.* **79**, 221 (1964).
80. R. Simha and J. M. Zimmerman, *J. Polymer Sci.* **42**, 309 (1960); **51**, 539 (1961).
81. A. Mitra and A. N. Saha, *Sci. Cult.* (*Calcutta*) **22**, 510 (1957).
82. H. Batzer and A. Moschle, *Makromol. Chem.* **22**, 195 (1957).
83. A. Buzagh, K. Udvarhelyi, and F. Horkay, *Kolloid-Z.* **157**, 53 (1958).
84. W. Kern, G. Dall'Asta, and H. Kämmerer, *Makromol. Chem.* **8**, 252 (1952).
85. K. A. Andrianov and A. A. Zhdanov, *Izvest. Akad. Nauk. SSSR, Otd. Khim. Nauk* p. 1590 (1959).
86. F. Howlett and A. R. Urquhart, *J. Textile Inst. Trans.* **37**, T89 (1946).
87. S. P. Wenediktow, W. A. Landyschewa, and S. A. Rogovin, *Khim. Prom.* p. 470 (1958).
88. D. R. Morey and J. W. Tamblyn, *J. Phys. & Colloid Chem.* **51**, 721 (1947).
88a. H. Ishii, *Chem. High Polymers* (*Tokyo*) **15**, 412 (1958).
89. K. Aejmelaeus and H. Sihtola, *Paper Timber* (*Helsinki*) **40**, 437 (1958).
90. K. Aejmelaeus, *Ann. Acad. Sci. Fennicae: Ser. A II 1956*, No. 75, 1.
91. G. A. Petropawlowski and N. I. Nikitin, *J. Appl. Chem. USSR* (*English Transl.*) **31**, 1862 (1958).
92. C. F. Bennett and T. E. Timell, *Svensk Papperstid.* **59**, 73 (1956).
93. T. Abe, K. Matsuzaki, A. Hatano, and H. Sobue, *Textile Res. J.* **25**, 254 (1955).
94. P. C. Scherer and J. G. Iacoviello, *Rayon & Synthetic Textiles* **32**, 47 (1951).
95. I. Croon and E. Flamm, *Svensk Papperstid.* **61**, 963 (1958).
96. K. F. Shigatsch, M. S. Finkelstein, I. M. Timochin, and I. A. Malinina, *Dokl. Akad. Nauk. SSSR* **123**, 289 (1958).
97. T. E. Timell, *Svensk Papperstid.* **56**, 311 and 483 (1953).
98. T. E. Timell and H. M. Spurlin, *Svensk Papperstid.* **55**, 700 (1952).
99. E. J. Lorand and E. A. Georgi, *J. Am. Chem. Soc.* **59**, 1166 (1937).
99a. E. W. Eisenbraun, *Tappi* **46**, 104 (1963).
100. R. E. Brooks, D. E. Strain, and A. McAlevy, *India Rubber World* **127**, 791 (1953).
101. I. Sakurada, Y. Sakaguchi, and S. Shima, *Chem. High Polymers* (*Tokyo*) **13**, 348 (1956).
102. A. Beresniewicz, *J. Polymer Sci.* **39**, 63 (1959).

D. Chemical Inhomogeneity and its Determination 375

103. S. R. Rafikov, G. N. Chelnokova, and P. N. Gribkova, *Polymer Sci. USSR*, (*English Transl.*) **1**, 135 (1960).
103a. I. Sakurada, Y. Sakaguchi, K. Fukami, and K. Takashima, *Chem. High Polymers* (*Tokyo*) **20**, 81 (1963).
104. D. C. Watts, C. Donninger, and E. P. Whitehead, *Biochem. J.* **81**, 4P (1961).
105. R. Nelson, *Tappi* **43**, 313 (1960).
106. H. A. McKenzie and R. G. Wake, *Australian J. Chem.* **12**, 712 (1959).
107. V. Kostka, B. Keil, and F. Sorm, *Collection Czech. Chem. Commun.* **24**, 2768 (1959).
108. S. A. Glikman and I. G. Shubtsowa, *Kolloidn. Zh.* **19**, 281 (1957).
109. R. G. Le Bel and D. A. I. Goring, *J. Polymer Sci.* **C2**, 29 (1963).
110. R. W. Hartley, E. A. Peterson, and H. A. Sober, *Biochemistry* **1**, 60 (1962).
111. T. H. J. Huisman, J. van de Brande, and C. A. Meyering, *Clin. Chim. Acta* **5**, 375 (1960).
112. N. Timasheff and R. Townend, *J. Am. Chem. Soc.* **82**, 3157 (1960).
112a. Z. Vodrazka and H. Holeysovska, *Collection Czech. Chem. Commun.* **29**, 1284 (1964).
112b. J. Sponar, I. Fric, S. Stokrova, and J. Kovarikova, *Collection Czech. Chem. Commun.* **28**, 1831 (1963).
112c. T. M. Enari and J. Mikola, *Suomen Kemistilehti* **B33**, 206 (1960); *European Brewery Conv., Proc. Congr.* **8**, 62 (1961).
112d. S. Wada, T. Iwamida, R. Iizima, and K. Yabe, *Chem. High Polymers* (*Tokyo*) **19**, 699 (1962).
112e. A. J. Rondier, *Bull. Soc. Chim. France* p. 976 (1961).
113. P. E. M. Allen, J. M. Downer, G. W. Hastings, H. W. Melville, P. Molyneux, and J. R. Urwin, *Nature* **177**, 910 (1956).
114. D. J. Angier, R. J. Ceresa, and W. F. Watson, *J. Polymer Sci.* **34**, 699 (1959).
115. G. M. Burnett, P. Meares, and C. Paton, *Trans. Faraday Soc.* **58**, 723 (1962).
115a. M. Leng and P. Rempp, *Compt. Rend.* **250**, 2720 (1960).
115b. D. Freyes, P. Rempp, and H. Benoit, *J. Polymer Sci.* **B2**, 217 (1964).
115c. M. Baer, *J. Polymer Sci.* **A2**, 417 (1964).
116. G. Champetier, M. Fontanille, and P. Sigwalt, *Compt. Rend.* **250**, 3653 (1960).
117. H. W. Melville and B. D. Stead, *J. Polymer Sci.* **16**, 505 (1956).
117a. V. G. Pokrikyan, V. A. Sergeev, and V. V. Korshak, *Izv. Akad. Nauk SSSR, Otd. Khim. Nauk* p. 1106 (1963).
118. R. K. Graham, J. R. Panchak, and M. J. Kampf, *J. Polymer Sci.* **44**, 411 (1960).
119. S. Schlick and M. Levy, *J. Phys. Chem.* **64**, 883 (1960).
120. R. J. Orr and H. L. Williams, *J. Am. Chem. Soc.* **79**, 3137 (1957).
121. C. Sadron, *Angew. Chem.* **75**, 472 (1963).
122. C. H. Bamford and E. F. T. White, *Trans. Faraday Soc.* **54**, 278 (1958).
123. J. Furukawa, T. Saegusa, and N. Mise, *Makromol. Chem.* **38**, 244 (1960).
124. G. Bier, A. Gumbold, and G. Lehmann, *Trans. Plastics Inst.* (*London*) **28**, 3 (1960).
125. G. Bier, G. Lehmann, and H. J. Leugering, *Makromol. Chem.* **44–46**, 347 (1961).
126. Y. Gallot, P. Rempp, and J. Parrod, *J. Polymer Sci.* **B1**, 329 (1963).
127. P. Hayden and R. Roberts, *Intern. J. Appl. Radiation Isotopes* **5**, 269 (1959).
128. G. Smets and M. Claesen, *J. Polymer Sci.* **8**, 289 (1952).
129. I. Mita, *J. Chim. Phys.* **59**, 530 (1962).
130. R. Hardy and P. E. M. Allen, *Makromol. Chem.* **42**, 33 (1960).
130a. W. K. Chen and H. Z. Friedländer, *J. Polymer Sci.* **C4**, 1195 (1964).
130b. T. Czvikovszki and J. Dobo, *IUPAC, Symp. Prague, 1965* p. 371.
131. I. A. Blanchette and L. E. Nielsen, *J. Polymer Sci.* **20**, 317 (1956).
132. A. Chapiro, P. Cordier, J. Jozefowicz, and J. Sebban-Danon, *IUPAC, Symp. Paris, 1963* p. 50.

133. J. Sebban-Danon, *J. Chim. Phys.* **58**, 246 and 263 (1961).
133a. J. Jozefowicz, *IUPAC, Symp. Prague, 1965* p. 579.
134. H. Kämmerer and F. Rocaboy, *Compt. Rend.* **256**, 4440 (1963).
135. S. G. Turley, *J. Polymer Sci.* **C1**, 101 (1963).
136. R. J. Ceresa, *J. Polymer Sci.* **53**, 9 (1961).
137. K. Hayakawa, C. C. Lin, K. Kawase, and T. Matsuda, *Chem. High Polymers (Tokyo)* **20**, 540 (1963).
138. H. Sumitomo and Y. Hachihama, *J. Chem. Soc. Japan, Ind. Chem. Sect.* **66**, 1508 (1963).
139. H. Sumitomo, S. Takamura, and Y. Hachihama, *J. Chem. Soc. Japan, Ind. Chem. Sect.* **66**, 269 (1963).
140. G. N. Richards, *J. Appl. Polymer Sci.* **5**, 539 (1961).
140a. H. A. Ende and V. Stannett, *J. Polymer Sci.* **A2**, 4047 (1964).
141. G. N. Richards, *J. Appl. Polymer Sci.* **5**, 558 (1961).
141a. G. S. Kolesnikov and G. T. Gurgenidze, *Vysokomolekul. Soedin.* **5**, 524 (1963).
141b. R. Feinauer, W. Funke, and K. Hamann, *Makromol. Chem.* **82**, 123 (1965).
142. G. Smets, J. Roovers, and W. van Humbeck, *J. Appl. Polymer Sci.* **5**, 149 (1961).
143. W. Kobryner and A. Banderet, *Compt. Rend.* **244**, 604 (1957); **245**, 689 (1957).
144. W. Kobryner and A. Banderet, *J. Polymer Sci.* **34**, 381 (1959).
145. W. Cooper, G. Vaughan, and R. W. Madden, *J. Appl. Polymer Sci.* **1**, 329 (1959).
146. W. Cooper and G. Vaughan, *J. Appl. Polymer Sci.* **1**, 254 (1959).
147. W. Cooper, G. Vaughan, D. E. Eaves, and R. W. Madden, *J. Polymer Sci.* **50**, 159 (1961).
148. I. Sakurada, S. Matuzawa, and Y. Kubota, *Makromol. Chem.* **69**, 115 (1963).
148a. B. A. Feit, A. Bar-Nun, M. Lahav, and A. Zilka, *J. Appl. Polymer Sci.* **8**, 1869 (1964).
149. G. Smets and A. Hertoghe, *Makromol. Chem.* **17**, 189 (1956).
150. F. D. Hartley, *J. Polymer Sci.* **34**, 397 (1959).
151. C. Schneider, J. Herz, and D. Hummel, *Makromol. Chem.* **51**, 182 (1962).
152. H. Sumitomo and Y. Hachihama, *Technol. Rept. Osaka Univ.* **8**, 157 (1958).
153. A. Chapiro and A. Jendrychowska-Bonamour, *J. Chim. Phys.* **60**, 1029 (1963).
153a. M. B. Huglin, *Polymer* **5**, 135 (1964).
153b. A. Y. Kulkarni and P. C. Mehta, *J. Polymer Sci.* **B1**, 509 (1963).
154. G. Kockelbergh and G. Smets, *J. Polymer Sci.* **33**, 227 (1958).
154a. S. E. Bresler, L. M. Pyrkov, and S. Y. Frenkel, *Vysokomolekul. Soedin.* **5**, 1315 (1963).
154b. F. Severini, M. Pegoraro, C. Tavazzani, and G. Aurello, *IUPAC, Symp. Prague, 1965* p. 338.
154c. G. S. Kolesnikov and G. T. Gurgenidze, *Izv. Akad. Nauk SSSR, Otd. Khim. Nauk* p. 2097 (1962).
155. A. Rarre and J. T. Khamis, *J. Polymer Sci.* **61**, 185 (1962).
156. D. Braun, H. Hintz, and W. Kern, *Makromol. Chem.* **65**, 1 (1963).
157. G. Natta, *Makromol. Chem.* **16**, 213 (1955).
158. G. Meyerhoff and M. J. R. Cantow, *J. Polymer Sci.* **34**, 503 (1959).
159. H. Higashi, T. Tanaka, M. Tanimura, and T. Egashira, *Makromol. Chem.* **43**, 245 (1961).
160. G. Bier and H. Krämer, *Makromol. Chem.* **18/19**, 151 (1956).
161. G. Bier and H. Krämer, *Kunststoffe* **46**, 498 (1956).
161a. M. Freeman and P. P. Manning, *J. Polymer Sci.* **A2**, 2017 (1964).
162. T. Kobayashi, *Bull. Chem. Soc. Japan* **35**, 726 (1962).
163. H. W. Melville, F. W. Peaker, and R. L. Vale, *Makromol. Chem.* **28**, 140 (1958).
164. P. Bosworth, C. R. Masson, H. W. Melville, and F. W. Peaker, *J. Polymer Sci.* **9**, 565 (1952).

D. Chemical Inhomogeneity and its Determination 377

164a. G. P. Belonovskaja, Z. D. Cernova, and L. A. Bessonova, *Zh. Prikl. Khim.* **37**, 2473 (1964).
165. Q. A. Trementozzi, *J. Polymer Sci.* **23**, 887 (1957).
166. S. W. Hawkins and H. Smith, *J. Polymer Sci.* **28**, 341 (1958).
166a. K. Shirayama, T. Okada, and S. Kita, *J. Polymer Sci.* **A3**, 907 (1965).
167. S. Newman, *J. Polymer Sci.* **47**, 111 (1960).
168. G. Natta, M. Pegoraro, and M. Peraldo, *Ric. Sci., Suppl.* **28**, 1473 (1958).
168a. A. Nakajima and H. Fujiwara, *Bull. Chem. Soc. Japan* **37**, 909 (1964).
168b. R. S. Porter, M. J. R. Cantow, and J. F. Johnson, *Am. Chem. Soc., Div. Org. Coatings Plastics Chem., Preprints* **25**, 162 (1965).
169. A. I. Jakubtschik and W. A. Filatowa, *J. Appl. Chem. USSR* (*English Transl.*) **32**, 1340 (1959).
170. S. Inoue, T. Tsuruta, and J. Furukawa, *Makromol. Chem.* **53**, 215 (1962).
170a. G. Allen, C. Booth, and M. N. Jones, *Polymer* **5**, 257 (1964).
171. P. T. Mora, *J. Polymer Sci.* **23**, 345 (1957).
172. P. T. Mora, J. W. Wood, R. Maury, and B. G. Young, *J. Am. Chem. Soc.* **80**, 693 (1958).
173. G. Mino, S. Kaizerman, and E. Rasmussen, *J. Polymer Sci.* **39**, 523 (1959).
174. P. J. Flory and F. S. Leutner, *J. Polymer Sci.* **3**, 880 (1948); **5**, 267 (1950).
175. K. Imai and M. Matsumoto, *Bull. Chem. Soc. Japan* **36**, 455 (1963).
176. H. Fujii, J. Furukawa, T. Saegusa, and A. Kawasaki, *Makromol. Chem.* **40**, 226 (1960).
177. M. Farina and G. Bressan, *Makromol. Chem.* **61**, 79 (1963).
178. G. Natta, G. Dall'Asta, G. Mazzanti, and G. Motroni, *Makromol. Chem.* **69**, 163 (1963).
179. A. Dobry, *J. Chim. Phys.* **42**, 109 (1945).
179a. R. Schöllner, *Plaste Kautschuk* **12**, 79 (1965).
179b. P. Longi, F. Bernardini, and L. Colombo, *Rend. Ist. Lombardo Sci.* **A93**, 134 (1959).
180. F. Burkhardt, H. Majer, and W. Kuhn, *Helv. Chim. Acta* **43**, 1192 (1960).
181. T. R. Paxton, *J. Appl. Polymer Sci.* **7**, 1499 (1963).
182. R. J. Kern, *J. Polymer Sci.* **21**, 19 (1956).
183. W. J. Langford and D. J. Vaughan, *Nature* **184**, 116 (1959).
184. V. E. Eskin, I. A. Baranovskaja, A. D. Litmanovich, and A. V. Topchiev, *Vysokomolekul. Soedin.* **6** (5) 896 (1964).
185. C. G. Overberger and N. Yamamoto, *Am. Chem. Soc. Meeting*, Phoenix, 1966, Vol. 7, p. 115.
186. K. Ito and Y. Yamashita, *J. Polymer Sci.* **B3**, 631 (1965).
187. I. A. Baranovskaja, A. D. Litmanovich, M. S. Protasova, and V. E. Eskin, *Vysokomolekul. Soedin.* **7**, 509 (1965).
188. J. Blackford and R. F. Robertson, *J. Polymer Sci.* **A3**, 1289 (1965).
189. B. D. Gesner, *J. Polymer Sci.* **A3**, 3825 (1965).
190. G. Mino, *J. Polymer Sci.* **22**, 369 (1956).
191. L. P. Ellinger, *Polymer* **6**, 549 (1965).
192. R. R. Garrett, *Rubber Plastics Age* **46**, 915 (1965).
193. G. Bucci and T. Simonazzi, *J. Polymer Sci.* **C7**, 203, (1964).
194. M. F. Shostakooskii, G. G. Skvortsova, and K. V. Zapunnaya, *Vysokomolekul. Soedin.* **5** (5), 767 (1963).
195. N. P. Dymarchuk, I. D. Petrovskaya, and S. L. Talmud, *Izv. Vysshikh Uchebn. Zavedenii, Khim i Khim Tekhnol.* **7** (2), 292 (1964).
196. D. M. P. Phillips and E. W. Johns, *Biochem. J.* **94**, 127 (1965).
197. R. Got, *Clin. chim. acta* **11**, 431 (1965).

198. J. Sponar, L. Pivec, P. Munk, and Z. Sormova, *Collection Czech. Chem. Commun.* **29** (1), 289 (1964).
199. M. Fontanille and P. Sigwalt, *Compt. Rend.* **251**, 2947 (1960).
200. S. Ohta, T. Saegusa, and J. Furukawa, *J. chem. Soc. Japan ind. Chem. Sect.* **67**, 947 (1965).
201. B. W. Bender, *J. Appl. Polymer Sci.* **9**, 2887 (1965).
202. Y. Gallot and M. Leng, *Compt. Rend.* **254**, 2334 (1962).
203. S. Polowinski and W. Reimschüssel, *Zeszyty Nauk Politechn. Lodz, Chem.* **14**, 87 (1964).
204. J. Pellon and K. J. Valan, *J. Appl. Polymer Sci.* **9**, 2955 (1965).
205. A. Yamada, Y. Yamamuro, and M. Yanagita, *Bull. chem. Soc. Japan* **35**, 609 (1962).
206. T. Nakata, T. Otsu and M. Imoto, *J. Polymer Sci* **A3**, 3383 (1965).
207. A. I. Yakubchik, B. I. Tikhomirov, and L. N. Mikhailova, *Vysokomolekul. Soedin.* **7**, 1562 (1965).
208. H. Sato, *J. Chem. Soc. Japan ind. Chem. Sect.* **65**, 385 (1962).
209. W. I. Bengongh and M. Onozuka, *Polymer* **6**, 625 (1965).
210. J. Curchod and T. Ve, *J. Appl. Polymer Sci.* **9**, 3541 (1965).
211. T. D. Perrine and P. F. Goolsby, *J. Polymer Sci.* **A3**, 3031 (1965).
212. P. Pino, G. Montagnoli, F. Ciardelli, and E. Benedetti, *Makromol. Chem.* **93**, 158 (1966).
213. G. Natta, G. Mazzanti, and G. Pregaglia, *DAS* 1 203 960 (1960).
214. J. B. Snell, *J. Polymer Sci*, **A3**, 2591 (1965).
215. K. Kamio, K. Meyersen, R. C. Schulz, and W. Kern, *Makromol. Chem.* **90**, 187 (1966).
216. H.-G. Elias and U. Gruber, *Makromol. Chem.* **78**, 72 (1964).

CHAPTER E

Treatment of Data

L. H. Tung
THE DOW CHEMICAL COMPANY, MIDLAND, MICHIGAN

I. Introduction... 379
II. Methods of Expressing Molecular Weight Distribution................ 380
   A. Graphical Representation...................................... 380
   B. Distribution Functions.. 382
   C. Molecular Weight Distribution Indices.......................... 386
III. Calculation of Molecular Weight Distribution from Fractionation Data.. 387
   A. Calculation through Integral Distribution....................... 388
   B. Direct Calculation of Differential Distribution..................... 395
   C. Calculation of Molecular Weight Distribution Indices.............. 398
IV. Calculation of Molecular Weight Distribution from Average Molecular Weight Measurements.. 400
V. Comparison of Methods... 402
   Appendix: Numerical Illustrations................................ 405
   References.. 412

I. Introduction

We shall discuss in this chapter the various methods of deriving molecular weight distribution from data obtained from conventional fractionation experiments. These experiments all involve the separation of the polymer sample into a finite number of fractions which in turn are characterized by some form or forms of molecular weight determination.

In addition we shall discuss the calculation of molecular weight distributions from data on average molecular weights of the unfractionated polymer. This latter type of data does not give as accurate information about molecular weight distribution as does data from an actual fractionation. Nevertheless, it has been used in many important discussions about molecular weight distribution in the literature. It seems worthwhile to know the limitations as well as some merits of the methods used in this type of data treatment. The experimental techniques, however, are not within the scope of this book. For this information one should consult other treatises; one edited by Allen [1] is an excellent source.

The chapter will begin with a section on methods of expressing molecular weight distribution. Although for some readers such preparation is not necessary, we shall thus avoid interruptions caused by defining these terms during the actual discussion of data treatment.

II. Methods of Expressing Molecular Weight Distribution

A. Graphical Representation

The easiest way of plotting the distribution of a system of a limited number of components is by a histogram such as the one shown in Fig. 1. A polymer chemist, however, is almost never confronted with such a simple system. He is faced with a system which comprises a nearly infinitely large number of components, each a member of the same homologous series but of a different molecular weight. It is convenient then to ignore the

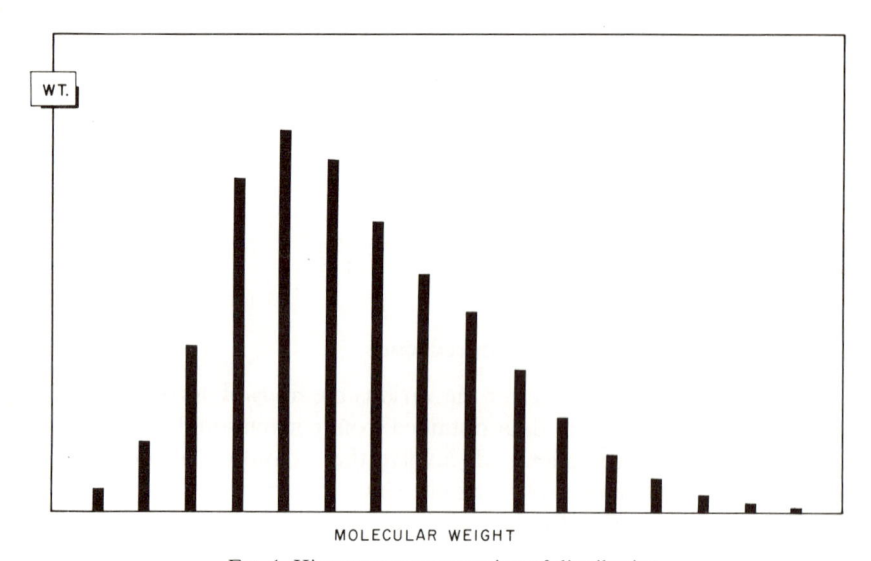

Fig. 1. Histogram representation of distribution.

discreteness of molecular weight and treat the distribution as a continuous one. A frequency function, defined as the amount of material per unit change of molecular weight, M, is used as the ordinate in the graph. A typical distribution curve, using weight fraction per unit change of molecular weight as the frequency function, is shown in Fig. 2. The amount of material in units of weight fraction having molecular weights between M_1 and M_2 is represented by the shaded area in the graph. The total area bounded by the curve and the abscissa is unity. Such a curve is called a differential weight distribution curve or simply a differential distribution

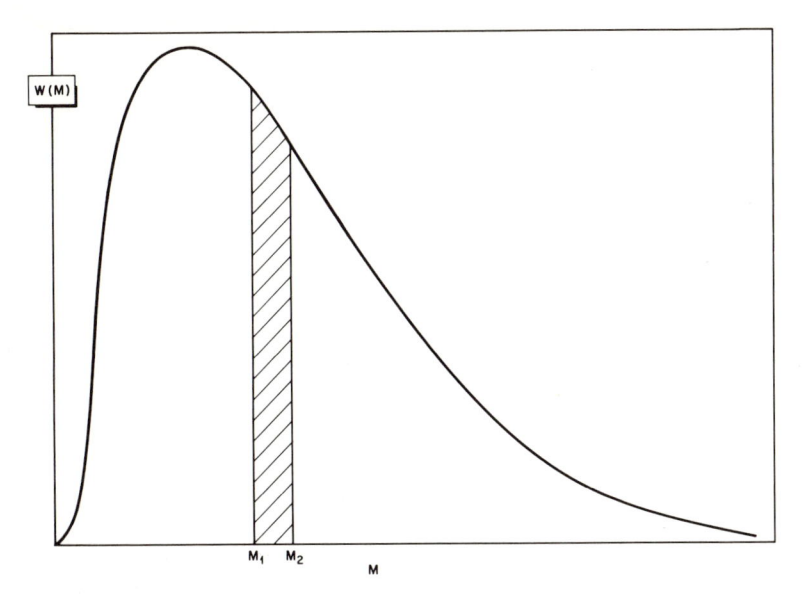

FIG. 2. Typical differential molecular weight distribution curve.

curve; the frequency function used is the differential distribution function $W(M)$. One can also use mole fraction per unit change of molecular weight as the frequency function and a curve plotted in such a manner is called the number distribution curve; the frequency function then is the number distribution function $N(M)$.

Another method of representing the molecular weight distribution graphically is by plotting the cumulative weight fraction versus molecular weight. The cumulative weight fraction corresponding to a molecular weight, say M_1, is the weight fraction of all molecular species having molecular weight smaller than or equal to M_1. Figure 3 is a typical curve on such a plot and the curve is called the integral distribution curve. The function representing the relation between the cumulative weight fraction and molecular weight is called the integral distribution function $I(M)$. One can obtain the differential distribution curve by graphically differentiating the integral distribution curve.

The following equations describe the relations between the various functions discussed:

$$dI(M)/dM = W(M) \tag{1}$$

$$I(M) = \int_0^M W(M)\,dM \tag{2}$$

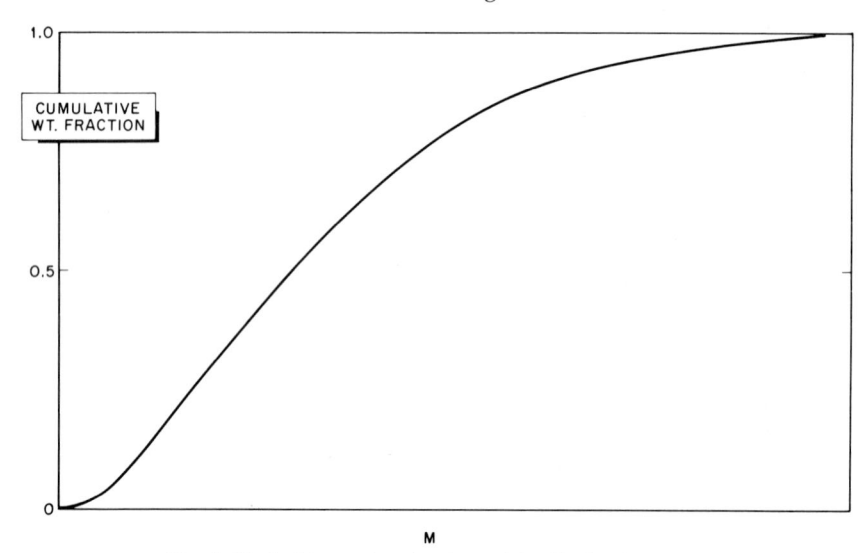

FIG. 3. Typical integral molecular weight distribution curve.

$$N(M) = \frac{W(M)}{M} \bigg/ \int_0^\infty \frac{W(M)}{M} \, dM \qquad (3)$$

$$\int_0^\infty W(M) \, dM = 1 \qquad (4)$$

$$I(\infty) = 1 \qquad (5)$$

$$\int_0^\infty N(M) \, dM = 1 \qquad (6)$$

B. Distribution Functions

The molecular weight distribution of polymers can sometimes be described by analytical functions of two or more parameters. Many molecular weight distribution functions have been proposed; some were derived from the kinetics of polymerization; others were devised empirically to fit the distribution curve. A few that are used in the following discussion of data treatment will be given here.

E. Treatment of Data

The first function of interest,

$$W(M) = \frac{(-\ln \alpha)^{b+2}}{\Gamma(b+2)} M^{b+1} \alpha^M \tag{7}$$

was derived by Schulz [2] for chain coupling vinyl polymerization. The function turned out, however, to fit many other vinyl polymers and condensation polymers. In Eq. (7), α and b are parameters which can be adjusted to fit the distribution of the polymer, and $\Gamma(b+2)$ is the gamma function of $b + 2$. The parameter b varies inversely with the breadth of the distribution. The other parameter α together with b determines the mean or average molecular weight of the distribution.

The many average molecular weights of a polymer whose distribution follows that of Eq. (7) can be expressed in terms of the two parameters, b and α. Their relations are

$$\overline{M}_n = \int_0^\infty M N(M)\, dM = 1 \bigg/ \int_0^\infty \frac{W(M)}{M}\, dM = \frac{(b+1)}{(-\ln \alpha)} \tag{8}$$

$$\overline{M}_w = \int_0^\infty M W(M)\, dM = \frac{b+2}{(-\ln \alpha)} \tag{9}$$

$$\overline{M}_z = \int_0^\infty M^2 W(M)\, dM \bigg/ \int_0^\infty M W(M)\, dM = \frac{b+3}{(-\ln \alpha)} \tag{10}$$

where $\overline{M}_n$ is the number average molecular weight; $\overline{M}_w$, the weight average molecular weight; and $\overline{M}_z$, the z average molecular weight.

Figure 4 shows four distribution curves with the same weight average molecular weight but different values of the parameter b. The number average molecular weights, $\overline{M}_n$, coincide with the peaks of these curves whereas the weight average molecular weights $\overline{M}_w$ are always greater than the molecular weights at the peaks of the distribution.

A second distribution function of interest is

$$W(M) = yz\, e^{-yM^z} M^{z-1} \tag{11}$$

Equation (11) is an empirical distribution function first used by Tung [3] to fit molecular weight distribution data of polymers. It also applies well to many condensation and vinyl polymers. The two adjustable parameters in Eq. (11) are y and z. As the parameters α and b in Eq. (7), z varies inversely

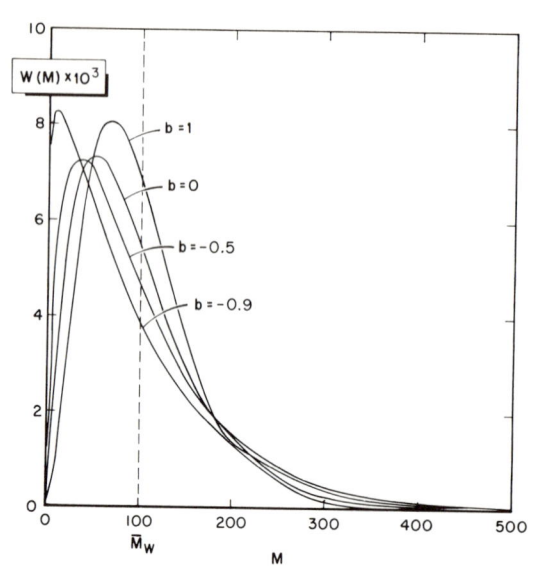

FIG. 4. Schulz distributions at constant $\bar{M}_w$: ($b = 1$, $\bar{M}_w/\bar{M}_n = 1.5$; $b = 0$, $\bar{M}_w/\bar{M}_n = 2$; $b = -0.5$, $\bar{M}_w/\bar{M}_n = 3$; $b = -0.9$, $\bar{M}_w/\bar{M}_n = 11$).

with the breadth of the distribution; y together with z determines the average molecular weights of the distribution. The similarity between Eqs. (7) and (11) has been compared by Green [4], who has found that curves calculated from Eqs. (7) and (11) are indistinguishable within experimental errors of practical fractionation technique. Equation (11), however, can be integrated analytically to the integral distribution form

$$I(M) = \int_0^M W(M)\,dM = 1 - e^{-yM^z} \tag{12}$$

which has certain advantages in treating polymer fractionation data [4,5].

The average molecular weights for those polymers which follow Eq. (11) are

$$\bar{M}_n = \left[y^{1/z}\Gamma\left(1 - \frac{1}{z}\right)\right]^{-1} \tag{13}$$

$$\bar{M}_w = y^{-1/z}\Gamma\left(1 + \frac{1}{z}\right) \tag{14}$$

$$\bar{M}_z = y^{-1/z}\Gamma\left(1 + \frac{2}{z}\right) \tag{15}$$

Equation (13) will not hold true when z is less or equal to 1. An explanation for such a case has been given by Howard [5].

A third distribution function of interest is the log-normal distribution

$$W(M) = \frac{1}{\beta\sqrt{\pi}} \frac{1}{M} \exp\left(-\frac{1}{\beta^2} \ln^2 \frac{M}{M_0}\right) \quad (16)$$

It was first used by Lansing and Kraemer [6] to describe polymer molecular weight distribution. Wesslau [7] later discovered that Eq. (16) applies to polyethylene polymerized by the Ziegler process [8]. The two adjustable parameters in Eq. (16) are β and M_0; β increases with the increase of the breadth of distribution; M_0 together with β determine the average molecular weights of the distribution.

The average molecular weights expressed in terms of β and M_0 are

$$\overline{M}_n = M_0 \, e^{-\beta^2/4} \quad (17)$$

$$\overline{M}_w = M_0 \, e^{\beta^2/4} \quad (18)$$

$$\overline{M}_z = M_0 \, e^{3\beta^2/4} \quad (19)$$

Figure 5 shows a set of distribution curves with the same $\overline{M}_w$ but with different values of β. These curves represent polymers having the same

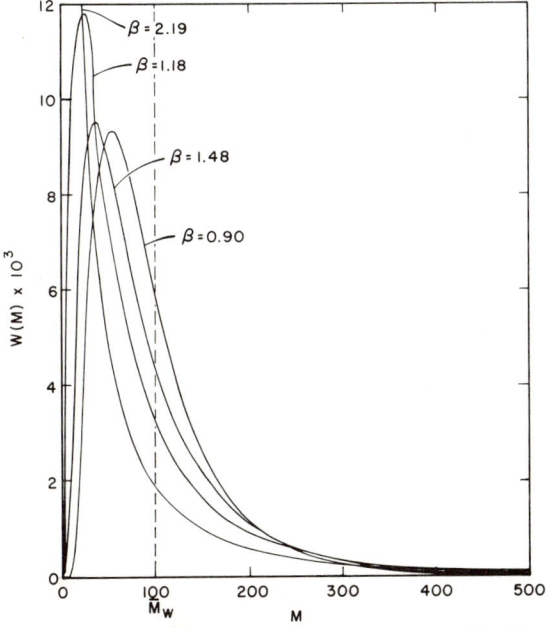

FIG. 5. Log-normal distributions at constant $\overline{M}_w$: $\beta = 0.90$, $\overline{M}_w/\overline{M}_n = 1.5$; $\beta = 1.18$, $\overline{M}_w/\overline{M}_n = 2$; $\beta = 1.48$, $\overline{M}_w/\overline{M}_n = 3$; $\beta = 2.19$, $\overline{M}_w/\overline{M}_n = 11$.

$\overline{M}_w/\overline{M}_n$ ratios as those in Fig. 4. It should be noted that for the log-normal distributions the molecular weight at the peak of the curves corresponds to

$$M_{max} = M_0 e^{-\beta^2/2}$$

which is smaller than both $\overline{M}_n$ and $\overline{M}_w$.

C. MOLECULAR WEIGHT DISTRIBUTION INDICES

Other than the distribution curves and functions described above, a single number or index has also been used to describe the molecular weight distribution of a polymer. We have seen that the distributions of a good number of polymers are describable by functions of two parameters, only one of which is connected directly with the breadth of the distribution. The parameters b, z, and β can therefore be used as indices for molecular weight distribution.

The more commonly used index, however, is the ratio of weight average molecular weight to number average molecular weight $\overline{M}_w/\overline{M}_n$. This ratio is unity for a polymer of uniform molecular weight and becomes larger as the distribution becomes broader.

Three other indices which have been used are

$$U = \frac{\overline{M}_w}{\overline{M}_n} - 1$$

$$\Delta = \left(\frac{\overline{M}_w}{\overline{M}_n} - 1\right)^{1/2}$$

$$g = \left(\frac{\overline{M}_z}{\overline{M}_w} - 1\right)^{1/2}$$

The index U was introduced by Schulz [9] who called it the inhomogeneity (Uneinheitlichkeit). U is related to the statistical standard deviation for number distribution S_n, which can be written as

$$S_n = \left[\int_0^\infty (M - \overline{M}_n)^2 N(M)\, dM\right]^{1/2} \qquad (20)$$

It follows then that

$$S_n^2 = \overline{M}_w \overline{M}_n - \overline{M}_n^2 \qquad (21)$$

Thus

$$U = \frac{\overline{M}_w}{\overline{M}_n} - 1 = \frac{S_n^2}{\overline{M}_n^2} \qquad (22)$$

E. TREATMENT OF DATA

The square root of U, Δ, which has been used by Lowry [10] as an index for molecular weight distribution is then

$$\Delta = S_n/\overline{M}_n \tag{23}$$

Similarly the g index, introduced by Hosemann and Schramek [11], is related to the standard deviation for weight distribution S_w in a similar way

$$g = S_w/\overline{M}_w \tag{24}$$

Hosemann and Schramek called g the polydispersity.

III. Calculation of Molecular Weight Distribution from Fractionation Data

Fractionation data can usually be arranged according to Table I. In Table I λ is the total number of fractions. The fractions are arranged in the order of ascending molecular weight. The weight fractions $w_1, w_2, \cdots w_\lambda$ are simply the weight of each fraction divided by the total weight of the sample. The average molecular weights of the fractions are usually determined through intrinsic viscosity measurements.

TABLE I

TYPICAL FRACTIONATION DATA

Fraction	Weight fraction	Average molecular weight
1	w_1	M_1
2	w_2	M_2
.	.	.
.	.	.
.	.	.
λ	w_λ	M_λ

One important feature of any fractionation experiment is that the fractions obtained are still not uniform in molecular weight distribution. Figure 6 shows the differential distribution curves of the original polymer and the fractions calculated from a hypothetical fractional precipitation experiment. The distributions of the fractions overlap quite extensively. This overlapping may be more or less, depending on the efficiency of the fractionation, but it is unavoidable in any practical method of separation. We see immediately that the data in Table I would be quite meaningless if plotted directly using w_i as the ordinate and M_i as the abscissa as in a histogram. The more appropriate approaches to this problem are described in the following sections.

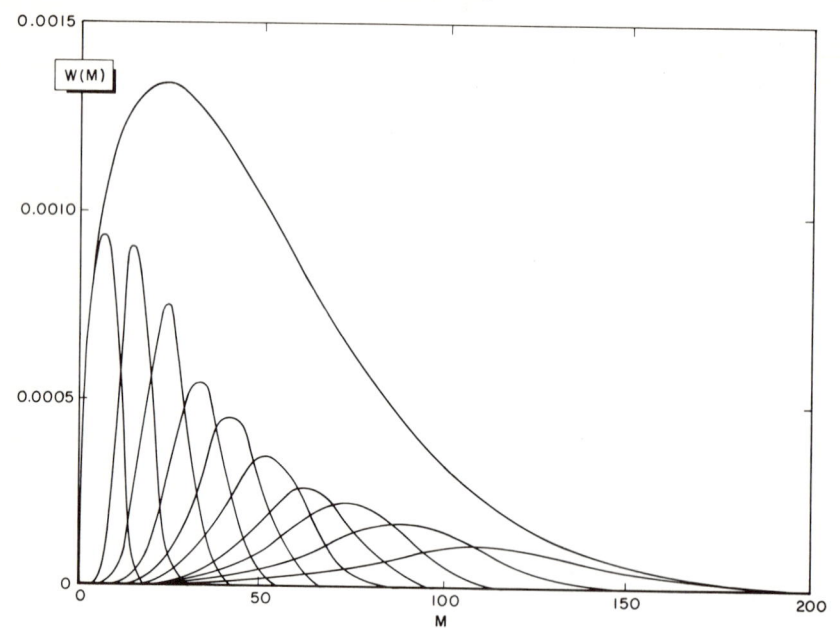

FIG. 6. Differential distributions of fractions computed from a hypothetic partial precipitation fractionation (Tung [12]).

A. Calculation through Integral Distribution

1. Schulz's Method

In Schulz's method [9] it is assumed that one-half of the weight of a fraction consists of molecules having molecular weights below the average molecular weight determined for that fraction and the other half above the average molecular weight. The cumulative weight fraction $C(M_i)$, for the ith fraction is then computed by adding one-half of its weight fraction w_i to the weight fractions of all previous fractions

$$C(M_i) = \tfrac{1}{2}w_i + \sum_{j=1}^{i-1} w_j \qquad (25)$$

The integral distribution curve is obtained by passing a smooth curve through the points on the plot of $C(M_i)$ versus M_i. Figure 7 is a typical experimental result plotted in this manner. Graphical differentiation is then used to obtain the differential distribution curve.

If in a fractionation experiment the sample is first separated into several primary fractions and each of the primary fractions is then refractionated into several more subfractions, the data can be treated by first determining the integral distribution curve for each of the primary fractions. The

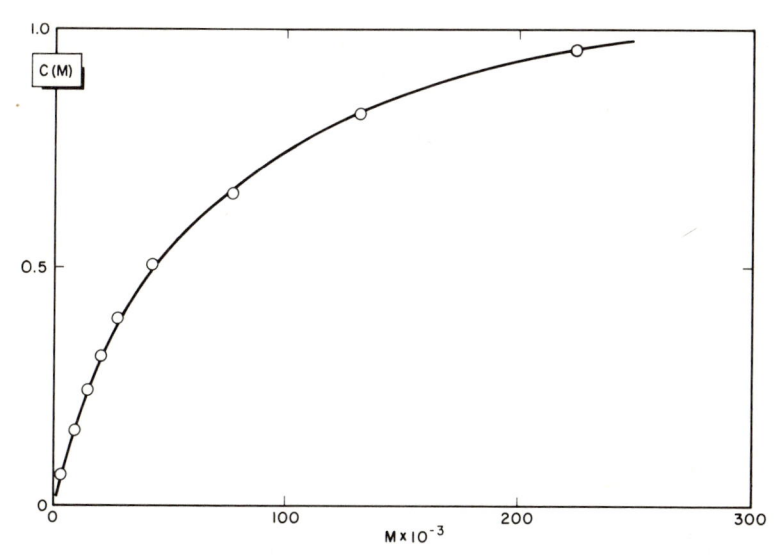

FIG. 7. Integral distribution curve of a high-density polyethylene (data of Lapsley and Pascoe [13]).

integral distribution for the entire sample is then the sum of the distributions of the primary fractions. Thus, if we let w_i represent the weight fraction of the primary fraction i, and w_{ij} the weight fractions of the subfraction, then

$$w_i = \sum_j w_{ij}$$

and

$$\sum_i \sum_j w_{ij} = 1 \qquad (26)$$

The cumulative weight for the ith primary fraction is then

$$C(M_{ij}) = \tfrac{1}{2} w_{ij} + \sum_{k=1}^{j-1} w_{ik} \qquad (27)$$

The summing of the integral distributions is done graphically, as shown in Fig. 8 (data of Davis and Tobias [14]).

2. Modification of Schulz's Method

Although, as will be shown in Section V, Schulz's method is quite satisfactory for treating fractionation data of high resolution, it cannot completely compensate for the overlapping shown in Fig. 6. Furthermore, the passing of a smooth curve through the experimental points involves some

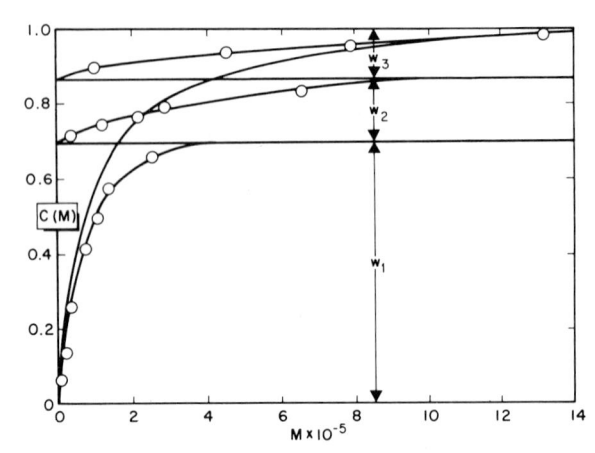

FIG. 8. Integral distribution curve of an isotactic polypropylene (data of Davis and Tobias [14]).

arbitrariness. Graphical differentiation then further amplifies such arbitrariness. To correct for these drawbacks a few modifications have been introduced.

a. *Correction for Overlapping.* At first glance the assumption used in Schulz's method as described by Eq. (25) seems to imply that the distribution curve of the ith fraction should not exceed the average molecular weights of its neighboring fractions [$(i + 1)$th, $(i - 1)$th]. The actual demand of this assumption is somewhat less restrictive. The overlapping of the distribution exceeding the average molecular weights of the neighboring fractions is in part compensated for by the similar overlapping of the distributions of other fractions. As shown in Fig. 9 the shaded area to the right of the M_i of the ith fraction is compensated for in part by the shaded area to the left. For the curves shown in Fig. 9 the cumulative weight fractions for the low molecular weight fractions are slightly lower than the value of the true integral distribution curve and those of the high molecular weight fractions, slightly higher than the true values. The largest discrepancy is at the highest molecular weight fraction where the overlapping to the right of M_λ is not compensated for at all. This deficiency was recognized by Schulz [9]. He recommended that the cumulative weight for the last fraction $C(M_\lambda)$ should be computed as

$$C(M_\lambda) = \sum_{j=1}^{\lambda - 1} w_j$$

instead of

$$C(M_\lambda) = \tfrac{1}{2} w_\lambda + \sum_{j=1}^{\lambda - 1} w_j$$

E. TREATMENT OF DATA

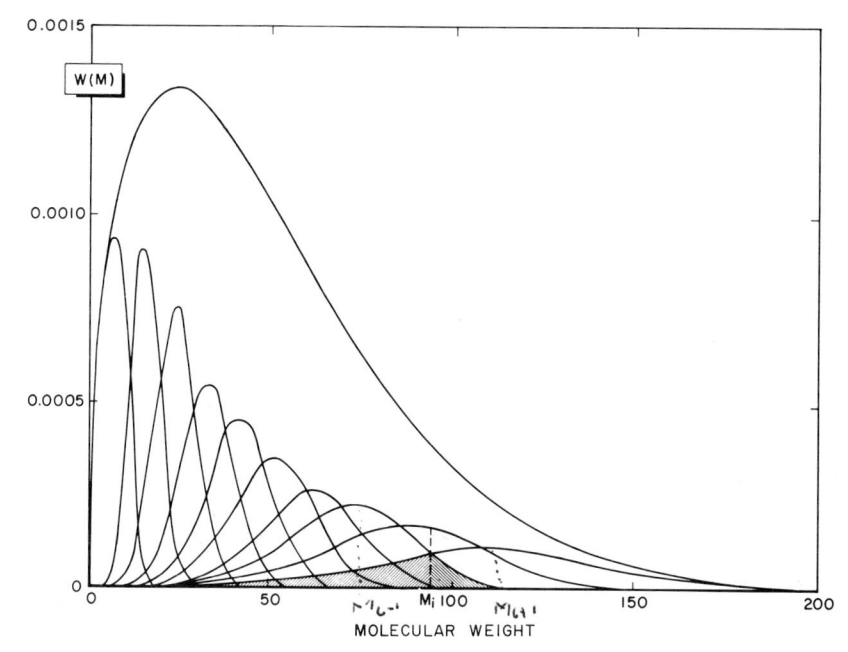

FIG. 9. Differential distributions of fractions computed from a hypothetic partial precipitation fractionation.

Haseley [15] later proposed another method to correct for the overlapping. He refractionated a few of the fractions, determined the integral distributions from the data of refractionation of these fractions by Schulz's method, and then estimated from these fractions the integral distributions of the rest of the fractions. The cumulative weight fraction for the ith fraction is then corrected by subtracting from it the sum of the weight fractions having molecular weight greater than M_i in all lower fractions and adding to it the same having molecular weight lower than M_i in the higher fractions. Thus

$$C(M_i)_{\text{corrected}} = C(M_i)_{\text{uncorrected}} - \sum_{j=1}^{i-1} w_j(M > M_i) + \sum_{j=i+1}^{\lambda} w_j(M < M_i) \quad (28)$$

where $w_j(M > M_i)$ represents the weight fraction in fraction j having molecular weight greater than M_i and $w_j(M < M_i)$ represents the similar quantity but having molecular weight lower than M_i.

b. *Elimination of Graphical Integration.* In practical fractionations, the experimental points often appear scattered from a smooth curve on the integral distribution plot. Such scattering is mainly the result of the nonuniformity of the fraction size and the fractionation efficiency. As shown in

Section II,2,a, the accuracy of the cumulative weight fractions depends on the compensation of the overlapping of the distribution of the fractions. If a large fraction is obtained next to a small one, the compensation of overlapping is influenced and experimental points will appear scattered. Similarly if a broader distribution fraction appears among neighboring narrower distribution fractions the compensation of the overlapping is also disturbed. The error involved in the determination of average molecular weights for these fractions can also contribute to the scattering of data. Rarely are we justified in treating this scattering of data points as actual bumps or dips in the integral distribution. A smooth curve is needed to pass through these points. A few methods have been devised to avoid the arbitrariness in drawing such curves and to eliminate graphical differentiation in the subsequent step of determining the differential distribution curve. These methods all involve the use of distribution functions to fit the data so that a straight line may be used to smooth out the experimental points.

The first method was introduced by Boyer [16]. He noticed that a good number of published molecular weight distribution data conform to the Schulz distribution function, Eq. (7). Boyer then picked appropriate values of the parameters α and b of Eq. (7) and numerically calculated the integral distribution function using Eqs. (2) and (7). A graph paper was then prepared for each set of α and b selected. The abscissa of the graph paper was divided arithmetically in molecular weight scale. Suitable intervals of $I(M)$ were then used for the ordinates so that data fitting Eq. (7) with matched α and b would form a straight line on the graph paper. He found that if degrees of polymerization is used instead of molecular weight M in Eq. (7), with the following combinations of α and b

$$\alpha = 0.99 \qquad b = 0, 1, 2$$
$$\alpha = 0.995 \qquad b = 1, 2, 3, 4$$

almost all the known molecular weight distributions that obeyed Eq. (7) were covered. For a few cases with values of b between 6 and 8 the data were found to fit the regular Gaussian probability graph paper.

Figure 10 shows a graph paper designed for $\alpha = 0.995$ and $b = 1$. The data corresponding to other values of b deviate markedly from a straight line. Thus by using these graph papers one can determine the parameters α and b and $W(M)$ can be directly calculated without going through graphical differentiation.

Theoretically Boyer's method can be used on data which fit other distribution functions. It is, however, especially suitable for Eq. (7) because the parameter α varies only within a limited range and the number of special graph papers required will not be unreasonably large.

E. Treatment of Data

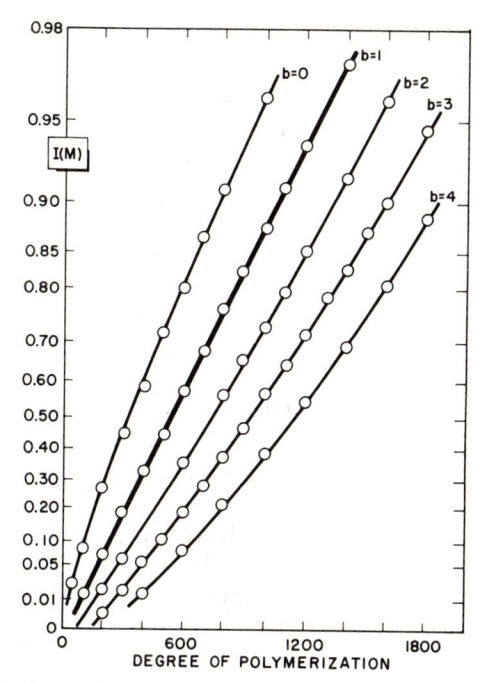

FIG. 10. Calculated cumulative weight fraction curves based on the Schulz distribution function at $\alpha = 0.995$ and $b = 0, 1, 2, 3$, and 4. Graph paper was designed for $\alpha = 0.995$ and $b = 1$ (Boyer [16]).

A second method to eliminate graphical differentiation was proposed by Mussa [17]. He noted that many two-parameter molecular weight distribution functions could be reduced to one-parameter functions if instead of M, M/M_+ was used as the independent variable, where M_+ is related to one of the two adjustable parameters. For instance, Eq. (7) can be written as

$$W\left(\frac{M}{M_+}\right) = \frac{1}{\Gamma(b+2)}\left(\frac{M}{M_+}\right)^{b+1} e^{-M/M_+} \qquad (29)$$

For selected values of b, the integral distribution $I(M/M_+)$ can be calculated by numerical integration. The fractions obtained from experiments are usually characterized by their intrinsic viscosities $[\eta]$. Since the cumulative weight fraction calculated by Schulz's method can be identified with $I(M/M_+)$ a relationship between $[\eta]$ for the fractions and M/M_+ can be established. For practically all polymers the relationship between $[\eta]$ and M is known to follow the Mark-Houwink equation

$$[\eta] = KM^a \qquad (30)$$

A plot of $[\eta]$ versus M/M_+ thus should assume a straight line on a log–log graph paper if the proper value of b is selected. The slope of such a plot should be identical to a in the Mark-Houwink equation and from the position of the straight line the parameter M_+ and hence also the parameter α of the original distribution function can be determined. The values of $I(M/M_+)$ calculated for a set of values of parameter b for Eq. (7) can be found in Mussa's original paper [17].

A third method utilizing distribution Eq. (11) was proposed by Tung [3]. The integral distribution equation (12) for that distribution can be rewritten as

$$\ln \frac{1}{1 - I(M)} = yM^z \qquad (31)$$

Here $I(M)$ can be identified with the cumulative weight fraction calculated from the data. A plot of $\log 1/[1 - I(M)]$ versus M on a log–log graph paper will then give a straight line. From the slope and position of the straight line the parameters y and z can be determined. Figure 11 shows

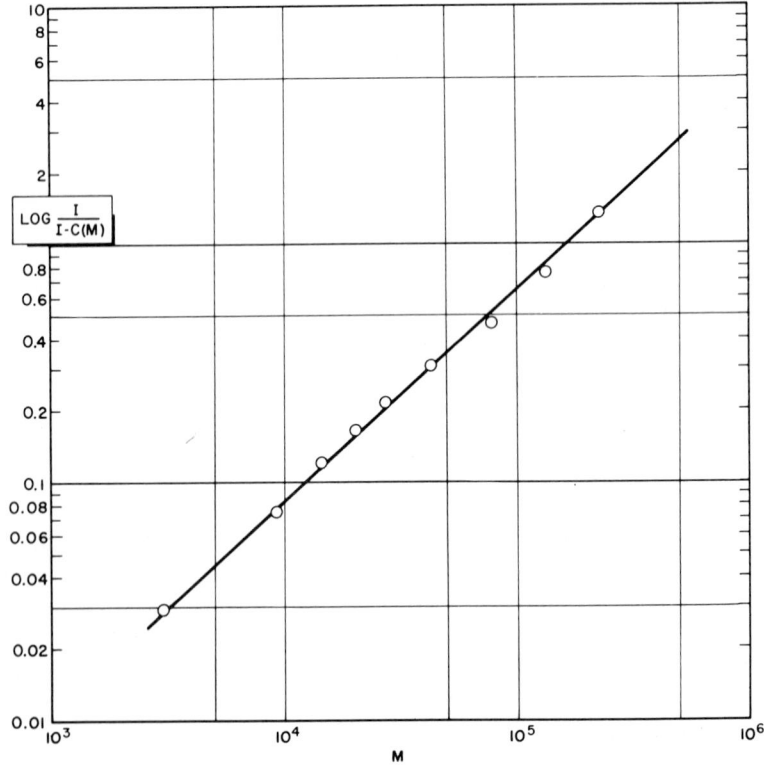

FIG. 11. Log $[1/1 - C(M)]$ versus M plot of the data of Lapsley and Pascoe [13].

the fractionation data of a high-density polyethylene from Lapsley and Pascoe [13] plotted in such a manner. A numerical example of this method is shown in Example 1 in the Appendix.

For polymers obeying the log-normal distribution function Eq. (16), Wesslau [7] has shown that a straight line can be obtained by plotting the cumulative weight fractions versus molecular weight on a log–probability graph paper. Figure 12 shows the fractionation data of a typical high-density polyethylene plotted in this manner. The parameters β and M_0 can be determined from the slope and position of the straight line. A numerical example of this method is shown in Example 2 in the Appendix.

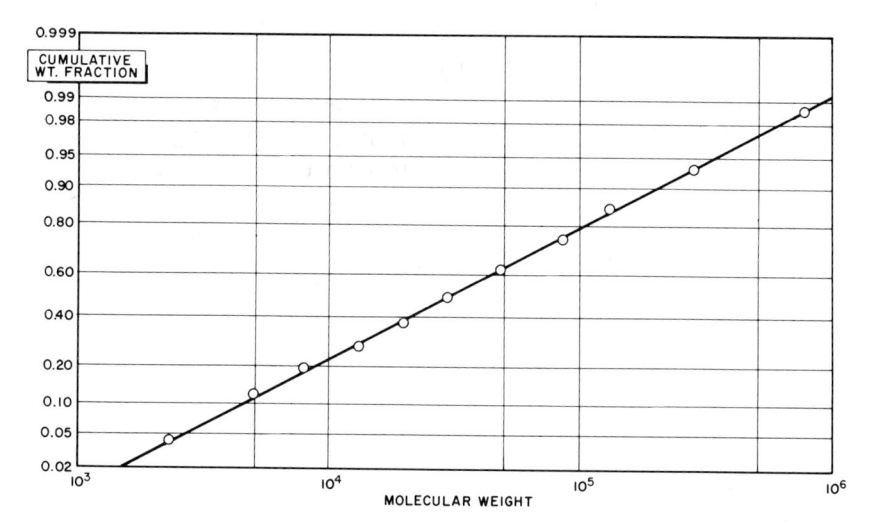

FIG. 12. Fractionation of a typical high-density polyethylene plotted on log-probability graph paper (Tung [12]).

B. Direct Calculation of Differential Distribution

1. Beall's Method

Instead of amending Schulz's method to correct for the overlapping of the distribution curves of the fractions, Beall [18], used a different approach to treat the fractionation data. In his method the distribution of the fractions is assumed to follow a two-parameter distribution function. The parameters of the distribution are determined through the average molecular weights measured for the fractions. The distribution of the entire polymer is then obtained by summing the distributions of all the fractions.

In Beall's original paper, he used a binomial distribution function to describe the distribution of the fractions. Since his work has been quoted by many authors we shall discuss the binomial distribution function in some detail.

As it is more convenient to describe the binomial distribution in terms of degree of polymerization, x, we shall in this section forego the use of molecular weight as the independent variable. The binomial distribution function for number distribution can be written as

$$N(x) = \frac{l!}{(l-x)!x!} q^{l-x} p^x \tag{32}$$

where l and p are two adjustable parameters and $q = 1 - p$. Equation (32) is the xth term in the binomial expansion of $(p + q)^l$. It has been used to describe the Bernoulli probability for exactly x successes out of l trials with a success probability p. Since $p + q = 1$,

$$\sum_{x=0}^{l} N(x) = \sum_{x=0}^{\infty} N(x) = 1 \tag{33}$$

The second part of Eq. (33) is true because $(l - x)! = 0$ when $x > l$. It can be shown that the first moment, μ_1, of $N(x)$, defined as

$$\sum_{0}^{\infty} x N(x)$$

and the second moment, μ_2, defined as

$$\sum_{0}^{\infty} x^2 N(x)$$

can be expressed as

$$\mu_1 = lp \tag{34}$$
$$\mu_2 = p^2 l(l-1) + \mu_1 \tag{35}$$

The number and weight average degree of polymerization are

$$\bar{x}_n = \mu_1 = lp \tag{36}$$
$$\bar{x}_w = \mu_2/\mu_1 = 1 + p(l-1) \tag{37}$$

When Eq. (32) is used to describe a polymer distribution the parameter l becomes a negative noninteger. In such case $N(x)$ can be written as

$$N(x) = \frac{\Gamma(x-l)}{x! \Gamma(-l)} q^{l-x} p^x \tag{38}$$

The relations described by Eqs. (33)–(37) are still true.

Beall determined μ_1 by direct measurement of the osmotic molecular weights of the fractions. He then assumed that the solubility of the polymer obeyed the equation

$$\chi_x = e^{a_1 x} \tag{39}$$

where χ_x is the fraction of polymer of degree of polymerization x in solution when partial precipitation took place. By using Eq. (39) in conjunction with the amount of polymer separated during a partial precipitation step he was able to calculate by successive approximation the second moment μ_2 for the fractions. The calculation was tedious. As it was later shown by Booth and Beason [19] that the values of μ_2 so calculated were not satisfactory, the procedure for the successive approximation is not presented here.

Knowing μ_1 and μ_2 one can determine the parameters l and p through Eqs. (34) and (35). The number distribution for each fraction is then known. If we use the subscript i to denote the distribution function for the ith fraction, the differential distribution function $W_i(x)$ is

$$W_i(x) = xN(x) \bigg/ \sum_{x=0}^{\infty} xN_i(x) = xN(x)/\mu_{1i} \tag{40}$$

and the differential distribution $W(x)$ for the entire polymer is

$$W(x) = \sum_{i=1}^{\lambda} w_i W_i(x) \tag{41}$$

2. Modification of Beall's Method

Beall's method as shown above does not impose any limit on the overlapping of the distribution of the fractions. It does not preassume the shape of the distribution curve of the whole polymer. The distribution can be bimodal or even discontinuous as long as the fractions obtained obey the binomial distribution function and the second moment of the fractions can be obtained through the successive approximation step.

The first assumption is easily satisfied. It was demonstrated by Koningsveld and Tuijnman [20] that the type of two-parameter function assumed for the fraction is not critical at all. The fractions obtained from most fractionations are narrow enough that any two-parameter function can adequately be used to describe their distribution. Koningsveld and Tuijnman even substituted artificial triangular and rectangular shape functions for the binomial function and found that the resulting distribution for the whole polymer did not deviate greatly from that calculated using the binomial distribution. Booth and Beason [19] in their calculation have shown that the final results calculated from using the binomial distribution and the Schulz's distribution function Eq. (7) are essentially identical.

The second assumption used by Beall is, however, not justified. As has already been mentioned in Section III,B,1, Booth and Beason's calculations showed that the successive approximation method used by Beall was not satisfactory. In fact their calculations have shown that the success of Beall's method depends on knowing the precise average molecular weights of the fractions. The successive approximation used by Beall actually gives results less representative of the true distribution than Schulz's method.

The above discussions indicate that a significant improvement in Beall's method can be made by determining the parameters of the distribution function assumed for the fractions directly through two average molecular weight measurements. The two average molecular weights are preferably the number average molecular weight measured from osmometry (for low molecular weight fractions ebulliometry or cryoscopy may prove to be necessary) and the weight average molecular weight from intrinsic viscosity measurements. If an accurate relationship between the intrinsic viscosity and molecular weight is not known light scattering measurements then should be used. The selection of a distribution function for the fraction is not critical. Equations (7), (11), and (16) are equally applicable. Because of the factorial term involved, the binomial distribution is somewhat cumbersome to use. If the whole polymer is known to follow any particular function, it is appropriate then to use the same function for the fractions. A numerical example for the modified Beall method is shown in Example 3 in the Appendix.

C. Calculation of Molecular Weight Distribution Indices

A simple way of calculating the molecular weight distribution indices, such as the $\overline{M}_w/\overline{M}_n$ ratio, is by summing the experimental results given in Table I. Thus from

$$\overline{M}_w = \sum_{i=1}^{\lambda} w_i M_i \tag{42}$$

$$\overline{M}_n = 1 \bigg/ \sum_{i=1}^{\lambda} \frac{w_i}{M_i} \tag{43}$$

the ratio $\overline{M}_w/\overline{M}_n$, and the indices U and Δ can be obtained. Similarly through Eq. (43), and

$$\overline{M}_z = \sum_{i=1}^{\lambda} w_i^2 M_i \bigg/ \sum_{i=1}^{\lambda} w_i M_i \tag{44}$$

the index g can be obtained.

The M_i values given in Table I are usually determined through intrinsic viscosity and are almost identical to the weight average molecular weights for the fractions. Equation (42) therefore should give the true weight average molecular weights. However, $\bar{M}_n$ obtained from Eq. (43) is always higher than the true number average molecular weight because of the polydispersity of the fractions. Since the lower molecular weight fractions contribute more to $\bar{M}_n$, the result will be improved significantly by making an absolute molecular weight determination for the lowest molecular weight fraction [12]. Similarly $\bar{M}_z$ obtained from Eq. (44) is always too low. Hosemann and Schramek [11] have discussed ways to calculate better values for the index g. Details of their method are not given here as g is a less frequently used index.

If one has fitted the experimental data to a distribution function these indices can be directly calculated from the parameters of the distribution function. From Schulz's distribution, Eq. (7)

$$\frac{\bar{M}_w}{\bar{M}_n} = \frac{b+2}{b+1}$$

$$U = \frac{1}{b+1}$$

$$\Delta = \left(\frac{1}{b+1}\right)^{1/2} \tag{45}$$

$$g = \left(\frac{1}{b+2}\right)^{1/2}$$

From distribution function equation (11)

$$\frac{\bar{M}_w}{\bar{M}_n} = \Gamma\left(1 + \frac{1}{z}\right)\Gamma\left(1 - \frac{1}{z}\right) = \frac{\pi/z}{\sin \pi/z}$$

$$U = \Gamma\left(1 + \frac{1}{z}\right)\Gamma\left(1 - \frac{1}{z}\right) - 1 = \frac{\pi/z}{\sin \pi/z} - 1 \tag{46}$$

$$\Delta = \left[\Gamma\left(1 + \frac{1}{z}\right)\Gamma\left(1 - \frac{1}{z}\right) - 1\right]^{1/2} = \left(\frac{\pi/z}{\sin \pi/z} - 1\right)^{1/2}$$

$$g = \left[\Gamma\left(1 + \frac{2}{z}\right) - \Gamma\left(1 + \frac{1}{z}\right)\right]^{1/2} / \left[\Gamma\left(1 + \frac{1}{z}\right)\right]^{1/2}$$

From the log-normal distribution equation (16)

$$\overline{M}_w/\overline{M}_n = e^{\beta^2/2}$$
$$U = e^{\beta^2/2} - 1$$
$$\Delta = \left(e^{\beta^2/2} - 1\right)^{1/2} \tag{47}$$
$$g = \Delta = \left(e^{\beta^2/2} - 1\right)^{1/2}$$

IV. Calculation of Molecular Weight Distribution from Average Molecular Weight Measurements

If one has the knowledge that the molecular weight distribution of a polymer follows a certain two-parameter distribution function, then the entire molecular weight distribution can be calculated from the measurement of two average molecular weights, e.g., number average molecular weight by osmometry and weight average molecular weight by light scattering. From these two average molecular weights one can calculate the parameters of the distribution function, hence the whole distribution, through relationships such as Eqs. (8) and (9). The molecular weight distribution indices can also be calculated directly from these measured molecular weights.

In these calculations, however, one cannot freely substitute the molecular weight determined from intrinsic viscosity for the weight average molecular weight as done in earlier sections for the fractions. The intrinsic viscosity–molecular weight relationship of polymers is describable by the Mark–Houwink equation

$$[\eta] = KM^a \tag{30}$$

The two constants K and a are usually determined by comparing the intrinsic viscosities of several narrow distribution fractions with the absolute weight average molecular weights determined for these fractions. When Eq. (30) is used to calculate the molecular weight of a broad distribution sample a viscosity average molecular weight $\overline{M}_v$ is obtained. Since

$$[\eta]_{av} = \int_0^\infty [\eta]W(M)\,dM \tag{48}$$

$$\overline{M}_v = \left(\frac{[\eta]_{av}}{K}\right)^{1/a} = \left[\int_0^\infty M^a W(M)\,dM\right]^{1/a} \tag{49}$$

Only when the exponent a is unity, does $\overline{M}_v = \overline{M}_w$. For polymers which assume a coiled configuration in solution, the exponent a falls between values 0.8 to 0.5, depending on the interaction between the solvent and the polymer. Hence the viscosity average molecular weight is significantly lower than the weight average molecular weight when the distribution is broad [21]. The relationship between $\overline{M}_v$ and the parameters of the distribution functions (7), (11), and (16) are, respectively,

$$\overline{M}_v = \frac{1}{-\ln \alpha} \left[\frac{\Gamma(a+b+2)}{\Gamma(b+2)} \right]^{1/a} \tag{50}$$

$$\overline{M}_v = y^{-1/z} \left[\Gamma\left(1 + \frac{a}{z}\right) \right]^{1/a} \tag{51}$$

$$\overline{M}_v = M_0 \, e^{a\beta^2/4} \tag{52}$$

Thus one of these equations instead of Eqs. (9), (14), or (18) should be used if the molecular weight is determined by intrinsic viscosity measurement instead of by light scattering.

We have seen that the exponent a in the Mark–Houwink equation depends on the interaction between the polymer molecule and the solvent in which the intrinsic viscosity measurements are made. The $\overline{M}_v$ measured in a good solvent therefore differs from that measured in a poor solvent. Frisch and Lundberg [22] suggested that the ratio

$$(\overline{M}_v)_{\text{solv. 1}} / (\overline{M}_v)_{\text{solv. 2}}$$

can also be used as an index for molecular weight distribution. But because the range of variation of the exponent a is so small such a ratio is not a very sensitive index. More on this can be found in Chapter C.5.

Distribution functions which contain more than two parameters have also been used to describe the molecular weight distribution. Generally a curve between two points can be fitted by a series containing infinite terms of orthogonal functions. This is discussed in detail in Chapter F. A series involving Laguerre polymonials was used by Wales *et al.* [23], Bamford and Tompa [24], and Miyake [25] to describe the distribution of polymer molecular weights. A series involving Hermite polynomials was used by Herdan [26]. The coefficients of these series can be related to the moments of the distribution. To evaluate the coefficients of higher terms higher moments of the distribution are required. Both Wales *et al.* and Herdan used $\overline{M}_z$ and $\overline{M}_{z+1}$ in addition to $\overline{M}_n$ and $\overline{M}_w$ to evaluate the third and fourth moments of the distribution. $\overline{M}_{z+1}$ is defined by

$$\overline{M}_{z+1} = \frac{\int_0^\infty M^3 W(M)\,dM}{\int_0^\infty M^2 W(M)\,dM} \tag{53}$$

Experimentally the z and $z + 1$ average molecular weights must be determined by ultracentrifugation and yet they cannot be determined, especially for $\overline{M}_{z+1}$, with a high degree of precision. It seems that if one has at his disposal an ultracentrifuge it is simpler to use sedimentation velocity as given in Chapter C.2 to determine the molecular weight distribution directly. The methods of Wales et al. and Herdan are therefore seldom used since the latter experimental technique becomes more developed.

V. Comparison of Methods

We have seen from discussions in Sections III and IV that the modified Beall method is the most precise way of treating polymer fractionation data. Determination of weight average molecular weights of the fractions through intrinsic viscosity measurements poses no great problem. The precise determination of the number average molecular weights by osmometry, ebulliometry, or cryoscopy is, however, time consuming and requires very careful experimentation. Consequently, very few published works have relied on the modified Beall method for treating fractionation data. The recent development of high-speed precision osmometers [27,28] may pave the way for more frequent use of this method.

If accurate absolute molecular weights for the fractions are not available, the less precise Schulz method is the preferred way of treating the data. Booth and Beason's calculations [19] have demonstrated that errors in molecular weights of the fractions greatly influence the result from Beall's method. They have also shown that the objection to Schulz's method is not as serious as believed by some authors. A set of recent calculations [12], based also on Flory–Huggins equation of polymer solubility as in Booth and Beason's work, but carried out to a higher degree of precision, has shown that when the distributions of the fractions are reasonably sharp Schulz's method is quite acceptable. The curve in Fig. 13 is the integral distribution of the starting polymer. The points are the calculated cumulative weight fractions of three hypothetic fractionation runs using the curve as the starting distribution. The result representing a run based on Flory's scheme of refractionation [29] agrees with the starting distribution curve within normal experimental errors. The conditions used for the calculation of the hypothetical fractionations are all experimentally feasible. A calculation simulating an actual fractionation further demonstrated that the results of these calculations correspond quite well to the efficiencies of actual experiments. It may be concluded then that if one lacks the tools or confidence in determining precise absolute molecular weights, he should improve the efficiency of fractionation and use Schulz's method to treat the data.

E. Treatment of Data

The methods described in Section III,A,2,b, are advantageous to use whenever they are applicable. One should, however, be very cautious about using the functions to extrapolate the distribution into high and low molecular weight regions beyond the data points for highest and lowest fractions. Only the modified Beall method gives reliable information in these regions.

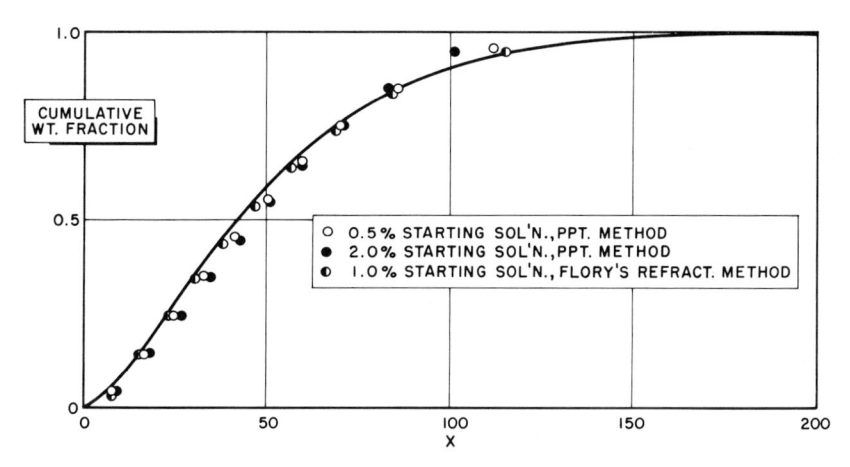

FIG. 13. Integral distributions of hypothetic fractionation experiments made at different fractionation efficiency (Tung [12]).

Figure 13 shows also that the correction proposed by Schulz (Section III,A,2,a) cannot be applied generally. If one substitutes

$$c(M_\lambda) = \sum_{i=1}^{\lambda-1} w_i$$

for

$$c(M_\lambda) = \tfrac{1}{2}w_\lambda + \sum_{j=1}^{\lambda-1} w_j$$

two of the three fractionation results will be overcorrected.

Similar considerations can be given to the calculation of molecular weight distribution indices. For instance, the calculation of $\overline{M}_w/\overline{M}_n$ ratio using the two-parameter distribution function, Eq. (7), (11), or (16) implicitly involves extrapolation of the distribution to molecular weight regions beyond the data points. The result, therefore, cannot be compared with the ratio obtained from the direct summation equations (42) and (43). The indices obtained from different distribution functions also cannot be compared with one another as each distribution function extrapolates differently into the unknown regions.

Distributions containing more than one peak or even discontinuities do not influence the treatment of fractionation data by the modified Beall method. If one should rely on the Schulz method, then graphical differentiation must be used to determine the differential distribution curve. A judgment is needed to distinguish whether the experimental points are truly reflecting inflections in the integral distribution curve or merely a scattering of data due to experimental errors. Figure 14 shows an integral distribution curve determined by Henry [30] on a polymer sample blended from a high and a low molecular weight fraction—a two-peak distribution.

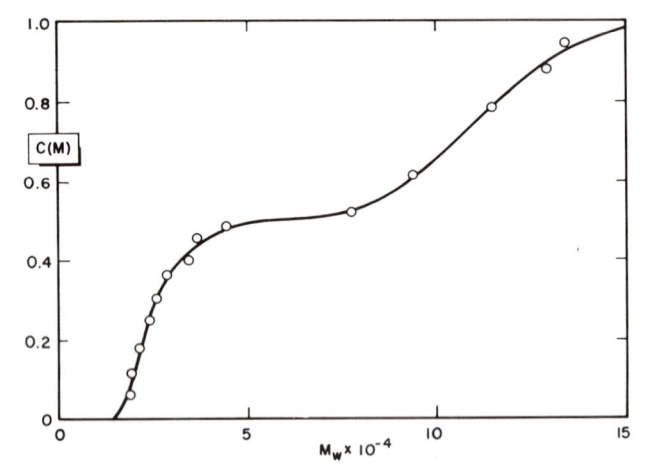

FIG. 14. Integral distribution of a blend of two high-density polyethylene fractions (data of Henry [30]).

It has already been mentioned (Section IV) that the use of multiparameter functions for calculating molecular weight distribution from experimentally determined average molecular weights is not practical. If one has *a priori* knowledge that a polymer sample follows a two-parameter distribution function, it is permissible then to determine the distribution through absolute molecular weight determinations of $\overline{M}_n$ and $\overline{M}_w$ (or $\overline{M}_v$). The determinations of the absolute molecular weights are, however, often more difficult for whole polymers than for the fractions. Thus unless one has confidence in the determination of absolute molecular weights, fractionation is still the more reliable approach even when only molecular weight distribution indices are sought for the samples.

Appendix: Numerical Illustrations

EXAMPLE 1

In this example the fractionation data of Lapsley and Pascoe [13] on a sample of high-density polyethylene are used to demonstrate the plotting of the integral distribution curve by the Schulz method. In addition we shall show the determination of the differential distribution curve through the use of Eq. (11). The data of Lapsley and Pascoe were obtained by a sand elution column (cf. Chapter B.2) and are shown in Table II.

TABLE II

FRACTIONATION DATA OF A HIGH-DENSITY POLYETHYLENE BY A SAND ELUTION COLUMN

Fraction	Weight fraction w	Cumulative weight fraction $C(M)$	Molecular weight $M \times 10^{-3}$
1	0.0212	0.0106	Insufficient sample
2	0.0888	0.0656	3.0
3	0.0960	0.1580	9.3
4	0.0722	0.2421	14.4
5	0.0722	0.3143	20.2
6	0.0870	0.3939	27.4
7	0.1368	0.5058	42.3
8	0.1624	0.6554	77.7
9	0.1750	0.8241	131.8
10	0.0884	0.9558	225.4

The cumulative weight fractions in the third column are calculated from Eq. (25) according to Schulz. Figure 7 (Section III,A,1.) shows the cumulative fractions plotted against the molecular weights of the fractions and the integral distribution curve drawn through the experimental points.

These data were found to fit the distribution function Eq. (11). Table III shows the calculated values of $1/[1 - C(M)]$ and $\log 1/[1 - C(M)]$. The plot of $\log 1/[1 - C(M)]$ versus M is shown in Fig. 11 (Section III,A,2,b) on a log–log graph paper.

To calculate the parameters y and z from Fig. 11, we start with Eq. (31).

$$\ln \frac{1}{1 - I(M)} = yM^z \qquad (31)$$

If $I(M)$ is identified with $C(M)$ and the natural logarithm is changed to

TABLE III

$1/[1 - C(M)]$ AND LOG $1/[1 - C(M)]$ FOR THE DATA IN TABLE II

Fraction	$1/[1 - C(M)]$	$\log 1/[1 - C(M)]$
1	—	—
2	1.070	0.0295
3	1.188	0.0747
4	1.319	0.1203
5	1.458	0.1638
6	1.650	0.2175
7	2.024	0.3062
8	2.902	0.4627
9	5.685	0.7547
10	22.62	1.354

logarithm to the base ten, we have

$$\log \log \frac{1}{1 - C(M)} = \log \frac{y}{2.303} + z \log M \tag{54}$$

Let M_1 be the molecular weight corresponding to $\log[1/1 - C(M)] = 1$ and M_2 be that corresponding to $\log[1/1 - C(M)] = 0.1$; then

$$\log \frac{y}{2.303} + z \log M_1 = 0 \tag{55}$$

$$\log \frac{y}{2.303} + z \log M_2 = -1 \tag{56}$$

Solving for y and z, we have

$$z = 1 \bigg/ \left(\log \frac{M_1}{M_2} \right) \tag{57}$$

$$y = 2.303/M_1^z \tag{58}$$

M_1 and M_2 read from Fig. 11 are 165,000 and 12,400, respectively. Substituting these values into Eq. (56) and (57), we obtain $z = 0.890$ and $y = 5.23 \times 10^{-5}$. The differential distribution $W(M)$ calculated from

$$W(M) = yz \, e^{-yM^z} M^{z-1} \tag{11}$$

is shown in Fig. 15.

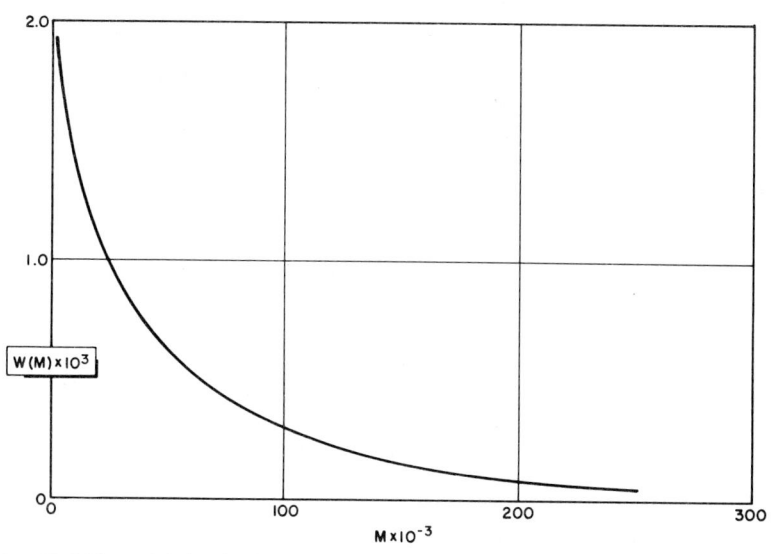

FIG. 15. Differential distribution of Lapsley and Pascoe's fractionation data on high-density polyethylene.

EXAMPLE 2

In this example the fractionation results of another high-density polyethylene sample are used to demonstrate the use of the log-normal distribution function, Eq. (16), in determining the differential distribution curve. The data shown in Table IV were obtained by partial precipitation using the Flory scheme of refractionation. The integral distribution curve of the polymer is shown in Fig. 16. The sample in this example is a typical high-density polyethylene which fits the log-normal distribution function, Eq. (16). The plot of these data on log-probability graph paper is shown in Fig. 12 (Section III,A,2,b).

To calculate the parameters β and M from Fig. 12 we start with the integral distribution function

$$I(M) = \frac{1}{\beta\sqrt{\pi}} \int_0^M \exp\left(-\frac{1}{\beta^2}\ln^2\frac{M}{M_0}\right)\frac{1}{M}\,dM \tag{59}$$

Let

$$u = \frac{\sqrt{2}}{\beta}\ln\frac{M}{M_0}$$

then

$$I(M) = \frac{1}{\sqrt{2\pi}} \int_{-\infty}^u e^{-u^2/2}\,du \tag{60}$$

TABLE IV

Fractionation Data of a High-Density Polyethylene by Partial Precipitation

Fraction	Weight fraction w	Cumulative weight fraction $C(M)$	Molecular weight $M \times 10^{-3}$
1	0.0863	0.0432	2.34
2	0.0734	0.1230	5.03
3	0.0808	0.2001	8.00
4	0.0792	0.2801	13.2
5	0.1159	0.3777	20.0
6	0.1224	0.4968	29.9
7	0.1321	0.6241	48.6
8	0.1244	0.7523	85.6
9	0.0868	0.8579	132
10	0.0720	0.9373	284
11	0.0267	0.9867	773

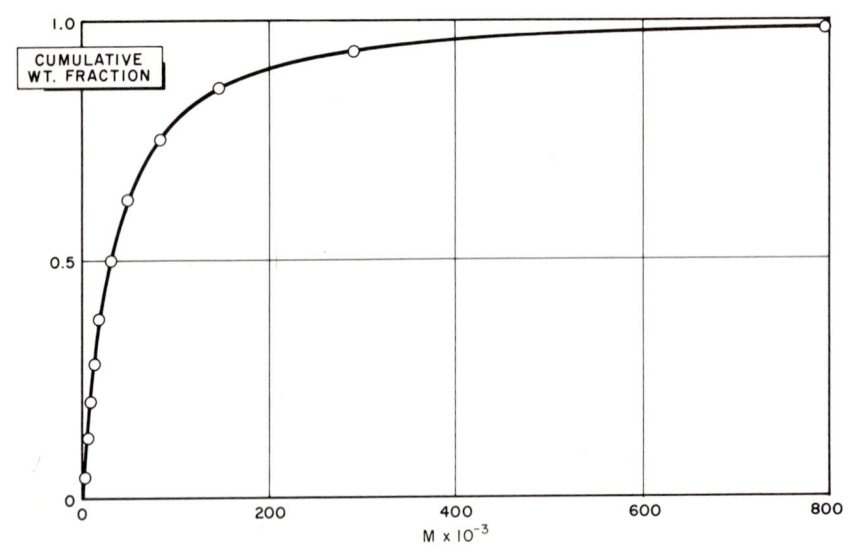

FIG. 16. Integral distribution of a typical high-density polyethylene (Tung [12]).

The integrand in Eq. (60) is now symmetrical with respect to the origin; it follows then that

$$\int_{-\infty}^{-u} e^{-u^2/2} \, du = \int_{u}^{\infty} e^{-u^2/2} \, du \tag{61}$$

We may now write

$$I(M) = \frac{1}{2\sqrt{2\pi}} \left(\int_{-\infty}^{\infty} e^{-u^2/2} \, du + \int_{-u}^{u} e^{-u^2/2} \, du \right) \qquad (62)$$

The value of the first integral in Eq. (62) is $\sqrt{2\pi}$, thus

$$I(M) = \frac{1}{2}\left(1 + \frac{1}{\sqrt{2\pi}} \int_{-u}^{u} e^{-u^2/2} \, du \right) \qquad (63)$$

When $u = 0$, $M = M_0$ and from Eq. (63) $I(M) = 0.5$. The value of the integral in Eq. (63) for other values of u can be found in many mathematical tables. Thus when $u = 1$,

$$I(M) = \tfrac{1}{2}(1 + 0.6827) = 0.8413$$

and since

$$u = (\sqrt{2}/\beta) \ln (M/M_0)$$

$$\beta = \sqrt{2} \ln (M/M_0)$$

Figure 12 gives 29,800 for M when $I(M) = 0.5$ and 131,000 for M when $I(M) = 0.8413$. From the above two conditions, we obtain $M_0 = 29,800$ and $\beta = 2.09$. The differential distribution curve calculated from

$$W(M) = \frac{1}{\beta\sqrt{\pi}} \frac{1}{M} \exp\left(-\frac{1}{\beta^2} \ln^2 \frac{M}{M_0}\right) \qquad (16)$$

is shown by the dotted curve in Fig. 17.

EXAMPLE 3

In this example we shall use the same fractionation result as in the previous example to demonstrate the Modified Beall Method of data treatment. The molecular weights reported in Table IV were determined through intrinsic viscosities and can therefore be taken as the weight average molecular weights for the fractions. However, in Table V only the lowest number average molecular weight was determined experimentally (by ebulliometry). The other values were obtained through a fractionation simulation computation as described by Tung [12]. Such an approach is never practical; we are merely borrowing these values for the sake of demonstrating numerically the modified Beall method.

We shall use the log-normal distribution function, as shown in Eq. (16),

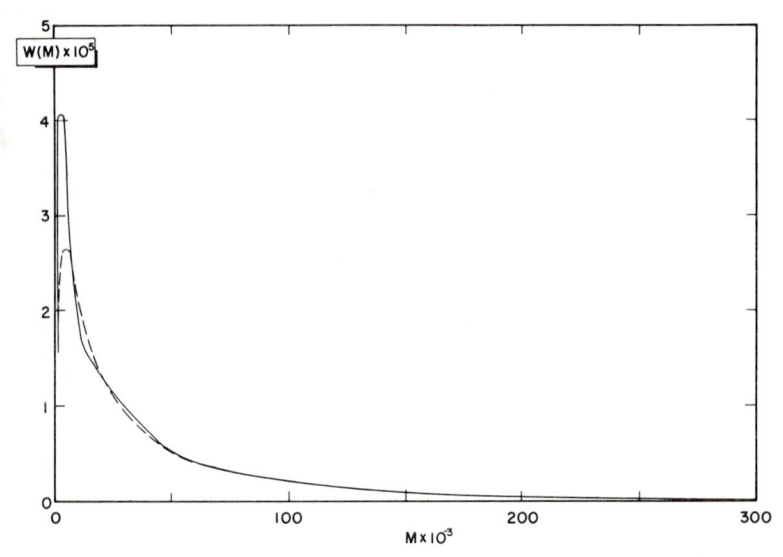

FIG. 17. Differential distribution curves of a typical high-density polyethylene: ---, by Schulz's method and log-normal distribution function; —— by modified Beall's method.

TABLE V

FRACTIONATION DATA OF A HIGH-DENSITY POLYETHYLENE BY PARTIAL PRECIPITATION[a]

Fraction	Weight fraction	Weight average molecular weight $\bar{M}_w \times 10^{-3}$	Number average molecular weight $\bar{M}_n \times 10^{-3}$	$\bar{M}_w/\bar{M}_n$
1	0.0863	2.34	1.95	1.20
2	0.0734	5.03	4.58	1.10
3	0.0808	8.00	7.21	1.11
4	0.0792	13.2	12.3	1.07
5	0.1159	20.0	18.5	1.08
6	0.1224	29.9	27.4	1.09
7	0.1321	48.6	43.4	1.12
8	0.1244	85.6	72.5	1.18
9	0.0868	132	112	1.18
10	0.0720	284	212	1.34
11	0.0267	773	430	1.80

[a] ($\bar{M}_n$ of the first fraction was by ebulliometry, the remainder were by fractionation simulation computation).

to fit the distributions of the fractions. Equations (17) and (18)

$$\overline{M}_n = M_0 \, e^{-\beta^2/4} \tag{17}$$

$$\overline{M}_w = M_0 \, e^{\beta^2/4} \tag{18}$$

are used to solve for the parameters β and M_0 for each of the fractions. The results are listed in Table VI.

TABLE VI

PARAMETERS β AND M_0 FOR THE FRACTIONS

Fraction	β	$M_0 \times 10^{-3}$
1	0.604	2.14
2	0.433	4.80
3	0.456	7.60
4	0.376	12.7
5	0.395	19.2
6	0.418	28.6
7	0.476	45.9
8	0.576	78.8
9	0.573	121.6
10	0.765	245
11	0.083	576

The differential distribution $W_i(M)$ for each fraction are computed by substituting the proper values of β and M_0 in Eq. (16). Equation (41)

$$W(M) = \sum_{i=1}^{\lambda} w_i W_i(M) \tag{41}$$

is then used to compute the differential distribution $W(M)$ for the whole polymer. The computation is illustrated below by the calculation of $W(M)$ for $M = 5000$.

Fraction	$W_i(M) \times 10^5$ at $M = 5000$	$w_i \times W_i(M) \times 10^5$ at $M = 5000$
1	2.59	0.222
2	25.8	1.896
3	1.068	0.863
4	0.0610	0.00484
5	—	—
		$\sum w_i W_i(M) = 2.986$

Table VII shows $W(M)$ calculated for other values of M. The differential distribution is the solid curve plotted in Fig. 17.

TABLE VII

DIFFERENTIAL DISTRIBUTION $W(M)$

$M \times 10^{-3}$	$W(M) \times 10^5$
1	1.661
2	4.065
3	2.988
5	2.986
7	2.184
10	1.644
20	1.401
30	0.961
50	0.508
100	0.204
200	0.052
300	0.022

ACKNOWLEDGMENT

The author wishes to thank Dr. R. F. Boyer for reviewing the manuscript.

REFERENCES

1. P. W. Allen, "Techniques of Polymer Characterization." Butterworth, London and Washington, D.C., 1959.
2. G. V. Schulz, *Z. Physik. Chem.* **B43**, 25 (1939).
3. L. H. Tung, *J. Polymer Sci.* **20**, 495 (1956).
4. J. H. S. Green, *Chem. & Ind.* (*London*) p. 924 (1959).
5. G. J. Howard, *J. Polymer Sci.* **59**, S4 (1962).
6. W. D. Lansing and E. O. Kraemer, *J. Am. Chem. Soc.* **57**, 1369 (1935).
7. H. Wesslau, *Makromol. Chem.* **20**, 111 (1956).
8. K. Ziegler, E. Holzkamp, H. Breil, and H. Martin, *Angew. Chem.* **67**, 541 (1955).
9. G. V. Schulz, *Z. Physik. Chem.* **B47**, 155 (1940).
10. G. G. Lowry, *J. Polymer Sci.* **B1**, 489 (1963).
11. R. Hosemann and W. Schramek, *J. Polymer Sci.* **59**, 29 (1962).
12. L. H. Tung, *J. Polymer Sci.* **61**, 449 (1962).
13. J. Lapsley and G. M. Pascoe, unpublished data (1962).
14. T. E. Davis and R. L. Tobias, *J. Polymer Sci.* **50**, 227 (1961).
15. E. A. Haseley, *J. Polymer Sci.* **35**, 309 (1959).
16. R. F. Boyer, *Ind. Eng. Chem. Anal. Ed.* **18**, 342 (1946).
17. C. Mussa, I.V., *J. Polymer Sci.* **26**, 67 (1957).
18. G. Beall, *J. Polymer Sci.* **4**, 483 (1949).
19. C. Booth and L. R. Beason, *J. Polymer Sci.* **42**, 81 (1960).

20. R. Koningsveld and C. A. F. Tuijnman, *J. Polymer Sci.* **39**, 445 (1959).
21. R. Chiang, *J. Polymer Sci.* **36**, 91 (1959).
22. H. L. Frisch and J. L. Lundberg, *J. Polymer Sci.* **37**, 123 (1959).
23. M. Wales, F. T. Adler, and K. E. Van Holde, *J. Phys. Colloid Chem.* **55**, 145 (1951).
24. C. H. Bamford and H. Tompa, *J. Polymer Sci.* **10**, 345 (1953).
25. A. Miyake, *J. Polymer Sci.* **45**, 232 (1960).
26. G. Herdan, *J. Polymer Sci.* **10**, 1 (1963).
27. R. E. Steele, W. E. Walker, D. E. Burge, and H. C. Ehrmantraut, *paper presented at Pittsburgh Conf. Anal. Chem. Appl. Spectr., 1963.*
28. F. B. Rolfson, *Anal. Chem.* **35**, 1303 (1963).
29. P. J. Flory, "Principles of Polymer Chemistry," Chapter VIII. Cornell Univ. Press, Ithaca, New York, 1953.
30. P. M. Henry, *J. Polymer Sci.* **36**, 3 (1959).

CHAPTER F

The Numerical Analysis and Kinetic Interpretation of Molecular Weight Distribution Data

F. C. Goodrich[1]

CHEVRON RESEARCH COMPANY, RICHMOND, CALIFORNIA

I. Introduction 415
   A. Distribution Functions and Their Moments 416
   B. Generating Functions 417
   List of Symbols 417
II. On the Prediction of Molecular Weight Distributions from Kinetic Schemes 418
   A. Condensation Polymers 419
   B. Free Radical Polymerizations 421
   C. Ionic Polymerizations 431
III. Numerical Methods in the Handling of Molecular Weight Distribution Data 438
   A. The Cumulative Distribution and Data Smoothing 439
   B. Laguerre Polynomials 440
   C. Finite Laguerre Expansions and Gaussian Integration 441
   D. The Choice of Scaling Parameters; Iterations; the Zeroth Moment 443
   E. A Numerical Recipe and a Worked Example 444
   F. Comparison Between Theory and Experiment: Thermally Polymerized Polystyrene 449
   Appendix to Chapter F 452
   References 459

I. Introduction

This chapter divides itself readily into two topics. The first, treated in Section II, which follows this, shows how postulated reaction sequences in polymerization kinetics may be used to predict the molecular weight distribution of the resulting polymer at various stages of the conversion of monomer to polymer. The second, treated in Section III, introduces a numerical method whereby experimentally obtained fractionation data may be used to calculate to good approximation both the moments of the distribution curve as well as the curve itself. The use of Section III does not depend upon a knowledge of Section II, and the reader interested only in the numerical analysis of his data without reference to the interpretation of the results in terms of chemical kinetics is advised, after reading I, A to turn directly to Section III.

[1] Present address: Department of Chemistry, Clarkson College of Technology, Potsdam, New York.

A. Distribution Functions and Their Moments

All synthetically produced high molecular weight compounds are obtained in a mixture of molecular sizes differing from each other in the number x of monomer units incorporated into the chain. The introduction of statistical methods is a natural response to this state of affairs, and it is convenient to introduce a number density distribution function $f(M)$ such that for a given sample of polymer $f(M)\,dM$ is the number of moles of molecules which possess a molecular weight lying between M and $M + dM$. The molecular weight M is related to the number of monomer units in the chain by $M = M_0 x$, where M_0 is the molecular weight of the monomer unit. The reader will observe that our notation implies that M will be treated as a continuous variable while x will take on only positive, integral values. This small inconsistency is of no importance, for no fractionation method is capable of distinguishing between high molecular weight molecules which differ by a single monomer unit. For the same reason, $f(M)$ is defined for $0 \le M \le \infty$, although molecules of either 0 or infinite molecular weight are clearly impossible.

The moments of the distribution $f(M)$ are the set of integrals

$$\mu_r = \int_0^\infty M^r f(M)\,dM. \qquad (1)$$

Because theoretical molecular size distributions evolve from kinetic hypotheses in the notation P_x = moles/liter of species of size x, an alternate interpretation of $f(M)$ is

$$f(M) = P_x/m_0 M_0 \qquad (2)$$

where m_0 is the initial monomer concentration in moles per liter. Correspondingly, we may write

$$\mu_r = (m_0)^{-1}(M_0)^r \sum_1^\infty x^r P_x \qquad (3)$$

and both Eqs. (1) and (3) will be used interchangeably. Important parameters which are used in polymer chemistry to characterize a given sample of polymer are the number average molecular weight $\overline{M}_n = \mu_1/\mu_0$, the weight average molecular weight $\overline{M}_w = \mu_2/\mu_1$, and the z average molecular weight $\overline{M}_z = \mu_3/\mu_2$.

A theorem in statistics, valid for conditions which are always met in polymer chemistry, states that a unique relationship exists between a distribution and its moments, so that a complete set of moments uniquely defines its parent distribution and *vice versa*. Thus, the derivation of a set of moments depending on a quantity q in the form

$$\mu_r = r!\,q^r$$

enables us to infer immediately that the parent distribution is

$$f(M) = q^{-1}\exp(-M/q).$$

This theorem is the basis of the method presented in Section III and we shall need to refer to it occasionally in Section II.

B. Generating Functions

A much used mathematical construct in elucidating complicated statistical problems is the generating function $G(y,t)$ defined by

$$G(y,t) = yP_1(t) + y^2 P_2(t) + \ldots = \sum_1^\infty y^x P_x(t) \qquad (4)$$

Here the $P_x(t)$ are a set of quantities which we shall later identify as the molar concentrations of species of polymer molecules, and the series is presumed convergent for all $0 \leq y \leq 1$. The ordering parameter y is without physical meaning, for all useful physical information is derived from the generating function by setting $y = 1$. Thus the reader may readily verify from (4) that

$$\left(y\frac{\partial}{\partial y}\right)^r G(y,t)\bigg|_{y=1} = \sum_1^\infty x^r P_x(t) \qquad (5)$$

Because sums of this sort are required in the evaluation of distribution moments [Eq. (3)], a knowledge of the generating function enables the investigator to calculate the moments of a distribution even without possessing an explicit formula for the distribution itself.

A second useful property of the generating function is its ability to order convolution sums. We shall require in our work expressions of the type

$$\sum_{j=1}^{x-1} P_j P_{x-j} \qquad (6)$$

known as convolution sums, and by squaring both sides of Eq. (4) and rearranging the result as a power series in y, Eq. (6) may be identified as the coefficient of y^x in $G^2(y,t)$.

LIST OF SYMBOLS

M	molecular weight
M_0	molecular weight of a monomer unit
$f(M)$	number density distribution function
$F(M)$	cumulative weight density distribution function
μ_r	rth moment of $f(M)$
P_x	concentration of dead polymer of size x
R_x	concentration of active polymer of size x
J_r	rth moment of radical distribution

m	monomer concentration
m_0	initial monomer concentration
C	initiator concentration
C_0	initial initiator concentration
S	solvent or chain transfer agent concentration
t	time
u	synthetic time
y	ordering parameter
$G(y, t)$	generating function
$H(y, t)$	generating function
p	fractional conversion of monomer to a condensation polymer
k	rate constant in polycondensation reactions
k_i	rate constant of initiation reaction
k_p	rate constant of propagation reaction
k_t	rate constant for combination reaction
k'_t	rate constant for disproportionation reaction
k_{tr}	rate constant for transfer reaction to monomer
k'_{tr}	rate constant for transfer reaction to solvent
I	initiation rate
f	initiator efficiency
a	scaling parameter
s	scaling parameter
$p_n^s(z)$	a normalized, associated, Laguerre polynomial
λ_k	a root of $p_5^s(z)$
h_{nr}	polynomial coefficients of $p_n^s(z)$
H	the matrix (h_{nr})
c_n	expansion coefficient
G	(g_{kn}), a matrix
Q	(q_{rk}), a matrix
ω_k	weights for the calculation of μ_0

II. On the Prediction of Molecular Weight Distributions from Kinetic Schemes

The distribution of molecular sizes which is the result of any synthetic polymerization is a consequence of the kinetic path followed by the reactants, and it is of interest to inquire as to the influence of various possible kinetic hypotheses upon the final molecular weight distribution. To hold the discussion within reasonable bounds, I shall restrict the discussion to homopolymers, but the methods used go over with only minor modifications to the treatment of copolymerization kinetics [1].

A useful simplification in all of this work lies in the assumption that chemical rate constants for reactive groups in a polymerization reaction are independent of the size of the molecule to which they are attached. This assumption is implicit in all of our work. I shall furthermore use a symbol such as P_x to stand for both a molecule and its molar concentration. This notation has become traditional and should lead to no confusion.

A. Condensation Polymers

The kinetically simplest polymerization reactions are those leading to the formation of condensation polymers, for a simple basic reaction resulting in the coupling of difunctional molecules is here repeated over and over. If P_1 be the monomer, then coupling takes place via the reaction,

$$P_x + P_z \xrightarrow{k} P_{x+z} \qquad (7)$$

and this mechanism leads to the kinetic equations

$$\frac{dP_1}{dt} = -kP_1 \sum_1^\infty P_j$$
$$\frac{dP_x}{dt} = \frac{1}{2}k \sum_1^{x-1} P_j P_{x-j} - kP_x \sum_1^\infty P_j \qquad x > 1 \qquad (8)$$

for which t is, of course, the time.

To solve Eqs. (8), we introduce a generating function

$$G(y, t) = \sum_1^\infty y^x P_x \qquad (9)$$

Multiplying the first of Eqs. (8) by y, the second by y^x, and summing over x from 1 to ∞, the entire set of differential equations is reduced to a single partial differentiation equation in $G(y, t)$

$$(\partial/\partial t)G(y, t) = -kG(1, t)G(y, t) + \tfrac{1}{2}kG^2(y, t) \qquad (10)$$

The meaning of the terms in this equation becomes clearer when we observe from Eq. (9) that

$$G(1, t) = \sum_1^\infty P_x \qquad (11)$$

is the total concentration of molecules of all sorts present at time t, and that the convolution sum

$$\sum_1^{x-1} P_j P_{x-j}$$

is the coefficient of y^x in $G^2(y, t)$.

In (10) set $y = 1$. There results

$$(\partial/\partial t)G(1, t) = -\tfrac{1}{2}kG^2(1, t)$$

and this equation may be integrated immediately under the initial condition that at $t = 0$, only monomer P_1 is present at concentration m_0,

$$G(1, t) = m_0/(1 + \tfrac{1}{2}m_0 kt) \tag{12}$$

Equation (12) has physical meaning, for through Eq. (11) it describes the total molar concentration of polymers of all types as a function of time. Substituting Eq. (12) into Eq. (10), we obtain a differential equation in $G(y, t)$ whose solution is

$$G(y, t) = ym_0[(1 - p)^2/(1 - py)] \tag{13}$$

in which

$$p = \tfrac{1}{2}m_0 kt/(1 + \tfrac{1}{2}m_0 kt) = 1 - G(1, t)/m_0$$

is the fraction of end groups which have disappeared by time t. To complete the problem, expand Eq. (13) as a power series in y and extract the coefficient of y^x, whence we have as the general solution to Eqs. (8)

$$P_x = m_0(1 - p)^2 p^{x-1} \tag{14}$$

Equation (14) is our first derived example of a molecular size distribution function, and we might through the substitution of $x = M/M_0$ rewrite it in the notation $f(M)$ adopted in the introductory section, Eq. (2). Before doing this, however, it is well to note from Eq. (14) that molecules of large size (x large) will be present in significant quantities only in the case that p is very near 1, which is to say when the reaction is practically complete. Secondly, if our major interest lies in the moments μ_r of the molecular size distribution rather than in an explicit expression for the distribution itself, then Eq. (14) is unnecessary, for the moments may be obtained directly from the generating function.

Applying Eq. (5) to the right-hand side of Eq. (13) and referring to Eq. (3), we find in the limit of p very close to 1

$$\mu_r \to (M_0)^r r! [1/(1 - p)]^{r-1} \tag{15}$$

But these quantities are the moments of the uniquely defined distribution function

$$f(M) = (M_0)^{-1}(1 - p)^2 \exp[-(1 - p)(M/M_0)] \tag{16}$$

a fact which might have been established directly from Eq. (14). A sketch of $f(M)$ and of $Mf(M)$ based on Eq. (16) is given in Fig. 1. The appearance of these curves is typical of the type of plot to which we shall devote increased attention in Section III.

F. Molecular Weight Distribution Data

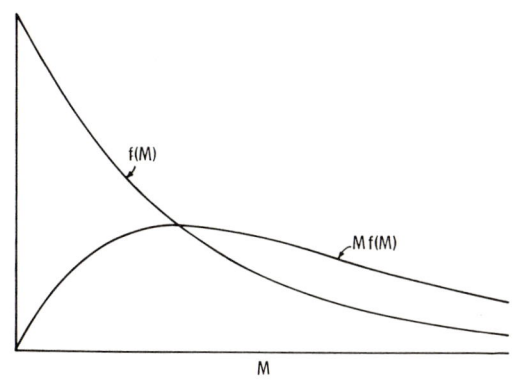

FIG. 1. Number density distribution $f(M)$ and weight distribution $Mf(M)$ for a condensation polymer.

The moments [Eq. (15)] obey a recurrence relation

$$\mu_r/\mu_{r-1} = M_0/(1-p)r$$

whence it follows that for condensation polymers, $\overline{M}_n$, $\overline{M}_w$, and $\overline{M}_z$ are in the ratio 1, 2, 3. Alternatively, a plot of $\log(\mu_r/r!)$ against r should be linear and of slope $\log[M_0/(1-p)]$. Such a plot, together with an independently determined value of p through titration of end groups or equivalent procedure, would serve to test experimentally the validity of the assumed mechanism [Eq. (7)].

B. Free Radical Polymerizations

1. Reaction Types

Addition polymerizations, whether of the free radical or ionic varieties, offer a far greater range of kinetic possibilities than do condensation polymerizations. Instead of the single type of reaction with its single rate constant which is characteristic of condensation polymers, the investigator may reasonably invoke as many as half a dozen different basic reactions with as many rate constants as contributing to the final molecular weight distribution of an addition polymer. Four basic reaction types have been identified:

(a) Initiation, in which primary radicals R_1 are introduced into the system at a rate I. This may occur either by decomposition of an initiator molecule, $C \to 2R_1$, or by thermal or photochemical activation of monomer, $m \to R_1$. Whichever of these mechanisms is selected as being appropriate to the system in hand, the rate constant describing the kinetics of initiation will be written k_i.

(b) Propagation, in which the primary and all subsequent radicals add monomer in a step-by-step fashion to build up a long radical chain,

$$R_x + m \xrightarrow{k_p} R_{x+1}$$

(c) Termination, in which two radical chains interact to produce a "dead" molecule P which is inert to further propagation steps. This may occur in either of two different ways, by combination

$$R_x + R_z \xrightarrow{k_t} P_{x+z}$$

or by disproportionation

$$R_x + R_z \xrightarrow{k'_t} P_x + P_z$$

(d) Transfer, in which a radical chain abstracts a hydrogen atom from either a monomer molecule m or a solvent or impurity molecule S, thereby producing a primary radical R_1 and a dead polymer P. For simplicity, the reactivity of a radical produced from either m or S will be assumed to be the same, so that in our kinetic equations we use the symbol R_1 for either

$$R_x + m \xrightarrow{k_{tr}} R_1 + P_x$$

$$R_x + S \xrightarrow{k'_{tr}} R_1 + P_x$$

It must be admitted that there exists also the possibility of a transfer reaction between a radical and a dead polymer P_x, thus reactivating the dead polymer and leading to the production of branched molecules; but the complicated kinetics [1,2] of reactions of this type lie outside the scope of this article.

2. A Simplified Kinetic Scheme: The Steady State Approximation

While the mathematical methods about to be introduced are fully capable of handling the complete hierarchy of kinetic equations implied by all the reactions listed above, it is convenient at this point to simplify the program by ignoring several of the possible types of reactions. This has the advantage of illustrating the method without at the same time obscuring the details with an overcomplicated notation. I shall leave the mechanism of the initiation reaction open, and suppose only that radicals R_1 are being introduced into the system at a rate I. Termination will be supposed to occur by combination only, and transfer reactions will be neglected. With these restrictions, the kinetic equations to be solved are

$$\frac{dR_1}{dt} = I - k_p m R_1 - k_t R_1 \sum_1^\infty R_j \tag{17a}$$

$$\frac{dR_x}{dt} = k_p m R_{x-1} - k_p m R_x - k_t R_x \sum_1^\infty R_j \qquad x > 1 \tag{17b}$$

F. Molecular Weight Distribution Data

$$\frac{dP_x}{dt} = \frac{1}{2}k_t \sum_{1}^{x-1} R_j R_{x-j} \tag{17c}$$

$$\frac{dm}{dt} = -I - k_p m \sum_{1}^{\infty} R_j \tag{17d}$$

Equation (17d) contains a term $-I$ on the right-hand side, suggesting that monomer is consumed in the initiation reaction. This may not necessarily be true, as for instance in the case of peroxide-initiated polymerizations; but the possible confusion will be short-lived, for we shall shortly consider arguments leading to the conclusion that the rate of decrease of monomer due to its possible involvement in the initiation reaction is negligible with respect to its rate of decrease due to the propagation step.

To solve Eqs. (17), introduce the generating function

$$H(y, t) = \sum_{1}^{\infty} y^x R_x \tag{18}$$

Then, multiplying Eq. (17a) by y, Eq. (17b) by y^x, and summing over x we have

$$(\partial/\partial t)H(y, t) = yI + k_p m(y - 1)H(y, t) - k_t H(1, t)H(y, t) \tag{19}$$

Equation (19) becomes particularly simple in the case $y = 1$,

$$(\partial/\partial t)H(1, t) = I - k_t H^2(1, t) \tag{20}$$

and the reader may better appreciate the physical meaning of this equation by noting from Eq. (18) that $H(1, t) = \Sigma R_x$ is the total concentration of radicals of all kinds. The equation has been integrated in closed form [3] only in the special cases $I = $ constant and $I = $ constant $\times \exp(-k_i t)$, but the information derived from these particular solutions will enable us to solve the general case to an excellent approximation. In Eq. (20) we set $I = $ constant and assume an initial radical concentration $H(1, 0) = 0$. Then

$$H(1, t) = (I/k_t)^{1/2} \tanh(Ik_t)^{1/2} t$$

This important relationship shows that starting from an initial value of 0, the radical concentration increases monotonically to a limiting value of $(I/k_t)^{1/2}$, and that for times $t \gg (Ik_t)^{-1/2}$, $(\partial/\partial t)H(1, t)$ is practically zero. Now reactions between free radicals generally proceed with extraordinary swiftness, and we are justified in supposing k_t to be a very large number. This being the case, it would appear reasonable to expect that the total radical concentration $H(1, t)$ very rapidly achieves its asymptotic value, so rapidly in fact that we may assume with negligible error that $H(1, t) = (I/k_t)^{1/2}$ over the entire course of the reaction. This assumption is the basis of steady-state calculations of molecular weight distribution functions. Under steady-state conditions the rate of creation of radicals through the initiation mechanism is equal to their rate of destruction through terminations. Although we have come to this conclusion by a study of the particular initiation rate $I = $ constant, the evolution of free radical polymerizations to steady-state conditions is not limited to this case, and throughout the rest of our work we shall write

$$H(1, t) = [I(t)/k_t]^{1/2} \tag{21}$$

even when I is a slowly varying function of the time. It is only necessary to postulate that $t \gg (Ik_t)^{-1/2}$ for all t of experimental interest, which is to say that t is at least as great as the time elapsed between the start of the polymerization and the withdrawal of the first sample for molecular weight fractionation studies. If this inequality is satisfied, then the radical concentration is practically instantaneously responsive to slow changes in the initiation rate.

To return to our problem, substitute Eq. (21) into Eq. (19). The resulting linear differential equation may be integrated for initial radical concentrations all zero, $H(y, 0) = 0$,

$$H(y, t) = y \int_0^t I \exp\left\{\int_{t'}^t [(y - 1)k_p m - (Ik_t)^{1/2}] \, dt''\right\} dt' \tag{22}$$

It is not immediately apparent that Eq. (22) is consistent with Eq. (21), for setting $y = 1$ in Eq. (22) we have

$$H(1, t) = \int_0^t I(t') \exp\left\{-\int_{t'}^t (Ik_t)^{1/2} \, dt''\right\} dt' \tag{23}$$

To reconcile the two expressions, recall that to achieve steady-state conditions, $(Ik_t)^{1/2}$ must be a large number. The exponential in the integrand of Eq. (23) will therefore be significantly different from zero only in the neighborhood of $t' = t$, so that the major contribution to the integral must come from this region. Expand

$$\int_{t'}^t (Ik_t)^{1/2} \, dt'' = [I(t)k_t]^{1/2}(t - t') + \cdots$$

$$I(t') = I(t) + \cdots$$

and reject all but the leading terms. Then in place of Eq. (23) we have

$$H(1, t) = I(t)\int_0^t \exp\{-[I(t)k_t]^{1/2}(t - t')\} \, dt' \to I(t)\int_0^\infty \exp\{-[I(t)k_t]^{1/2}\tau\} \, d\tau$$

$$= \left[\frac{I(t)}{k_t}\right]^{1/2}$$

F. Molecular Weight Distribution Data

which is to be compared with Eq. (21). The approximations involved in these calculations form a standard method in the theory of asymptotic expansions where they are known as Watson's Lemma [4]. Exactly similar reasoning leads to the asymptotic formula

$$J_r \equiv \int_0^t I(t') \left\{ \int_{t'}^t k_p m \, dt'' \right\}^r \exp\left\{ -\int_{t'}^t (Ik_t)^{1/2} \, dt'' \right\} dt' \qquad (24)$$

$$\to r! \left[\frac{I(t)}{k_t} \right]^{1/2} \left\{ \frac{k_p m(t)}{[I(t)k_t]^{1/2}} \right\}^r$$

a result which we shall shortly require.

At this point we have the alternative possibilities of expanding Eq. (22) into a power series in y and in identifying R_x with the coefficient of y^x, or in developing the moments of R_x via the formula

$$\sum_1^\infty x^r R_x = \left. \left(y \frac{\partial}{\partial y} \right)^r H(y, t) \right|_{y=1} \qquad (25)$$

I choose the latter course, although either can be made to yield the same results. Substituting Eq. (22) into Eq. (25), we have in succession

$$\sum_1^\infty x R_x = J_0 + J_1$$

$$\sum_1^\infty x^2 R_x = J_0 + 3J_1 + J_2$$

$$\vdots$$

and I shall anticipate a result to be proved shortly to the effect that $J_r \gg J_{r-1}$, so that in general

$$\sum_1^\infty x^r R_x \to J_r$$

But the moments J_r are those of the uniquely defined distribution

$$R_x = [I(t)/k_p m(t)] \exp\{-x[I(t)k_t]^{1/2}/k_p m(t)\} \qquad (26)$$

and this we shall take to be the steady state solution of Eqs. (17a) and (17b). Turning now to Eq. (17d), we harken to chemical intuition which tells us that for the production of very large molecules the quantity of monomer consumed by propagation reactions must far exceed that consumed by initiations. The first term I on the right in Eq. (17d) must therefore be negligible with respect to the second, and we write instead of Eq. (17d)

$$\frac{dm}{dt} = -k_p m \sum_1^\infty R_j = -k_p m H(1, t) = -k_p m \left(\frac{I}{k_t} \right)^{1/2} \qquad (27)$$

But if I is negligible with respect to $k_p m (I/k_t)^{1/2}$, then we must also have $k_p m \gg (Ik_t)^{1/2}$, whence it follows from an examination of (24) that the condition $J_r \gg J_{r-1}$ upon which Eq. (26) was based is satisfied.

It is worth pausing at this point to survey the results so far achieved. The derivation of the radical distribution [Eq. (26)] depends on the existence in our polymerizing system of three widely different time scales. The coarsest of these time scales is t, which is of the order of the time required for the experimenter to perform his manipulations in the laboratory, withdrawing samples for analysis, etc. The second time scale is $(Ik_t)^{-1/2}$, which is the time required for the radical concentration starting from zero to build up steady-state conditions. Finest of all time scales is $(k_p m)^{-1}$, the time required for the addition of monomer to the growing radical chain. If these time scales are ordered by $t \gg (Ik_t)^{-1/2} \gg (k_p m)^{-1}$, then the radical distribution, responding practically instantaneously to slow changes in monomer concentration and initiation rate, is given by Eq. (26).

Our work is now almost complete, and we have only to integrate Eq. (17c) to obtain an expression for the distribution of dead polymer P_x. This might be done by making use of the convolution sum property [Eq. (6)] of generating functions, but it is simpler and under our present approximations just as accurate to replace the sum on the right in Eq. (17c) by an integral

$$\frac{dP_x}{dt} = \frac{1}{2} k_t \int_0^x R_j R_{x-j} \, dj$$

whence from (26)

$$dP_x/dt = \tfrac{1}{2} I [(Ik_t)^{1/2}/k_p m]^2 x \exp\{-x(Ik_t)^{1/2}/k_p m\} \tag{28}$$

Equivalently the moments Eqs. (1), (3) of the dead polymer distribution are defined by

$$\mu_r = \int_0^\infty M^r f(M) \, dM \to (m_0)^{-1} (M_0)^r \int_0^\infty x^r P_x \, dx$$

so that

$$d\mu_r/dt = \tfrac{1}{2}(r+1)! (m_0)^{-1} (M_0)^r I [k_p m/(Ik_t)^{1/2}]^r \tag{29}$$

Both Eqs. (28) and (29) are perfectly general whatever the initiation mechanism, so long as I is a slowly varying function of time.

To make further progress, an initiation mechanism must be chosen, and as a sample calculation I shall suppose that I is maintained constant. This could be accomplished, for instance, through use of an excess of a slowly decomposing peroxide initiator. Turning back to Eq. (27), we have for I constant

$$m = m_0 \exp\{-k_p (I/k_t)^{1/2} t\} \tag{30}$$

F. Molecular Weight Distribution Data

Substituting Eq. (30) into Eq. (29) and integrating over time from 0 to t,

$$\mu_0 = \frac{1}{2}\left(\frac{I}{m_0}\right)t$$

$$\mu_r = \frac{1}{2}\frac{(r+1)!}{r}(M_0)^r\left[\frac{k_p m_0}{(Ik_t)^{1/2}}\right]^{r-1}\left[1 - \exp\left\{-rk_p\left(\frac{I}{k_t}\right)^{1/2}t\right\}\right] \qquad r > 0$$

For comparison with experimental data, it is usually more convenient to write these last formulas in terms of the fraction of monomer m/m_0 left unconverted by time t,

$$\mu_0 = \frac{1}{2}\left[\frac{k_p m_0}{(Ik_t)^{1/2}}\right]^{-1}\ln\left(\frac{m_0}{m}\right)$$

$$\mu_r = \frac{1}{2}\frac{(r+1)!}{r}(M_0)^r\left[\frac{k_p m_0}{(Ik_t)^{1/2}}\right]^{r-1}\left[1 - \left(\frac{m}{m_0}\right)^r\right] \qquad r > 0$$

and these moments may be shown from Eq. (28) to be those of the molecular weight distribution function first obtained by Herrington and Robertson [5]

$$f(M) = \frac{1}{2}(M_0)^{-1}\left[\frac{k_p m_0}{(Ik_t)^{1/2}}\right]^{-2} \times \left\{\left[1 + \frac{M_0}{M}\frac{k_p m_0}{(Ik_t)^{1/2}}\right]\right.$$

$$\times \exp\left[-\frac{M}{M_0}\frac{(Ik_t)^{1/2}}{k_p m_0}\right] - \left[\frac{M_0}{M}\frac{k_p m_0}{(Ik_t)^{1/2}} + \frac{m_0}{m}\right]$$

$$\left.\times \exp\left[-\frac{M}{M_0}\frac{(Ik_t)^{1/2}}{k_p m_0}\frac{m_0}{m}\right]\right\}$$

The more complicated case in which the peroxide initiator decays exponentially with time has been labeled "dead end polymerization" by Tobolsky [6–8]. It may be treated in a similar way; but the general formulas are complex, and I shall quote only the first three moments. Denoting the initiator by C, its initial concentration by C_0, and supposing that

$$C \xrightarrow{k_i} 2R_1$$

then

$$C/C_0 = \exp(-k_i t)$$

$$I = 2fk_i C_0 \exp(-k_i t)$$

where f is the catalyst efficiency [9]—the fraction of primary radicals which actually initiate chains. There follows from Eq. (27),

$$m/m_0 = \exp\{-2k_p(2fC_0/k_i k_t)^{1/2}[1 - \exp(-k_i t/2)]\}$$

Substituting these into Eq. (29), integrating from 0 to t, and expressing the results in terms of m/m_0, we obtain

$$\mu_0 = \left(\frac{k_i k_t f C_0}{2}\right)^{1/2} (k_p m_0)^{-1} \ln\left(\frac{m_0}{m}\right) \times \left[1 - \left(\frac{k_i k_t}{2 f C_0}\right)^{1/2} (4k_p)^{-1} \ln\left(\frac{m_0}{m}\right)\right]$$

$$\mu_1 = M_0(1 - m/m_0)$$

$$\mu_2 = 6(M_0)^2 \left(\frac{k_p^2 m_0}{k_i k_t}\right) \exp\left\{-4k_p\left(\frac{2fC_0}{k_i k_t}\right)^{1/2}\right\} \times \left[\overline{\text{Ei}}\left\{4k_p\left(\frac{2fC_0}{k_i k_t}\right)^{1/2}\right\}\right.$$

$$\left. - \overline{\text{Ei}}\left\{4k_p\left(\frac{2fC_0}{k_i k_t}\right)^{1/2} - 2\ln\left(\frac{m_0}{m}\right)\right\}\right]$$

where $\overline{\text{Ei}}(z)$ is the exponential integral

$$\overline{\text{Ei}}(z) = \int_{-\infty}^{z} t^{-1} e^t \, dt$$

tabulated by Jahnke and Emde [10]. The list of moments could be continued, but they do not appear to be of sufficient interest to warrant it.

For both of the above problems the initiation rate was taken to be zeroth order in the monomer concentration. Our machinery, consisting of Eqs. (27) and (28) for the size distribution and Eqs. (27) and (29) for the moments, is equally applicable to a monomer-dependent initiation rate; and we might in particular take I to be first order in m,

$$I = k_i m$$

Then from Eq. (27)

$$m/m_0 = [1 + \tfrac{1}{2}k_p(k_i m_0/k_t)^{1/2} t]^{-2}$$

and from Eq. (29)

$$\mu_r = r!(M_0)^r [k_p(m_0/k_i k_t)^{1/2}]^{r-1}\{1 - (m/m_0)^{(1/2)(r+1)}\}$$

Correspondingly, if I is second order in m,

$$I = k_i m^2$$

then

$$m/m_0 = [1 + k_p m_0 (k_i/k_t)^{1/2} t]^{-1}$$

and

$$\mu_r = \tfrac{1}{2}(r+1)!(M_0)^r [k_p/(k_i k_t)^{1/2}]^{r-1}(1 - m/m_0) \tag{31}$$

3. The General Kinetic Scheme: Tabulation of Results

Complicated as is the appearance of some of the equations of Section II,B,2, they nevertheless represent simplifications of the more general hierarchy of kinetic possibilities listed in Section II,B,1. Without troubling the reader with details, I shall simply state that with neglect of none of the possible types of reactions, there corresponds to Eq. (28) for the size distribution

$$\frac{dP_x}{dt} = \left\{k_{tr}m + k'_{tr}S + k'_t\left[\frac{I}{(k_t + k'_t)}\right]^{1/2}\right\}\left[\frac{I}{(k_t + k'_t)}\right]^{1/2}$$

$$\times \left\{\frac{k_{tr}m + k'_{tr}S + [I(k_t + k'_t)]^{1/2}}{k_p m}\right\}$$

$$\times \exp\left\{-x\frac{k_{tr}m + k'_{tr}S + [I(k_t + k'_t)]^{1/2}}{k_p m}\right\} \qquad (32)$$

$$+ \frac{1}{2}\left(\frac{Ik_t}{k_t + k'_t}\right)\left\{\frac{k_{tr}m + k'_{tr}S + [I(k_t + k'_t)]^{1/2}}{k_p m}\right\}^2$$

$$\times x\exp\left\{-x\frac{k_{tr}m + k'_{tr}S + [I(k_t + k'_t)]^{1/2}}{k_p m}\right\}$$

and to Eq. (29) for the moments

$$\frac{d\mu_r}{dt} = (m_0)^{-1}(M_0)^r\left\{k_{tr}m + k'_{tr}S + k'_t\left[\frac{I}{(k_t + k'_t)}\right]^{1/2}\right\}$$

$$\times \left[\frac{I}{(k_t + k'_t)}\right]^{1/2}$$

$$\times r!\left\{\frac{k_p m}{k_{tr}m + k'_{tr}S + [I(k_t + k'_t)]^{1/2}}\right\}^r \qquad (33)$$

$$+ (m_0)^{-1}(M_0)^r\left[\frac{Ik_t}{(k_t + k'_t)}\right]\frac{1}{2}(r + 1)!$$

$$\times \left\{\frac{k_p m}{k_{tr}m + k'_{tr}S + [I(k_t + k'_t)]^{1/2}}\right\}^r$$

The approximations upon which these equations are based are the same as those leading to Eqs. (28) and (29). That is, for the achievement of steady state conditions and the production of long chains, we must have an ordering of time scales $t \gg [I(k_t + k'_t)]^{-1/2} \gg (k_p m)^{-1}$. Furthermore, because transfer reactions are not included in the general scheme, and because large molecules will result only if the propagation velocity greatly exceeds the transfer velocity, we must also have $(k_{tr}m)^{-1} \gg (k_p m)^{-1}$; $(k'_{tr}S)^{-1} \gg (k_p m)^{-1}$. These approximations lead to

$$dm/dt = -k_p m[I/(k_t + k'_t)]^{1/2} \tag{34}$$

$$S = \text{constant}$$

as replacing Eq. (27), and when once an initiation mechanism has been chosen, integration of Eq. (34) and substitution of the result into Eqs. (32) and (33) yield formal quadratures for the size distribution P_x and the moments μ_r. The necessary integrations are generally routine and usually tedious. A sample of some of the results is given below for the special case $k_{tr} = 0$, i.e., there is no transfer to monomer, but we permit transfer to S; and termination may be by either disproportionation or combination or both.

a. *Zeroth order initiation*: $I = \text{constant}$.

$$m/m_0 = \exp\{-k_p[I/(k_t + k'_t)]^{1/2} t\}$$

$$\mu_0 = \left\{ \frac{k'_{tr}S}{k_p m_0} + \frac{(k'_t + k_t/2)}{k_p m_0} \left[\frac{I}{(k_t + k'_t)} \right]^{1/2} \right\} \ln\left(\frac{m_0}{m}\right)$$

$$\mu_1 = M_0(1 - m/m_0)$$

$$\mu_2 = (M_0)^2 k_p m_0 \frac{k'_{tr}S + (k'_t + 3k_t/2)[I/(k_t + k'_t)]^{1/2}}{\{k'_{tr}S + [I(k_t + k'_t)]^{1/2}\}^2} \left[1 - \left(\frac{m}{m_0}\right)^2\right]$$

b. *First order initiation*: $I = k_i m$.

$$m/m_0 = \{1 + \tfrac{1}{2} k_p[k_i m_0/(k_t + k'_t)]^{1/2} t\}^{-2}$$

$$\mu_0 = \left(\frac{k'_{tr}S}{k_p m_0}\right) \ln\left(\frac{m_0}{m}\right) + \left[\frac{(2k'_t + k_t)}{k_p m_0}\right] \left[\frac{k_i m_0}{(k_t + k'_t)}\right]^{1/2}$$

$$\times \left[1 - \left(\frac{m}{m_0}\right)^{1/2}\right]$$

$$\mu_1 = M_0(1 - m/m_0)$$

$$\mu_2 = (M_0)^2 \frac{4k_p(k'_{tr}S)^3}{m_0 k_i^2 (k_t + k'_t)^2}$$

$$\times \bigg|_{q_2}^{q_1} \left\{\tfrac{1}{2}(1 + q)^2 - 3q + 3\ln(1 + q) + (1 + q)^{-1}\right\}$$

$$+ (M_0)^2 \frac{(4k'_t + 6k_t)k_p(k'_{tr}S)^3}{m_0 k_i^2 (k_t + k'_t)^3}$$

$$\times \bigg|_{q_2}^{q_1} \left\{\tfrac{1}{3}(1 + q)^3 - 2(1 + q)^2 + 6q - 4\ln(1 + q) - (1 + q)^{-1}\right\}$$

F. Molecular Weight Distribution Data

In the above equation for μ_2, the quantity in curly brackets preceded by a vertical line is to be evaluated between the upper limit $q_1 = [k_i m_0(k_t + k'_t)]^{1/2}/k'_{tr}S$ and the lower limit $q_2 = q_1(m/m_0)^{1/2}$ in the usual fashion after performing an integration.

c. *Second order initiation*: $I = k_i m^2$.

$$m/m_0 = \{1 + k_p m_0 [k_i/(k_t + k'_t)]^{1/2} t\}^{-1}$$

$$\mu_0 = \left(\frac{k'_{tr}S}{k_p m_0}\right) \ln\left(\frac{m_0}{m}\right) + \left[\frac{(k'_t + k_t/2)}{k_p}\right]\left[\frac{k_i}{(k_t + k'_t)}\right]^{1/2} \times \left(1 - \frac{m}{m_0}\right)$$

$$\mu_1 = M_0(1 - m/m_0)$$

$$\mu_2 = (M_0)^2 \frac{2k_p k'_{tr} S}{m_0 k_i (k_t + k'_t)} \bigg|_{q_2}^{q_1} \{\ln(1 + q) + (1 + q)^{-1}\}$$

$$+ (M_0)^2 \frac{k_p k'_{tr} S(2k'_t + 3k_t)}{m_0 k_i (k_t + k'_t)^2} \bigg|_{q_2}^{q_1} \{q - 2\ln(1 + q) - (1 + q)^{-1}\}$$

for which

$$q_1 = \frac{m_0[k_i(k_t + k'_t)]^{1/2}}{k'_{tr}S} \quad \text{and} \quad q_2 = q_1\left(\frac{m}{m_0}\right)$$

The reader will readily appreciate after these examples how complicated the various special cases can become. Despite their messy appearance, however, the labor involved in carrying out the necessary integrations and in evaluating numerically the resulting formulas is not excessive. Other results of this nature are tabulated in the literature [1,2,11–18].

C. Ionic Polymerizations

1. *Living Polymers*

The sequence of elemental reactions by which an electrically charged species may add monomer to build up a long chain is kinetically similar to that for free radical polymerizations. The investigator is again presented with an initiation reaction in which primary ions R_1 are introduced into the system at a rate I moles liter^{-1} second^{-1}. These and all successive ions add monomer m in a series of propagation reactions

$$R_x + m \xrightarrow{k_p} R_{x+1}$$

Furthermore, chain transfer reactions

$$R_x + S \to P_x + R_1$$

which produce a dead polymer P_x by interaction of R_x with a chain transfer agent S, are common, as are termination reactions yielding uncharged products [19]. If the steady-state hypothesis is valid, the kinetics of these most typical ionic polymerizations may be handled along the lines developed for free radical polymerizations in Section II,B. Of peculiar interest to the kineticist, however, is the fact that ionic polymerizations are known for which the termination reaction is absent. Such polymerizations cease only upon the exhaustion of the available supply of monomer, and the addition of fresh monomer to the system will cause the still active ends of the unterminated chains to resume their growth. Szwarc has coined the appropriate name of "living polymers" for such systems, and it is to them that we shall direct our attention in the next several paragraphs. From the mathematical point of view the interesting feature of living polymers is the fact that the absence of a termination reaction renders invalid the steady-state hypothesis upon which all of the calculations in Section II,B were based, and as a result the distributions predicted are of unusual shape. I shall not dwell too much upon these matters, for not many data on the kinetics of formation of living polymers exist.

2. A Pulse Initiation

The simplest possible kinetic hypothesis leading to the formation of a living polymer was considered by Flory [20] many years ago. We suppose that at time $t = 0$ there is a single pulse of initiations leading to the creation of R_1^0 moles per liter of active species, and that for the entire remainder of the polymerization the initiation rate is zero. The sequence of propagation steps leads to the set of differential equations

$$\begin{aligned} dR_1/dt &= -k_p m R_1 \\ dR_x/dt &= k_p m (R_{x+1} - R_x) \qquad x > 1 \\ dm/dt &= -k_p m \sum_1^\infty R_j \end{aligned} \qquad (35)$$

and these are to be integrated under the initial conditions $R_1 = R_1^0$; $R_x = 0$, $x > 1$; $m = m_0$ at $t = 0$. As was the case for condensation and free radical polymerizations, the use of generating functions is useful here and we define

$$H(y, t) = \sum_1^\infty y^x R_x \qquad (36)$$

F. Molecular Weight Distribution Data

whence we have in place of Eqs. (35)

$$(\partial/\partial t)H(y, t) = (y - 1)k_p mH(y, t)$$

$$dm/dt = -k_p mH(1, t)$$

$$\left.\begin{array}{l} H(y, 0) = yR_1^0 \\ m = m_0 \end{array}\right\} \text{at } t = 0$$

Setting $y = 1$, there results

$$H(1, t) = R_1^0$$

so that

$$m/m_0 = \exp(-k_p R_1^0 t)$$

and

$$H(y, t) = yR_1^0 \exp[(y - 1)u] \tag{37}$$

for which

$$u = \int_0^t k_p m \, dt = \left(\frac{m_0}{R_1^0}\right)\left(1 - \frac{m}{m_0}\right)$$

The size distribution R_x may be obtained by expanding Eq. (37) into a power series in y and extracting the coefficient of y^x,

$$R_{x+1} = R_1^0 (x!)^{-1} u^x e^{-u} \tag{38}$$

Equation (38) is the familiar Poisson distribution, and it states that the living polymer resulting from a pulse initiation is practically monodisperse of molecular weight $M = M_0(m_0/R_1^0)(1 - m/m_0)$. To make this point even clearer it is useful to replace the Poisson function [Eq. (38)] by its Gaussian approximation, valid for large x,

$$R_x \to R_1^0 (2\pi x)^{-1/2} \exp[-(u - x)^2/2x]$$

so that R_x is sharply peaked at $x = u$ and furthermore 95% of all chains are contained in the size region $u - 2\sqrt{u} \leq x \leq u + 2\sqrt{u}$. Thus, for example, a peak size of $x = 10,000$ has 95% of all chains lying between lengths 9,800 and 10,200. Such distributions have been experimentally observed [21,22].

3. More General Initiation Mechanisms

The Poisson distribution of molecular sizes which results from a pulse initiation is an interesting special case, but we cannot in general expect that all initiation reactions will be completed within the first few moments of the polymerization process; and a more complete kinetic description of the formation of living polymers is contained in the equations

$$dR_1/dt = I - k_p m R_1 \tag{39a}$$

$$dR_x/dt = k_p m(R_{x-1} - R_x) \quad x > 1 \tag{39b}$$

$$dm/dt = -I - k_p m \sum_1^\infty R_j \tag{39c}$$

together with $R_x = 0$; $m = m_0$ at $t = 0$. As was the case with Eq. (17d), the reader will note that the right-hand side of Eq. (39c) contains a term $-I$ implying that monomer is consumed by the initiation reaction. This may not necessarily be the case, but whether it is or not is of small consequence, for long chains can arise only if the amount of monomer consumed by propagation reactions greatly exceeds whatever monomer may be used up in initiations, so that $-I$ in Eq. (39c) is usually negligible.

Introducing the generating function [Eq. (36)], the set of differential Eqs. (39) is contained in

$$(\partial/\partial t)H(y, t) = yI + (y - 1)k_p m H(y, t)$$

$$dm/dt = -I - k_p m H(1, t) \to -k_p m H(1, t) \tag{40}$$

$$\left.\begin{array}{l} H(y, 0) = 0 \\ \\ m = m_0 \end{array}\right\} \text{at } t = 0$$

The first of these may be integrated immediately,

$$H(y, t) = y \int_0^t I(t') \exp\left\{(y - 1)\int_{t'}^t k_p m \, dt''\right\} dt' \tag{41}$$

and it is convenient at this point to simplify the notation by changing the variable of integration in Eq. (41) to synthetic time u defined by

$$u = \int_0^t k_p m \, dt'' \qquad v = \int_0^{t'} k_p m \, dt'' \tag{42}$$

whence

$$H(y, t) = y \int_0^u \left(\frac{I}{k_p m}\right) \exp\{(y - 1)(u - v)\} \, dv \tag{43}$$

Picking out the coefficient of y^x, we have for the size distribution

$$R_{x+1} = (x!)^{-1} \int_0^u \left(\frac{I}{k_p m}\right)(u - v)^x \exp[-(u - v)] \, dv \tag{44}$$

Equation (44) is not in a form which is suitable for computation, and because we are primarily interested in R_x when x is a large number, exact evaluation of the integral by conventional techniques is impracticable. There exists, however, an approximate method whose precision increases with x and which is based upon the consideration discussed in Section II,C,2 that the Poisson factor $(x!)^{-1}(u - v)^x \exp[-(u - v)]$ which occurs in Eq. (44) attains a sharp and narrow maximum in the neighborhood of $v = u - x$. If this point is included in the range of integration, practically the entire contribution to the integral will come from the immediate neighborhood of this maximum. If $v = u - x$ falls outside the range of integration, the value of the integral becomes negligibly small. Approximate integrations based upon a pronounced local maximum in an integrand are known as steepest descents [23], and lead in this case to the good approximation

$$R_x \to \frac{I(u - x)}{k_p m(u - x)} \frac{1}{2}\left\{1 + \text{erf}\left[\frac{(u - x)}{(2x)^{1/2}}\right]\right\} \tag{45}$$

in which erf z is the error function defined by

$$\text{erf } z = \left(\frac{2}{\sqrt{\pi}}\right)\int_0^z \exp(-z^2)\, dz$$

It is tabulated by Jahnke and Emde [24].

Equation (45) states that if I and m are known explicitly *as functions of synthetic time* u, then the size distribution R_x can be calculated. The quantity

$$\tfrac{1}{2}\{1 + \text{erf}[(u - x)/(2x)^{1/2}]\}$$

which occurs in Eq. (45) acts as a cutoff factor for the distribution, for if $x \le u - 2\sqrt{u}$, the cutoff factor is 1, while if $x \ge u + 2\sqrt{u}$, the cutoff factor is essentially zero. We should therefore expect in our distribution of living polymers that practically all chains have length x less than u.

To illustrate the usefulness of Eq. (45), consider an initiation which is first order in monomer concentration,

$$I = k_i m \tag{46}$$

A glance at Eq. (45) immediately tells us the general shape of the size distribution R_x, for

$$R_x = k_i/k_p$$

for all $x \le u - 2\sqrt{u}$, after which the distribution plunges steeply to 0 (see Fig. 2). It remains only to relate u and t so that the distribution may be

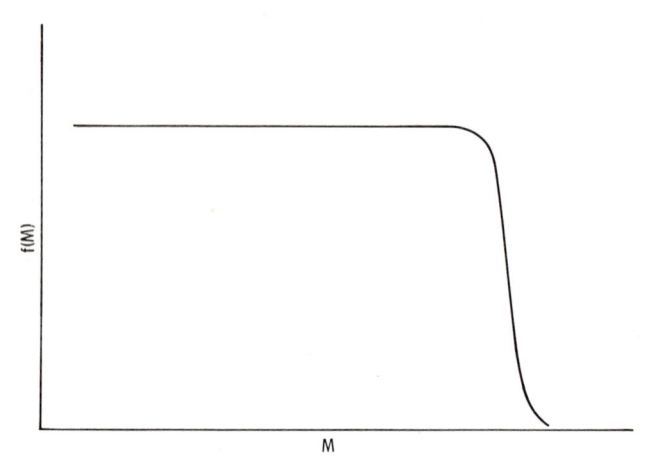

FIG. 2. Number density distribution $f(M)$ for living polymers: initiation first order in monomer.

known as a function of time, or alternatively as a function of monomer consumption. From Eqs. (40), (41), and (46) we have

$$\frac{dm}{dt} = k_p m H(1, t)$$
$$= -k_p m \int_0^t I \, dt' = -k_p m \int_0^t k_i m \, dt'$$

Dividing through both sides by $k_p m$ and introducing synthetic time $du = k_p m \, dt$,

$$\frac{dm}{du} = -\left(\frac{k_i}{k_p}\right) \int_0^u dv = -\left(\frac{k_i}{k_p}\right) u$$

so that in terms of synthetic time,

$$m = m_0 - \tfrac{1}{2}(k_i/k_p) u^2$$

By inversion of the integral [Eq. (42)], it follows that

$$t = \int_0^u \frac{du}{k_p m}$$
$$= \left(\frac{2}{k_i k_p m_0}\right)^{1/2} \operatorname{arctanh}\left[\left(\frac{k_i}{2 k_p m_0}\right)^{1/2} u\right]$$

so that explicitly

$$u = (2 k_p m_0 / k_i)^{1/2} \tanh[(k_i k_p m_0 / 2)^{1/2} t]$$

F. Molecular Weight Distribution Data

and

$$m/m_0 = 1 - \tanh^2[(k_i k_p m_0/2)^{1/2} t]$$

At $t \to \infty$ when the supply of monomer has been exhausted,

$$u \to (2k_p m_0/k_i)^{1/2}$$

and this should be the maximum chain length x which can be produced in a living polymer by a first order initiation process.

4. Additional Results

To bring our subject to a close, I shall list without derivation some additional results for living polymers.

1. *Initiation zero order in monomer:* $I = $ constant. Then

$$u = m_0(\pi k_p/2I)^{1/2} \operatorname{erf}[(k_p I/2)^{1/2} t]$$

$$m/m_0 = \exp[-(k_p I/2) t^2]$$

A parametric representation of m as a function of u is calculable from these equations, and when the result is substituted into Eq. (45) the size distribution takes on the form sketched in Fig. 3 for the special case $t \to \infty$, maximum chain length $\to m_0(\pi k_p/2I)^{1/2}$. The final part of the polymerization is

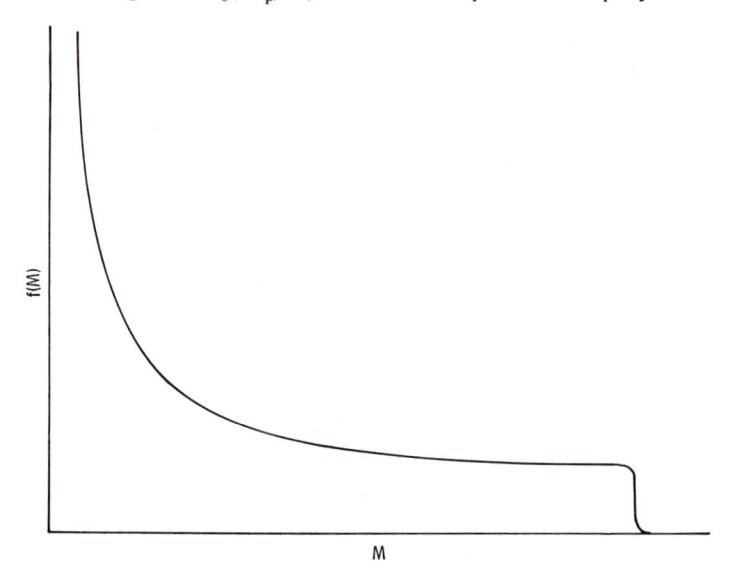

FIG. 3. Number density distribution $f(M)$ for living polymers: initiation zero order in monomer.

apparently occupied in the production of short chains, hence the steep rise in this curve near the origin. While qualitatively correct, this apparent singularity near $x = 0$ is a result of our assumption that the quantity of monomer consumed by propagations exceeds that consumed by initiations. If the initiation rate is maintained constant and the polymerization allowed to proceed so far as to deplete seriously the available supply of monomer, then this assumption must fail, leading to appreciable errors in the distribution near $x = 0$.

b. *Initiation second order in monomer*: $I = k_i m^2$. For this case

$$u = 2(k_p/k_i)^{1/2} \left[\arctan\{\exp[m_0(k_i k_p)^{1/2} t]\} - \pi/4\right]$$

$$m/m_0 = \cos[(k_i/k_p)^{1/2} u]$$

$$= 2 \exp[m_0(k_i k_p)^{1/2} t]/\{1 + \exp[2m_0(k_i k_p)^{1/2} t]\}$$

$$R_x = (k_i/k_p) \cos[(k_i/k_p)^{1/2}(u - x)] \times \tfrac{1}{2}\{1 + \text{erf}[(u - x)/(2x)^{1/2}]\}$$

and at $t \to \infty$, maximum chain length $= (\pi/2)(k_p/k_i)^{1/2}$, the distribution has the shape drawn in Fig. 4.

Further problems involving the molecular weight distributions of living polymers have been treated elsewhere [23,25–28,28a,b].

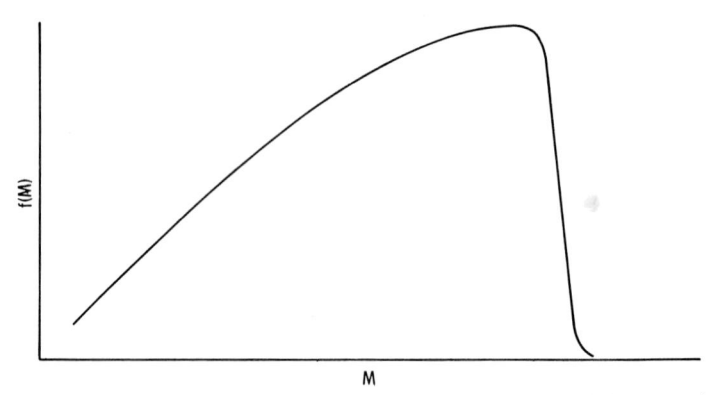

FIG. 4. Number density distribution $f(M)$ for living polymers: initiation second order in monomer.

III. Numerical Methods in the Handling of Molecular Weight Distribution Data

The objectives of Section II were in the main theoretical and the mathematical analysis of a postulated sequence of chemical reactions led to the prediction of a size distribution P_x giving the concentration of chains containing x monomer units as a function of time. Our approach in Section III

F. MOLECULAR WEIGHT DISTRIBUTION DATA

will be quite different, for we shall be interested in developing numerical methods leading to the accurate computation of both the molecular weight distribution $f(M)$ and of its moments μ_r from fractionation data. When this has been achieved, contact between the theoretical prediction P_x and the experimentally observed $f(M)$ might be made through Eq. (2) and the substitution $x = M/M_0$, in which M_0 is the molecular weight of a single monomer unit. This most direct method will actually prove to be less convenient in practice than a comparison between moments μ_r calculated from experimental data via Eq. (1) and those predicted theoretically from Eq. (3).

The numerical procedure to be described in the following pages has not heretofore been published in detail [28c]. There is a consequent disproportionate neglect of other approaches to the same problem, for which the reader is referred to the review by Tung in Chapter E of this volume.

A. THE CUMULATIVE DISTRIBUTION AND DATA SMOOTHING

The number density distribution function $f(M)$ is not a direct result of a series of fractionation experiments. The direct result of the calculations described in Chapter E is the cumulative distribution

$$F(M) = \int_0^M Mf(M)\, dM$$

which for any given sample measures the total weight $F(M)$ of polymer which possesses a molecular weight equal to or less than M. It is conventional to normalize $F(M)$ so that $F(M) \to 1$ as $M \to \infty$, and in this case $F(M)$ is the weight fraction of all chains whose molecular weight does not exceed M. An example of such a cumulative distribution is given in Fig. 5.

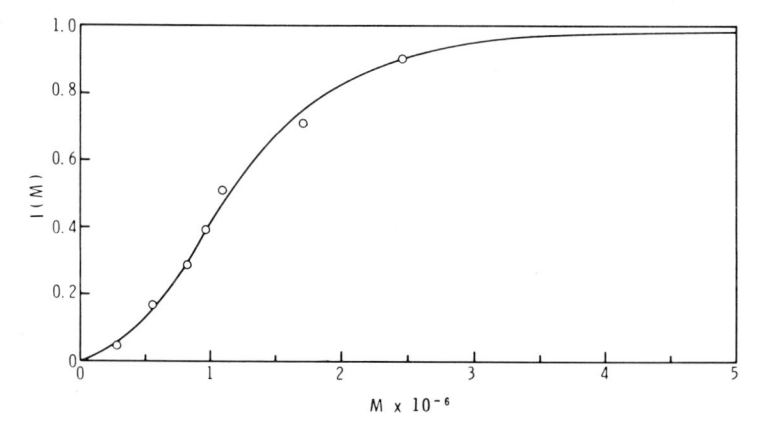

FIG. 5. Polystyrene thermally polymerized at 60.0°C to a monomer conversion of 77.2%.

The data are for polystyrene thermally polymerized at 60°C and are drawn from the doctoral dissertation of Cantow [29].

As is common with all experimental data, the points do not lie on a perfectly smooth curve, but it is essential to the method about to be introduced that some sort of decision be made as to the best curve which can be drawn through the points starting with $F(0) = 0$ and proceeding asymptotically to $F(\infty) = 1$. From the statistical point of view, appropriate data smoothing is performed by any of the various least-squares techniques, of which the Fourier method of Lanczos [30] is recommended. Generally speaking, however, such techniques are not adapted to hand computation, and I prefer to bypass the whole problem by recourse to the chemist's traditional method of data smoothing "by eye," which is to say that with French curve and pencil he draws the best curve through the points which his intuitive feel for the data may direct.

A few further remarks upon the limitations of the data are in order. Although we know that $F(M) \to 1$ asymptotically as $M \to \infty$, it is inevitable that we shall be able to measure $F(M)$ experimentally only over a finite range of M. The calculation of the moments

$$\mu_r = \int_0^\infty M^r f(M)\, dM \qquad (47)$$

requires on the contrary an extension of the data over the infinite range, and a result of our computational schedule must then be a suitable extrapolation technique out to $M \to \infty$. No such extrapolation technique, however, can generate information which is not present in the data, and our ignorance of $F(M)$ at large values of M forces us to abandon any hope of an accurate estimate of the higher moments, all of which depend increasingly on accurately known values of $F(M)$ for M large.

B. Laguerre Polynomials

Our numerical analysis of molecular weight distribution data starts with an expansion in normalized, associated, Laguerre[2] polynomials $p_n^s(z)$,

$$Mf(M) = a(aM)^s \exp(-aM) \sum_0^\infty c_n p_n^s(aM) \qquad (48)$$

Such expansions are common in statistics, where they are known as Poisson–Charlier series [32]. Here $a > 0$ and $s > -1$ are scaling parameters

---

[2] The normalized, associated, Laguerre polynomials $p_n^s(z)$ are proportional to the associated Laguerre polynomials $L_n^s(z)$ as conventionally defined by Morse and Feshbach [31].

$$p_n^s(z) = \{(n!)^{1/2}/[\Gamma(n+s+1)]^{3/2}\} L_n^s(z)$$

F. Molecular Weight Distribution Data

which may be chosen at pleasure, and $f(M)$ shall be said to be determined when the unknown coefficients c_n are determined. The normalized, associated, Laguerre polynomials are of the form

$$p_n^s(z) = \sum_{r=0}^{n} h_{nr} z^r \qquad (49)$$

and the first several members of the set are

$$\begin{aligned} p_0^s(z) &= \Gamma^{-1/2}(s+1) \\ p_1^s(z) &= \Gamma^{-1/2}(s+2)(s+1-z) \\ p_2^s(z) &= 2^{-1/2}\Gamma^{-1/2}(s+3)[(s+2)(s+1) - 2(s+2)z + z^2] \end{aligned} \qquad (50)$$

in which $\Gamma(s)$ is the gamma function. By a well-known theorem, the coefficients c_n are related to the distribution $f(M)$ by

$$c_n = \int_0^\infty M f(M) p_n^s(aM) \, dM \qquad (51)$$

and the usual procedure in these expansions is to compute the c_n from Eq. (51) for an analytically known $f(M)$, following which their numerical values are inserted into Eq. (48). The reader should note from Eq. (49) that what this implies in practice is a knowledge of all the μ_r, for substituting Eq. (49) into Eq. (51),

$$c_n = \sum_{r=0}^{n} h_{nr} a^r \mu_{r+1} \qquad (52)$$

so that each c_n is a linear combination of all moments from μ_1 to μ_{n+1} inclusive.

In the present instance, of course, we know neither $f(M)$ nor its moments. The set of linear equations [Eq. (52)] is, nevertheless, useful; for our approach in the following will be to develop an approximate algebraic method of calculating the c_n directly from $F(M)$ and in using the numbers so obtained to compute the moments by inversion of Eq. (52). In finding the c_n we may be said also to have found $f(M)$, for $f(M)$ is then determined by Eq. (48).

C. Finite Laguerre Expansions and Gaussian Integration

At this point I introduce the essential approximation which will limit all our subsequent calculations. It is to truncate the series Eq. (48) after the fifth term, so that we shall require the c_n only out to $n = 4$. Note that this presumes that $f(M)$ is to within experimental accuracy determined by the five moments $\mu_1, \cdots, \mu_5$, and that we abandon any hope of obtaining efficient estimates of higher moments. In place of Eq. (48) we thus have approximately

$$Mf(M) \sim a(aM)^s \exp(-aM) \sum_{n=0}^{4} c_n p_n^s(aM) \qquad (53)$$

and integrating both sides of Eq. (53) from 0 to M there results

$$\int_0^M Mf(M)\,dM = F(M) \sim \sum_{n=0}^{4} c_n a \int_0^M (aM)^s \exp(-aM) p_n^s(aM)\,dM \qquad (54)$$

This equation can be simplified, but before doing so let us observe that there are five terms on the right, so that we can make left- and right-hand sides agree exactly in at least five distinct points M_k ($k = 1, \cdots, 5$) by inversion of a set of linear equations in the c_n. The goodness of fit of the finite expansion from Eq. (53) to $F(M)$ for points M other than the M_k will vary with the choice of the M_k, and it is to our advantage to choose the M_k so as to reduce the residual error to a minimum. This problem has been solved by Gauss [33], who has shown that the optimum choice of data points is to take $M_k = \lambda_k/a$, where the λ_k are the five roots of the associated Laguerre polynomial of fifth degree, i.e.,

$$p_5^s(\lambda_k) = 0 \qquad k = 1, \cdots, 5$$

These roots may be shown always to be real, positive, and distinct.

In Eq. (54) I therefore set M equal to each $M_k = \lambda_k/a$ in turn and take advantage of known properties [31] of the normalized, associated, Laguerre polynomials to derive a set of five linear equations in five unknown c_n

$$F(M_k) = \sum_{n=0}^{4} g_{kn} c_n \qquad k = 1, \cdots, 5$$

in which the matrix $G = (g_{kn})$ is defined by

$$g_{k0} = \Gamma^{-1/2}(s+1) \int_0^{\lambda_k} z^s e^{-z}\,dz$$

$$g_{kn} = n^{-1/2}(\lambda_k)^{s+1} \exp(-\lambda_k) p_{n-1}^{s+1}(\lambda_k) \qquad n \geq 1$$

These matrix elements may all be computed independently of the $F(M_k)$, and the matrix G inverted and pretabulated. The investigator need thus only operate with G^{-1} on the five data points $F(M_k)$ according to the usual rules of matrix algebra in order to obtain the five coefficients c_n; $n = 0, \cdots, 4$.

Up to this point, I have concentrated on finding the coefficients c_n by finite operations on a discrete set of data. The practical result of these calculations is to differentiate the experimental curve $F(M)$, for the differential or weight distribution curve $Mf(M)$ may be synthesized by substituting the numerical values of c_n into Eq. (53). For theoretical purposes,

F. Molecular Weight Distribution Data

however, the μ_n are the more interesting quantities, and it would be preferable to have our method give the moments directly without the intermediate step of calculating the coefficients c_n. The necessary adjustment in our calculation schedule is easy, for from Eq. (52), the c's are related to the μ's by operation with a matrix $H = (h_{nr})$ which is the table of coefficients of the normalized, associated, Laguerre polynomials [Eq. (49)]. A little matrix algebra shows that if the c_n are to be calculated by operating on five data points $F(M_k)$ with the matrix G^{-1}, then the quantities $a^r \mu_{r+1}$ may be obtained by operating on the same data points with the matrix $H^{-1}G^{-1} \equiv Q$, which can also be pretabulated. This is the schedule which has been chosen, and the first matrix listed in each table in the Appendix is $Q = (q_{rk})$, suitable for direct calculation of the moments from five data points $F(M_k)$,

$$a^r \mu_{r+1} = \sum_1^5 q_{rk} F(M_k) \tag{55}$$

Also tabulated is the matrix H, so that should a plot of either the weight distribution $Mf(M)$ or the number density distribution $f(M)$ be required, the c_n may be calculated from the $a^r \mu_{r+1}$ by matrix multiplication with H [Eq. (52)], following which substitution into Eq. (53) and use of tables of the normalized, associated, Laguerre functions,

$$z^s e^{-z} p_n^s(z)$$

permits the construction of $Mf(M)$.

D. The Choice of Scaling Parameters; Iterations; the Zeroth Moment

So far I have said nothing about the choice of the scaling parameters a and s. Because we are using a finite series, we shall obtain the best results if it is made as rapidly convergent as possible; and this may be achieved if we select the values of a and s in such a way as to make c_1 and c_2 small. The reason for this is that the function $(aM)^s \exp(-aM)$ which appears as a factor on the right-hand side of Eq. (53) will in this way be made to fit as closely as possible the curve $Mf(M)$, and that the resulting corrections to this fit which arise from the subsequent series of associated, Laguerre polynomials will be made as small as possible.

A glance at Eqs. (50) and (52) shows that c_1 and c_2 vanish if

$$s = \frac{2 - \mu_3 \mu_1/(\mu_2)^2}{\mu_3 \mu_1/(\mu_2)^2 - 1} \qquad a = \frac{(s+1)\mu_1}{\mu_2} \tag{56}$$

An immediate objection to the practical realization of this choice is that the first three moments μ_1, μ_2, μ_3 must apparently be known before we can begin the calculation of a and s; but the difficulty is not a serious one, for a little experience will enable the user of this method to guess suitable starting approximations for a and s, to calculate with their use the first three moments, and then to use these first approximations to the true moments as the basis for a more refined choice of a and s through Eqs. (56). Because s is restricted to those discrete values listed in the tables, it will be possible to satisfy the first of Eqs. (56) only approximately; but the second may be satisfied exactly once s has been chosen. It is recommended that for the first iteration, the value $s = 1$ be selected together with any value of a which scales the data points $M_k = \lambda_k/a$ into the experimental range. If the reader will have the patience to follow through the steps of the numerical example worked in Section III,E, he will obtain a better grasp of the course of these iterations than could be given here by a descriptive essay.

The finite Laguerre expansion and its optimization through Gaussian integration and efficient choice of scaling parameters can be carried through to yield estimates of the five experimental moments $\mu_1, \cdots, \mu_5$. The zeroth moment

$$\mu_0 = \int_0^\infty f(M)\, dM$$

is, however, also of practical interest, for it enters into the definition of the number average molecular weight $\overline{M}_n = \mu_1/\mu_0$. By dividing both sides of Eq. (53) by M, integrating from 0 to ∞, and simplifying the result from known properties [31] of the associated Laguerre polynomials,

$$a^{-1}\mu_0 = \Gamma(s) \sum_0^4 c_n \left[\frac{n!}{\Gamma(n + s + 1)}\right]^{1/2}$$

so that μ_0 is expressible as a linear combination of the c_n. By matrix methods it can hence also be expressed as a linear combination of the data points $F(M_k)$, and this fact has led to the tabulation of the single row matrix ω_k in terms which

$$a^{-1}\mu_0 = \sum_1^5 \omega_k F(M_k) \tag{57}$$

The ω_k might have been appended to the tabulated square matrices Q as a zeroth row, but because the calculation of μ_0 should be performed only after the iterations leading to a proper choice of scaling parameters, I judged such a juxtaposition to be too confusing.

E. A Numerical Recipe and a Worked Example

Whatever may have been the reader's impression of the complexity of the above operations, the method is quite simple in practice. Needed are a desk calculator and an hour or more of time.

F. MOLECULAR WEIGHT DISTRIBUTION DATA

The Appendix is divided into a total of seven tables, one for each value of s starting with $s = 0$ and proceeding by steps of 0.5 to $s = 3$. The first row lists the five roots λ_k of $p_5^s(z)$. Next follows the 5×5 matrix Q and immediately thereafter (except in the case $s = 0$) the five quantities ω_k to be used for calculating μ_0. Finally, after the 5×5 triangular matrix H, is a brief tabulation of the first five normalized, associated, Laguerre functions

$$z^s e^{-z} p_n^s(z)$$

for values of z between 0 and 10.

Working instructions for the use of these tables are the following:

(a) The cumulative distribution $F(M)$ is plotted on finely graded graph paper and a smooth curve drawn through the points in the manner of Fig. 5. I shall assume that $F(M)$ has been normalized in such a way that $F(M) \to 1$ as $M \to \infty$.

(b) Set $s = 1$ and turn to the table so labeled in the Appendix. The five roots $\lambda_1, \cdots, \lambda_5$ of $p_5^1(z)$ are listed across the top of the page, and we are to choose a in such a way that the five numbers $M_k = \lambda_k/a$ fall roughly into the observed range of the data. For high polymers this will generally mean $a = 10^{-4}$ to $a = 10^{-6}$. As an example, the initial choice of $a = 10^{-6}$ is appropriate for Fig. 5. Calculate the M_k and read graphically the corresponding ordinates $F(M_k)$. For Fig. 5 this results in the following tabulation.

$s = 1; a = 10^{-6}$

$M_k (\times 10^{-6})$	$F(M_k)$
0.617	0.182
2.113	0.855
4.611	0.980
8.399	1
14.260	1

The reader should note that for the two points $M_4 = 8.339 \times 10^6$ and $M_5 = 14.260 \times 10^6$ which fall off the range of the graph, I have assumed ordinates $F(M_4) = F(M_5) = 1$.

(c) Next calculate μ_1 from the first row of the matrix Q by letting it operate on the column of quantities $F(M_k)$ according to the usual multiply–add rule of matrix multiplication, Eq. (55). In the above example this means

$$\mu_1 = (0.0099700)(0.182) - (0.0084367)(0.855) + (0.016853)(0.980)$$
$$- (0.075332)(1) + (1.0633)(1) = 1.000$$

This is a not unexpected result, for normalization of $F(M)$ to 1 as $M \to \infty$

means by definition that $\mu_1 = 1$, so that this part of the calculation is little more than a check on the accuracy of our procedure.

(d) By operating on the $F(M_k)$ with the second row of the matrix Q we obtain $a\mu_2$

$$a\mu_2 = -(0.86178)(0.182) - (2.1219)(0.855) - (2.7633)(0.980)$$
$$- (5.9993)(1) + (12.029)(1) = 1.351$$

whence it follows that $\mu_2 = 1.351 \times 10^6$, and we might continue this procedure through the remaining three rows of Q to obtain successively $a^2\mu_3$, $a^3\mu_4$, and $a^4\mu_5$. It is essential to the accuracy of the higher moments, however, that the calculation be broken off at this point to rescale the data. We retain $s = 1$, but modify a from Eq. (56) and our preliminary estimates of the first two moments,

$$a = (s + 1)\mu_1/\mu_2 = 2(1.000)/(1.351 \times 10^6) = 1.480 \times 10^{-6}$$

(e) The data points are rescaled $M_k = \lambda_k/a = \lambda_k/(1.480 \times 10^{-6})$ and new ordinates $F(M_k)$ are read from the graph.

$s = 1; a = 1.480 \times 10^{-6}$

$M_k (\times 10^{-6})$	$F(M_k)$
0.417	0.100
1.428	0.643
3.116	0.963
5.675	1
9.635	1

As in (c) and (d), the first two moments are calculated from these new data by operating on the $F(M_k)$ with the first two rows of Q. I find $\mu_1 = 1.000$ and $\mu_2 = 1.297 \times 10^6$, so that a third estimate of a is

$$a = 2(1.000)/(1.297 \times 10^6) = 1.542 \times 10^{-6}$$

When the M_k are rescaled to this choice of a and the first two moments recalculated, we obtain substantial agreement with the results of the second iteration,

$$\mu_1 = 1.000 \qquad \mu_2 = 1.294 \times 10^6$$

The calculation is therefore continued to the third moment by operating on $F(M_k)$ with the third row of Q, using the identical multiply–add rule which has governed all previous computations of this sort. I find $a^2\mu_3 = 5.52$, so that

$$\mu_3 = (5.52)/(1.542 \times 10^{-6})^2 = 2.32 \times 10^{12}$$

F. Molecular Weight Distribution Data

(f) Having successfully scaled a, we must now scale s, and a revised estimate for this quantity is calculated through Eq. (56) using our latest estimates for the first three moments,

$$\mu_3\mu_1/(\mu_2)^2 = (2.32)(1.000)/(1.294)^2 = 1.386$$
$$s = (2 - 1.386)/(1.386 - 1) = 1.59 \sim 1.5$$

Ideally we should thus choose $s = 1.59$, but because we are restricted by the tables to integer and half-integer values of s, we turn instead to the tables for $s = 1.5$. For our new a we have

$$a = (s + 1)\mu_1/\mu_2 = (2.5)(1.000)/(1.294 \times 10^6) = 1.932 \times 10^{-6}$$

(g) The calculations from this point on are substantially the same as those heretofore, except that now we must scale the data $M_k = \lambda_k/a$ and operate with the matrix Q tabulated for $s = 1.5$. As in (c) and (d), the first iterations for $s = 1.5$ are to calculate the first two moments only and to rescale a holding s fixed. With a satisfactorily readjusted, calculate μ_3 and check s via Eq. (56). If the new s lies closer to 1.5 than to any other tabulated value, the scaling iterations are complete. If not, turn to the table which best approximates the new s and continue. In the present case I end up with the table

$s = 1.5; a = 1.905 \times 10^{-6}$

$M_k(\times 10^{-6})$	$F(M_k)$
0.429	0.105
1.298	0.582
2.686	0.932
4.748	0.995
7.900	1

The complete matrix Q is now permitted to act on the column of $F(M_k)$ using the multiply–add law of Eq. (55). There results[3]

$\mu_1 = 1.000$
$a\mu_2 = 2.498 \qquad \mu_2 = 1.311 \times 10^6$
$a^2\mu_3 = 8.91 \qquad \mu_3 = 2.455 \times 10^{12}$
$a^3\mu_4 = 42.6 \qquad \mu_4 = 6.16 \times 10^{18}$
$a^4\mu_5 = 265 \qquad \mu_5 = 20.1 \times 10^{24}$

[3] The normalization convention of $\mu_1 = 1$ here differs from that adopted in Section III where the theoretical moments are normalized to $\mu_1 = M_0(1 - m/m_0)$. This has no effect on the molecular weight averages $\overline{M}_n$, $\overline{M}_w$, $\overline{M}_z$, but requires a renormalization of either the experimental or theoretical moments if they are to be compared on an absolute scale.

Finally, to calculate μ_0 from formula (57), we need the row matrix ω_k appropriate for $s = 1.5$, and carrying out this final multiply–add operation,

$$\mu_0 = 1.281 \times 10^{-6}$$

From these results we readily derive the molecular weight averages

$$\overline{M}_n = \mu_1/\mu_0 = 0.781 \times 10^6$$
$$\overline{M}_w = \mu_2/\mu_1 = 1.31 \times 10^6$$
$$\overline{M}_z = \mu_3/\mu_2 = 1.87 \times 10^6$$

(h) For many purposes the calculations may well stop here. If it is required, however, to obtain a plot of the weight distribution $Mf(M)$, then the five c_n are needed, and these may be computed from the column of quantities $a^r \mu_{r+1}$; $r = 0, \cdots, 4$; by operating on them with the matrix H. Thus, using Eq. (52) and successive rows of H,

$$c_0 = (0.86733)(1.000) = 0.8673$$

$$c_1 = (1.3714)(1.000) - (0.54854)(2.498) = 0.0012$$

and similarly

$$c_2 = 0.0366 \qquad c_3 = -0.0546 \qquad c_4 = 0.0321$$

Substituting these quantities into Eq. (53) and remembering that $a = 1.905 \times 10^{-6}$, we have through the tables of normalized, associated Laguerre functions $z^{1.5} e^{-z} p_n^{1.5}(z)$ a complete formula for the weight distribution. Both $Mf(M)$ and $f(M)$ constructed in this manner are plotted in Fig. 6.

(i) A word of warning connected with the use of these tables is in order. If the sample of polymer fractionated is rich in low molecular weight material, the calculations will probably end up with a low value of s. For $0 < s < 1$, the computational procedure is normal; but a glance at Eq. (53) shows that the number density distribution curve $f(M)$ will become infinite at the origin. Such an infinity cannot be accepted literally, of course, and indicates only that the scaling parameters give an optimal fit to the true moments of the distribution by integrating a curve which is asymptotically infinite at $M = 0$. If $s = 0$, the situation is even more curious, for $f(M)$ goes so rapidly to ∞ at $M = 0$ that the zeroth moment

$$\mu_0 = \int_0^\infty f(M)\,dM$$

does not exist, and thus no ω_k are tabulated for $s = 0$. This fact does not hinder the calculation of all higher moments for $s = 0$ nor does it destroy the user's ability to construct plots of $Mf(M)$ and $f(M)$. If, however, the

F. MOLECULAR WEIGHT DISTRIBUTION DATA

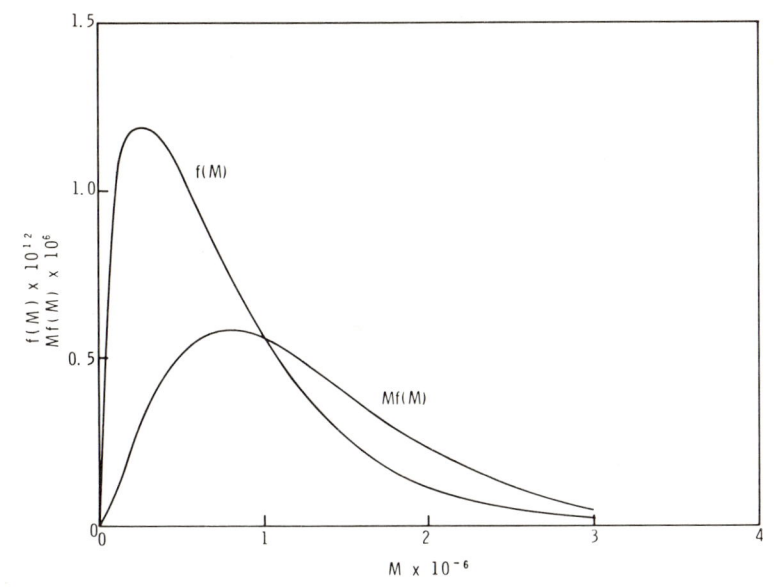

FIG. 6. Number density distribution $f(M)$ and weight distribution $Mf(M)$ computed from the data of Fig. 1.

zeroth moment be required, it can be estimated by rescaling the data to $s = 0.5$ and calculating μ_0 only. For the higher moments, those calculated for the choice $s = 0$ should be retained.

F. COMPARISON BETWEEN THEORY AND EXPERIMENT: THERMALLY POLYMERIZED POLYSTYRENE

According to Flory [34], thermal initiation of styrene polymerization is second order in monomer concentration, $I = k_i m^2$. In the absence of transfer reactions and assuming termination by combination, the theoretical moments of the resulting molecular weight distribution are given by Eq. (31), which predicts that $\log[\mu_r/(r + 1)!]$ should be linear in r and of slope $\log[M_0 k_p/(k_i k_t)^{1/2}]$.

The data of Fig. 5 were obtained for styrene thermally polymerized at 60°C to a 77.2% conversion of monomer. Using the distribution moments μ_r derived for these data in Section III,E, the excellent straight line of Fig. 7 is obtained; and from Flory's estimate of $(k_p)^2 k_t = 1 \times 10^{-3}$ liter/mole^{-1} second^{-1}, the further value of $k_i = 5 \times 10^{-13}$ liter/mole^{-1} second^{-1} may be calculated. At 100°C Flory quotes a value $k_i = 4 \times 10^{-11}$ liter/mole^{-1} second^{-1}, whence an activation energy of $\Delta H_i^{\ddagger} = 27$ kcal mole^{-1} for the initiation reaction may be inferred. On other grounds, Flory has estimated for this reaction an activation energy of $\Delta H_i^{\ddagger} = 29$ kcal mole^{-1}.

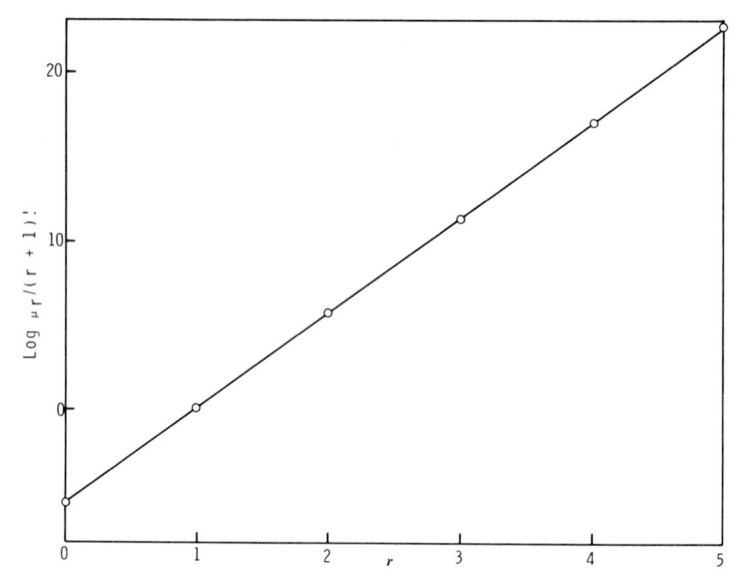

FIG. 7. Log $\mu_r/(r + 1)!$ plotted against r for thermally polymerized polystyrene.

The agreement between theory and experiment would thus appear to be good, but it is only fair to point out that the logarithmic plot of Fig. 7 smooths over a consistent departure of the data from theory. From Eq. (31) one may calculate four theoretical values of the dimensionless quantity $\mu_{r+1}\mu_{r-1}/(\mu_r)^2$ to be compared with experiment. Cantow's polystyrene is

	$\mu_{r+1}\mu_{r-1}/(\mu_r)^2$	
r	Theory	Experiment
1	1.50	1.68
2	1.33	1.43
3	1.25	1.34
4	1.20	1.30

thus slightly more heterogeneous than predicted, a result probably of transfer reactions leading to branched molecules.

Further calculations demonstrate that initiation mechanisms which are zero or first order in monomer may not be reconciled with the observed distribution moments, but the data do turn out to be as consistent with a third order initiation mechanism $I = k_i m^3$ as they are with a second. A

F. Molecular Weight Distribution Data

third order initiation has been proposed by Mayo [35]; but without distribution data on thermal polymerization at other temperatures, it is impossible to say whether or not Mayo's proposal leads to an acceptable value of the activation energy.

By way of a closing remark, confidence in the accuracy of the Gauss–Laguerre integration procedure developed in Section III is increased by experiments on artificial distribution "data" in which a cumulative curve $F(M)$ is synthesized from a known mathematical function $f(M)$ whose moments may be calculated analytically. So long as $Mf(M)$ contains only a single maximum, the Gauss–Laguerre method gives good estimates to the exact moments. This is particularly true if a properly scaled s happens to fall very close to one of the tabulated values, for then all moments are given with negligible error out to a maximum error of perhaps 10% in μ_5. In the least favorable case when a properly scaled s falls midway between tabulated values, μ_4 may be off by 10% with a 50% error in μ_5. In any case, the errors introduced by the algebraic procedure are likely to be considerably less than the experimental errors introduced into $F(M)$ by the irregularities of the data.

Appendix to Chapter F

$$s = 0.0$$

Roots λ_k

| 0.26356 | 1.41340 | 3.59644 | 7.08581 | 12.64080 |

Matrix Q

0.0046129	−0.0054982	0.013349	−0.067953	1.0580
−0.60241	−1.7299	−2.5476	−5.4451	10.367
0.88101	−6.1085	−16.335	−79.411	101.65
19.363	−33.067	−51.023	−937.36	1012.7
299.80	−375.90	352.38	−10559	10446

Matrix H

1.0000				
1.0000	−1.0000			
1.0000	−2.0000	0.5000		
1.0000	−3.0000	1.5000	−0.16667	
1.0000	−4.0000	3.0000	−0.66667	0.041667

$$s = 0.0$$

Associated Laguerre Functions $e^{-z}p_n^0(z)$

z	$n = 0$	$n = 1$	$n = 2$	$n = 3$	$n = 4$
0	1.00000	1.00000	1.00000	1.00000	1.00000
0.2	0.81873	0.65498	0.50761	0.37552	0.25768
0.4	0.67032	0.40219	0.18769	0.19663	−0.10832
0.6	0.54881	0.21952	−0.01098	−0.16245	−0.25169
0.8	0.44933	0.08987	−0.12581	−0.23605	−0.27151
1.0	0.36788	0.00000	−0.18394	−0.24525	−0.22992
1.2	0.30119	−0.06024	−0.20481	−0.21927	−0.16433
1.4	0.24660	−0.09864	−0.20221	−0.17689	−0.09599
1.6	0.20190	−0.12114	−0.18574	−0.12975	−0.03586
1.8	0.16530	−0.13224	−0.16199	−0.08463	0.01147
2.0	0.13534	−0.13534	−0.13534	−0.04511	0.04511
2.5	0.08208	−0.12313	−0.07182	0.02223	0.07888
3.0	0.04979	−0.09957	−0.02489	0.04979	0.06846
3.5	0.03020	−0.07549	0.00377	0.05222	0.04286
4.0	0.01832	−0.05495	0.01832	0.04274	0.01832
5.0	0.00674	−0.02695	0.02358	0.01797	−0.00870
6.0	0.00248	−0.01239	0.01735	0.00248	−0.01239
7.0	0.00091	−0.00547	0.01049	−0.00334	−0.00786
8.0	0.00034	−0.00235	0.00570	−0.00414	−0.00324
9.0	0.00012	−0.00099	0.00290	−0.00321	−0.00057
10.0	0.00005	−0.00041	0.00141	−0.00207	0.00050

F. Molecular Weight Distribution Data

$s = 0.5$

Roots λ_k				
0.43140	1.75976	4.10450	7.74670	13.45768
Matrix Q				
0.0071239	−0.0069572	0.015146	−0.071820	1.0608
−0.74339	−1.9293	−2.6599	−5.7258	11.205
1.3817	−8.4420	−19.483	−90.694	118.51
35.006	−51.451	−72.390	−1162.0	1272.1
577.85	−603.99	420.05	−14135	14088
Matrix ω_k				
7.2469	−0.4397	0.7667	−0.9659	0.7461
Matrix H				
1.0623				
1.3010	−0.86733			
1.4545	−1.9391	0.38788		
1.5711	−3.1422	1.2569	−0.11970	
1.6664	−4.4437	2.6662	−0.50785	0.028214

$s = 0.5$

Associated Laguerre Functions $z^{0.5} e^{-z} p_n^{0.5}(z)$

z	$n = 0$	$n = 1$	$n = 2$	$n = 3$	$n = 4$
0	0	0	0	0	0
0.2	0.38894	0.41284	0.39624	0.36321	0.32231
0.4	0.45034	0.40447	0.31408	0.21522	0.12028
0.6	0.45157	0.33184	0.18303	0.04778	−0.06208
0.8	0.42691	0.24400	0.06080	−0.08020	−0.17308
1.0	0.39078	0.15954	−0.03567	−0.15963	−0.21732
1.2	0.35048	0.08585	−0.10366	−0.19680	−0.21306
1.4	0.30994	0.02531	−0.14600	−0.20219	−0.17920
1.6	0.27128	−0.02215	−0.16741	−0.18620	−0.13108
1.8	0.23558	−0.05770	−0.17290	−0.15761	−0.07969
2.0	0.20331	−0.08300	−0.16703	−0.12314	−0.03206
2.5	0.13787	−0.11257	−0.12586	−0.03884	0.05035
3.0	0.09160	−0.11219	−0.07526	0.01935	0.07801
3.5	0.06001	−0.09800	−0.03287	0.04734	0.06973
4.0	0.03891	−0.07943	−0.00355	0.05317	0.04657
5.0	0.01600	−0.04574	0.02192	0.03494	0.00385
6.0	0.00645	−0.02370	0.02296	0.01281	−0.01301
7.0	0.00256	−0.01151	0.01661	0.00025	−0.01266
8.0	0.00101	−0.00535	0.01021	−0.00419	−0.00731
9.0	0.00039	−0.00241	0.00571	−0.00450	−0.00277
10.0	0.00015	−0.00106	0.00299	−0.00343	−0.00027

$s = 1$

Roots λ_k				
0.61704	2.11298	4.61090	8.39907	14.26010

Matrix Q				
0.0099700	−0.0084367	0.016853	−0.075332	1.0633
−0.86178	−2.1219	−2.7633	−5.9993	12.029
2.0064	−11.100	−22.772	−102.39	136.34
56.658	−75.296	−98.243	−1413.3	1566.8
993.29	−915.57	475.51	−18445	18521

Matrix ω_k				
4.1764	0.1240	0.3776	−0.2954	0.3286

Matrix H				
1.0000				
1.4142	−0.70711			
1.7321	−1.7318	0.28868		
2.0000	−3.0000	1.0000	−0.083333	
2.2361	−4.4721	2.2361	−0.37268	0.018634

$s = 1$

Associated Laguerre Functions $ze^{-z}p_n^1(z)$

z	$n=0$	$n=1$	$n=2$	$n=3$	$n=4$
0	0	0	0	0	0
0.2	0.16375	0.20841	0.22878	0.23569	0.23385
0.4	0.26813	0.30335	0.29103	0.25597	0.20957
0.6	0.32929	0.32598	0.26236	0.17847	0.09210
0.8	0.35946	0.30501	0.19093	0.07093	−0.03370
1.0	0.36788	0.26013	0.10620	−0.03066	−0.13025
1.2	0.36143	0.20446	0.02504	−0.10988	−0.18647
1.4	0.34524	0.14647	−0.04385	−0.16180	−0.20482
1.6	0.32303	0.09137	−0.09698	−0.18779	−0.19362
1.8	0.29754	0.04208	−0.13399	−0.19221	−0.16268
2.0	0.27067	0.00000	−0.15627	−0.18045	−0.12105
2.5	0.20521	−0.07255	−0.16291	−0.11329	−0.01314
3.0	0.14936	−0.10561	−0.12935	−0.03734	0.05845
3.5	0.10569	−0.11210	−0.08390	0.01872	0.08382
4.0	0.07326	−0.10361	−0.04230	0.04884	0.07645
5.0	0.03369	−0.07147	0.00973	0.05334	0.02825
6.0	0.01487	−0.04207	0.02576	0.02974	−0.00665
7.0	0.00638	−0.02257	0.02396	0.00904	−0.01653
8.0	0.00268	−0.01139	0.01704	−0.00179	−0.01320
9.0	0.00111	−0.00550	0.01058	−0.00528	−0.00702
10.0	0.00045	−0.00257	0.00603	−0.00515	−0.00237

F. Molecular Weight Distribution Data

$$s = 1.5$$

Roots λ_k					
0.81765	2.47236	5.11612	9.04415	15.04988	
Matrix Q					
0.013092	−0.0099217	0.018475	−0.078546	1.0655	
−0.96088	−2.3090	−2.8590	−6.2664	12.843	
2.7702	−14.074	−26.190	−114.48	155.13	
85.181	−105.23	−128.83	−1691.8	1898.3	
1580.2	−1327.6	511.45	−23559	23836	
Matrix ω_k					
2.3701	0.2225	0.2124	−0.0790	0.1749	
Matrix H					
0.86733					
1.3714	−0.54854				
1.8141	−1.4511	0.20733			
2.2219	−2.6662	0.76178	−0.056428		
2.6054	−4.1686	1.7865	−0.26467	0.012031	

$$s = 1.5$$

Associated Laguerre Functions $z^{1.5}e^{-z}p_n^{1.5}(z)$

z	$n = 0$	$n = 1$	$n = 2$	$n = 3$	$n = 4$
0	0	0	0	0	0
0.2	0.06351	0.09239	0.11220	0.12585	0.13482
0.4	0.14708	0.19535	0.21482	0.21598	0.20471
0.6	0.22122	0.26584	0.25965	0.22552	0.17644
0.8	0.27886	0.29982	0.25264	0.17604	0.09108
1.0	0.31907	0.30270	0.20975	0.09601	−0.01079
1.2	0.34340	0.28234	0.14694	0.00865	−0.10165
1.4	0.35429	0.24648	0.07707	−0.07051	−0.16711
1.6	0.35440	0.20173	0.00932	−0.13283	−0.20271
1.8	0.34623	0.15328	−0.05049	−0.17496	−0.21037
2.0	0.33200	0.10499	−0.09920	−0.19710	−0.19543
2.5	0.28142	0.00000	−0.16818	−0.18309	−0.10247
3.0	0.22438	−0.07096	−0.17432	−0.11496	0.00175
3.5	0.17150	−0.10846	−0.14348	−0.03905	0.07077
4.0	0.12708	−0.12056	−0.09873	0.01964	0.09618
5.0	0.06534	−0.10331	−0.01952	0.06642	0.06486
6.0	0.03160	−0.06994	0.02077	0.05319	0.01208
7.0	0.01465	−0.04169	0.03064	0.02585	−0.01573
8.0	0.00658	−0.02290	0.02636	0.00573	−0.02004
9.0	0.00289	−0.01188	0.01848	−0.00402	−0.01405
10.0	0.00125	−0.00591	0.01153	−0.00673	−0.00688

$s = 2.0$

Roots λ_k				
1.03111	2.83728	5.62049	9.68291	15.82847

Matrix Q				
0.016441	−0.011402	0.020018	−0.081504	1.0676
−1.0429	−2.4915	−2.9479	−6.5278	13.647
3.6873	−17.357	−29.729	−126.96	174.86
121.41	−141.87	−164.40	−1998.1	2268.1
2375.5	−1858.3	519.96	−29552	30127

Matrix ω_k				
1.5559	0.2220	0.1482	−0.0122	0.1224

Matrix H				
0.70711				
1.2247	−0.40825			
1.7321	−1.1545	0.14434		
2.2361	−2.2361	0.55902	−0.037268	
2.7386	−3.6515	1.3693	−0.18257	0.0076073

$s = 2.0$

Associated Laguerre Functions $z^2 e^{-z} p_n^2(z)$

z	$n = 0$	$n = 1$	$n = 2$	$n = 3$	$n = 4$
0	0	0	0	0	0
0.2	0.02316	0.03744	0.04935	0.05931	0.06752
0.4	0.07584	0.11384	0.13870	0.15323	0.15933
0.6	0.13970	0.19358	0.21559	0.21488	0.19801
0.8	0.20334	0.25828	0.25901	0.22600	0.17353
1.0	0.26013	0.30037	0.26549	0.19194	0.10355
1.2	0.30669	0.31872	0.24039	0.12724	0.01254
1.4	0.34177	0.31571	0.19255	0.04784	−0.07800
1.6	0.36547	0.29541	0.13130	−0.03267	−0.15315
1.8	0.37870	0.26237	0.06493	−0.10443	−0.20481
2.0	0.38279	0.22100	0.00000	−0.16140	−0.23061
2.5	0.36277	0.10472	−0.12959	−0.22704	−0.19880
3.0	0.31684	0.00000	−0.19403	−0.20039	−0.09204
3.5	0.26157	−0.07551	−0.20022	−0.12580	0.01706
4.0	0.20722	−0.11964	−0.16919	−0.04368	0.08917
5.0	0.11911	−0.13754	−0.07294	0.06278	0.10892
6.0	0.06310	−0.10929	0.00000	0.07981	0.04888
7.0	0.03160	−0.07297	0.03225	0.05329	−0.00374
8.0	0.01518	−0.04382	0.03719	0.02240	−0.02482
9.0	0.00707	−0.02449	0.03030	0.00224	−0.02395
10.0	0.00321	−0.01297	0.02097	−0.00677	−0.01520

F. Molecular Weight Distribution Data

$$s = 2.5$$

Roots λ_k				
1.25585	3.20715	6.12419	10.31614	16.59708
Matrix Q				
0.019971	−0.012870	0.021489	−0.084242	1.0695
−1.1092	−2.6701	−3.0307	−6.7840	14.442
4.7702	−20.942	−33.381	−139.81	195.53
166.17	−185.80	−205.15	−2333.1	2677.8
3418.4	−2527.3	492.55	−36501	37491
Matrix ω_k				
1.1110	0.2015	0.1154	0.0121	0.0990
Matrix H				
0.54854				
1.0262	−0.29321			
1.5393	−0.87950	0.097736		
2.0843	−1.7865	0.39701	−0.024061	
2.6570	−3.0365	1.0122	−0.12269	0.0047188

$$s = 2.5$$

Associated Laguerre Functions $z^{2.5}e^{-z}p_n^{2.5}(z)$

z	$n = 0$	$n = 1$	$n = 2$	$n = 3$	$n = 4$
0	0	0	0	0	0
0.2	0.00803	0.01417	0.02003	0.02552	0.03060
0.4	0.03721	0.06166	0.08161	0.09711	0.10830
0.6	0.08395	0.13013	0.16019	0.17601	0.17960
0.8	0.14109	0.20362	0.23103	0.23067	0.20954
1.0	0.20180	0.26966	0.27865	0.24674	0.18933
1.2	0.26062	0.32041	0.29673	0.22357	0.12754
1.4	0.31370	0.35213	0.28562	0.16885	0.04070
1.6	0.35863	0.36422	0.24984	0.09390	−0.05356
1.8	0.39415	0.35816	0.19594	0.01042	−0.14034
2.0	0.41995	0.33671	0.13094	−0.07138	−0.20930
2.5	0.44496	0.23784	−0.03964	−0.22445	−0.27655
3.0	0.42573	0.11378	−0.17067	−0.27311	−0.21218
3.5	0.37962	0.00000	−0.23674	−0.23312	−0.08572
4.0	0.32150	−0.08592	−0.24345	−0.14631	0.03613
5.0	0.20662	−0.16566	−0.15646	0.02606	0.14763
6.0	0.11990	−0.16022	−0.04807	0.10058	0.10733
7.0	0.06485	−0.12132	0.02022	0.09209	0.02904
8.0	0.03331	−0.08012	0.04600	0.05351	−0.02090
9.0	0.01645	−0.04836	0.04616	0.01867	−0.03494
10.0	0.00788	−0.02736	0.03613	−0.00203	−0.02858

$s = 3.0$

Roots λ_k				
1.49056	3.58150	6.62733	10.94447	17.35670

Matrix Q				
0.023649	−0.014320	0.022891	−0.086787	1.0712
−1.1612	−2.8452	−3.1080	−7.0356	15.230
6.0302	−24.822	−37.138	−153.04	217.11
220.25	−237.61	−251.30	−2697.2	3128.9
4750.8	−3355.7	420.10	−44483	46027

Matrix ω_k				
0.8384	0.1787	0.0955	0.0215	0.0863

Matrix H				
0.40825				
0.81650	−0.20412			
1.2910	−0.64540	0.064550		
1.8257	−1.3693	0.27386	−0.015215	
2.4152	−2.4152	0.72457	−0.080508	0.0028753

$s = 3.0$

Associated Laguerre Functions $z^3 e^{-z} p_n^3(z)$

z	$n = 0$	$n = 1$	$n = 2$	$n = 3$	$n = 4$
0	0	0	0	0	0
0.2	0.00267	0.00508	0.00763	0.01024	0.01284
0.4	0.01751	0.03152	0.04475	0.05667	0.06692
0.6	0.04840	0.08227	0.10988	0.13033	0.14343
0.8	0.09392	0.15027	0.18770	0.20654	0.20860
1.0	0.15019	0.22528	0.26121	0.26306	0.23799
1.2	0.21248	0.29747	0.31714	0.28659	0.22233
1.4	0.27625	0.35912	0.34768	0.27319	0.16524
1.6	0.33761	0.40513	0.35018	0.22627	0.07843
1.8	0.39356	0.43292	0.32607	0.15383	−0.02306
2.0	0.44200	0.44200	0.27955	0.06589	−0.12452
2.5	0.52361	0.39271	0.10349	−0.15855	−0.30770
3.0	0.54879	0.27439	−0.08677	−0.30678	−0.33626
3.5	0.52856	0.13214	−0.22983	−0.34226	−0.23616
4.0	0.47855	0.00000	−0.30266	−0.28535	−0.08089
5.0	0.34384	−0.17192	−0.27183	−0.06407	0.15741
6.0	0.21858	−0.21858	−0.13824	0.09775	0.18473
7.0	0.12769	−0.19154	−0.02019	0.13800	0.09443
8.0	0.07012	−0.14024	0.04435	0.10453	0.00395
9.0	0.03673	−0.09182	0.06388	0.05338	−0.04113
10.0	0.01853	−0.05560	0.05861	0.01382	−0.04699

REFERENCES

1. C. H. Bamford, W. G. Barb, A. D. Jenkins, and P. F. Onyon, "The Kinetics of Vinyl Polymerization by Radical Mechanisms," Chapter 7. Academic Press, New York, 1958.
2. C. H. Bamford and H. Tompa, *J. Polymer Sci.* **10**, 345 (1953).
3. J. C. W. Chien, *J. Am. Chem. Soc.* **81**, 86 (1959).
4. G. N. Watson, "Theory of Bessel Functions," p. 236. Cambridge Univ. Press, London and New York, 1952.
5. E. F. G. Herrington and A. Robertson, *Trans. Faraday Soc.* **38**, 490 (1942).
6. A. V. Tobolsky, *J. Am. Chem. Soc.* **80**, 5927 (1958).
7. R. H. Gobran, M. B. Berenbaum, and A. V. Tobolsky, *J. Polymer Sci.* **46**, 431 (1960).
8. A. V. Tobolsky, R. H. Gobran, R. Böhme, and R. Schaffhauser, *J. Phys. Chem.* **67**, 2336 (1963).
9. P. J. Flory, "Principles of Polymer Chemistry," p. 113. Cornell Univ. Press, Ithaca, New York, 1953.
10. E. Jahnke and F. Emde, "Funktionentafeln" (Reprint), pp. 6–9. Dover, New York, 1945.
11. E. F. G. Herrington and A. Robertson, *Trans. Faraday Soc.* **40**, 236 (1944).
12. G. Gee and H. W. Melville, *Trans. Faraday Soc.* **40**, 240 (1944).
13. W. F. Watson, *Trans. Faraday Soc.* **49**, 842 (1953).
14. E. Katchalski, I. Shalitin, and M. Gehatia, *J. Am. Chem. Soc.* **77**, 1925 (1955).
15. E. Katchalski, M. Gehatia, and M. Sela, *J. Am. Chem. Soc.* **77**, 6175 (1955).
16. R. J. Roe, *Polymer* **2**, 60 (1961).
17. T. A. Orofino, *Polymer* **2**, 295 (1961).
18. R. Zeman and N. R. Amundson, *A.I.Ch.E. (Am. Inst. Chem. Engrs.) J.* **9**, 297 (1963).
19. M. Szwarc, *Fortschr. Hochpolymer-Forsch.* **2**, 275 (1960).
20. P. J. Flory, *J. Am. Chem. Soc.* **62**, 1561 (1940).
21. H. W. McCormick, *J. Polymer Sci.* **36**, 341 (1959); **41**, 327 (1959).
22. A. F. Sirianni, D. J. Worsfold, and S. Bywater, *Trans. Faraday Soc.* **55**, 2124 (1959).
23. F. C. Goodrich, *J. Chem. Phys.* **35**, 2101 (1961).
24. E. Jahnke and F. Emde, "Funktionentafeln" (Reprint), p. 24. Dover, New York, 1945.
25. L. Gold, *J. Chem. Phys.* **28**, 91 (1958).
26. W. T. Kyner, R. M. Radock, and M. Wales, *J. Chem. Phys.* **30**, 363 (1959).
27. M. Litt and M. Szwarc, *J. Polymer Sci.* **42**, 159 (1960).
28. B. D. Coleman, F. Gornick, and G. Weiss, *J. Chem. Phys.* **39**, 3233 (1963).
28a. V. S. Nanda, *J. Chem. Phys.* **39**, 1363 (1963).
28b. V. S. Nanda, *Trans. Faraday Soc.* **60**, 949 (1964).
28c. F. C. Goodrich and M. J. R. Cantow, *J. Polymer Sci.* **C8**, 269 (1965).
29. M. J. R. Cantow, Dissertation, Mainz (1959), "Polymerization 'H'," pp. 19 and 22.
30. C. Lanczos, "Applied Analysis," pp. 331–344. Prentice-Hall, Englewood Cliffs, New Jersey, 1956.
31. P. M. Morse and H. Feshbach, "Methods of Theoretical Physics," pp. 784–785. McGraw-Hill, New York, 1953.
32. M. G. Kendall, "The Advanced Theory of Statistics," Vol. I, p. 34. Griffin, London, 1948.
33. C. Lanczos, "Applied Analysis," pp. 371–378 and 396–414. Prentice-Hall, Englewood Cliffs, New Jersey, 1956.
34. P. J. Flory, "Principles of Polymer Chemistry," pp. 129–132. Cornell Univ. Press, Ithaca, New York, 1953.
35. F. R. Mayo, *J. Am. Chem. Soc.* **75**, 6133 (1953).

CHAPTER G

Additional Methods of Fractionation

Manfred J. R. Cantow
CHEVRON RESEARCH COMPANY, RICHMOND, CALIFORNIA

Text	461
References	464
Appendix	465
References to Appendix	488

In this chapter, various methods of polymer fractionation will be discussed which have not yet found wide application. However, some of these procedures show considerable promise for future work. The most direct approach for the evaluation of a molecular weight distribution is to observe the individual molecules and to establish the distribution by a measuring and counting procedure. Husemann and Ruska [1] were the first to apply the electron microscope to the study of iodine-substituted glycogen. Good agreement with osmotic molecular weights was obtained. Boyer and Heidenreich [2] added a precipitating agent, propanol, to a very dilute solution of polychlorostyrene and evaporated a drop on a silica screen. The observed frequency distribution of particle sizes was used to calculate average molecular weights assuming that the particle size varies as the square root of molecular weight. The calculated values were found to be from four to five times greater than values measured by an independent method. Siegel et al. [3] sprayed solutions of polystyrene fractions in a poor solvent, cyclohexane, onto collodion films. Using the bulk density of polystyrene, the average molecular weight was calculated from the average volume of the spherical particles. Excellent agreement was found in the molecular weight range from 4.8×10^5 to 2.5×10^6. Schoon and van der Bie [4], as well as Schulz and Mula [5] applied the same method to the study of brominated Hevea latices. Richardson [6] recommends electron microscopy for molecular weights of the order of a million or more because other methods may have diminishing accuracy in that region. He prepared 0.0001% solutions of polystyrene in a good solvent and added a precipitant which had to have a lower vapor pressure than the solvent. The mixture was sprayed on a carbon film backed by mica. Metal shadowing and floating was carried out in the usual way. The particles observed were considered to be true spheres, and their diameters were calculated from their shadow lengths. In this manner, a sharp fraction with a viscosity average molecular weight of 1,500,000 was calculated to have a number average of 1,250,000 and a weight average of 1,340,000. Even greater accuracy is expected at higher molecular weights.

Another approach to the problem of polymer fractionation is the use of membranes. Rosenberg and Beckmann [7] employed sintered metal disks to fractionate mixtures of homologous amylopectins and also a degraded corn amylopectin. The expected efficiency of the method was calculated for a previously characterized sample of GRS rubber. It is concluded that the diffusion method has an advantage over fractional precipitation for cases where very large fractions of high molecular weight, free of low molecular weight material, are desired. Stavermann [8] pointed out the possibility of obtaining accurate data on polymer distributions by measuring simultaneously the apparent osmotic pressure and the permeation of various solutions of a polymer through membranes of different permeability. Gardon and Mason [9] prepared fractions of lignin sulfonates by ultrafiltration through membranes of different pore sizes. Wilkie et al. [10] used the same process to separate dextrane into fractions ranging from number average molecular weights of 11,000 up to 190,000. Green et al. [11] formed membranes by casting solutions of methoxymethyl nylon and fractionated a sample of polystyrene with a very broad distribution (M_w = 387,000 and M_n = 40,000). The process required considerable time; the fractions obtained ranged from M_n = 21,000 to 289,000, but were not of narrow distribution. Hellfritz et al. [12,13] recommend membrane fractionation for industrial polymers. Hoffmann and Unbehend [14] find that the diffusion coefficient of polymers in membranes depends to a great extent upon the molecular weight of the diffusing material. Molecular weight distributions can be determined from the time dependence of the osmotic pressure of a polymer solution and the thickness and degree of swelling of the membrane. The chemical nature of the sample is of little influence. This makes the method applicable to the investigation of copolymers. Best resolution is obtained at the low end of the distribution, up to molecular weights around 30,000.

Zone melting was applied to polymer fractionation by Peaker and Robb [15]. The method is based on the presence of different solute concentrations in a solid- and a liquid-solution phase at equilibrium. A segregation coefficient is defined as the ratio of the volume fractions of similar polymer species in the two different phases. This coefficient is found to vary as a function of the molecular weights of these species. The above authors fractionated a sample of polystyrene (M_w = 200,000 and M_n = 80,000) by this method. Ten fractions were obtained and characterized qualitatively by turbidimetric titration. A regular change in initial precipitation points was observed. Loconti and Cahill [16] investigated the effects of freezing rate and concentration on the system polystyrene-benzene. The number of zones passed, the zone lengths, and the zone travel rate were introduced as further variables by Ruskin and Parravano [17]. These authors applied the method to polystyrene in cyclohexane. A variation of zone refining was introduced by Eldib [18]: Zone precipitation is carried out in a solvent that does not have to solidify during the process. As in zone melting, it is desirable to use a poor solvent. The method was employed to fractionate waxes in *sec*-butyl acetate.

Bardwell and Winkler [19] reported that addition of a cross-linking agent to a rubber results in preferential cross-linking of high molecular weight material. During gelation, the molecular weight of the soluble portion decreases steadily. Selective cross-linking thus offers a possible basis for determining molecular weight distributions. The method was attempted by Golub [20] on GRS latices. The results were not satisfactory. Bueche and Harding [21] improved the procedure and obtained good results on natural and synthetic rubber. The dependence of sol–gel ratio as a function of cross-link density and one molecular weight average are required. Rehner [22] assumes that the molecular weight distribution follows Tung's empirical function (see Chapter E). With this assumption, only two pairs of data on sol–gel ratio and cross-link density are required. The method was tested with reasonable success on rubber and polystyrene. A similar procedure was proposed by Kobelt and Stemmler [23]. Calderon and Scott [24] applied the method to polyisoprene.

Scherer and Testerman [25] measured the dielectric dispersion of cellulose nitrate solutions in acetone over a frequency range of 100–500 kc/sec. The shape of the dispersion curves was brought into a relation with the shape of the corresponding integral molecular weight distribution.

Tobolsky and Murakami [26] investigated the effect of polydispersity on stress relaxation for polystyrenes. It is concluded that these measurements can be used to obtain weight average molecular weights and also to provide a qualitative measure of the heterogeneity.

McCall *et al.* [27] measured nuclear magnetic relaxation times for unfractionated branched polyethylenes with varying widths of molecular weight distribution. A definite correlation between the shape of the relaxation curve and polydispersity was found. A direct conversion was not carried out.

Imai and Matsumoto [28] report fractionation of polyvinyl alcohol with respect to stereoregularity by foaming an aqueous solution of the polymer. Fractions were obtained by repeated shaking of the solution and removal of the foam layer. Schulz and Nordt [29] attempted extraction of a polymer solution with a liquid immiscible with the solvent. The partition coefficient of the polymer between the two phases is again a function of solute molecular weight. However, the method provided unsatisfactory results for the system polyoxyglycol-chloroform-benzene. Almin et al. [30,31] added the principle of countercurrent distribution to the procedure. Polyoxyethylene glycols have thus been successfully fractionated in the system trichloroethylene–chloroform–water. In an experimental arrangement described by Craig et al. [32] up to 8,000 extractions per hour are performed automatically.

REFERENCES

1. E. Husemann and H. Ruska, *Naturwissenschaften* **28**, 534 (1940); *J. Prakt. Chem.* **156**, 1 (1940).
2. R. F. Boyer and R. D. Heidenreich, *J. Appl. Phys.* **16**, 621 (1945).
3. B. M. Siegel, D. H. Johnson, and H. Mark, *J. Polymer Sci.* **5**, 111 (1950).
4. T. G. Schoon and G. J. van der Bie, *J. Polymer Sci.* **16**, 63 (1955).
5. G. V. Schulz and A. Mula, *Proc. Natl. Rubber Res. Conf.*, Kuala Lumpur, 1960, p. 602. Rubber Res. Inst. Malaya, 1961.
6. M. J. Richardson, *J. Polymer Sci.* **C3**, 21 (1963).
7. J. L. Rosenberg and C. O. Beckmann, *J. Colloid Sci.* **3**, 483 (1948).
8. A. J. Stavermann, *Ind. Chim. Belge* **18**, 235 (1953).
9. J. L. Gardon and S. G. Mason, *Can. J. Chem.* **33**, 1477 (1955).
10. K. C. B. Wilkie, J. K. N. Jones, B. J. Excell, and R. E. Semple, *Can. J. Chem.* **35**, 795 (1957).
11. J. H. S. Green, H. T. Hookway, and M. F. Vaughan, *Chem. & Ind. (London)* **36**, 862 (1958).
12. H. Hellfritz, H. Krämer, and W. Schmieder, *Kunststoffe* **49**, 391 (1959).
13. H. Hellfritz, *Kunststoffe* **50**, 502 (1960).
14. M. Hoffmann and M. Unbehend, *Makromol. Chem.* **88**, 256 (1965).
15. F. W. Peaker and J. C. Robb, *Nature* **182**, 1591 (1958).
16. J. D. Loconti and J. W. Cahill, *J. Polymer Sci.* **A1**, 3163 (1963).
17. A. M. Ruskin and G. Parravano, *J. Appl. Polymer Sci.* **8**, 565 (1964).
18. I. A. Eldib, *Ind. Eng. Chem., Process Design Develop.* **1**, 2 (1962).
19. J. Bardwell and C. A. Winkler, *Can. J. Res.* **B27**, 128 (1946).
20. M. A. Golub, *J. Polymer Sci.* **10**, 591 (1952).
21. F. Bueche and S. W. Harding, *J. Appl. Polymer Sci.* **2**, 273 (1959).
22. J. Rehner, Jr., *J. Appl. Polymer Sci.* **4**, 95 (1960).
23. D. Kobelt and H. D. Stemmler, *Kautschuk Gummi* **16**, 195 (1963).
24. N. Calderon and K. W. Scott, *J. Polymer Sci.* **A3**, 551 (1965).
25. P. C. Scherer and M. K. Testerman, *J. Polymer Sci.* **7**, 549 (1951).
26. A. V. Tobolsky and K. Murakami, *J. Polymer Sci.* **47**, 55 (1960).
27. D. W. McCall, D. C. Douglas, and E. W. Anderson, *J. Polymer Sci.* **59**, 301 (1962).
28. K. Imai and M. Matsumoto, *Bull. Chem. Soc. Japan* **36**, 455 (1963).
29. G. V. Schulz and E. Nordt, *J. Prakt. Chem.* **155**, 115 (1940).
30. K. E. Almin and B. Sitenberg, *Acta Chem. Scand.* **11**, 936 (1957).
31. K. E. Almin and R. Lundberg, *Acta Chem. Scand.* **13**, 1274 (1959).
32. L. C. Craig, W. Hausmann, E. H. Ahrens, and E. Harfenist, *Anal. Chem.* **23**, 1236 (1951).

Appendix to Chapter G

The following tabulation contains information on the fractionation of a large variety of polymers. The method employed is indicated by a symbol referring to the respective chapter in this book. Information on additional polymers may be found in separate tables in Chapters C.1 on turbidimetric titration and D on fractionation of mixtures and copolymers.

POLYOLEFINS

Polymer	Method	Solvent; Solvent–Nonsolvent	Temperature (°C)	Remarks	Reference
Hexene-1-polysulfone	B.1	Acetone–methanol	15–25	—	211
Low-M.W. polyolefins	B.2	—	—	—	156
	B.4	—	—	—	156
Polyalkane	B.3	Benzene–ethanol	20–60	—	216
Poly-1-butene	B.2	Xylene–cellosolve	—	—	229
	B.2	Ligroin–ethanol	60	—	252
Polychlorotrifluoroethylene	B.1	Dichlorobenzotrifluoride–diethyl phthalate	—	—	103
	B.1	Dichlorobenzotrifluoride–diethyl phthalate	150	—	223
Polyethylene	B.1	Xylene–n-propanol	90	—	12
	B.1	2-Ethylhexanol–decalin	—	—	222
	B.1	Xylene–polyethylene glycol	80	—	311
	B.1	Toluene–n-propanol	—	—	316
	B.1	Xylene–triethylene glycol	110	—	344
	B.1	—	—	—	410
	B.1	Xylene–triethylene glycol	130	—	418
	B.1	Xylene–triethylene glycol	130	—	419
	B.1	Toluene–n-propanol	—	—	422
	B.1	2-Ethylhexanol–decalin	—	—	443
	B.1	2-Ethylhexanol–decalin	—	—	444
	B.2	Touene	50–80	—	8
	B.2	Toluene	50–80	—	97
	B.2	Toluene	50–80	—	98
	B.2	Xylene–cellosolve	127	—	139
	B.2	Xylene–n-butanol	120	—	144
	B.2	Xylene	—	—	190
	B.2	p-Xylene–cellosolve	126.5	Large-scale	193

G. Additional Methods of Fractionation

Polymer		Solvent system	Temperature	Notes	Ref.
	B.2	p-Xylene–cellosolve	165		202
	B.2	Xylene–ethyl cellosolve	127		228
	B.2	Xylene–ethyl cellosolve	127		229
	B.2	Decalin–ethylene glycol, n-hexanol	139		246
	B.2	p-Xylene			247
	B.2	Toluene			316
	B.2	Toluene–polyethylene glycol	80		320
	B.2	Toluene–polyethylene glycol	80		321
	B.2	Xylene–polyethylene glycol			323
	B.2	Xylene–triethylene glycol	130		418
	B.2	Xylene–cellosolve	127		444
	B.3	Tetralin–butyl cellosolve	110–160		83
	B.3	Petroleum ether		Low mol. wt.	155
	B.3	Tetralin–butyl cellosolve	110–160		169
	B.3	Tetralin–butyl cellosolve	110–160		170
	C.2	α-Bromonaphthalene	110		275
	C.2	α-Bromonaphthalene	120		303
	C.5				294
	C.5				373
Polyethylene tetrasulfide	B.3	Dioxane			157
Polyisobutylene	B.1	Benzene–acetone			134
	B.1	Benzene–acetone			137
	B.1	Benzene–acetone			138
	B.1	Benzene–acetone			250
	B.1	2,4,4-Trimethylpentene–n-butanol			269
	B.3	Benzene–acetone	28–50		71
	B.3	Benzene–butanone	40–90		177
	B.3	Toluene–methanol			254
	B.3	2-Methylheptane			269
	B.3	Mixed xylenes–n-propanol	10–60		329
	C.2	2,2,4-Trimethylpentane			434
	C.5				149
	C.5	—			265

POLYOLEFINS—continued

Polymer	Method	Solvent; Solvent–Nonsolvent	Temperature (°C)	Remarks	Reference
Polypropylene	C.5	Benzene	—	—	380
	B.1	Xylene–polyethylene glycol	—	—	136
	B.2	o-Dichlorobenzene	—	—	93
	B.2	Xylene–cellosolve	165	—	202
	B.2	Butyl cellosolve–butyl carbitol	165	—	203
	B.2	Tetralin-o-dimethyl phthalate	155	—	284
	B.2	o-Dichlorobenzene–methyl carbitol	168	—	284
	B.2	Hydrocarbons with increasing b.p.	—	Acc. to crystallinity	315
	B.2	Isopropyl ether	—		317
	B.2	9 Solvents	—	Acc. to crystallinity	319
	B.2	Tetralin–diethylene glycol monoethyl ether	178	—	330
	B.2	Tetralin–Carbowax 200	140	—	348
	B.2	n-Heptane	99	Extract. of amorph.	355
	B.2	Hydrocarbons–2-butoxyethanol	—	—	388
	B.2	Kerosene–butyl carbitol	—	—	447
	B.2	Solvents of increasing b.p.	—	—	451
	B.3	Tetralin–butyl carbitol	140–180	—	83
	B.3	Tetralin–butyl carbitol	140–180	—	171

G. Additional Methods of Fractionation

Aliphatic Vinylpolymers

Polymer	Method	Solvent; Solvent–Nonsolvent	Temperature (°C)	Remarks	Reference
Poly(vinyl acetate)	B.1	Methanol–water	35	—	39
	B.1	Acetone–water	—	—	40
	B.1	Dioxane–isopropanol	—	—	121
	B.1	Benzene–isopropanol	—	—	121
	B.1	Acetone–water	—	—	314
	B.1	Acetone–n-hexane	—	—	374
	B.1	Acetone–methanol, water	—	—	433
	B.2	Methyl acetate–petroleum ether	—	—	146
	B.2	—	—	—	154
	B.2	Methyl isobutyl ketone	—	—	255
	B.2	Acetone	—	—	81
	B.5	Water	—	—	259
	B.5	Water	—	—	260
	B.5	Various solvents	—	—	263
	C.1	Acetone–water	—	—	305
	C.5	—	—	—	148
Poly(vinyl acetal)	B.1	—	—	—	105
	B.2	Benzene	10–24	—	105
Poly(vinyl alcohol)	B.1	Water–acetone	—	—	101
	B.1	Water–acetone	—	—	104
	B.2	Water–methanol	65	—	35
	B.2	Water–n-propanol	65	—	145
	B.5	Water	—	—	143
	B.5	Water	—	—	257
	C.3	—	—	—	142
Polyvinylbutyral	B.1	Isopropanol, benzene–water	—	—	200
Poly(vinyl chloride)	B.1	Tetrahydrofuran–water	50	—	23

ALIPHATIC VINYLPOLYMERS—continued

Polymer	Method	Solvent; Solvent–Nonsolvent	Temperature (°C)	Remarks	Reference
Poly(vinyl chloride)	B.1	Cyclohexanone–methanol	—	—	95
	B.1	Cyclohexanone–n-butanol	—	—	106
	B.1	Tetrahydrofuran–water	—	—	141
	B.1	Tetrahydrofuran–water	—	—	283
	B.1	Cyclohexanone–ethylene glycol	—	—	283
	B.1	Chlorobenzene, cyclohexane–acetone, methanol	—	—	291
	B.1	Tetrahydrofuran–water	—	—	304
	B.1	Chlorobenzene–gasoline	—	Acc. to Cl content	356
	B.1	Acetone, chlorobenzene–methanol	—	—	356
	B.2	Methyl isobutyl ketone	—	—	255
	B.2	Tetrahydrofuran–water	44	—	400
	B.5	Cyclohexanone	—	—	176
	B.5	Cyclohexanone	—	—	260
	C.1	Cyclohexanone–heptane, carbon tetrachloride	—	—	325
Poly-α-chlorovinylacetic acid	B.1	Methanol–water	—	—	239
Poly(vinyl isobutyl ether)	B.1	Toluene, butanone–ethanol	—	—	365

POLYPHENYLS

Polymer	Method	Solvent; Solvent–Nonsolvent	Temperature (°C)	Remarks	Reference
Polyphenyls	B.2	Dioxane–Water	63	—	378

G. Additional Methods of Fractionation

Aromatic Vinylpolymers

Polymer	Method	Solvent; Solvent–Nonsolvent	Temperature (°C)	Remarks	Reference
Na-poly(-p-styrene sulfonate)	B.1	4N aqueous NaI—N aqueous NaI	—	—	271
Poly(ammonium-p-styrene sulfonate)	B.2	—	—	—	184
Poly-α-methylstyrene	B.2	Toluene–methanol	30	—	61
	C.1	—	—	—	342
	C.2	Cyclohexane	35	—	276
Polystyrene	B.1	Butanone–ethanol	—	—	41
	B.1	Benzene–methanol	—	—	49
	B.1	Butanone–methanol	—	—	72
	B.1	Butanone–butanol, water	—	—	82
	B.1	Ethyl acetate–ethanol	—	—	140
	B.1	Butanone–butanol, water	—	—	159
	B.1	Butanone–methanol	—	—	173
	B.1	Toluene–methanol	—	—	178
	B.1	Toluene–petroleum ether	—	—	179
	B.1	Butanone–methanol	—	—	212
	B.1	Toluene–methanol	—	Grafted PST	218
	B.1	Butanone–butanol	—	—	225
	B.1	Butanone–butanol, water	—	—	285
	B.1	Toluene–methanol	—	—	318
	B.1	Butanone–propanol	—	—	341
	B.1	Toluene–n-decane	—	—	341
	B.1	Chloroform–methanol	—	—	392
	B.1	Butanone–ethanol	—	—	405
	B.1	Butanone–methanol	—	—	427
	B.2	Benzene–methanol	—	—	166
	B.2	Toluene–butanol	—	—	166
	B.2	Chloroform–butanol	—	—	166

AROMATIC VINYLPOLYMERS—continued

Polymer	Method	Solvent; Solvent–Nonsolvent	Temperature (°C)	Remarks	Reference
Polystyrene	B.2	Butanone–butanol	—	—	166
	B.2	Benzene–methanol	—	—	182
	B.2	Butanone–methanol	25	Isotactic	228
	B.2	Butanone–ethanol	—	—	366
	B.2	Butanone–ethanol	15	—	368
	B.2	Various solvents	—	—	454
	B.3	Butanone–ethanol	—	—	19
	B.3	Butanone–ethanol	10–60	—	216
	B.3	Toluene–ethanol	—	—	289
	B.3	Benzene–ethanol	12–65	Large-scale	333
	B.3	Butanone–ethanol	40–70	—	367
	B.3	Butanone–ethanol	10–60	—	368
	B.3	Butanone–ethanol	40–70	—	369
	B.3	Butanone–ethanol	—	—	405
	B.4	Toluene	—	On PMM-gel	100
	B.4	Toluene	—	On PST-gel	262
	B.4	Toluene	—	On PST-gel	302
	B.5	Toluene	—	—	96
	B.5	Toluene	—	—	122
	B.5	o-Xylene	—	—	122
	B.5	Styrene	—	—	122
	B.5	Ethylbenzene	—	—	122
	B.5	Dioxane	—	—	122
	B.5	Pyridine	—	—	122
	B.5	Toluene	—	—	201
	B.5	Butyl acetate	—	—	201
	B.5	Toluene	—	—	256

G. Additional Methods of Fractionation 473

B.5	Toluene	—	—	—	260
B.5	Toluene	—	—	—	261
B.5	Toluene	—	—	—	408
B.5	Butanone	—	—	—	408
C.1	Benzene–isopropanol	—	—	—	120
C.1	Benzene–methanol	—	—	—	162
C.1	Benzene–methanol	—	—	—	186
C.1	Butanone–acetone, water	—	—	—	192
C.1	Benzene–methanol	—	—	—	282
C.1		—	—	—	342
C.1	Butanone–isopropanol	—	—	—	424
C.2	Cyclohexane	35	—	—	58
C.2	Water	—	Latex	—	68
C.2	Cyclohexane	35	Sed. balance	—	69
C.2	Methyl cyclohexane	75	Sed. balance	—	69
C.2	Butanone	20	—	—	165
C.2	Cyclohexanol, carbon tetrachloride	—	Gradient sedim.	—	195
C.2	Ethyl acetate, butanone	—	—	—	224
C.2	Cyclohexane	35	—	—	274
C.2	Chloroform	—	—	—	389
C.2	Butanone	—	—	—	435
C.2	Cyclohexane	34	—	—	437
C.2	Butanone	—	—	—	450
C.4	Benzene–methanol	24.5	—	—	50
C.5		—	—	—	148
C.5		—	Living PST	—	194
C.5		—	—	—	353
C.5		—	—	—	354
C.5		—	—	—	361
C.5		—	—	—	414

Heterocyclic Vinylpolymers

Polymer	Method	Solvent; Solvent–Nonsolvent	Temperature (°C)	Remarks	Reference
Polyvinylpyridine	B.1	Methanol–toluene	15	—	36
	B.1	Nitromethane–benzene	—	—	132
	B.1	tert-Butyl alcohol–benzene	—	—	132
	B.2	tert-Butyl alcohol–benzene	—	—	397
Polyvinylpyrrolidone	B.1	Water–acetone	—	—	102
	B.1	Water–acetone	—	—	150
	B.1	Water–acetone	—	—	213
	B.1	Water–acetone	—	—	370
	B.2	Water–organic acids	—	—	383
	B.5	Water	—	—	256
	B.5	Water	—	—	258
	B.5	Water	—	—	259
	B.5	Water	—	—	260
	B.5	Water	—	—	261
	B.5	Ethanol	—	—	261
	C.1	Water–sodium sulfate soln.	—	—	67
	C.1	Water–sodium sulfate soln.	—	—	371
	C.2	Water	—	—	293
	C.3	Water	—	—	293

G. Additional Methods of Fractionation

Diene Polymers

Polymer	Method	Solvent; Solvent–Nonsolvent	Temperature (°C)	Remarks	Reference
Polybutadiene	B.1	Benzene–methanol	—	—	52
	B.1	Benzene–methanol	—	—	54
	B.1	Benzene–acetone	28	—	84
	B.1	Toluene–methanol	—	—	108
	B.1	Benzene–acetone, dioxane	—	—	152
	B.1	Benzene–acetone	—	—	428
	B.3	Benzene–ethanol	15–60	—	86
	B.3	Diisobutylene–isooctane	40–90	*cis*–1,4	207
	C.2	Octane	—	—	53
	C.2	Cyclohexane	—	—	54
	C.2	Cyclohexene	—	—	54
	C.2	Cyclohexene	—	—	54
	C.2	Theta solvents	—	*cis*–1,4	336
Polychloroprene	B.1	Benzene–acetone	—	—	188
	B.1	Benzene–methanol	—	—	298
	B.1	Benzene–methanol	—	—	299
	B.1	Benzene–methanol	—	—	300
	B.5	Benzene	—	—	244
	B.5	Benzene	—	—	249
Polyisoprene	B.1	Toluene–methanol	—	*cis*–1,4	32
	B.1	Benzene–ethanol	28	*trans*	84
	B.1	Acetone–ethanol	—	*cis*	278
	B.1	Toluene–*n*-butanol	—	—	338
	B.1	Benzene–*n*-butanol	—	—	338
Polytrichlorobutadiene	B.1	Benzene–white spirits	—	—	331

ACRYLIC POLYMERS

Polymer	Method	Solvent; Solvent–Nonsolvent	Temperature (°C)	Remarks	Reference
Poly(acrolein-thiophenolmercaptals)	B.1	Benzene–methanol	—	—	377
Polyacrylamide	B.1	Water–methanol	30	—	31
	B.1	Water–methanol	—	—	376
Polyacrylic acid	B.5	Water	—	—	257
Polyacrylonitrile	B.1	Dimethylformamide–heptane	60	—	45
	B.1	Dimethylformamide–heptane	60	—	80
	B.1	Dimethylformamide–ligroin	55	—	164
	B.1	Dimethylformamide–heptane, ether	—	—	198
	B.1	Dimethylformamide–heptane	60	—	204
	B.1	Dimethyl sulfoxide–toluene	—	—	235
	B.1	Hydroxyacetonitrile, ethanol–benzene	—	—	236
	B.1	Dimethyl sulfoxide–toluene	35	—	237
	B.1	Dimethylformamide–lauric alcohol	—	—	402
	B.2	Dimethylformamide–heptane	—	—	251
	B.5	Dimethylformamide	—	—	260
	B.5	Dimethylformamide	—	—	261
	B.5	Dimethylformamide	—	—	264
	C.5	Dimethylformamide	—	—	110
	C.5	Dimethylformamide	—	—	111
	C.5	Dimethylformamide	—	—	113
	C.5	—	—	—	115
Poly(butyl methacrylate)	B.1	Acetone–methanol	—	—	78
	B.5	Benzene	—	—	243
	B.5	Benzene	—	—	245
Poly(cyclohexyl methacrylate)	B.1	Dioxane–methanol	—	—	286
Poly(1,1-dihydroperfluorobutyl acrylate)	B.1	Benzotrifluoride–methanol	—	—	347
Poly(ethyl methacrylate)	B.1	Acetone–acetone, water	—	—	77

G. Additional Methods of Fractionation 477

Polymer	Code	Solvents	Value	Ref.
Poly(n-hexyl methacrylate)	B.1	Acetone–ethanol	—	79
Polymethacrylic acid	B.1	Methanol–ether	—	221
	B.1	Methanol–methyl isobutyl ketone	—	445
Poly(methyl acrylate)	B.1	Acetone–water, methanol	—	174
Poly(methyl methacrylate)	B.1	Acetone–acetone, water	—	30
	B.1	Acetone–hexane	—	42
	B.1	Chloroform, acetone–petroleum ether	25	88
	B.1	Benzene–cyclohexane	—	125
	B.1	Acetone–petroleum ether	—	125
	B.1	Benzene–cyclohexane	—	127
	B.1	Benzene–cyclohexane	—	232
	B.1	Benzene–n-hexane	—	248
	B.1	Benzene, chloroform–petroleum ether	—	290
	B.1	Chloroform–petroleum ether	—	417
	B.1	Butanone, ethanol–cyclohexane	—	442
	B.2	—	—	99
	B.2	Toluene–petroleum ether	—	119
	B.2	Butanone, ethanol–cyclohexane	—	442
	B.3	Butanone–cyclohexane	10–60	441
	B.3	Butanone, ethanol–cyclohexane	—	442
	B.5	Acetone	—	143
	B.5	Benzene	—	242
	B.5	Toluene	—	260
	B.5	Toluene	—	261
	C.1	Acetone–water	—	183
	C.2		—	126
	C.2		—	128
	C.2		—	232
	C.4		—	233
	C.5		—	168
	C.5		—	439
Poly(N-phenyl-methacrylamide)	B.1	Acetone–benzene	—	360

POLYETHERS

Polymer	Method	Solvent; Solvent–Nonsolvent	Temperature (°C)	Remarks	Reference
Polydiketenes	B.1	Acetone–dioxane	—	—	226
Polyethylene oxide	B.1	Chloroform–hexane	—	—	18
	B.1	—	—	—	309
	B.3	—	—	Thin layer	60
Polyhydroxyethers	B.2	Chloroform–hexane	—	—	312
Poly(oxybenzyl ether)	B.1	Benzene–petroleum ether	22	—	217
Poly(oxymethylene diacetate)	B.1	—	—	—	219
Poly(propylene oxide)	B.1	Isopropanol–water	74	—	4
	B.1	Isooctane	—	By cooling	5

ALDEHYDE RESINS

Polymer	Method	Solvent; Solvent–Nonsolvent	Temperature (°C)	Remarks	Reference
p-Cresol-formaldehyde	B.1	Methanol, tetrahydrofurane–water	—	—	231
Phenol-formaldehyde	B.1	Various solvent–nonsolvent systems	—	—	430
	B.2	Ethanol–saline solutions	—	—	63
	B.2	Various solvent–nonsolvent systems	—	—	64
	B.2	Benzene–acetic acid	15	On paper	131
	B.2	Benzene–acetic acid	—	On paper	332
Urea-formaldehyde	B.1	Ethanol, water–methanol	—	—	431
	B.2	Butanone–methyl butyl ketone, water	—	On paper	107

G. Additional Methods of Fractionation

Polymer	Method	Solvent; Solvent–Nonsolvent	Temperature (°C)	Remarks	Reference
Adipic acid + ethylene glycol	B.1	Butanol–petroleum ether	—	—	75
	B.1	Benzene–petroleum ether	—	—	343
α,α'-Dibutylsebacic acid + hexanediol	B.1	Benzene–methanol	—	—	25
	B.2	Benzene–methanol	—	—	25
Hydroxy-11-undecanoic acid	B.1	Benzene–methanol	35	—	268
Isophthalic acid + maleic anhydride + propylene glycol	B.3	Acetone–n-heptane	27–50	—	70
Phthalic anhydride + glycerol	B.1	Butanone, acetone–methanol, water	—	—	47
	B.1	Acetone–water	—	—	133
Sebacic, adipic, tartaric acids + 1,6-hexanediol, 1,10-decanediol	B.1	Benzene–methanol	—	—	24
	B.1	Benzene–methanol	—	—	448
	B.2	Benzene–methanol	—	—	24
	B.2	Benzene–methanol	—	—	448
Succinic, pimelic acids + hexanediol	B.1	Benzene–methanol	—	—	26
Sebacic acid + hexanediol + hexane triol	B.1	Benzene–methanol	—	—	26
Succinic acid + hexanediol	B.3	Butanone–cyclohexane	25–60	—	180
Tartaric acid + hexanediol	B.2	Benzene	—	—	230
Toluene diisocyanate–oxypropylene glycol	B.2	Benzene–isooctane	34	—	346
Polycarbonates	B.1	Methylene chloride–methanol	—	—	158
	B.1	Phenol–cyclohexane	70	—	270
Poly(ethylene terephthalate)	B.1	Phenol, tetrachloroethane–ligroin	100	—	241
	B.1	Dimethylformamide	—	By cooling	357
	B.1	m-Cresol–ligroin	50	—	423
	B.2	Phenol, tetrachloroethane–nonane	95	—	423
	B.2	Various solvent–nonsolvent systems	—	—	185
	B.3	Trifluoroacetic acid–chloroform	—	—	420
Various polyesters	B.2	—	—	On paper	59
					11

POLYAMIDES

Polymer	Method	Solvent; Solvent–Nonsolvent	Temperature (°C)	Remarks	Reference
Poly(ε-caprolactam)	B.1	m-Cresol–petroleum ether	—	—	167
	B.1	m-Cresol–cyclohexane	—	—	215
	B.1	Phenol–water	70	—	449
	B.2	Phenol	—	—	196
	B.2	Various solvent–nonsolvent systems	—	—	420
	B.2	Phenol–ethyleneglycol, water	—	—	421
	B.3	Formic acid–water	—	—	15
	B.5	Formic acid	—	—	260
	C.1	m-Cresol–cyclohexane	25	—	206
Poly(γ-benzyl-L-glutamate)	B.2	Formic acid, ethanol	—	—	295
	B.3	Phenol, ethanol–ethanol, water	—	—	295
Polyhexamethyleneadipamide	B.1	m-Cresol–cyclohexane	—	—	205
	B.1	Phenol–water	—	—	409
Polyhexamethyleneadipamide + ε-Caprolactam	B.1	Phenol–water	80	—	27
Various polyamides	B.1	Cresol, benzene–ligroin	—	—	358
	C.2	Methanol	—	—	55

480   *Manfred J. R. Cantow*

SILICONES

Polymer	Method	Solvent; Solvent–Nonsolvent	Temperature (°C)	Remarks	Reference
Silicones	B.2	Methanol–carbon tetrachloride–cyclohexane	—	—	22
	B.3	Ethanol–methanol	—	—	22
Monoallyl derivatives	B.2	Acetone–ethanol–heptane	—	—	416

NATURAL POLYMERS

Polymer	Method	Solvent; Solvent–Nonsolvent	Temperature (°C)	Remarks	Reference
Amylose	B.1	Acetone–acetone, water	—	—	209
Asphaltenes	B.2	Benzene	—	—	172
	B.4	Benzene	—	—	7
Cellulose	B.1	Cuene–propanol	—	—	391
	B.1	60% Sulfuric acid–water	—	—	393
	B.2	Cuene	—	—	310
	B.2	Cuoxam	0	—	387
	B.3	—	—	Review	74
	C.2	—	—	Microcrystals	197
	C.5	—	—	—	113
	C.5	Cuoxam	—	—	266
	C.5	Cuoxam	—	—	287
	C.5	Cuoxam	—	—	288
Cellulose acetate	B.1	Acetone–ethanol	—	—	16
	B.1	Acetone–water	—	—	175
	B.1	Various solvent–nonsolvent systems	—	—	308
	B.1	Acetone–ethanol	—	—	334
	B.1	Acetone–heptane, acetone	—	—	362
	B.1	Acetone–ethanol	—	—	396
	B.1	Acetone, water–water	—	—	401
	B.1	Acetone–ethanol	—	—	404
	B.2	Acetone	—	—	292
	B.5	Ethylene chloride	—	—	260
	B.5	Ethylene chloride	—	—	261
	B.5	Ethylene chloride	—	—	264
	C.1	Butanol, ethanol–ethanol	—	—	44
	C.2	Acetone	—	—	436
	C.4	Aqueous NaOH–acetone	—	—	87
	C.5	Cyclohexanone	—	—	359

G. Additional Methods of Fractionation 483

Cellulose acetate–butyrate	B.1	Acetone–isopropyl ether	—	306
	B.1	Acetone–water, acetone	—	406
Cellulose nitrate	C.1	Acetone–ethanol, water	—	307
	B.1	Acetone–acetone, heptane	—	1
	B.1	Acetone–water	—	2
	B.1	Acetone–water	—	3
	B.1	Acetone–acetone, water	—	14
	B.1	Acetone–n-hexane	—	94
	B.1	Acetone–petroleum ether	—	123
	B.1	Acetone, hexane–hexane	—	181
	B.1	Acetone–hexane	—	208
	B.1	Acetone–water	0	272
	B.1	Acetone–water	21	273
	B.1	Acetone–water	—	279
	B.1	Acetone–water	—	280
	B.1	Acetone–water	—	281
	B.1	Acetone, water–water	25	296
	B.1	Acetone–acetone, water	—	313
	B.1	Acetone, water–water	—	350
	B.1	Ethyl acetate–n-heptane	—	363
	B.1	Acetone–acetone, water	—	375
	B.1	Acetone, water–water	—	411
	B.1	Acetone–water	25	413
	B.1	Acetone–water	—	438
	B.1	Acetone–acetone, water	—	455
	B.1	Acetone–acetone, water	—	456
	B.2	Methyl acetate–ethanol	—	57
	B.2	Ethyl acetate–ethanol	—	123
	B.2	Ethyl acetate–ethanol	—	199
	C.1	Acetone–methanol, water	—	327
	C.2	Acetone	—	326
	C.4	Acetone–water	—	345
	C.4	Acetone–water	—	407
	C.5	Butyl acetate	—	112

NATURAL POLYMERS—continued

Polymer	Method	Solvent; Solvent–Nonsolvent	Temperature (°C)	Remarks	Reference
Cellulose tributyrate	C.5	—	—	—	114
	D	—	—	—	2
Ethyl cellulose	B.1	Acetone–water	—	—	253
Hemicelluloses	B.1	Ethyl acetate, acetone–water	—	—	364
Methyl cellulose	B.3	—	—	—	62
Viscose	B.5	Water	—	—	143
	C.5	—	—	—	381
	C.5	—	—	—	382
Desoxyribonucleic acid	B.2	Aqueous NaCl	—	—	34
	C.2	Aqueous NaCl	—	—	386
Dextran	B.1	Water–methanol	—	—	13
	B.1	Water–ethanol	—	—	384
	B.1	Water–methanol	—	—	458
	C.2	Water–methanol	—	—	109
Fibrinogen	C.2	Phosphate buffer	—	—	124
Gelatin	B.2	Water–ethanol	—	—	340
	B.2	Water–ethanol	—	—	399
Glycogen	C.2	Water	—	—	56
Lactoglobulin	B.1	Aqueous ammonium sulfate	—	—	415
Lignin sulfonates	B.1	Water–ethanol	—	—	129
	B.2	Water–ethanol	—	—	129
	B.2	Aqueous NaCl–ethanol	—	—	297
	C.3	Aqueous NaCl, ethanol	—	—	297
Pectins	B.1	Dioxane–petroleum ether	—	—	328
	C.2	Aqueous alkali	—	—	398
Polyarabinose	B.1	Acetone–petroleum ether	—	—	163
	B.2	Acetone–petroleum ether	—	—	163
Polymerized oils	B.2	Various alcohols	—	—	38

G. Additional Methods of Fractionation

Polypeptides	B.1	Polyacrylic acids at various pH	—	446
	B.3	Methanol, ethanol–cyclohexane	10–60	339
Polysarcosine dimethylamide	B.3	Water–dioxane	15–65	73
Ribonucleic acid	B.2	Aqueous NaCl	23	51
Rubber	—		Review	33
	B.1	Benzene–methanol	—	46
	B.1	Benzene–isopropanol	—	46
	B.1	Chloroform–acetone	—	65
	B.1	Toluene–methanol	5	432
	B.1		—	440
	B.2	Ether	—	91
	B.2	n-Hexane	—	189
	B.2	Toluene–methanol	—	214
	B.3	Cyclohexanone	—	20
	B.3	Chloroform–methanol, water	22	37
	B.3	Benzene–ethanol	—	66
	C.2		—	372
	C.2		—	429
	C.5	Toluene	—	113
	C.5		—	116
	C.5		—	117
	C.5		—	118
Chlorinated Rubber	B.1	Toluene–methanol	—	6
Serum albumin	B.1	Toluene–methanol	—	349
	B.5	Phosphate buffer	—	234
	C.3	Water	—	92
Silk fibroin	B.1	Aqueous LiCNS–water	—	390
Sodium alginate	B.1	Water–aqueous NaCl	—	187
	B.1	Aqueous $MnCl_2$–aqueous $CaCl_2$	—	277
Starch	B.1	Thymol–n-butanol	—	90
	B.2	Aqueous NaOH–n-butanol	—	21
	B.2	Water	—	28
	C.2	Aqueous NaOH	—	29

COPOLYMERS

Polymer	Method	Solvent; Solvent–Nonsolvent	Temperature (°C)	Remarks	Reference
Acetaldehyde-co-propylene oxide	B.2	Methanol–chloroform	—	—	147
Acrylonitrile-co-methyl methacrylate	B.1	Dimethylformamide–n-hexane, ether	—	—	89
Acrylonitrile-co-vinyl acetate-co-methylvinylpyridine	B.1	Dimethylformamide–n-hexane, ether	—	—	89
Acrylonitrile-co-vinyl chloride	B.1	Acetone–methanol	—	—	379
Butadiene-co-styrene	B.1	Toluene–methanol	20	—	48
	B.1	Toluene–ethanol	—	—	153
	B.1	Toluene–methanol	—	—	214
	B.1	Toluene–isopropanol	—	—	351
	B.1	Chloroform–isopropanol	—	—	351
	B.1	Benzene–methanol	—	—	453
	B.2	Benzene	—	—	324
	B.2	Benzene	—	—	161
Butadiene-co-vinyl isopropyl ether	B.1	Benzene–methanol	—	—	426
Cumarone-co-indene	B.1	Benzene, ethyl acetate–methanol	—	—	457
Divinylbenzene-co-styrene	B.1	Acetone, dioxane–methanol	—	—	412
	C.2	Theta solvents	—	—	335
	C.2	n-Octane	21	—	337
	C.3	—	21	—	337
1-Dodecane-co-1-octadecane	B.1	Benzene–ethanol	—	—	135
	B.3	Benzene–ethanol	23–73	—	135
Ethylene-co-propylene	C.1	Heptane–n-propanol	—	—	151
p-Laurylamide-co-styrene	B.1	Butanone–butanol	—	—	322
Maleic anhydride-co-styrene	B.1	Butanone–cyclohexane	—	—	130
Methyl methacrylate-co-styrene	B.1	Acetone–methanol	—	—	9
	B.1	Benzene, chlorobenzene–petroleum ether	—	—	9
	B.1	Toluene–hexane, methanol	—	—	267

G. Additional Methods of Fractionation

Styrene-co-1,3-vinyltoluene	B.1	Butanone–diisopropyl ether	—	403
Vinyl acetate-co-vinylidene cyanide	B.1	Butanone–butanol	—	301
Cellulose acetate + methyl methacrylate	B.1	Nitromethane–methanol, water	50	452
Polyacrylonitrile + styrene	C.1	Acetone–water	—	210
Poly(butadiene-co-styrene) + epoxide	B.1	Chloroform–methanol	—	385
Polybutyl methacrylate + styrene	C.1	Chloroform–methanol	—	220
Polybutyl methacrylate + vinyl acetate	B.1	Benzene–methanol	—	191
Polychloromethylstyrene + isobutene	B.1	Acetone–methanol, water	—	191
Polyisobutylene + styrene	B.1	Benzene, butanone–methanol	—	240
	B.1	Cyclohexane–propanol	25	76
	B.3	Cyclohexane–isopropanol	25–60	76
Polymethyl methacrylate + styrene	B.1	Benzene, chlorobenzene–methanol	—	17
	B.1	Benzene–methanol	—	191
	C.1	Butanone–isopropanol	—	425
Polymethyl methacrylate + vinyl acetate	B.1	Acetone, water–methanol	—	191
Polystyrene + vinyl acetate	B.1	Butanone–methanol	—	394
Polyvinyl acetate + butyl methacrylate	B.1	Acetone–methanol, water	—	191
Polyvinyl acetate + methyl methacrylate	B.1	Acetone–methanol, water	—	191
Polyvinyl acetate + styrene	B.1	Benzene–petroleum ether	—	191
Polyvinyl benzoate + methyl methacrylate	B.1	Acetone–methanol	—	395
	B.2	Petroleum ether–acetone	—	10
	B.2	Acetone	—	238
Rubber + methyl methacrylate	B.3	Petroleum naphtha–benzene	18–50	85

REFERENCES TO APPENDIX

1. K. Aejmelaeus, *Ann. Acad. Sci. Fennicae: Ser. A II*, 63 (1956).
2. K. Aejmelaeus and H. Sihtola, *Paperi ja Puu* **40**, 437 (1958).
3. J. C. Aggarwala and J. L. McCarthy, *J. Indian Chem. Soc.* **26**, 11 (1949).
4. G. Allen, C. Booth, and M. N. Jones, *Polymer* **5**, 195 (1964).
5. G. Allen, C. Booth, and M. N. Jones, *Polymer* **5**, 257 (1964).
6. R. Allirot, *Compt. Rend.* **231**, 1065 (1950).
7. K. H. Altgelt, *J. Appl. Polymer Sci.* **9**, 3389 (1965).
8. F. R. Anderson, *J. Polymer Sci.* **C8**, 275 (1965).
9. D. J. Angier and R. J. Ceresa, *J. Polymer Sci.* **34**, 699 (1959).
10. D. J. Angier and D. T. Turner, *J. Polymer Sci.* **28**, 265 (1958).
11. I. Arendt and H. J. Schenck, *Kunststoffe* **48**, 111 (1958).
12. R. S. Aries and A. P. Sachs, *J. Polymer Sci.* **21**, 551 (1956).
13. L. H. Arond and H. P. Frank, *J. Phys. Chem.* **58**, 953 (1954).
14. H. Asaoka and A. Suzuki, *J. Soc. Textile Cellulose Ind., Japan* **11**, 32 (1955).
15. C. W. Ayers, *Analyst* **78**, 382 (1953).
16. W. J. Badgley and H. Mark, *J. Phys. Chem.* **51**, 58 (1947).
17. M. Baer, *J. Polymer Sci.* **A2**, 417 (1964).
18. F. E. Bailey, G. M. Powell, and K. L. Smith, *Ind. Eng. Chem.* **50**, 8 (1958).
19. C. A. Baker and R. J. Williams, *J. Chem. Soc.* p. 2352 (1956).
20. T. F. Banigan, *Science* **117**, 249 (1953).
21. W. Banks, C. T. Greenwood, and J. Thomson, *Makromol. Chem.* **31**, 197 (1959).
22. D. W. Bannister, C. S. Phillips, and R. J. Williams, *Anal. Chem.* **26**, 1451 (1954).
23. H. Batzer and A. Nisch, *Makromol. Chem.* **22**, 131 (1957).
24. H. Batzer, *Makromol. Chem.* **5**, 66 (1950).
25. H. Batzer and F. Wiloth, *Makromol. Chem.* **8**, 55 and 111 (1952).
26. H. Batzer, *Makromol. Chem.* **12**, 145 (1954).
27. H. Batzer and A. Moschle, *Makromol. Chem.* **22**, 195 (1957).
28. H. Baum, G. A. Gilbert, and H. L. Wood, *J. Chem. Soc.* p. 4047 (1955).
29. H. Baum and G. A. Gilbert, *J. Colloid Sci.* **11**, 428 (1956).
30. J. H. Baxendale, S. Bywater, and M. G. Evans, *Trans. Faraday Soc.* **42**, 675 (1946).
31. B. Baysal, G. Adler, and D. Ballantine, *J. Polymer Sci.* **B1**, 257 (1963).
32. W. H. Beattie and C. Booth, *J. Appl. Polymer Sci.* **7**, 507 (1963).
33. R. Belmas, *Rev. Gen. Caoutchouc* **27**, 152 (1950).
34. A. Bendich, J. R. Fresco, and H. S. Rosenkranz, *J. Am. Chem. Soc.* **77**, 3671 (1955).
35. A. Beresniewicz, *J. Polymer Sci.* **35**, 321 (1959).
36. J. B. Berkowitz, M. Yamin, and R. M. Fuoss, *J. Polymer Sci.* **28**, 69 (1958).
37. D. E. Bernal, *Rev. General Caoutchouc* **32**, 889 (1955).
38. I. M. Bernstein, *J. Phys. & Colloid Chem.* **52**, 613 (1948).
39. G. C. Berry and R. G. Craig, *Polymer* **5**, 19 (1964).
40. J. C. Bevington, G. M. Guzmán, and H. W. Melville, *Proc. Roy. Soc.* **A221**, 437 (1954).
41. J. P. Bianchi, F. P. Price, and B. H. Zimm, *J. Polymer Sci.* **25**, 27 (1957).
42. F. W. Billmeyer, Jr., and W. H. Stockmayer, *J. Polymer Sci.* **5**, 121 (1950).
43. F. W. Billmeyer, Jr., *J. Polymer Sci.* **C8**, 161 (1965).
44. F. Bischoff and V. Desreux, *Bull. Soc. Chim. Belges* **60**, 137 (1951).
45. J. Bisschops, *J. Polymer Sci.* **17**, 81 (1955).
46. G. F. Bloomfield, *Rubber Chem. Technol.* **24**, 737 (1951).
47. E. G. Bobalek, S. S. Levy, and C. C. Lee, *J. Appl. Polymer Sci.* **8**, 625 (1964).
48. C. Booth and L. R. Beason, *J. Polymer Sci.* **42**, 93 (1960).
49. R. F. Boyer, *J. Polymer Sci.* **8**, 73 (1952).

50. R. F. Boyer, *J. Polymer Sci.* **9**, 197 (1952).
51. D. F. Bradley and A. Rich, *J. Am. Chem. Soc.* **78**, 5898 (1956).
52. S. E. Bresler, I. Y. Poddubnyi, and S. Y. Frenkel, *Zh. Tekhn. Fiz.* **23**, 1521 (1953).
53. S. E. Bresler and S. Y. Frenkel, *Zh. Tekhn. Fiz.* **23**, 1502 (1953).
54. S. E. Bresler, I. Y. Poddubnyi, and S. Y. Frenkel, *Rubber Chem. Technol.* **30**, 5C7 (1957).
55. S. E. Bresler, V. V. Korshak, and S. A. Pavlova, *Dokl. Akad. Nauk. SSSR* **87**, 961 (1952).
56. W. B. Bridgman, *J. Am. Chem. Soc.* **64**, 2349 (1942).
57. M. C. Brooks and R. M. Badger, *J. Am. Chem. Soc.* **72**, 1705 (1950).
58. F. M. Brower and H. W. McCormick, *J. Polymer Sci.* **A1**, 1749 (1963).
59. S. D. Bruck, *J. Polymer Sci.* **32**, 519 (1958).
60. K. Buerger, *Z. Anal. Chem.* **196**, 259 (1963).
61. D. E. Burge and D. B. Bruss, *J. Polymer Sci.* **A1**, 1927 (1963).
62. A. Buurman, *Textile Res. J.* **23**, 888 (1953).
63. A. Buzagh, K. Udvarhelyi, and F. Horkay, *Kolloid-Z.* **154**, 130 (1957).
64. A. Buzagh, K. Udvarhelyi, and F. Horkay, *Kolloid-Z.* **157**, 53 (1958).
65. S. Bywater and P. Johnson, *Trans. Faraday Soc.* **47**, 195 (1951).
66. G. Cajelli, *Rubber Chem. Technol.* **12**, 762 (1939).
67. H. Campbell, P. O. Kane, and I. G. Ottewill, *J. Polymer Sci.* **12**, 611 (1954).
68. H.-J. Cantow, *Makromol. Chem.* **70**, 130 (1964).
69. H.-J. Cantow, *Makromol. Chem.* **30**, 81 (1959).
70. M. J. R. Cantow, R. S. Porter, and J. F. Johnson, *J. Appl. Polymer Sci.* **8**, 2963 (1964).
71. M. J. R. Cantow, R. S. Porter, and J. F. Johnson, *J. Polymer Sci.* **C1**, 187 (1963).
72. M. J. R. Cantow, G. Meyerhoff, and G. V. Schulz, *Makromol. Chem.* **49**, 1 (1961).
73. S. R. Caplan, *J. Polymer Sci.* **35**, 409 (1959).
74. C. M. Conrad, *Ind. Eng. Chem.* **45**, 2511 (1953).
75. J. Cepelak, *Chem. Prumysl* **6**, 106 (1956).
76. A. Chapiro, J. Jozefowicz, and J. Sebban-Danon, *J. Polymer Sci.* **C4**, 491 (1963).
77. S. N. Chinai and R. J. Samuels, *J. Polymer Sci.* **19**, 463 (1956).
78. S. N. Chinai and R. A. Guzzi, *J. Polymer Sci.* **21**, 417 (1956).
79. S. N. Chinai, *J. Polymer Sci.* **25**, 413 (1957).
80. G. Ciampa and H. Schwindt, *Chim. Ind. (Milan)* **37**, 169 (1955).
81. S. Claesson, *Discussions Faraday Soc.* **7**, 321 (1949).
82. D. Cleverdon and D. Laker, *J. Appl. Chem.* **1**, 6 (1951).
83. R. L. Combs, D. F. Slonaker, and J. T. Summers, *Am. Chem. Soc., Div. Org. Coatings Plastics Chem., Preprints* **21**, 249 (1961).
84. W. Cooper, D. E. Eaves, and G. Vaughan, *J. Polymer Sci.* **59**, 241 (1962).
85. W. Cooper, G. Vaughan, and R. W. Madden, *J. Appl. Polymer Sci.* **1**, 329 (1959).
86. W. Cooper, G. Vaughan, and J. Yardley, *J. Polymer Sci.* **59**, S2 (1962).
87. S. Coppick, O. A. Battista, and M. R. Lytton, *Ind. Eng. Chem.* **42**, 2533 (1950).
88. B. J. Cottam, D. M. Wiles, and S. Bywater, *Can. J. Chem.* **41**, 1905 (1963).
89. G. R. Cotten and W. C. Schneider, *J. Appl. Polymer Sci.* **7**, 1243 (1963).
90. J. M. Cowie and C. T. Greenwood, *J. Chem. Soc.* p. 4640 (1957).
91. B. S. Das and P. K. Choudhury, *Indian J. Appl. Chem.* **22**, 73 (1959).
92. M. Daune, H. Benoit, and C. Sadron, *J. Polymer Sci.* **16**, 483 (1955).
93. T. E. Davis and R. L. Tobias, *J. Polymer Sci.* **50**, 227 (1961).
94. W. E. Davis, *J. Am. Chem. Soc.* **69**, 1453 (1947).
95. L. de Brouckere, E. Bidaine, and A. van der Heyden, *Bull. Soc. Chim. Belges* **58**, 418 (1949).
96. P. Debye and A. M. Bueche, in "High Polymer Physics" (H. A. Robinson, ed.), p. 497. Remsen Press Division, New York, 1948.

97. V. Desreux and M. C. Spiegels, *Bull. Soc. Chim. Belges* **59**, 476 (1950).
98. V. Desreux, *Rec. Trav. Chim.* **68**, 789 (1949).
99. V. Desreux, *Bull. Soc. Chim. Belges* **57**, 416 (1948).
100. H. Determann, G. Lueben, and T. Wieland, *Makromol. Chem.* **73**, 168 (1964).
101. K. Dialer, K. Vogler, and F. Patat, *Helv. Chim. Acta* **35**, 869 (1952).
102. K. Dialer and K. Vogler, *Makromol. Chem.* **6**, 191 (1951).
103. H. A. Dieu, *J. Polymer Sci.* **37**, 173 (1959).
104. H. A. Dieu, *J. Polymer Sci.* **12**, 417 (1954).
105. A. Dobry, *J. Chim. Phys.* **35**, 392 (1938).
106. P. Doty, H. L. Wagner, and S. Singer, *J. Phys. Chem.* **51**, 32 (1947).
107. K. Dusek, *Chem. Listy* **50**, 1948 (1956).
108. K. C. Eberly and B. L. Johnson, *J. Polymer Sci.* **3**, 283 (1948).
109. K. H. Ebert, M. Brosche, and K. F. Elgert, *Makromol. Chem.* **72**, 191 (1964).
110. K. Edelmann, *Faserforsch. Textiltech.* **6**, 269 (1955).
111. K. Edelmann, *Kolloid-Z.* **145**, 92 (1956).
112. K. Edelmann, *Faserforsch. Textiltech.* **5**, 59 (1954).
113. K. Edelmann, *Rubber Chem. Technol.* **30**, 470 (1957).
114. K. Edelmann, *Faserforsch. Textiltech.* **8**, 184 (1957).
115. K. Edelmann, *Faserforsch. Textiltech.* **5**, 325 (1954).
116. K. Edelmann, *Rheol. Acta* **1**, 53 (1958).
117. K. Edelmann, *Gummi Asbest* **11**, 251 (1958).
118. K. Edelmann, *Gummi Asbest* **12**, 66 (1959).
119. E. J. Elgood, N. S. Heath, and D. H. Solomon, *J. Appl. Polymer Sci.* **8**, 881 (1964).
120. H.-G. Elias and U. Gruber, *Makromol. Chem.* **78**, 72 (1964).
121. H.-G. Elias and F. Patat, *Makromol. Chem.* **23**, 13 (1957).
122. A. H. Emery and H. G. Drickamer, *J. Chem. Phys.* **23**, 2252 (1953).
123. C. Emery and W. E. Cohen, *Australian J. Appl. Sci.* **2**, 473 (1951).
124. H. A. Ende and G. V. Schulz, *Z. Physik. Chem. (Frankfurt)* [N.F.] **33**, 143 (1962).
125. A. F. V. Eriksson, *Acta Chem. Scand.* **7**, 377 (1953).
126. A. F. V. Eriksson, *Acta Chem. Scand.* **7**, 623 (1953).
127. A. F. V. Eriksson, *Acta Chem. Scand.* **3**, 1 (1949).
128. A. F. V. Eriksson, *Acta Chem. Scand.* **10**, 360 and 378 (1956).
129. V. F. Felicetta, A. Ahola, and J. L. McCarthy, *J. Am. Chem. Soc.* **78**, 1899 (1956).
130. J. D. Ferry, D. C. Udy, and W. F. Chi, *J. Colloid Sci.* **6**, 429 (1951).
131. S. R. Finn and J. W. James, *J. Appl. Chem.* **6**, 466 (1956).
132. E. B. Fitzgerald and R. M. Fuoss, *Ind. Eng. Chem.* **42**, 1603 (1950).
133. J. R. Fletcher, L. Polgar, and D. H. Solomon, *J. Appl. Polymer Sci.* **8**, 663 (1964).
134. P. J. Flory, *J. Am. Chem. Soc.* **65**, 372 (1943).
135. D. L. Flowers, W. A. Hewett, and R. D. Mullineaux, *J. Polymer Sci.* **A2**, 2305 (1964).
136. C. M. Fontana, *J. Phys. Chem.* **63**, 1167 (1959).
137. T. G. Fox and P. J. Flory, *J. Am. Chem. Soc.* **70**, 2384 (1948).
138. T. G. Fox and P. J. Flory, *J. Phys. Chem.* **53**, 197 (1949).
139. P. S. Francis, R. C. Cooke, Jr., and J. H. Elliott, *J. Polymer Sci.* **31**, 453 (1958).
140. H. P. Frank and J. W. Breitenbach, *J. Polymer Sci.* **6**, 609 (1951).
141. M. Freeman and P. P. Manning, *J. Polymer Sci.* **A2**, 2017 (1964).
142. L. Freund and M. Daune, *J. Polymer Sci.* **29**, 161 (1958).
143. H. Fritzemeier and J. J. Hermans, *Bull. Soc. Chim. Belges* **57**, 136 (1948).
144. O. Fuchs, *Z. Elektrochem.* **60**, 229 (1956).
145. O. Fuchs, *Makromol. Chem.* **7**, 259 (1952).
146. O. Fuchs, *Makromol. Chem.* **5**, 245 (1951).

G. Additional Methods of Fractionation 491

147. H. Fujii, T. Saegusa, and J. Furukawa, *Makromol. Chem.* **63**, 147 (1963).
148. H. Fujita and K. Ninomiya, *J. Polymer Sci.* **24**, 233 (1957).
149. H. Fujita and K. Ninomiya, *J. Phys. Chem.* **61**, 814 (1957).
150. A. Gallo, *Chim. Ind. (Milan)* **35**, 487 (1953).
151. L. W. Gamble, W. T. Wipke, and T. Lane, *J. Appl. Polymer Sci.* **9**, 1503 (1965).
152. V. Garten and W. Becker, *Makromol. Chem.* **3**, 78 (1949).
153. G. Gavoret and M. Magat, *J. Chim. Phys.* **44**, 90 (1947).
154. G. Gavoret and J. Duclaux, *J. Chim. Phys.* **42**, 41 (1945).
155. G. Geiseler, H. Herold, and F. Runge, *Erdoel Kohle* **7**, 357 (1954).
156. G. Geiseler and H. P. Baumann, *Z. Elektrochem.* **62**, 209 (1958).
157. A. N. Genkin, T. P. Nasonova, and I. Y. Poddubnyi, *Vysokomolekul. Soedin.* **4**, 1088 (1962).
158. G. Glöckner, *Plaste Kautschuk* **10**, 154 (1963).
159. A. I. Goldberg, W. P. Hohenstein, and H. Mark, *J. Polymer Sci.* **2**, 503 (1947).
160. M. A. Golub, *J. Polymer Sci.* **11**, 281 (1953).
161. M. A. Golub, *J. Polymer Sci.* **11**, 583 (1953).
162. G. Gooberman, *J. Polymer Sci.* **40**, 469 (1959).
163. A. E. Goodban and H. S. Owens, *J. Polymer Sci.* **23**, 825 (1957).
164. A. Gordienko, *Faserforsch. Textiltech.* **4**, 499 (1953).
165. N. Gralèn and G. Lagermalm, *J. Phys. Chem.* **56**, 514 (1952).
166. J. H. S. Green and M. F. Vaughan, *Chem. & Ind. (London)* p. 829 (1958).
167. W. Griehl and H. Luckert, *J. Polymer Sci.* **30**, 399 (1958).
168. H. Groeblinghoff, Ph.D. Thesis, Graz University (1961).
169. J. E. Guillet, *J. Polymer Sci.* **47**, 307 (1960).
170. J. E. Guillet, R. L. Combs, and D. F. Slonaker, *J. Polymer Sci.* **47**, 307 (1960).
171. J. E. Guillet, R. L. Combs, and H. W. Coover, Jr., *SPE (Soc. Plastics Engrs.) Trans.* **2**, 164 (1964).
172. E. Gundermann, *Chem. Tech. (Berlin)* **7**, 678 (1955).
173. G. M. Guzmán, *J. Polymer Sci.* **19**, 519 (1956).
174. G. M. Guzmán, *Anales Real Soc. Espan. Fis. Quim. (Madrid)* **B50**, 631 (1954).
175. G. M. Guzmán and J. M. Fatou, *Anales Real Soc. Espan. Fiz. Quim. (Madrid)* **B53**, 669 (1957).
176. G. M. Guzmán and J. M. Fatou, *Anales Real Soc. Espan. Fis. Quim. (Madrid)* **B54**, 609 (1958).
177. W. F. Haddon, R. S. Porter, and J. F. Johnson, *J. Appl. Polymer Sci.* **8**, 1371 (1964).
178. W. Hahn, W. Müller, and R. W. Webber, *Makromol. Chem.* **21**, 131 (1956).
179. J. Hannus and G. Smets, *Bull. Soc. Chim. Belges* **60**, 76 (1951).
180. C. M. Hansen and G. A. Sather, *J. Appl. Polymer Sci.* **8**, 2479 (1964).
181. W. G. Harland, *J. Textile Inst. Trans.* **46**, 483 (1955).
182. R. E. Harrington and B. H. Zimm, *Am. Chem. Soc., Div. Polymer Chem., Preprints* **6**, 346 (1965).
183. I. Harris and R. G. J. Miller, *J. Polymer Sci.* **7**, 337 (1951).
184. N. Hartler, *Acta Chem. Scand.* **11**, 1162 (1957).
185. E. A. Haseley, *J. Polymer Sci.* **35**, 309 (1959).
186. G. W. Hastings, D. W. Ovenall, and F. W. Peaker, *Nature* **177**, 1091 (1956).
187. A. Haug, *Acta Chem. Scand.* **13**, 601 (1959).
188. E. A. Hauser and D. S. Le Beau, *J. Phys. Chem.* **54**, 256 (1950).
189. E. A. Hauser and D. S. Le Beau, *Rubber Chem. Technol.* **20**, 70 (1947).
190. S. W. Hawkins and H. Smith, *J. Polymer Sci.* **28**, 341 (1958).
191. P. Hayden and R. Roberts, *Intern. J. Appl. Radiation Isotopes* **5**, 269 (1959).

192. J. Hengstenberg, *Z. Elektrochem.* **60**, 236 (1956).
193. P. M. Henry, *J. Polymer Sci.* **36**, 3 (1959).
194. M. Hermann, Ph.D. Thesis, Graz University (1962).
195. J. J. Hermans and H. A. Ende, *J. Polymer Sci.* **C1**, 161 (1963).
196. P. H. Hermans, D. Heikens, and P. F. van Velden, *J. Polymer Sci.* **16**, 451 (1955).
197. J. Hermans, *J. Polymer Sci.* **C2**, 129 (1963).
198. P. Herrent, *J. Polymer Sci.* **8**, 346 (1952).
199. E. Heuser, W. Shockley, and R. Kjellgreen, *Tappi* **33**, 92 (1950).
200. S. M. Hirshfield and E. L. Allen, *J. Polymer Sci.* **39**, 554 (1959).
201. J. D. Hoffmann and B. H. Zimm, *J. Polymer Sci.* **15**, 405 (1955).
202. R. H. Horowitz, *Am. Chem. Soc., Div. Polymer Chem., Preprints* **3**, 167 (1962).
203. R. H. Horowitz, *Am. Chem. Soc., Div. Polymer Chem., Preprints* **4**, 689 (1963).
204. R. C. Houtz, *Textile Res. J.* **20**, 786 (1950).
205. G. J. Howard, *J. Polymer Sci.* **37**, 310 (1959).
206. G. J. Howard, *J. Polymer Sci.* **A1**, 2667 (1963).
207. J. M. Hulme and L. A. McLeod, *Polymer* **3**, 153 (1962).
208. M. L. Hunt, H. A. Scheraga, and P. J. Flory, *J. Phys. Chem.* **60**, 1278 (1956).
209. E. Husemann, *Makromol. Chem.* **26**, 181 and 199 (1958).
210. F. Ide, R. Handa, and K. Nakatsuka, *Chem. High Polymers* (*Tokyo*) **21**, 57 (1964).
211. K. J. Ivin, H. A. Ende, and G. Meyerhoff, *Polymer* **3**, 129 (1962).
212. H. G. Jellinek and G. White, *J. Polymer Sci.* **6**, 757 (1951).
213. B. Jirgensons, *J. Polymer Sci.* **8**, 519 (1952).
214. B. L. Johnson, *Ind. Eng. Chem.* **40**, 351 (1948).
215. J. Juilfs, *Kolloid-Z.* **141**, 88 (1955).
216. J. L. Jungnickel and F. J. Weiss, *J. Polymer Sci.* **49**, 437 (1961).
217. H. Kämmerer, W. Kern, and W. Heuser, *J. Polymer Sci.* **28**, 331 (1958).
218. H. Kämmerer and F. Rocaboy, *Compt. Rend.* **256**, 4440 (1963).
219. H. Kakiuchi and W. Fukuda, *J. Chem. Soc. Japan, Ind. Chem. Sect.* **66**, 964 (1963).
220. V. A. Kargin, N. A. Plate, and A. S. Dobrynina, *Colloid J. USSR* (*English Transl.*) **20**, 332 (1958).
221. A. Katchalsky and H. Eisenberg, *J. Polymer Sci.* **6**, 145 (1951).
222. H. S. Kaufman and E. K. Walsh, *J. Polymer Sci.* **26**, 124 (1957).
223. H. S. Kaufman and E. Solomon, *Ind. Eng. Chem.* **45**, 1779 (1953).
224. K. Kawahara, *Makromol. Chem.* **73**, 1 (1964).
225. K. Kawahara, *Ann. Rept. Inst. Fiber Res. Japan* **9**, 30 (1955).
226. A. Kawasaki, J. Furukawa, and T. Tsuruta, *Markromol. Chem.* **42**, 25 (1960).
227. A. Keller and A. O'Connor, *Polymer* **1**, 163 (1960).
228. A. S. Kenyon and I. O. Salyer, *J. Polymer Sci.* **43**, 427 (1960).
229. A. S. Kenyon, I. O. Salyer, and D. R. Brown, *J. Polymer Sci.* **C8**, 205 (1965).
230. W. Kern, H. Schmidt, and H. Steinwehr, *Makromol. Chem.* **16**, 74 (1955).
231. W. Kern, H. Kämmerer, and G. Dall'Asta, *Makromol. Chem.* **6**, 206 (1951).
232. P. O. Kinell, *Acta Chem. Scand.* **1**, 832 (1947).
233. P. O. Kinell, *Acta Chem. Scand.* **1**, 335 (1947).
234. J. G. Kirkwood and R. A. Brown, *J. Am. Chem. Soc.* **74**, 1056 (1952).
235. H. Kobayashi and Y. Fujisaki, *J. Polymer Sci.* **B1**, 15 (1963).
236. H. Kobayashi, *J. Polymer Sci.* **26**, 230 (1957).
237. H. Kobayashi, K. Sasaguri, and T. Amano, *J. Polymer Sci.* **A2**, 313 (1964).
238. W. Kobryner and A. Banderet, *Compt. Rend.* **244**, 604; **245**, 689 (1957).
239. R. Kocher and C. Sadron, *Makromol. Chem.* **10**, 172 (1953).
240. G. Kockelbergh and G. Smets, *J. Polymer Sci.* **33**, 227 (1958).

G. Additional Methods of Fractionation 493

241. H. M. Koepp and H. Werner, *Makromol. Chem.* **32**, 79 (1959).
242. I. Kössler and J. Krejsa, *J. Polymer Sci.* **57**, 509 (1962).
243. I. Kössler and J. Krejsa, *J. Polymer Sci.* **29**, 69 (1958).
244. I. Kössler and M. Stolka, *J. Polymer Sci.* **44**, 213 (1960).
245. I. Kössler and J. Krejsa, *J. Polymer Sci.* **35**, 308 (1959).
246. V. Kokle and F. W. Billmeyer, Jr., *J. Polymer Sci.* **C8**, 217 (1965).
247. R. Koningsveld and A. J. Pennings, *Rec. Trav. Chim.* **83**, 552 (1964).
248. S. Krause and E. Cohn-Ginsberg, *J. Polymer Sci.* **A2**, 1393 (1964).
249. J. Krejsa, *Markromol. Chem.* **33**, 244 (1959).
250. W. R. Krigbaum and P. J. Flory, *J. Am. Chem. Soc.* **75**, 1775 (1953).
251. W. R. Krigbaum and A. M. Kotliar, *J. Polymer Sci.* **32**, 323 (1958).
252. W. R. Krigbaum, J. E. Kurz, and P. Smith, *J. Phys. Chem.* **65**, 1984 (1961).
253. R. F. Landel and J. D. Ferry, *J. Phys. Chem.* **59**, 658 (1955).
254. I. Landler, *Compt. Rend.* **225**, 629 (1947).
255. W. J. Langford and D. J. Vaughan, *Nature* **184**, 116 (1959).
256. G. Langhammer, H. Pfennig, and K. Quitzsch, *Z. Elektrochem.* **62**, 458 (1958).
257. G. Langhammer, *Naturwissenschaften* **41**, 552 (1954).
258. G. Langhammer and K. Quitzsch, *Z. Elektrochem.* **65**, 706 (1961).
259. G. Langhammer, *Kolloid-Z.* **146**, 44 (1956).
260. G. Langhammer, *Svensk Kem. Tidskr.* **69**, 328 (1957).
261. G. Langhammer, *J. Polymer Sci.* **29**, 505 (1958).
262. G. Langhammer and K. Quitzsch, *Makromol. Chem.* **43**, 160 (1961).
263. G. Langhammer and H. Foerster, *Z. Physik. Chem. (Frankfurt)* [N.F.] **15**, 212 (1958).
264. G. Langhammer, *Makromol. Chem.* **21**, 74 (1956).
265. H. Leaderman, R. G. Smith, and L. C. Williams, *J. Polymer Sci.* **36**, 233 (1959).
266. F. Linsert, Ph.D. Thesis, University of Köln (1950).
267. A. D. Litmanovich, V. Y. Shtern, and A. V. Topchiev, *Neftekhimiya* **3**, 217 (1963).
268. F. Lombard, *Makromol. Chem.* **8**, 201 (1952).
269. B. V. Losikov, N. I. Kaverina, and A. A. Fedyantseva, *Khim. i Tekhnol. Topl.* **3**, 51 (1956).
270. S. Lunak and M. Bohdanecky, *Collection Czech. Chem. Commun.* **30**, 2756 (1965).
271. C. A. Marshall and R. A. Mock, *J. Polymer Sci.* **17**, 591 (1955).
272. M. Marx-Figini, *J. Polymer Sci.* **30**, 119 (1958).
273. M. Marx-Figini, *Makromol. Chem.* **32**, 233 (1959).
274. H. W. McCormick, *J. Polymer Sci.* **36**, 341 (1959).
275. H. W. McCormick, *J. Polymer Sci.* **A1**, 103 (1963).
276. H. W. McCormick, *J. Polymer Sci.* **41**, 327 (1959).
277. R. H. McDowell, *Chem. & Ind. (London)* p. 1401 (1958).
278. J. W. Meeks, T. F. Banigan, and R. W. Planck, *India Rubber World* **122**, 301 (1950).
279. A. M. Meffroy-Biget, *Compt. Rend.* **240**, 1707 (1955).
280. A. M. Meffroy-Biget, *Bull. Soc. Chim. France* p. 458 (1954).
281. P. C. Mehta and E. Pacsu, *Textile Res. J.* **19**, 699 (1949).
282. H. W. Melville and B. D. Stead, *J. Polymer Sci.* **16**, 505 (1956).
283. Z. Menčik, *Chem. Zvesti* **9**, 165 (1955).
284. R. A. Mendelson, *J. Polymer Sci.* **A1**, 2361 (1963).
285. E. H. Merz and R. W. Raetz, *J. Polymer Sci.* **5**, 587 (1950).
286. E. H. Merz, *J. Polymer Sci.* **3**, 790 (1948).
287. W. Meskat, *Chem.-Ing.-Tech.* **24**, 333 (1952).
288. W. Meskat, *Dechema Monograph.* **25**, 9 (1955).
289. G. Meyerhoff and J. Romatowski, *Makromol. Chem.* **74**, 222 (1964).

290. G. Meyerhoff, *Makromol. Chem.* **12**, 45 (1954).
291. N. V. Mikhailov and S. G. Zelikman, *Colloid. J. USSR (English Transl.)* **18**, 717 (1956).
292. B. Miller and E. Pacsu, *J. Polymer Sci.* **41**, 97 (1959).
293. L. E. Miller and F. A. Hamm, *J. Phys. Chem.* **57**, 110 (1953).
294. D. R. Mills, G. E. Moore, and D. W. Pugh, *Tech. Papers, Reg. Tech. Conf., Soc. Plastics Engrs.* **6**, 10 (1960).
295. J. C. Mitchell, A. E. Woodward, and P. Doty, *J. Am. Chem. Soc.* **79**, 3955 (1957).
296. R. L. Mitchell, *Ind. Eng. Chem.* **45**, 2526 (1953).
297. J. Moacanin, H. Nelson, and E. Back, *J. Am. Chem. Soc.* **81**, 2054 (1959).
298. W. E. Mochel, J. B. Nichols, and C. J. Mighton, *J. Am. Chem. Soc.* **70**, 2185 (1948).
299. W. E. Mochel and J. B. Nichols, *J. Am. Chem. Soc.* **71**, 3435 (1949).
300. W. E. Mochel and J. B. Nichols, *Ind. Eng. Chem.* **43**, 154 (1951).
301. R. A. Mock, C. A. Marshall, and V. D. Floria, *J. Polymer Sci.* **11**, 447 (1953).
302. J. C. Moore, *J. Polymer Sci.* **A2**, 835 (1964).
303. L. D. Moore, Jr., G. R. Greear, and J. O. Sharp, *J. Polymer Sci.* **59**, 339 (1962).
304. W. R. Moore and R. J. Hutchinson, *Nature* **200**, 1095 (1964).
305. D. R. Morey, E. W. Taylor, and G. P. Waugh, *J. Colloid Sci.* **6**, 470 (1951).
306. D. R. Morey and J. W. Tamblyn, *J. Phys. Chem.* **51**, 721 (1947).
307. D. R. Morey and J. W. Tamblyn, *J. Appl. Phys.* **16**, 419 (1945).
308. D. R. Morey and J. W. Tamblyn, *J. Phys. & Colloid Chem.* **50**, 12 (1946).
309. T. M. Moshkina and A. N. Pudovik, *Vysokomolekul. Soedin.* **5**, 1106 (1963).
310. W. A. Muller and L. N. Rogers, *Ind. Eng. Chem.* **45**, 2522 (1953).
311. C. Mussa, *J. Polymer Sci.* **28**, 587 (1958).
312. G. E. Myers and J. R. Dagon, *J. Polymer Sci.* **A2**, 2631 (1964).
313. H. Nakahara and M. Shihanda, *J. Soc. Textile Cellulose Ind., Japan* **8**, 438 (1952).
314. A. Nakajima and I. Sakurada, *Chem. High Polymers (Tokyo)* **11**, 11 (1954).
315. A. Nakajima and H. Fujiwara, *Bull. Chem. Soc. Japan* **37**, 909 (1964).
316. A. Nasini and C. Mussa, *Makromol. Chem.* **22**, 59 (1957).
317. G. Natta, M. Pegoraro, and M. Peraldo, *Ric. Sci.* **28**, 1473 (1958).
318. G. Natta, F. Danusso, and G. Moraglio, *Makromol. Chem.* **20**, 37 (1956).
319. S. Newman, *J. Polymer Sci.* **47**, 111 (1960).
320. L. Nicolas, *Compt. Rend.* **236**, 809 (1953).
321. L. Nicolas, *Compt. Rend.* **242**, 2720 (1956).
322. N. T. Notley, *J. Polymer Sci.* **A1**, 227 (1963).
323. H. Okamoto, *J. Polymer Sci.* **A2**, 3451 (1964).
324. R. J. Orr and H. L. Williams, *J. Am. Chem. Soc.* **79**, 3137 (1957).
325. A. Oth and V. Desreux, *Bull. Soc. Chim. Belges.* **63**, 261 (1954).
326. A. Oth and V. Desreux, *Ric. Sci.* **25**, 447 (1955).
327. A. Oth, *Bull. Soc. Chim. Belges.* **58**, 285 (1949).
328. H. S. Owens, J. C. Miers, and W. D. Maclay, *J. Colloid Sci.* **3**, 277 (1948).
329. C. J. Panton, P. H. Plesch, and P. P. Rutherford, *J. Chem. Soc.* p. 2586 (1964).
330. P. Parrini, F. Sebastiano, and G. Messina, *Makromol. Chem.* **38**, 27 (1960).
331. S. A. Pavlova, T. A. Soboleva, and A. P. Suprun, *Vysokomolekul. Soedin.* **6**, 122 (1964).
332. H. G. Peer, *Rec. Trav. Chim.* **78**, 631 (1959).
333. D. C. Pepper and P. P. Rutherford, *J. Appl. Polymer Sci.* **2**, 100 (1959).
334. H. J. Philipp and C. F. Bjork, *J. Polymer Sci.* **6**, 383 and 549 (1951).
335. I. Y. Poddubnyi, V. A. Grechanovskii, and M. I. Mosevitskii, *Vysokomolekul. Soedin.* **5**, 1042 (1963).
336. I. Y. Poddubnyi, V. A. Grechanovskii, and M. I. Mosevitskii, *Vysokomolekul. Soedin.* **5**, 1049 (1963).

G. Additional Methods of Fractionation

337. I. Y. Poddubnyi, V. A. Grechanovskii, and M. I. Mosevitskii, *Vysokomolekul. Soedin.* **7**, 1042 (1963).
338. D. J. Pollock, L. J. Elyash, and T. W. De Witt, *J. Polymer Sci.* **15**, 87 and 336 (1955).
339. N. T. Pope, T. J. R. Weakley, and R. J. P. Williams, *J. Chem. Soc.* p. 3442 (1959).
340. J. Pouradier and A. M. Venet, *J. Chim. Phys.* **47**, 11 (1950).
341. P. O. Powers, *Ind. Eng. Chem.* **42**, 2558 (1950).
342. N. A. Pravikova, Y. U. Ryabova, and P. Vyrskii, *Vysokomolekul. Soedin.* **5**, 1165 (1963).
343. S. R. Rafikov, V. V. Korshak, and G. N. Chelnokova, *Bull. Acad. Sci. USSR, Div. Chem. Sci.* (*English Transl.*) p. 642 (1948).
344. N. K. Raman and J. J. Hermans, *J. Polymer Sci.* **35**, 71 (1959).
345. B. G. Ranby, O. W. Woltersdorf, and O. A. Battista, *Svensk. Paperstid.* **60**, 373 (1957).
346. N. S. Rapp and J. D. Ingham, *J. Polymer Sci.* **A2**, 689 (1964).
347. G. B. Rathmann and F. A. Bovey, *J. Polymer Sci.* **15**, 544 (1955).
348. O. Redlich, A. L. Jacobson, and W. H. McFadden, *J. Polymer Sci.* **A1**, 393 (1963).
349. M. Riou and R. Pibarot, *Rev. Gen. Caoutchouc.* **27**, 596 (1950).
350. W. E. Roseveare and L. Poore, *Ind. Eng. Chem.* **45**, 2518 (1953).
351. L. S. Rosik and B. Krabal, *Chem. Prumysl.* **9**, 377 (1959).
352. J. F. Rudd and E. F. Gurnee, *J. Polymer Sci.* **A1**, 2857 (1963).
353. J. F. Rudd, *J. Polymer Sci.* **44**, 459 (1960).
354. J. F. Rudd, *J. Polymer Sci.* **60**, S9 (1962).
355. C. A. Russell, *J. Appl. Polymer Sci.* **4**, 219 (1960).
356. B. N. Rutowski and W. W. Tschebotarwski, *Kunststoffe* **41**, 230 (1951).
357. F. Rybnikar, *Chem. Listy* **50**, 1190 (1956).
358. F. Rybnikar, *Collection Czech. Chem. Commun.* **21**, 1101 (1956).
359. C. Sadron and H. Mosimann, *J. Phys. Radium* **9**, 384 (1938).
360. M. I. Savitskaya and S. Y. Frenkel, *Zh. Fiz. Khim.* **32**, 1063 (1958).
361. K. H. Schäfer, Ph.D. Thesis, Graz University (1959).
362. P. C. Scherer and R. B. Thompson, *Rayon Synthetic Textiles* **31**, 51 (1950).
363. P. C. Scherer and B. P. Rouse, *Rayon Synthetic Textiles* **29**, 55 and 85 (1948).
364. P. C. Scherer and R. D. McNeer, *Rayon Synthetic Textiles* **30**, 56 (1949).
365. E. Schildknecht, S. Gross, and H. Davidson, *Ind. Eng. Chem.* **40**, 2104 (1948).
366. N. S. Schneider, J. D. Loconti, and L. G. Holmes, *J. Appl. Polymer Sci.* **3**, 251 (1960).
367. N. S. Schneider, L. G. Holmes, and C. F. Miyal, *J. Polymer Sci.* **37**, 551 (1959).
368. N. S. Schneider, J. D. Loconti, and L. G. Holmes, *J. Appl. Polymer Sci.* **5**, 354 (1961).
369. N. S. Schneider and L. G. Holmes, *J. Polymer Sci.* **38**, 552 (1959).
370. W. Scholtan, *Makromol. Chem.* **24**, 83 (1957).
371. W. Scholtan, *Makromol. Chem.* **24**, 104 (1957).
372. T. G. Schoon and G. J. van der Bie, *J. Polymer Sci.* **16**, 63 (1955).
373. H. P. Schreiber, E. B. Bagley, and D. C. West, *Polymer* **4**, 355 and 365 (1963).
374. A. R. Schultz, *J. Am. Chem. Soc.* **76**, 3422 (1954).
375. G. V. Schulz and M. Marx, *Makromol. Chem.* **14**, 52 (1954).
376. R. C. Schulz, G. Renner, and A. Henglein, *Makromol. Chem.* **12**, 20 (1954).
377. R. C. Schulz, E. Müller, and W. Kern, *Makromol. Chem.* **30**, 39 (1959).
378. W. W. Schulz and W. C. Purdy, *Anal. Chem.* **35**, 2044 (1963).
379. J. Schurz, T. Steiner, and H. Streitzig, *Makromol. Chem.* **23**, 141 (1957).
380. J. Schurz, T. Steiner, and M. Hermann, *Gummi, Asbest, Kunststoffe* **14**, 1122 (1961).
381. J. Schurz, *Papier* **9**, 45 (1955).
382. J. Schurz, *Kolloid-Z.* **138**, 149 (1954).
383. B. Sebille and J. Neel, *J. Chim. Phys.* **60**, 475 (1963).

384. F. R. Senti, *J. Polymer Sci.* **17**, 527 (1955).
385. J. Shimura, I. Mita, and H. Kambe, *J. Polymer Sci.* **B2**, 403 (1964).
386. K. V. Shooter and J. A. Butler, *J. Polymer Sci.* **23**, 705 (1957).
387. N. V. Shulyatikova and D. I. Mandel'baum, *Zh. Prikl. Khim.* **24**, 264 (1951).
388. S. Shyluk, *J. Polymer Sci.* **62**, 317 (1962).
389. R. Signer and H. Gross, *Helv. Chim. Acta* **17**, 726 (1934).
390. R. Signer and R. Glanzmann, *Makromol. Chem.* **5**, 257 (1951).
391. H. Sihtola, E. Kaila, and L. Laamanen, *J. Polymer Sci.* **23**, 809 (1957).
392. H. Sihtola, E. Kaila, and N. Virkola, *Makromol. Chem.* **11**, 70 (1953).
393. C. C. Simionescu and E. Alistru, *Faserforsch. Textiltech.* **7**, 171 (1956).
394. G. Smets and M. Claesen, *J. Polymer Sci.* **8**, 289 (1952).
395. G. Smets and A. Hertoghe, *Makromol. Chem.* **17**, 189 (1956).
396. A. M. Sookne and M. Harris, *Ind. Eng. Chem.* **37**, 475 and 478 (1945).
397. P. P. Spiegelman and G. Parravano, *J. Polymer Sci.* **A2**, 2245 (1964).
398. C. J. Stacy and J. F. Foster, *J. Polymer Sci.* **25**, 39 (1957).
399. G. Stainsby, *Discussions Faraday Soc.* **18**, 288 (1954).
400. H. Staudinger and M. Haberle, *Makromol. Chem.* **9**, 48 (1952).
401. H. Staudinger and T. Eichen, *Makromol. Chem.* **10**, 235 (1953).
402. R. Stefani, M. Chevreton, and C. Eyraud, *Compt. Rend.* **248**, 2006 (1959).
403. W. H. Stockmayer, L. D. Moore, Jr., and M. Fixman, *J. Polymer Sci.* **16**, 517 (1955).
404. D. L. Swanson and J. W. Williams, *J. Appl. Phys.* **26**, 810 (1955).
405. G. Talamini and G. Vidotto, *Chim. Ind. (Milan)* **45**, 548 (1963).
406. J. W. Tamblyn, D. R. Morey, and R. H. Wagner, *Ind. Eng. Chem.* **37**, 573 (1945).
407. J. E. Tasman and A. J. Corey, *Pulp Paper Mag. Can.* **48**, 166 (1947).
408. D. L. Taylor, *J. Polymer Sci.* **A2**, 611 (1964).
409. G. B. Taylor, *J. Am. Chem. Soc.* **60**, 639 (1947).
410. W. C. Taylor and L. H. Tung, *J. Polymer Sci.* **B1**, 157 (1963).
411. B. B. Thomas and W. J. Alexander, *J. Polymer Sci.* **15**, 361 (1955).
412. C. D. Thurmond and B. H. Zimm, *J. Polymer Sci.* **8**, 477 (1952).
413. T. E. Timell, *Ind. Eng. Chem.* **47**, 2166 (1955).
414. A. V. Tobolsky, A. Mercurio, and K. Murakami, *J. Colloid Sci.* **13**, 196 (1958).
415. M. P. Tombs, *Biochem. J.* **67**, 517 (1957).
416. A. V. Topchiev, *Izv. Akad. Nauk. SSSR, Otd. Khim. Nauk.* p. 269 (1963).
417. E. Trommsdorff, H. Kohle, and P. Lagally, *Makromol. Chem.* **1**, 169 (1948).
418. L. H. Tung, *J. Polymer Sci.* **20**, 495 (1956).
419. L. H. Tung, *J. Polymer Sci.* **24**, 333 (1956).
420. E. Turska and L. Utracki, *J. Appl. Polymer Sci.* **2**, 46 (1959).
421. E. Turska and M. Laczkowski, *J. Polymer Sci.* **23**, 285 (1957).
422. K. Ueberreiter, H. J. Orthmann, and G. Sorge, *Makromol. Chem.* **8**, 21 (1952).
423. K. Ueberreiter and T. Götze, *Makromol. Chem.* **29**, 61 (1959).
424. J. R. Urwin, D. O. Jordan and R. A. Mills, *Makromol. Chem.* **72**, 53 (1964).
425. J. R. Urwin and J. M. Stearne, *Makromol. Chem.* **78**, 194 (1964).
426. S. N. Ushakov, S. P. Mitsengendler, and N. V. Krasulina, *Bull. Acad. Sci. USSR, Div. Chem. Sci. (English Transl.)* **3**, 366 (1957).
427. I. Valyi, A. G. Janssen, and H. Mark, *J. Phys. Chem.* **49**, 461 (1945).
428. G. Vaughan, D. E. Eaves, and W. Cooper, *Polymer* **2**, 235 (1961).
429. G. Verhaar, *India Rubber World* **126**, 636 and 644 (1952).
430. R. E. Vogel, *Kunststoffe* **42**, 17 (1952).
431. R. E. Vogel, *Kunststoffe* **44**, 335 (1954).
432. H. L. Wagner and P. J. Flory, *J. Am. Chem. Soc.* **74**, 195 (1952).

G. Additional Methods of Fractionation

433. R. H. Wagner, *J. Polymer Sci.* **2**, 21 (1947).
434. M. Wales, F. T. Adler, and K. E. van Holde, *J. Phys. & Colloid Chem.* **55**, 145 (1951).
435. M. Wales, J. W. Williams, and J. O. Thompson, *J. Phys. & Colloid Chem.* **52**, 983 (1948).
436. M. Wales and D. L. Swanson, *J. Phys. & Colloid Chem.* **55**, 203 (1951).
437. M. Wales and S. J. Rehfeld, *J. Polymer Sci.* **62**, 179 (1962).
438. H. A. Wannow and F. Thormann, *Kolloid-Z.* **112**, 94 (1949).
439. J. M. Watkins, R. D. Spangler, and E. C. McKannan, *J. Appl. Phys.* **27**, 685 (1956).
440. W. F. Watson, *J. Polymer Sci.* **13**, 595 (1954).
441. W. F. Watson, *Trans. Faraday Soc.* **49**, 842 and 1369 (1953).
442. T. J. R. Weakley, R. J. P. Williams, and J. D. Wilson, *J. Chem. Soc. (London)* p. 3963 (1960).
443. H. Wesslau, *Makromol. Chem.* **20**, 111 (1956).
444. H. Wesslau, *Makromol. Chem.* **26**, 96 and 102 (1958)
445. N. M. Wiederhorn and A. R. Brown, *J. Polymer Sci.* **8**, 651 (1952).
446. T. Wieland, *Makromol. Chem.* **10**, 136 (1953).
447. P. W. O. Wijga, J. van Schooten, and J. Boerma, *Makromol. Chem.* **36**, 115 (1960).
448. F. Wiloth, *Makromol. Chem.* **8**, 111 (1952).
449. F. Wiloth, *Makromol. Chem.* **14**, 156 (1954).
450. N. Yamada, *Kobunshi Kagaku* **19**, 358 (1962).
451. K. Yamaguchi, H. Kojima, and A. Takahashi, *Intern. Chem. Eng.* **5**, 169 (1965).
452. J. A. Yanko, *J. Polymer Sci.* **22**, 153 (1956).
453. J. A. Yanko, *J. Polymer Sci.* **3**, 576 (1948).
454. S. J. Yeh and H. L. Frisch, *J. Polymer Sci.* **27**, 149 (1958).
455. F. Zapf, *Makromol. Chem.* **3**, 164 (1949).
456. F. Zapf, *Makromol. Chem.* **10**, 61 (1953).
457. A. C. Zettelmoyer and E. T. Pieski, *Ind. Eng. Chem.* **45**, 165 (1953).
458. M. Zief, G. Brunner, and J. Metzendorff, *Ind. Eng. Chem.* **48**, 119 (1956).

Author Index

Numbers in parentheses are reference numbers and indicate that an author's work is referred to although his name is not cited in the text. Numbers in italic show the page on which the complete reference is listed.

A

Abe, T., 365(93), *374*
Ackers, G. K., 127(170, 171), 137, 142(199), 152(170, 171), 153, *177, 178*
Acres, G. J. K., 351(11), 366(11), 367(11), *371*
Adams, H. E., 192, 241, *246*
Adler, F. T., 278(65), 279(65), *284*, 401(23), *413*, 467(434), *497*
Adler, G., 356(31), 367(31), *372*, 476(31), *488*
Adler, J., 126(92), *175*
Aejmelaeus, K., 365(89, 90), *374*, 483(1, 2), 484(2), *488*
Agfa, A. G., 192(20), 201(20), 204(20), 205(20), 210(20), 211(20), 238, *247*
Aggarwala, J. C., 483(3), *488*
Ahola, A., 484(129), *490*
Ahrens, E. H., 464(32), *464*
Aijmelaeus, K., 351(14), 365(14), *371*
Akutin, M. S., 200(39), 241(39), *247*
Albert, W., 363(62), *373*
Albertsson, P. A., 124(3), *173*
Albrecht, A. C., 263(22), *283*
Aldoshin, V. G., 364(68b), *373*
Alexander, W. J., 483(411), *496*
Alfrey, T., Jr., 148, *178*
Alistru, E., 482(393), *496*
Allen, E. L., 469(200), *492*
Allen, G., 369(170a), *377*, 478(4, 5), *488*
Allen, P. W., 379, *412*
Allen, P. E. M., 192, 218, 229, 230, 231, 235, 236, 237, *247*, 248, 366(113), 367(130), 368(130), *375*
Allirot, R., 485(6), *488*
Allison, A. C., 152(190a), *177*
Allison, J. B., 86, *93*
Alm, R. S., 98, *120*
Almin, K. E., 464, *464*
Altgelt, K. H., 127(163), 129(163), 130, 139(191), 146(163, 248), 153(163), 154(191), 162(163, 248), 167, *177, 179*, 482(7), *488*
Amano, T., 476(237), *492*
Ames, B. N., 126(21), 129(21), 145(21), *174*

Amundson, N. R., 431(18), *459*
Anderer, F. A., 126(31), 127(31), *174*
Anderson, E. W., 463(27), *464*
Anderson, F. R., 466(8), *488*
Anderson, R. E., 364(68), *373*
Andrews, P., 127(168, 169), *177*
Andrews, R. D., 333(64), *339*
Andrianov, K. A., 365(85), *374*
Anfinsen, C. B., 126(62), 129(62), *174*
Angier, D. J., 366(114), 367(114), *375*, 486(9), 487(10), *488*
Anyas-Weisz, L., 127(182), 145(182), *177*
Appel, P., 275(61), *284*
Appel, W., 126(89), *175*
Archibald, W. J., 274, *283*, 305, *306*
Arendt, I., 479(11), *488*
Aries, R. S., 466(12), *488*
Arond, L. H., 62(41), *66*, 484(13), *488*
Asaoka, H., 483(14), *488*
Aspberg, K., 126(132), 127(132), *176*
Aug, T. L., *373*
Aurello, G., 368(154b) *376*
Auricchio, S., 126(145), *176*
Ayers, C. W., 480(15), *488*

B

Back, E., 288(13), 294(13), 295(13), 302(13), 305, 484(297), *494*
Badger, R. M., 85, *93*, 483(57), *489*
Badgley, W. J., 482(16), *488*
Baer, M., 366(115c), *375*, 487(17), *488*
Bagdasaryan, R. V., 241, *249*
Bagley, E. B., 4(5), *41*, 331(51, 52), *338*, 467(373), *495*
Bailey, E. D., 256(7), *282*, 287(4), *305*
Bailey, F. E., 478(18), *488*
Baker, C. A., 40(71), *42*, 83, 90, *93*, 95, 98(1), 102(1), 104, 106, 110(1), 113, *120*, 169, *179*, 472(19), *488*
Balwin, R. L., 254, 262, 263, 264, 266, 267, 270(49), 271(52), 279, 282(81), *282, 283*, *284*

Ballantine, D., 356(31), 367(31), *372*, 476(31), *488*
Ballman, R. L., 4(8), *41*, 320(16), 332(60e), *337*, *338*
Bamford, C. H., 79, *92*, 366(122), *375*, 401, *413*, 418(1), 422(1, 2), 431(1, 2), *459*
Banderet, A., 367(143, 144), *376*, 487(238), *492*
Banigan, T. F., 475(278), 485(20), *488*, *493*
Banks, W., 485(21), *488*
Bannister, D. W., 481(22), *488*
Baranovskaja, I. A., 370(184, 187) *377*
Barb, W. G., 418(1), 422(1), 431(1), *459*
Bardwell, J., 463, *464*
Barlow, C. F., 126(114), 129(114), *176*
Bar-Nun, A., 367(148a), 368(148a), *376*
Bassett, E., 126(64), *175*
Battista, O. A., 308(3), 309(3, 12), 311, *315*, *316*, 482(87), 483(345), *489*, *495*
Batzer, H., 365(82), *374*, 469(23), 479(24, 25, 26), 480(27), *488*
Baum, A., 145(211), *178*
Baum, H., 485(28, 29), *488*
Bauman, W. C., 128, 139(184), 148(184, 224), *177*, *178*
Baumann, H. P., 466(156), *491*
Baxendale, J. H., 477(30), *488*
Baysal, B., 476(31), *488*
Beall, G., 39(68), *42*, 308, 311, *315*, 395, *412*
Beams, J. W., 280(73), *284*
Beason, L. R., 45(10), 46(10), *66*, 397, 402, *412*, 486(48), *488*
Beattie, W. H., 62(43), *66*, 192, 198, 216, 231, 232, *247*, 475(32), *488*
Becker, G. W., 334(67), *339*
Becker, W., 475(152), *491*
Beckmann, C. O., 462, *464*
Beiser, S. M., 126(64), *175*
Beiss, U., 126(79), *175*
Beling, C. G., 126(69), 162, *175*, *179*
Belmas, R., 485(33), *488*
Belonovskaja, G. P., 369(164a), *377*
Bender, B. W., 370(201), *378*
Bendich, A., 484(34), *488*
Benedetti, E., 371(212), *378*
Bengongh, W. I., 371(209), *378*
Bengtsson, C., 127(151), 129(151), *177*
Bennett, C. F., 365(92), *374*
Bennich, H., 126(23), 127(23), 160, 161, *174*, *179*

Benoit, H., 302, *306*, 355(24, 25), 363(24, 25, 42), 366(115b), *372*, *375*, 485(92), *489*
Berenbaum, M. B., 427(7), *459*
Beresniewicz, A., 365(102), *374*, 469(35), *488*
Berestneva, Z. Ya., 192(18a), 230(18a), 231(18a), *247*
Berger, H. L., 172, *179*
Bergsnov-Hansen, B., 17, 26, *41*
Berkowitz, J. B., 45(5), *65*, 474(36), *488*
Bernal, D. E., 485(37), *488*
Bernardini, F., 369(179b), 370(179b), *377*
Bernstein, I. M., 484(38), *484*
Berry, G. C., 469(39), *488*
Bessonova, L. A., 369(164a), *377*
Bevington, J. C., 469(40), *488*
Bianche, J. P., 471(41), *488*
Bidaine, E., 470(95), *489*
Bier, G., 366(124, 125), 369(160, 161), *375*, *376*
Bill, A., 126(60), *174*
Billick, I. H., 268(42), 269(42), *283*
Billmeyer, F. W., Jr., 40, *42*, 276(62), *284*, 308, *316*, 467(246), 477(42), *488*, *493*
Bischoff, F., 482(44), *488*
Bismuth, F., 126(35), *174*
Bisschops (Bischoff), J., 192(6, 8), 201(6), 203(6), 204, 205(6, 7), 208, 237, 240, *246*, *247*, 476(45), *488*
Björk, I., 126(32), *174*
Björk, W., 126(13, 19). 128(13), *173*, *174*
Bjork, C. F., 482(334), *482*
Blackford, J., 363(48a), 370(188), *373*, *377*
Blair, J. E., 269, *283*
Blanchette, I. A., 367(131), *375*
Bloemendal, H., 126(70), 127(70), *175*
Bloomfield, G. F., 485(46), *488*
Bobalek, G. E., 479(47), *488*
Bock, R. M., 99(14), *120*
Böhme, R., 334(70), 335(70), *339*, 427(8), *459*
Boehmke, G., 246, *249*
Boeke, P. J., 331(54), *338*
Boerma, J., 25(52), *42*, 75(16), 87(52), 89(52), 92, 93, 468(*447*), *497*
Bohdanecky, M., 479(270), *493*
Bohn, L., 355(23), 363(23), *372*
Boman, H. G., 127(149), 129(149), 152, *176*
Booth, C., 45(10), 46(10), 62(43), *66*, 351(15), 369(15, 170a), *371*, *377*, 397, 402, *412*, 475(32), 478(4, 5), 486(48), *488*

Author Index 501

Bortnick, N. M., 148(288), 155(237), 156(240), *178, 179*
Bosch, L., 126(70), 127(70), *175*
Bosworth, P., 369(164), *376*
Botty, M. C., 364(68), *373*
Bourrillon, R., 126(126), *176*
Bovey, F. A., 476(347), *495*
Boyer, R. F., 45(9), 49, 52, 57, *66*, 308, *315*, 392, 393, *412*, 461, *464*, 471(49), 473(50), *488, 489*
Boyer-Kawenoki, F., 146(215), *178*
Bradley, D. F., 485(51), *489*
Brady, A. P., 17, 26, *41*
Braun, D., 369(156), *376*
Breazeale, F., 319(12), *337*
Breil, H., 385(8), *412*
Breitenbach, J. W., 319(5), *336, 337*, 471(140), *490*
Bresler, S. E., 126(40, 109), 127(40), 129(109), *174, 176*, 281, *284*, 368(154a), *376*, 475(52, 53, 54), 480(55), *489*
Bressan, G., 369(177), *377*
Brewer, P. I., 126(101), 127(157a, 164, 165), 129(101, 157a, 164, 165), 130, 134, 154 (164, 165), *175, 177*
Brey, W. S., 356(36), 369(36), *372*
Bridgman, W. B., 267(37), *283*, 484(56), *489*
Brink, N. G., 126(90), *175*
Broda, A., 26(59, 60), 40, *42*, 309, *316*
Broda, E., 145(211), *178*
Broman, L., 167, *179*
Brooks, M. C., 85, *93*, 483(57), *489*
Brooks, R. E., 365(100), *374*
Brosche, M., 484(109), *490*
Brower, F. M., 4(12), *41*, 332(60i), *338*, 473(58), *489*
Brown, A. R., 477(445), *497*
Brown, D. R., 83(36a), *93*, 120(46), *121*, 466(229), 467(229), *492*
Brown, R. A., 485(234), *492*
Bruck, S. D., 479(59), *489*
Brunner, G., 484(458), *497*
Bruss, D. B., 471(61), *489*
Bucci, G., 370(193), *377*
Buchdahl, R., 281, 282, *284*, 355(28), 369(28), *372*
Bucovaz, E. T., 126(65), 129(65), *175*
Bueche, A. B., 20, *42*
Bueche, A. M., 181, 184, 185, 186, 187, *189*, 362(41c), *372*, 472(96), *489*

Bueche, F., 3, 4(4), *41*, 334, 335(77), *339*, 463, *464*
Buerger, K., 478(60), *489*
Bungenberg de Jong, H. G., 78, *92*
Burge, D. E., 402(27), *413*, 471(61), *489*
Burkhardt, F., 370(180), *377*
Burnett, G. M., 366(115), *375*
Bushuk, W., 355(24), 363(24, 42), *372*
Busse, W. F., 137, *178*
Butler, A. V., 257(9), 273(9), *282*
Butler, G. B., 356(36), 369(36), *372*
Butler, J. A., 484(386), *496*
Buurman, A., 484(62), *489*
Buzagh, A., 365(83), *374*, 478(63, 64), *489*
Bywater, S., 433(22), *459*, 477(30, 88), 485(65), *488, 489*

C

Cahill, J. W., 463, *464*
Cajelli, G., 489(66), *489*
Calderon, N., 463, *464*
Caldwell, J. R., 364(72), *374*
Campbell, H., 200(37), 235, 239, *247*, 474(67), *489*
Cantoni, G. L., 126(105), *175*
Cantow, H.-J., 192, 201(22), 203(22), 204(22), 206(22), 231, 233, *247*, 272(55), *283*, 292(16), 293(16), 299(16), 304(27), 305, *305, 306*, 357, 359, 360(41), 361, *372*, 473 (68, 69), *489*
Cantow, M. J. R., 56, *66*, 83, 84(34, 39), 87(39), 91(35), *92*, *93*, 97(7), 98(7), 99(16), 100(16), 102(7, 16), 103(16), 105(7), 114(7), 119(7), 120(7, 44, 45), *120, 121*, 173, *179*, 369(158, 168b), *376, 377*, 439(28c), 440, *459*, 467(71), 471(72), 479(70), *489*
Caplan, S. R., 41, *42*, 104, 105, 107(22), 108(22), 109, 111(22), *121*, 485(73), *489*
Carbonaro, A., 364(71b), *374*
Carlstrom, A. A., 102(20), *121*
Carpenter, D. K., 282(80), *284*
Centola, G., 363(59), *373*
Čepelák, J., 242, *249*, 479(75), *489*
Ceresa, R. J., 366(114), 367(114, 136), 368(136), *375, 376*, 486(9), *488*
Cerf, R., 335, *339*
Cernescu, N., 127, *177*
Cernia, E., 234, 236, 240, *248*

Cernova, Z. D., 369(164a), *377*
Chadwick, M., 126(78), *175*
Champetier, G., 366(116), *375*
Chandan, R. C., 126(84), *175*
Chapiro, A., 111(31), *121*, 367(132), 368(153), *375, 376*, 487(76), *489*
Chelnokova, G. N., 365(103), *375*, 479(343), *495*
Chen, W. K., 367(130a), *375*
Cherkin, A., 98(11), *120*
Chevreton, M., 476(402), *496*
Chi, W. F., 486(130), *490*
Chiang, R., 401(21), *413*
Chien, J. C. W., 423(3), *459*
Ch'ien, J. V., 110(30), *121*
Chinai, S., 328(48), *338*
Chinai, S. N., 62(42), *66*, 476(77, 78), 477(79), *489*
Choudhury, P. K., 485(91), *489*
Chu, S. N., 110(30), *121*
Ciampa, G., 476(80), *489*
Ciampelli, F., 364(69a), *373*
Ciardelli, F., 352(19), 369(19), 371(212), *372, 378*
Claesen, M., 366(128), 367(128), 368(128), *375,* 487(394), *496*
Claesson, J., 127, 145(179, 180), *177*
Claesson, S., 127, 145(179, 180, 181), *177*, 192, 205(15), 219, 223, 226, 240, *247, 248,* 288(11), *305,* 469(81), *489*
Clark, O. K., 172, *179*
Cleverdon, D., 471(82), *489*
Climie, I. E., 200(34), 237, *247*
Coen, A., 332(60i), *338*
Cohen, W. E., 483(123), *490*
Cohn-Ginsberg, E., 477(248), *493*
Coleman, B. D., 362(41f), *372*, 438(28), *459*
Coleman, J. E., 126(122), *176*
Colobert, L., 126(30), 127(30), 129(30), *174*
Colombo, L., 369(179b), 370(179b), *377*
Combs, R. L., 84(38, 43), 87(38), 89(38), *93,* 102(18), 104(18), 105(18), 106(26), 107(18), 108(18), 109(18), 110(18), 111(26), 116(18, 26), 117(18), 118(18, *120,* 467(83, 170), 468(83, 171), *489, 491*
Condliffe, P. G., 163(252), *179*
Connell, G. E., 126(130), *176*
Conrad, C. M., 482(74), *489*
Cooke, R. C., Jr., 83(37), 84(37), 85(37), 86(37), 87(37), 88(37), *93,* 466(139), *490*

Cooper, W., 89, 90, *93,* 104, 106, 111(21), 116, *121,* 364(70), 367(145, 146, 147), 369(147), *373, 376,* 475(84, 86, 428), 487(85), *489, 496*
Coover, H. W., Jr., 84(38, 43), 87(38), 89(38), *93,* 102(18), 104(18), 105(18), 106(26), 107(18), 108(18), 109(18), 110(18), 111(26), 116(18, 26), 117(18), 118(18), *121,* 468(171), *491*
Coppick, S., 308, 309, 311, *315,* 482(87), *489*
Cordier, P., 111(31), *121,* 367(132), *375*
Corey, A. J., 309, *316,* 483(407), *496*
Cornelius, C. E., 126(44), *174*
Cornillott, P., 126(126), *176*
Corradini, P., 75(20), 85(20), 91(20), *92*
Cortis-Jones, B., 126(112), 127(166), 129(166), 130, *176, 177*
Cottam, B. J., 332(60i), *338,* 477(88), *489*
Cotten, G. R., 364(67), *373,* 486(89), *489*
Cowie, J. M., 485(90), *489*
Cowling, E. B., 126(88), *175*
Cox, W. P., 320(16), *337*
Cragg, L. H., 74(10), *92,* 363(55), 367(55), *373*
Craig, L. C., 126(18, 74), 137, *174, 175, 178,* 464, *464*
Craig, R. G., 469(39), *488*
Crestfield, A. M., 126(116), *176*
Croon, I., 365(95), *374*
Cruft, H. J., 126(139), *176*
Curchod, J., 371(210), *378*
Curdel, A., 127(175), *177*
Curtain, C. C., 126(72), *175*
Cyperovic, A. S., 352(17), 366(17), *372*
Czvikovszki, T., 367(130b), *375*

D

Dagon, J. R., 478(312), *494*
Daisley, K. W., 127(22), *174*
Dall'Asta, G., 364(69b, 69c), 365(84), 369(178), *373, 374, 377,* 478(231), *492*
Dalton, F. L., 351(11), 366(11), 367(11), *371*
Danusso, F., 56(36), *66,* 471(318), *494*
Das, B. S., 485(91), *489*
Daum, U., 336(82a), *339*
Daune, M., 287(5, 6, 7), 288(5, 6, 7), 294, 296, 301, 302(23, 24), 303, 304(29), 305, *305, 306,* 469(142), 485(92), *489, 490*
Davidson, H., 470(365), *495*

Author Index 503

Davis, J. W., 126(65), 129(65), *175*
Davis, T. E., 78, 79(29), 86(29), 89, *92*, 389, 390, *412*, 468(93), *489*
Davis, W. E., 483(94), *489*
Davydova, V. P., 246(77a), *249*
De, T. W., 475(338), *495*
De Brouckere, L., 470(95), *489*
Debye, P., 181, 184, 185, 186, 187, *189*, 472(96), *489*
Decker-Freyss, D., 363(43a), *372*
Declerck-Raskin, M., 126(111), 129(111), *176*
de Groot, S. R., 182, 183(9), *190*
Demian, N., 363(55c), *373*
DeMoor, P., 126(111), 129(111), *176*
Desnuelle, P., 126(43), *174*
Desreux, V., 74, 79, 85, 86, 89, *92*, 115(42), *121*, 192, 201(6, 7), 203, 204, 205(6, 7), 206, 208, 215, 234, 240, *246*, 266(34), 268, *283*, 466(97, 98), 470(325), 477(99), 482(44), 483(326), *488*, *490*, *494*
Determann, H., 124, 126(33, 113), 127(162), 129(5, 162), 133(162), 159(5), 169, *173*, *174*, *176*, *177*, 472(100), *490*
Detoro, F. E., 364(68), *373*
Deuel, H., 127, 145(182, 183), *177*
Deusser, P., 41(73), *42*, 71, *92*, 112(38), *121*
Dexheimer, H., 353(21), 370(21), *372*
Dialer, K., 56, *66*, 469(101), 474(102), *490*
Dieu, H. A., 466(103), 469(104), *490*
Dimonie, M., 363(55c), *373*
Dinglinger, A., 53(29), 54(29), *66*
Dirheimer, G., 126(30), 127(30), 129(30), *174*
Djurtoft, R., 126(106), 127(106), *175*
Dobo, J., 367(130b), *375*
Dobry, A., 146(215), *178*, 369(179), *377*, 469(105), *490*
Dobrynina, A. S., 200(38), 241(38), *247*, 365(79), *374*, 487(220), *492*
Domareva, N. M., 332(60i), *338*
Donaldson, K. O., 98(12), 99(15), *120*
Donninger, C., 366(104), *375*
Doty, P., 470(106), 480(295), *490*, *494*
Douglas, D. C., 463(27), *464*
Downer, J. M., 235(65), *248*, 366(113), *375*
Dreval, V. E., 320(18), *337*
Dreyer, W. J., 129(188), *177*
Drickamer, H. G., 184, 185, *190*, 472(122), *490*
Duclaux, J., 469(154), *491*
Dunn, A. S., 192(13), 236(13), *247*
Dunn, M. S., 98(11), *120*

Dusek, K., 478(107), *490*
Dymarchuk, N. P., 370(195), *377*

E

Eaves, D. E., 367(147), 369(147), *376*, 475(84, 428), *489*, *496*
Eberly, K. C., 475(108), *490*
Ebersbach, H. W., 364(64), *373*
Ebert, K. H., 484(109), *490*
Edelmann, K., 320(19), 325, *337*, 476(110, 111, 113, 115), 482(113), 483(112), 484(114), 485(113, 116, 117, 118), *490*
Edmunson, A. B., 126(136), *176*
Edwards, M. R., 126(77), *175*
Egashira, T., 369(159), *376*
Ehrmantraut, H. C., 402(27), *413*
Eichen, T., 482(401), *496*
Eirich, F. R., 240, *249*, 318(1), 322(1), *336*
Eisenberg, H., 477(221), *492*
Eisenbraun, E. W., 365(99a), *374*
Eldib, I. A., 463, *464*
Elgert, K. F., 484(109), *490*
Elgood, E. J., 477(119), *490*
Elias, H.-G., 220, 229, *248*, 371(216), *378*, 469(121), 473(120), *490*
Ellinger, L. P., 370(191), *377*
Elliott, J. H., 83(37), 84(37), 85(37), 86(37), 87(37), 88(37), *93*, 466(139), *490*
Elmquist, A., 126(58), 129(58), *174*
Elyash, L. J., 332(60g), *338*, 475(338), *495*
Emde, F., 428, 435, *459*
Emery, A. H., 472(122), *490*
Emery, A. H., Jr., 184(11), 185, *190*
Emery, C., 483(123)., *490*
Enari, T. M., 126(142), *176*, 366(112c), *375*
Ende, H. A., 63(44), *66*, 281(76, 78), 282(79), *284*, 355(28), 363(53a), 367(140a), 369(28), *372*, *373*, *376*, 466(211), 473(195), 484(124), *490*, *492*
Enomoto, S., 355(27), 363(27), *372*
Epstein, W. V., 126(46, 97, 119), *174*, *175*, *176*
Eriksson, A. F. V., 264, 265, 270, *283*, 477(125, 126, 127, 128), *490*
Erlander, S. R., 275, *283*
Eskin, V. E., 370(184, 187), *377*
Ettre, L. S., 141, *178*
Evans, M. G., 477(30), *488*
Evans, W., 335(77), *339*

Evans, W. W., 333(65), *339*
Excell, B. J., 462(10), *464*
Eyraud, C., 476(402), *496*
Eyring, H., 330, *338*

F

Farina, M., 369(177), *377*
Fatou, J. M., 182, 187, 188, *190*, 470(176), 482(175), *491*
Faucher, J. A., 330, *338*
Faxen, H., 261, *283*
Fedyantseva, A. A., 467(269), *493*
Feinauer, R., 367(141b), *376*
Feit, B. A., 367(148a), 368(148a), *376*
Felicetta, V. F., 288(12, 13), 294, 295(12, 13), 302(12, 13), 304(12), *305*, 484(129), *490*
Felton, C., 364(68), *373*
Ferguson, J., 332(60d), *338*
Ferry, J. D., 331(56), 332(61), 333(65), 334, 335(77), *339*, 484(253), 486(130), *490*, *493*
Feshbach, H., 440, 442(31), 444(31, *459*
Filatowa, W. A., 369(169), *377*
Finkelstein, M. S., 365(96), *374*
Finlayson, J. S., 352(20), 366(20), *372*
Finn, S. R., 478(131), *490*
Firemark, H., 126(114), 129(114), *176*
Fischer, E., 365(79b), *374*
Fischer, S., 156(242), *179*
Fitzgerald, E. B., 474(132), *490*
Fitzgerald, E. R., 334(67), *339*
Fixman, M., 363(43), *372*, 487(403), *496*
Flamm, E., 365(95), *374*
Fletcher, J. R., 479(133), *490*
Flodin, P., 124, 126(10, 98, 99, 100, 131, 132, 133), 127(132), 128(131, 187), 129(187), 131(1), 133(133), 134, 135, 139, 141, 142, 143, 150(1), 162, 165, *173*, *175*, *176*, *177*
Floria, V. D., 363(51), *373*, 487(301), *494*
Flory, P. J., 2(2), 9, 11, 12, 15(35), 17(2), 18, 21, 22, 23(44), 24(48, 49), 26, 28, *41*, *42*, 44, 52, 54, 55, 63, 64, *65*, *66*, 70, 76, 87, *92*, 107, 108, *121*, 146(219), *178*, 263(18), *283*, 298(19), 305(19), *306*, 369(174), *377*, 402, *413*, 427, 432, 449, *459*, 467(134, 137, 138, 250), 483(208), 485(432), *490*, *492*, *496*
Flowers, D. L., 96(3), 97, 98, 114, 115(3), *120*, 486(135), *490*

Foerster, H., 469(263), *493*
Fomina, A. S., 320(18), *337*
Fontana, C. M., 468(136), *490*
Fontanille, M., 366(116), 370(199), *375*, *378*
Foster, J. F., 275, *283*, 484(398), *496*
Fothergill, F. E., 126(28), *174*
Fox, T. G., 3(3), *41*, 263(19), *283*, 356(34), 362(41f), 369(34), *372*, 467(137, 138), *490*
Francis, P. S., 83, 84(37), 85(37), 86, 87(37), 88, *93*, 466(139), *490*
Frank, H. P., 62(41), *66*, 319(4, 5), *336*, 471(140), 484(13), *488*, *490*
Frazer, W. J., 332(60g), *338*
Freedland, R. A., 126(44), *174*
Freeman, M., 369(161a), *376*, 470(141), *490*
Frenkel, S. Y., 281(77), 364(68b), 368(154a), *373*, *376*, *284*, 475(52, 53, 54), *489*, 477(360), *495*
Fresco, J. R., 484(34), *488*
Freund, L., 287(5, 6, 7), 288(5, 6, 7), 294, 296, 301, 302(24), 303, 304(5, 6, 7, 29, 30), 305(5, 6, 7, 25, 30), *305*, 306, 469(142), *490*
Frey, K., 43(1), *65*
Freyes, D., 366(115b), *375*
Fric, I., 366(112b), *375*
Friedberg, F., 98(12), *120*
Friedländer, H. Z., 367(130a), *375*
Frisch, H. L., 319, *336*, 401, *413*, 472(454), *497*
Frisch, N., 156(242), *179*
Fritzemeier, H., 182, 187, *189*, 469(143), 477(143), 484(143), *490*
Fuchs, O., 76, 77, *92*, 341(1), 345(6), 346(6), 347(8), 353(21), 354(22), 357, 359, 360(41), 361, 363(6), 364(8), 365(6), 369(8, 22), 370(21), *371*, *372*, 466(144), 469(145, 146), *490*
Fuchs, S., 126(115), *176*
Fujii, H., 356(32), 364(32), 369(176), *372*, *377*, 486(147), *491*
Fujii, T., 356(32), 364(32), *372*
Fujine, T., 355(29), 367(29), *372*
Fujisaki, Y., 50, 58(38), 66, 476(235), *492*
Fujita, H., 252, 259(1), 262, 263, 268, 273(1), 278, 281, *282*, *283*, *284*, 333, 334(68), *339*, 467(149), 469(148) 473(148), *491*
Fujiwara, H., 76, *92*, 369(168a), *377*, 468(315), *494*
Fukami, K., 365(103a), *375*
Fukuda, W., 478(219), *492*
Funke, W., 367(141b), *376*

AUTHOR INDEX

Fuoss, R. M., 45(5), 65, 474(36, 132), 488
Furukawa, J., 356(32), 364(32, 73, 75, 76), 365(79c), 366(123), 369(170, 176), 370(200), 372, 374, 375, 377, 378, 478(226), 486(147), 491, 492

G

Gässler, G., 202(48), 248
Gaillard, B. D. E., 356(33), 369(33), 372
Galic, I. P., 352(17), 366(17), 372
Gallo, A., 474(150), 491
Gallot, Y., 366(126), 367(126), 370(202), 375, 378
Gamble, L. W., 192, 201(25), 204(25), 243, 247, 364(69), 373, 486(151), 491
Gard, J., 356(31), 367(31), 372
Gardon, J. L., 462, 464
Garrett, R. R., 370(192), 377
Garry, B. G., 126(21), 129(21), 145(21), 174
Garten, V., 475(152), 491
Gavoret, G., 469(154), 486(153), 491
Gawronska, B., 26(59), 40(59), 42
Gee, G., 49, 66, 431(12), 459
Gehatia, M., 431(14, 15), 459
Geil, P. H., 4(14), 24(47), 41, 42
Geiseler, G., 466(156), 467(155), 491
Gelotte, B., 124, 126(42, 59, 99), 127(146), 128(59), 129(42, 59), 145, 169(42), 173, 174, 175, 176
Genkin, A. N., 467(157), 491
George, W. H. S., 126(73, 110), 175, 176
Georgi, E. A., 365(99), 374
Gernert, J. F., 84, 87, 93, 99(16), 100(16), 102(16), 103(16), 120
Gesner, B. D., 370(189), 377
Geymer, D. O., 16, 26, 41, 42
Gibbs, J. W., 9, 41
Gibbs, P. A., 127(150), 176
Giera, A., 145(212), 178
Giesekus, H., 192, 195, 201(19), 202(19), 204(19), 205(19), 206, 210, 228, 234, 236, 237, 239, 240, 241, 243, 244, 245, 247, 248, 249
Gilbert, G. A., 485(28, 29), 488
Gillespie, J. M., 352(18), 366(18), 372
Gillis, J., 300, 301(21), 303, 306
Ginsburg, A., 275(61), 284
Givol, D., 126(115), 176

Glanzmann, R., 485(390), 496
Glass, G. B. J., 126(118), 176
Glaudemans, C. P. J., 356(30), 367(30), 372
Glick, J. H., Jr., 126(124), 127(124), 176
Glikman, S. A., 351(16), 366(16, 108), 372, 375
Glöckner, G., 231, 238, 248, 479(158), 491
Glueckauf, E., 141(205), 143, 178
Gobran, R. H., 427(7, 8), 459
Goettler, L. A., 326, 337
Götze, T., 48(15), 66
Gold, L., 438(25), 459
Goldberg, A. I., 471(159), 491
Goldsmith, H. L., 137, 178
Golub, M. A., 309, 316, 363(46), 373, 463, 464, 486(161), 491
Golubenkova, L. I., 200(39), 241(39), 247
Gooberman, G., 200(43), 217, 220, 226, 231, 232, 248, 473(162), 491
Goodban, A. E., 484(163), 491
Goodman, M., 200(35), 239, 247
Goodrich, F. C., 435(23), 438(23), 439(28c), 459
Goolsby, P. F., 371(211), 378
Gordienko, A., 200, 239, 248, 476(164), 491
Gordon, G. K., 156, 179
Goring, D. A. I., 366(109), 375
Gornick, F., 438(28), 459
Gosting, L. J., 270, 283
Got, R., 126(126), 176, 370(197), 337
Gotze, T., 479(423), 496
Graevskaya, R. A., 126(40, 109), 127(40), 129(109), 174, 176
Grafflin, M. W., 341(3), 362(3), 371
Graham, J. P., 231, 248
Graham, R. K., 366(118), 375
Gralén, N., 181, 189, 264, 265, 266, 270(50), 283, 291, 300, 303(15), 305, 473(165), 491
Granath, K. A., 126(92a, 131, 133), 128(131), 133(133), 139, 175, 176
Grandine, L. D., 334(67), 339
Grasbeck, R., 126(45), 174
Gratch, S., 3(3), 41
Graul, E. H., 126(55), 174
Graydon, W. F., 362(41g), 372
Grechanovskii, V. A., 363(49), 373, 475(336), 486(335, 337), 494, 495
Greear, A., 467(303), 494
Greear, G. R., 90(54), 93, 113(40), 117(40), 118(40), 121, 271(51), 283

Green, J. H. S., 98, 118, *120*, 127(156), 129(156), *177*, 384, *412*, 462, *464*, 471(166), 472(166), *491*
Greenwood, C. T., 485(21, 90), *488*, *489*
Greenwood, F. C., 126(39), *174*
Gregory, F. D., 126(25), 127(25), *174*
Gribkova, P. N., 365(103), *375*
Griehl, W., 200(45), 239(45), *248*, 480(167), *491*
Grodon, M. A., 126(77), *175*
Gröblinghoff, H., 328(42, 43), *337*, *338*, 477(168), *491*
Grohn, H., 200(33), 234, *247*
Gronwall, A., 192(9), 204(9), 228(9), 240(9), *247*
Gross, E., 126(144), 127(144), *176*
Gross, H., 267(38), *283*, 300(22), *306*, 473(389), *496*
Gross, S., 470(365), *495*
Grossman, P. U. A., 334(75), *339*
Gruber, U., 229, *248*, 371(216), *378*, 473(120), *490*
Guglielmino, P., 335(77a), *339*
Guidotti, G., 127(152), *177*
Guillemin, R., 126(49), *174*
Guillet, J. E., 84(38, 43), 87, 89, 102(18), 104(18), 105(18), 106, 107, 108(18), 109, 110(18), 111(26), 116(18), 117(18), 118(18), *121*, 467(169, 170), 468(171), *491*
Gumbold, A., 366(124), *375*
Gundermann, E., 482(172), *491*
Gurgenidze, G. T., 367(141a), 368(154c), *376*
Gurnee, E. F., 336, *339*
Gutweiler, K., 365(79b), *374*
Guzeev, V. V., 201(47), 203(47), 204(47), 205(47), *248*
Guzmán, G. M., 74, *92*, 182, 187, 188, *190*, 469(40), 470(176), 471(173), 477(174), 482(175), *488*, *491*
Guzzi, R. A., 62(42), *66*, 476(78), *489*

H

Haahti, E., 166(254), *179*
Haber, E., 126(62), 129(62), *174*
Haberle, M., 470(400), *496*
Hachihama, Y., 367(138, 139), 368(152), *376*
Haddon, W. F., Jr., 110(34), *121*, 467(177), *491*

Hahn, W., 471(178), *491*
Hale, D. K., 139(204), 155(231), *178*, *179*
Hall, H. K., 63(46), 64(46), *66*
Hall, R. W., 45(11), 62, *66*, 74, 77, 89, *92*, 193(26), 200(26), *247*
Haller, W., 147(219b), 153, 154(219b), *178*
Hallows, B. G., 126(90), *175*
Ham, J. S., 184, *190*
Hamann, K., 367(141b), *376*
Hamilton, P. B., 155, 162, *179*
Hamm, F. A., 290, 303, *305*, 474(293), *494*
Hammerschlag, H., 74(10), *92*
Hána, L., 126(50), *174*
Handa, R., 487(210), *492*
Handschuh, D., 126(31), 127(31), *174*
Hanna, R. J., 363(58), *373*
Hannus, J., 471(179), *491*
Hansen, C. M., 111(32), 113, *121*, 479(180), *491*
Hanson, A. W., 363(53), 364(53), *373*
Hanson, L. A., 126(15), 127(151), 129(151), *173*, *177*
Harding, S. W., 463, *464*
Hardy, R., 192(18), 218(18), 229(18), 230(18), 231(18), 235(18), 236(18), 237(18), *247*, *248*, 367(130), 368(130), *375*
Hardy, T. L., 126(47), 129(47), *174*
Harfenist, E., 464(32), *464*
Harland, W. G., 46(12), *66*, 483(181), *491*
Harmon, D. J., 127(160), 129(160), *177*
Harrington, R. E., 75, *92*, 472(182), *491*
Harrington, W. F., 260(10), 277(10), *282*
Harris, I., 200, 236, 237, *248*, 477(183), *491*
Harris, M., 482(396), *496*
Harrison, G. D., 192, 234, *247*
Hartler, N., 471(184), *491*
Hartley, F. D., 200(42), 235, *248*, 368(150), *376*
Hartley, R. W., 366(110), *375*
Harwood, H. J., 356, 362(41j), *372*, *373*
Haseley, E. A., 39(67), *42*, 391, *412*, 479(185), *491*
Hastings, G. W., 192, 230, 231, 235(65), *247*, *248*, 366(113), *375*, 473(186), *491*
Hatano, A., 365(93), *374*
Haug, A., 485(187), *491*
Hauser, E. A., 475(188), 485(189), *491*
Hausmann, W., 464(32), *464*
Haward, R. N., 332(60d), *338*
Hawkins, S. W., 25, 26(53), *42*, 109(36), *121*, 369(166), *377*, 466(190), *491*

Author Index

Hayakawa, K., 367(137), *376*
Hayashi, I., 356(38), 363(38, 51a), *372, 373*
Hayden, P., 366(127), 367(127), 368(127), *375,* 487(191), *491*
Heath, N. S., 477(119), *490*
Hecht, G., 228(57), 236(57), 240(57), *248*
Heftmann, E., 159(244), *179*
Heidenreich, R. D., 461, *464*
Heikens, D., 480(196), *492*
Heirwegh, K., 126(111), 129(111), *176*
Helfferich, F., 135, 136, *178*
Hellfritz, H., 341(1), *371*, 462, *464*
Hellman, M. Y., 319(8), *336*
Hendrickson, J. G., 127(158), 129(158), 139(158), 153(158), 170, 173(158, 265), *177, 179*
Henglein, A., 476(376), *495*
Hengstenberg, J., 192, 193(12), 196, 231, 232, 234, 235, *247*, 473(192), *492*
Henry, P. M., 80(33), 82, 83, 84, 91(33), *92,* 113, *121*, 404, *413*, 466(193), *492*
Herdan, G., 401, *413*
Heremans, J. F., 126(111), 129(111), *176*
Hermann, H.-D., 365(79a, 79b), *374*
Hermann, M., 328(39, 45), *337, 338*, 468(380), 473(194), *492, 495*
Hermans, J., 482(197), *492*
Hermans, J. J., 182, 187, *189*, 281(76), 282, *284*, 362(41a), 363(53a), *372, 373*, 466(344), 469(143), 473(195), 477(143), 484(143), *490, 492, 495*
Hermans, P. H., 480(196), *492*
Herold, H., 467(155), *491*
Herren, C. L., 184, *190*
Herrent, P., 476(198), *492*
Herrington, E. F. G., 427, 431(11), *459*
Hertoghe, A., 367(149), 368(149), *376,* 487(395), *496*
Herz, J., 363(43a), 368(151), *372, 376*
Hess, K., 323(24), 326(24), *337*
Heusch, R., 246, *249*
Heuser, E., 483(199), *492*
Heuser, W., 478(217), *492*
Hewett, W. A., 96(3), 97(3), 98(3), 114(3), 115(3), *120*, 486(135), *490*
Hexner, P. E., 280, *284*
Higashi, H., 369(159), *376*
Hildebrand, J. H., 19(38), *42*, 146(218), *178*
Hill, E. H., 364(72), *374*
Hill, R. F., 126(14), *173*

Hill, R. J., 126(14), 127(152), *177*
Hint, H., 192(9), 204(9), 228(9), 240(9), *247*
Hintz, H., 369(156), *376*
Hirai, N., 330(48), *338*
Hirooka, H., 84(42), 87(42), *93*
Hirshfield, S. M., 469(200), *492*
Hjertén, S., 126(108), 127(149), 129(149, 189), 152, *176, 177*
Högman, C. F., 126(48), *174*
Hoffman, D. T., Jr., 184(11), *190*
Hoffman, J. D., 182, 184, 185, *190*, 472(201), *492*
Hoffmann, M., 195, 196, 230, 231, 234, 235, 236, 237, 238, 241, 242(29), *247, 248*, 462, *464*
Hohenstein, W. P., 471(159), *491*
Hojima, H., 76, *92*
Holeysovska, H., 366(112a), *375*
Holland, V. F., 5(15), *41*
Holmes, L. G., 71, 85(48), 87(3), 90(48), *92, 93*, 97(4), 98(4), 102(4), 106(24), 107(28), 108(24, 28), 110(4, 33), 113, 115(24), *120, 121*, 472(366, 367, 368, 369), *495*
Holzkamp, E., 385(8), *412*
Hook, E. O., 364(74a), *374*
Hookway, H. T., 462(11), *464*
Horkay, F., 365(83), *374*, 478(63, 64), *489*
Horowitz, R. H., 84(40), 86, *93*, 467(202), 468(202, 203), *492*
Hosemann, R., 387, 399, *412*
Hotta, J., 363(51c), *373*
Houtz, R. C., 476(204), *492*
Howard, G. J., 201, 202, 228, 239, *248*, 384, *412*, 480(205, 206), *492*
Howlett, F., 365(86), *374*
Huggins, M. L., 11, 12(32), 21, 23(27, 43), 24(50), *41, 42*
Hughes, E. C., 364(74a), *374*
Huglin, M. B., 368(153a), *376*
Huisman, T. H. J., 366(111), *375*
Hulme, J. M., 97, 104, 106, 107, 111(5), *120*, 475(207), *492*
Hummel, D., 368(151), *376*
Hummel, J. P., 129(188), *177*
Humphrey, J. H., 152(190a), *177*
Hunt, M. L., 483(208), *492*
Hunter, W. M., 126(39), *174*
Husemann, E., 461, *464*, 482(209), *492*
Hutchinson, R. J., 470(304), *494*
Huu-Binh, H., 200(33), 234, *247*

AUTHOR INDEX

Hwa, J. C. H., 155(232), *179*

I

Iacoviello, J. G., 365(94), *374*
Ichikawa, R. I., 356(38), 363(38), *372*
Ide, F., 487(210), *492*
Iizima, R., 366(112d), *375*
Ilchenko, P. A., 332(60i), *338*
Imai, H., 365(79c), *374*
Imai, K., 369(175), *377*, 463, *464*
Imoto, M., 355(29), 367(29), 371(206), *372, 378*
Inamoto, Y., 332(60i), *338*
Ingelman, B., 192(9), 204(9), 228(9), 240(9), *247*
Ingham, J. D., 85, *93*, 479(346), *495*
Ingram, V. M., 127(148), *176*
Inoue, S., 369(170), *377*
Isgur, I. E., 363(49a), *373*
Ishii, H., 365(88a), *374*
Ishikura, H., 126(71, 123), 127(71), *175, 176*
Iskhakov, O. A., 364(68c), *373*
Ito, K., 370(186), *377*
Ivin, K. J., 63(44), *66*, 466(211), *492*
Iwamida, T., 366(112d), *375*
Iwatsubo, M., 127(175), *177*

J

Jackson, J. B., 24(48), *42*
Jacob, M., 287(6, 7), 288(6, 7), 294, 296(6, 7), 301(6, 7), 304(6, 7, 26, 30), 305(6, 7, 26, 30), *305, 306*
Jacobson, A. L., 78(30), 79(30), *92*, 468(348), *495*
Jacobsson, L., 126(34, 41), *174*
Jahnke, E., 428, 435, *459*
Jahnke, K., 228(58), 236(58), 240(58), *248*
Jakubtschik, A. I., 364(74), 369(169), *374, 377*
James, A. T., 166(253), *179*
James, E. A., 5(16), 23(16), *41*
James, J. W., 478(131), *490*
Janeschitz-Kriepl, H., 336(82a), *339*
Janssen, A. G., 471(427), *496*
Jaques, R., 126(37), *174*
Jellinek, H. G., 471(212), *492*
Jendrychowska-Bonamour, A., 368(153), *376*
Jenkins, A. D., 418(1), 422(1), 431(1), *459*
Jentoft, R. E., 102(20), *121*
Jirgensons, B., 474(213), *492*

Johansson, B. G., 126(15, 120), 127(151), 129(151), 169, *173, 176, 177, 179*
Johns, E. W., 370(196), *337*
Johnson, B. L., 475(108), 485(214), 486(214), *490, 492*
Johnson, D. H., 461(3), *464*
Johnson, J. F., 83(34, 35), 84(34, 39), 87(39), 91(35), *92, 93*, 97(7), 98(7), 99(16), 100(16), 102(7, 16), 103(16), 105(7), 110(34), 114(7), 119(7), 120(7, 44, 45), *120, 121, 173, 179*, 332(60b), *338*, 369(168b), *377*, 467(71, 177), 479(70), *489, 491*
Johnson, M., 335(77), *339*
Johnson, M. F., 333(65), *339*
Johnson, P., 485(65), *489*
Jones, J. K. N., 462(10), *464*
Jones, M. N., 351(15), 369(15, 170a), *371, 377*, 478(4, 5), *488*
Jones, R. W., 333(62), *338*
Jordan, D. O., 207(50), 216(50), 231(50), 232(50), *248*, 473(424), *496*
Jordan, I., 333(65), *339*
Jordan, L., 335(77), *339*
Jorgensen, L., 308, *315*
Jozefowicz, J., 111(31), *121*, 367(132, 133a), *375, 376*, 487(76), *489*
Juilfs, J., 480(215), *492*
Jullander, I., 264, 266, *283*
Jungnickel, J. L., 90(55), *93*, 97, 98(2), 102(2), 104, 106, 110(2), 111(2), *120*, 466(216), 472(216), *492*
Jutisz, M., 126(49), *174*

K

Kämmerer, H., 365(84), *374*, 367(134), *376*, 471(218), 478(217, 231), *492*
Kaila, E., 471(392), 482(391), *496*
Kaizerman, S., 369(173), *377*
Kakei, M., 126(118), *176*
Kakiuchi, H., 478(219), *492*
Kamata, T., 65(47), *66*
Kambara, S., 364(71a), *374*
Kambe, H., 363(55a), *373*, 487(385), *496*
Kamide, K., 332(60i), *338*
Kamio, K., 371(215), *378*
Kampf, M. J., 366(118), *375*
Kanda, H., 84(42), 87(42), *93*
Kane, P. O., 200(37), 235(37), 239(37), *247*, 474(67), *489*

AUTHOR INDEX

Kargin, V. A., 200(38), 241, *247*, 365(79), *374*, 487(220), *492*
Karlsson, R., 126(45), *174*
Katchalski, E., 431(14, 15), *459*
Katchalsky, A., 477(221), *492*
Kaufman, H. S., 466(222, 223), *492*
Kaverina, N. I., 467(269), *493*
Kawaguchi, H., 65(47), *66*
Kawahara, K., 45(7, 8), 59, 60, 61, 65(8), *65*, *66*, 471(225), 473(224), *492*
Kawai, T., 24, 25, *42*, 52(23), *66*
Kawasaki, A., 369(176), *377*, 478(226), *492*
Kawase, K., 367(137), *376*
Ke, B., 356(39), *372*
Kedem, O., 300, 301(21), 303, *306*
Kegeles, G., 275, *283*
Keil, B., 366(107), *375*
Keith, H. D., 5, *41*
Keller, A., 25, *42*, 75, 91(21), *92*
Kendall, M. G., 440(32), *459*
Kenn, R. S., 362(41b), *372*
Kenyon, A. S., 72, 74, 83, 84(6), 86(6), 88, 89(6), *92*, *93*, 106(25), 120(46), *121*, 466(229), 467(228, 229), 472(228), *492*
Kern, R. J., 370(182), *377*
Kern, W., 365(84), 369(156), 371(215), *374*, *376*, *378*, 476(377), 478(217, 231), 479(230), *492*, *495*
Kerten, T. E., 166(255), *179*
Ketley, A. D., 364(76a), *374*
Khamis, J. T., 368(155), *376*
Kiehne, H., 342(5), *371*
Kilb, R. W., 20, *42*, 362(41c), *372*
Killander, F., 126(26), *174*
Killander, J., 126(48, 98, 99, 100), 133, 134, *174*, *175*, *177*
Killmann, E., 85(47), *93*
Kim, C. S. Y., 364(74a), *374*
Kim, W. K., 330(48), *338*
Kin, L., 4(12), *41*, 332(60i), *338*
Kind, T. P., 126(141), *176*
Kinell, P. O., 265, 266, *283*, 477(232, 233), *492*
King, T. E., 126(54), *174*
Kinsinger, J. B., 112(37), *121*
Kirichenko, V. A., 246(77a), *249*
Kirkwood, J. G., 485(234), *492*
Kisliuk, R. L., 126(94), *175*
Kita, S., 369(166a), *377*
Kjellgreen, R., 483(199), *492*
Kjellin, K., 167, *179*

Klainer, S. M., 275, *283*
Klee, W. A., 126(105), *175*
Klein, J., 219, 221, *248*
Klenine, S., 304(26), 305(26), *306*
Knight, G. W., 173, *179*
Knudsen, P. J., 126(68), *175*
Kobayashi, F., 365(77), *374*
Kobayashi, H., 50, 57, 58(38), *66*, 476(235, 236, 237), *492*
Kobayashi, T., 369(162), *376*
Kobelt, D., 463, *464*
Kobryner, W., 367(143, 144), *376*, 487(238), *492*
Kocher, R., 470(239), *492*
Kockelbergh, G., 368(154), *376*, 487(240), *492*
Koefoed, J., 126(68), *175*
Koepp, H. M., 479(241), *493*
Koessler, I., 475(244), 476(243, 245), 477(242), *493*
Kössler, I., 181(2), 186, 187, 189, *189*, *190*
Kohle, H., 477(417), *496*
Kojima, H., 76(21), *92*, 468(451), *497*
Kokle, V., 467(246), *493*
Kolesnikov, G. S., 367(141a), 368(154c), *376*
Konigsberg, W., 126(14), 127(152), *173*, *177*
Koningsveld, R., 25, 26(54a), *42*, 91, *93*, 397, *413*, 467(247), *493*
Korshak, V. V., 366(117a), *375*, 479(343), 480(55), *489*, *495*
Kosai, K., 363(51c), *373*
Koshland, D. E., 126(16), *173*
Kostka, V., 366(107), *375*
Kotera, A., 65(47), *66*
Kotliar, A. M., 476(251), *493*
Kovarikova, J., 366(112b), *375*
Kovarskaya, B. M., 200(39), 241, *247*
Krabal, B., 363(48), *373*, 486(351), *495*
Krämer, H., 369(160, 161), *376*, 462(12), *464*
Kraemer, E. O., 266, 278(33), *283*, 293, 305, 385, *412*
Krantz, A. B., 126(59), 128(59), 129(59), *174*
Krasovec, F., 363(57), *373*
Krasulina, N. V., 364(71), *374*, 486(426), *496*
Krause, S., 362(41d), *372*, 477(248), *493*
Krejsa, J., 181(2), 186, 187, 189, *189*, *190*, 475(249), 476(243, 245), 477(242), *493*
Kressmann, T. R. E., 139(203), 148(203, 229), 155(203, 236), 156(203), *178*, *179*
Krigbaum, W. R., 16, 26, *41*, *42*, 75, 84(45), 85, *92*, *93*, 97, 98(6), *120*, 263(18), 282,

283, *284*, 466(252), 467(250), 476(251), *493*
Krozer, S., 200(32), 237, 247
Krueger, P. M., 166(255), *179*
Kruyt, H. R., 78, *92*
Kubota, Y., 367(148), *376*
Kuboyama, M., 126(54), *174*
Kudryavtseva, L. G., 363(44a), *372*
Kuhn, W., 370(180), *377*
Kulkarni, A. Y., 368(153b), *376*
Kun, K. A., 156(241), *179*
Kunin, R., 155(233), 156(240), *179*
Kurata, M., 5(20), *41*, 263(21), *283*
Kurz, J. E., 75(15), 83(36a), 84(45), 85, *92*, *93*, 97, 98(6), 120(46), *120*, *121*, 466(252), *493*
Kuyama, S., 126(76), *175*
Kuznetsov, E. V., 364(68c), *373*
Kyner, W. T., 438(26), *459*

L

Laamanen, L., 482(391), *496*
Laczkowski, M., 480(421), *496*
Lagally, P., 477(417), *496*
Lagermalm, G., 265, 270(50), *283*, 473(165), *491*
Lahav, M., 367(148a), 368(148a), *376*
Laker, D., 471(82), *489*
Lakshmanan, T. K., 102(17), *120*
Lamm, O., 255, *282*
Lamp, J., 332(60f), *338*
Lanczos, C., 440, 442(33), *459*
Landel, R. F., 484(253), *493*
Landler, I., 145(213), *178*, 467(254), *493*
Landyschewa, W. A., 365(87), *374*
Lane, T., 192(25), 201(25), 204(25), 243(25), *247*, 364(69), *373*, 486(151), *491*
Lang, M., 355(25), 363(25), *372*
Langford, W. J., 370(183), *377*, 469(255), 470(255), *493*
Langhammer, G., 146, *178*, 181(1), 182(6, 7), 184, 185, 186, 187, 188, 189, *189*, *190*, 469(257, 259, 260, 263), 470(260), 472(256, 262), 473(260, 261), 474(256, 258, 259, 260, 261), 476(257, 260, 261, 264), 477(260, 261), 480(260), 482(260, 261, 264), *493*
Lansing, W. D., 266, 278(33), *283*, 293, *305*, 385, *412*

Lapresle, C., 126(51), *174*
Lapsley, J., 389, 394, 395, 405, *412*
Lathe, G. H., 127(167), 128, 152(167, 186), *177*
Latov, V. K., 236, *248*
Laurent, E. P., 136, *178*
Laurent, T. C., 133, 134, 136, *177*, *178*
Lea, D. J., 126(140), 129(140), 150, *176*
Leaderman, H., 4, *41*, 331, 333, *338*, 467(265), *493*
Le Beau, D. S., 475(188), 485(189), *491*
Le Bel, R. G., 366(109), *375*
Lee, C. C., 479(47), *488*
Lee, Y. C., 126(81), 127(81), *175*
Lehmann, G., 366(124, 125), *375*
Leng, M., 366(115a), 370(202), *375*, *378*
Lesnini, D. G., 162(249), *179*
Leugering, H. J., 341(1), 366(125), *371*, *375*
Leutner, F. S., 369(174), *377*
Levantovskaya, I. I., 200(39), 241(39), *247*
Levi, G. R., 145(212), *178*
Levin, A., 362(41l), *372*
Levy, M., 366(119), *375*
Levy, S. S., 479(47), *488*
Li, C. H., 126(36), 127(36), *174*
Lieberman, S., 102(17), *120*
Lin, C. C., 367(137), *376*
Lindenmeyer, P. H., 5(15), *41*
Lindner, E. B., 126(58, 104), 129(58), 163(104), 169(104), *174*, *175*
Lindquist, B., 126(121, 135), 128, *176*, *177*
Ling, Nan-Sing, 99(14), *120*
Linklater, A. M., 278(67), *284*
Lins, H., 228(58), 236(58), 240(58), *248*
Linsert, F., 325, 326, *337*, 482(266), *493*
Lipp, W. J., 126(27), *174*
Lisowski, F., 126(93), 129(93), *175*
Lissitzky, S., 126(17, 35), 145(214), 163(214), *173*, *174*, *178*
Litmanovich, A. D., 362(41h, 41k), 363(44, 44a), 370(184, 187), *372*, *377*, 486(267), *493*
Litt, M., 438(27), *459*
Liu, W. K., 126(36), 127(36), *174*
Lloyd, W. G., 148, *178*
Loconti, J. D., 71, 85(48), 87(3), 90(48), *92*, *93*, 97(4), 98(4), 102(4), 106(24), 107(28), 108(24, 28), 110(4), 115(24), *120*, *121*, 463, *464*, 472(366, 368), *495*
Lombard, F., 479(268), *493*

Author Index

Longi, P., 75(18), 91(18), 92, 369(179b), 370(179b), 377
Lontie, R., 126(137), 176
Lorand, E. J., 365(99), 374
Lorenzi, G. P., 352(19), 369(19), 372
Loshack, S., 3(3), 41
Losikov, B. V., 467(269), 493
Lowry, G. G., 318(3), 336, 387, 412
Lu, C. Y., 362(41g), 372
Luchkina, V. M., 364(68c), 373
Luckert, H., 480(167), 491
Lueben, G., 127(161, 162), 129(161, 162), 133(162), 177, 472(100), 490
Lütje, H., 184(15), 190
Lunak, S., 479(270), 493
Lundberg, J. L., 319(7, 8), 336, 401, 413
Lundberg, R., 464(31), 464
Lundblad, G., 126(96), 175
Luongo, J. P., 75, 76, 92
Lytton, M. R., 308(3), 309(3), 311, 315, 482(87), 489

M

McAlevy, A., 365(100), 374
McCall, D. W., 463, 464
McCarthy, J. L., 288(12, 13), 294, 295(12, 13), 302(12, 13), 304(12), 305, 483(3), 484(129), 488, 490
McCord, R. A., 331(54), 338
McCormick, H. W., 4(12), 41, 271(53), 272(53, 54, 56), 273(56), 283, 305, 306, 332(60i), 338, 433(21), 459, 467(275), 471(276), 473(58, 274), 489, 492, 493
McDowell, R. H., 51, 66, 485(277), 493
McFadden, W. H., 78(30), 79(30), 92, 468(348), 495
McGarvey, F. X., 155(233), 179
McKannan, E. C., 334(74), 339, 477(439), 477
McKelvey, J. M., 332(57), 338
McKenzie, H. A., 366(106), 375
Maclay, W. D., 484(328), 494
McLeod, L. A., 97(5), 104, 106, 107, 111(5), 120, 475(207), 492
McLoughlin, J. R., 333(64), 339
McNally, J. G., 192, 241, 246
McNeer, R. D., 484(364), 495
Madden, R. W., 104(21), 111(21), 121, 367(145, 147), 369(147), 376, 487(85), 489

Magat, M., 486(153), 491
Mahoney, J. F., 341(2), 371
Majer, H., 370(180), 377
Major, J. R., 192(18), 218(18), 229(18), 230(18), 231(18), 235(18), 236(18), 237(18), 247
Maley, L. E., 127(159), 129(159), 130, 160, 161, 177
Malinina, I. A., 365(96), 374
Mammen, E., 126(87), 175
Mandel'baum, D. I., 482(387), 496
Mandelkern, L., 63(46), 64(46), 66, 263(18, 19, 20), 278, 283, 284
Manderkern, L., 24(46), 42
Manning, P. P., 369(161a), 376, 470(141), 490
Mansford, K. R. L., 126(47), 129(47), 174
Manson, J. A., 363(55), 367(55), 373
Marakhonov, I. A., 332(60i), 338
Marchal, E., 336(86), 339
Marchal, J., 336(86), 339
Mark, H., 145(210), 178, 310, 316, 336(85), 339, 461(3), 464, 471(159, 427), 482(16), 488, 491, 496
Marks, G. S., 127(153), 177
Maron, S. H., 11(31), 41
Marr, W. E., 139(203), 148(203), 155(203), 178
Marsden, N., 126(60), 174
Marshall, C. A., 51, 66, 363(51), 373, 471(271), 487(301), 493, 494
Marshall, L. M., 98(12), 99(15), 120
Marshall, R. D., 127(153), 177
Martin, H., 385(8), 412
Martin, R. G., 126(21), 129(21), 145(21), 174
Martinez, F. E., 98(11), 120
Martinovich, R. J., 331, 338
Marx, R., 126(79), 175
Marx-Figini, M., 483(272, 273, 375), 493, 495
Mason, S. G., 137, 178, 462, 464
Masson, C. R., 369(164), 376
Mašura, V., 326, 337
Matheka, H. D., 126(67), 175
Matheson, A. J., 332(60f), 338
Mathieson, A. R., 200(44), 231, 232, 248
Matsuda, T., 367(137), 376
Matsumoto, M., 26(57, 58), 39(57), 40, 42, 349(10), 365(10), 369(175), 371, 377, 463, 464
Matsumoto, S., 52(26), 53(26), 66
Matsuya, K., 365(77), 374
Matsuzaki, K., 365(93), 374

Matuzawa, S., 367(148), *376*
Maury, R., 369(172), *377*
Mayo, F. R., 451, *459*
Mazzanti, G., 75(18), 91(18), *92*, 364(69b, 69c), 369(178), 371(213), *373*, *377*, *378*
Meares, P., 366(115), *375*
Meeks, J. W., 475(278), *493*
Meffroy-Biget, A. M., 54, 55, *66*, 483(279, 280), *493*
Mehta, P.C., 368(153b), *376*, 483(281), *493*
Meitzner, E. F., 148(226), 156(240, 242), *178* *179*
Melkonyan, L. G., 241, *249*
Melville, H. W., 192, 201, 205(14), 229, 231, 232, 235(65), 236, *247*, *248*, 366(113, 117), 369(117, 163, 164), *375*, *376*, 431(12), *459*, 469(40), 473(282), *488*, *493*
Menčik, Z., 50, 51, *66*, 226, 228, *248*, 470(283), *493*
Mendelson, R. A., 25, 26(54), 39(54), *42*, 86, *93*, 115(43), *121*, 468(284), *493*
Menefee, E., 330(46), 334, *338*
Mercurio, A., 334(70), 335(70), *339*, 473(414), *496*
Merz, E. H., 43(3), *65*, 471(285), 476(286), *493*
Meselson, M., 280, 281(74), 282(74), *284*
Meskat, W., 318(1), 322(1), 325, 326, *337*, 482(287, 288), *493*
Messina, G., 84(44), 88(44), *93*, 468(330), *494*
Metz, D. J., 356(31), 367(31), *372*
Metzendorff, J., 484(458), *497*
Meyer, K. H., 73, 75(7), *92*
Meyerhoff, G., 56(34), 63(44), *66*, 98(13), *120*, 184, 185, *190*, 260(12), *282*, 288(9, 10), *305*, 369(158), *376*, 466(211), 471(72), 472(289), 477(290), *489*, *492*, *493*, *494*
Meyering, C. A., 366(111), *375*
Meyersen, K., 371(215), *378*
Michailow, N. W., 363(60), *373*
Michl, K. H., 364(64), *373*
Michon, J., 126(126), *176*
Miers, J. C., 484(328), *494*
Mighton, C. J., 475(298), *494*
Mijal, C. F., 97(4), 98(4), 102(4), 110(4), *120*
Mikes, J. K., 148, *178*
Mikhailov, N. V., 58, *66*, 470(291), *494*
Mikhailova, L. N., 371(207), *378*
Mikola, J., 366(112c), *375*

Millar, J. R., 139(203), 148(229), 155(203), 156(203), *178*
Miller, B., 482(292), *494*
Miller, L. E., 290, 303, *305*, 474(293), *494*
Miller, R. G. J., 200, 236, 237, *248*, 477(183), *491*
Miller, W. L., 356(36), 369(36), *372*
Millin, D. J., 126(143), *176*
Mills, D. R., 326, 331, 332(60c), *337*, *338*, 467(294), *494*
Mills, R. A., 207(50), 216(50), 231(50), 232(50), *248*, 473(424), *496*
Mino, G., 363(55b), 369(173), 370(190), *373*, *377*
Minoura, Y., 355(29), 367(29), *372*
Miranda, F., 126(17), 145, 163(214), *173*, *178*
Mise, N., 366(123), *375*
Mita, I., 363(55a), 367(129), *373*, *375*, 487(385), *496*
Mitchell, J. C., 480(295), *494*
Mitchell, R. L., 49(16), *66*, 483(296), *494*
Mitra, A., 365(81), *374*
Mitsengendler, S. P., 364(71), *374*, 486(426), *496*
Miyake, A., 401, *413*
Miyal, C. F., 472(367), *495*
Miyamichi, K., 363(54), *373*
Miyazawa, Y., 65(47, *66*)
Moacanin, J., 288, 294, 295, 302, 304, *305*, 484(297), *494*
Mochel, W. E., 475(298, 299, 300), *494*
Mock, R. A., 51, *66*, 363(51), *373*, 471(271), 487(301), *493*, *494*
Molyneux, P., 192(18), 218(18), 229(18), 230(18), 231(18), 235(18, 65), 236(18), 237(18), *247*, *248*, 366(113), *375*
Monfoort, C. H., 126(52), *174*
Montagnoli, G., 371(212), *378*
Montgomery, R., 126(81), 127(81), *175*
Moore, G. E., 326(33), 331(53), 332(58, 60c), *337*, *338*, 467(294), *494*
Moore, J. C., 127(157, 158), 129(157, 158), 130, 139(158), 148, 153(157, 158), 154(230), 157, 170, 171(157), 173(158, 265), *177*, *178*, *179*, 472(302), *494*
Moore, L. D., 363(43), *372*, 467(303), 487(403), *494*, *496*
Moore, L. D., Jr., 90(54), *93*, 113(40), 117, 118, *121*, 171(51), *283*
Moore, S., 126(116), *176*

Moore, W. R., 470(304), *494*
Mora, P. T., 369(171, 172), *377*
Moraglio, G., 56(36), *66*, 347(7), *371*, 471(318), *494*
Morey, D. R., 50, *66*, 192, 191, 201(2), 205(2), 219, 222, 234, 239, *246*, *247*, *248*, 365(88), *374*, 469(305), *482*(308), 483(306, 307, 406), *494*, *496*
Morikava, H., 200(40a), 242(40a), *248*
Morini, M., 155(234), *179*
Morozov, V. I., 201(47), 203(47), 204(47), 205(47), *248*
Morris, C. J. O. R., 124, 150, 152(4), *173*
Morris, P., 124, 150, 152(4), *173*
Morse, P. M., 440, 442(31), 444(31), *459*
Mosbach, R., 126(108), 152, *176*
Moschle, A., 365(82), *374*, 480(27), *488*
Mosevitskii, M. I., 363(49), *373*, 475(336), 486(335, 337), *494*, *495*
Moshkina, T. M., 478(309), *494*
Mosimann, H., 268, *283*, 335(78), *339*, 482(359), *495*
Motroni, G., 369(178), *377*
Mueller, E., 476(377), *495*
Müller, K., 126(129), 145(129), *176*
Müller, O., 238, *248*
Müller, W., 471(178), *491*
Münster, A., 19(40), *42*, 52(25), *66*
Mukoyama, S., 351(13), *371*
Mula, A., 461, *464*
Muller, W. A., 482(310), *494*
Mullineaux, R. D., 96(3), 97(3), 98(3), 114(3), 115(3), *120*, 486(135), *490*
Munk. P., 370(198), *378*
Murahashi, S., 363(51c), *373*
Murakami, K., 334(70), 335(70), *339*, 463, *464*, 473(414), *496*
Murray, M., 126(78), *175*
Mussa, C., 65(48), *66*, 78, 79(26), 86, 89, *92*, 234, 236, 240, *248*, 319, 335, *337*, *339*, 393, 394(17), *412*, 466(311, 316), 467(316), *494*
Myagchenkov, V. A., 364(68c), *373*
Myers, G. E., 478(312), *494*

N

Nachtigall, K., 184, *190*
Nairn, R. C., 126(28), *174*
Nakaguchi, K., 84(42), 87(42), *93*
Nakahara, H., 483(313), *494*
Nakajima, A., 76, *92*, 369(168a), *377*, 468(315), 469(314), *494*
Nakajima, N., 332, *338*
Nakamura, A., 200(40a), 242(40a), *248*
Nakata, T., 371(206), *378*
Nakatsuka, K., 487(210), *492*
Nanda, V. S., 438(28a, 28b), *459*
Nasini, A., 65(48), *66*, 78, 79(26), 86, 89, *92*
Nasonova, T. P., 467(157), *491*
Nassini, A., 466(316), 467(316), *494*
Natta, G., 56, *66*, 75, 85, 91, *92*, 352(19), 365(69a, 69b, 69c, 71b), 369(19, 157, 168, 178), 371(213), *372*, *373*, *374*, *376*, *377*, *378*, 468(317), 471(318), *494*
Nazarian, G. M., 280(72), *284*
Neel, J., 474(383), *495*
Nelson, H., 288(13), 294(13), 295(13), 302(13), *305*, 484(297), *494*
Nelson, R., 366(105), *375*
Neshima, T., 365(79c), *374*
Nestler, L., 146, *178*
Neuberger, A., 127(153), *177*
Neukom, H., 127(183), 145(183), *177*
Newman, S., 282(80), *284*, 369(167), *377*, 468(319), *494*
Nicholas, M. L., 45(6), 65(6), *65*
Nichols, J. B., 256(7), 264(24), *282*, *283*, 287(4), *305*, 475(298, 299, 300), *494*
Nicolas, L., 78, 79(28), 86, *92*, 467(320, 321), *494*
Nielsen, L. E., 364(63), 367(131), *373*, *375*
Nikitin, N. I., 365(91), *374*
Nikkari, T., 166(254), *179*
Nikolaeva, I. I., 332(60i), *338*
Ninomiya, K., 333, 334(68), *339*, 467(149), 469(148), 473(148), *491*
Nisch, A., 469(23), *488*
Nishioka, A., 356(35), 369(35), *372*
Niwinska, T., 26(59, 60), 40(59, 60), *42*, 309(9, 10), *316*
Nomura, H., 65(47), *66*
Nordin, P., 126(134), *176*
Nordt, E., 463, *464*
Norman, P. S., 126(141), *176*
Notley, N. T., 363(56), *373*, 486(322), *494*
Nozakura, S., 363(52), *373*
Nultsch, W., 126(107), *175*
Nummi, M., 126(142), *176*
Nystrom, E., 154(230b), *179*

O

O'Connor, A., 75, 91(21), *92*
Ogata, Y., 355(29), 367(29), *372*
Ogston, A. G., 133, *177*
Ohta, S., 370(200), *378*
Ohuo, K., 332(60i), *338*
Ohyanagi, Y., 26(57), 39(57), *42*, 52(26), 53(26), *66*
Okada, R., 45(8), 65(8), *65*
Okada, T., 369(166a), *377*
Okamoto, H., 19(39), 26(61, 62, 63), 27, 39(62, 63), 40, *42*, 47, 48, *66*, 467(323), *494*
Oline, J. A., 148(226), *178*, 156(242), *179*
Onozuka, M., 371(209), *378*
Onyon, P. F., 319, *336*, 418(1), 422(1), 431(1), *459*
Orofino, T. A., 431(17), *459*
Orr, R. J., 366(120), *375*, 486(324), *494*
Orthmann, H. J., 48(14), *66*, 466(422), *496*
Ostling, G., 125(11), 129(11), *173*
Oth, A., 74, 79(13), 85, *92*, 192(5, 7, 8), 201(5, 7), 203(7), 205(5, 7), 206, 215, 234, 237, 240, *246*, *247*, 470(325), 483(326, 327), *494*
Oth, J., 266(34), 268, *283*
Otsu, T., 371(206), *378*
Ott, E., 341(3), 362, *371*
Ottewill, I. G., 200(37), 235(37), 239(37), *247*, 474(67), *489*
Ovenall, D. W., 192(17), 230(17), 231(17), *247*, 473(186), *491*
Overberger, C. G., 363(52, 54), 370(185), *373*, *377*
Owens, H. S., 484(163, 328), *491*, *494*

P

Pacsu, E., 482(292), 483(282), *493*, *494*
Padden, P. J., 5, *41*
Pakshver, E. A., 320(17), *337*
Palmstierna, H., 126(63), *174*
Panchak, J. R., 366(118), *375*
Panton, C. J., 110(35), *121*, 467(329), *494*
Papkoff, H., 126(36), 127(36), *174*, *177*
Parravano, G., 463, *464*, 474(397), *496*
Parrini, P., 84(44), 88(44), *93*, 468(330), *494*
Parrod, J., 366(126), 367(126), *375*
Pascoe, G. M., 389, 394, 395, 405, *412*

Pasquon, I., 75(20), 85(20), 91(20), *92*, 364(69a, 69b, 69c), *373*
Passaglia, E., 356(30), 367(30), *372*
Pasternak, R. A., 280, *284*
Patat, F., 56(35), *66*, 85, *93*, 217, 219(52a, 52b), 221, *248*, 469(101, 121), *490*
Paton, C., 366(115), *375*
Pavlova, S. A., 475(331), 480(55), *489*, *494*
Paxton, T. R., 370(181), *377*
Peaker, F. W., 192(17), 193(27), 230(17), 231(17), 234, *247*, *248*, 341(4), 369(163, 164), *371*, *376*, 462, *464*, 473(186), *491*
Pedersen, K. O., 126(95), 128(95), 145(95), *175*, 256(6), *282*
Peebles, L. H., 281(78), 282(79), *284*, 355(28), 369(28), *372*
Peer, H. G., 478(332), *494*
Pegoraro, M., 75(19, 20), 85(19, 20), 91(19, 20), *92*, 368(154b), 369(168), *376*, *377*, 468(317), *494*
Peller, L., 254, *282*
Pellon, J., 371(204), *378*
Pennings, A. J., 25, 26(54a, 54b), *42*, 91, *93*, 467(247), *493*
Pepper, D. C., 70, 84(2), 98(8), 102(8), 107, 108, *120*, 472(333), *494*
Pepper, K. W., 139, *178*
Peraldo, M., 75(19, 20), 85(19, 20), 91(19, 20), *92*, 369(168), *377*, 468(317), *494*
Perrine, T. D., 371(211), *378*
Peterlin, A., 335, *339*
Petersen, D. F., 126(125), 127(125), *176*
Peterson, E. A., 366(110), *375*
Peticolas, W. L., 4(11), *41*, 330, 334, *338*, *339*
Petraglia, G., 332(60i), *338*
Petropawlowski, G. A., 365(91), *374*
Petrovskaya, I. D., 370(195), *377*
Pettersson, G., 126(88), *175*
Pfennig, H., 181(1), 184(1), 185(1), 186(1), 187(1), 188(1), *189*, 472(256), 474(256), *493*
Philipp, H. J., 482(334), *494*
Philippoff, W., 322(23), 323, 326, *337*
Phillips, A. W., 127(150), *176*
Phillips, C. S., 481(22), *488*
Phillips, D. M. P., 370(196), *337*
Philpot, J., 287, *305*
Pibarot, R., 485(349), *495*
Pickett, H. E., 173, *179*
Pierce, F. V., 126(24), *174*

Author Index

Pierce, J. G., 127(148), *176*
Pieski, E. T., 486(457), *497*
Pino, P., 352(19), 369(19), 371(212), *372,378*
Pivec, L., 370(198), *378*
Planck, R. W., 475(278), *493*
Plate, N. A., 200(38), 241(38), *247*, 365(79), *374*, 487(220), *492*
Plesch, P. H., 110(35), *121*, 467(329), *494*
Poddubnyi, I. Y., 363(49), *373*, 467(157), 475(52, 54, 336), 486(335, 337), *489, 491, 494, 495*
Pokrikyan, V. G., 366(117a), *375*
Polgar, L., 479(133), *490*
Pollock, D. J., 475(338), *495*
Polowinski, S., 26(59, 60), 40(59, 60), *42*, 309(9, 10), *316*, 370(203), *378*
Polson, A., 126(138), 129(138), 152(138, 153), *176*
Pontén, F., 126(26), *174*
Poore, L., 483(350), *495*
Pope, N. T., 1, 111(29), *121*, 485(339), *495*
Porath, J., 124(3), 126(10, 12, 13, 58, 75, 83, 85, 86, 88, 102, 104), 128(12, 13, 102), 129(58, 102), 133, 145, 150, 160, 161, 163(104), 169(104), *173, 174, 175, 177, 179*
Porri, L., 364(71b), *374*
Porter, R. R., 126(38), *174*
Porter, R. S., 83(34, 35), 84(34, 39), 87(39), 91(35), *92, 93*, 97(7), 98(7), 99(16), 100(16), 102(7, 16), 103(16), 105(7), 110(34), 114(7), 119(7), 120(7, 44, 45), *120, 121, 179*, 332(60b), *338*, 369(168b), *377*, 467(71, 177), 479(70), *489, 491*
Pouradier, J., 484(340), *495*
Powell, E., 351(15), 369(15), *371*
Powell, G. M., 478(18), *488*
Powers, P. O., 192, 241, *246*, 471(341), *495*
Pramer, D., 126(76), *175*
Prati, G., 363(59), *373*
Pravda, Z., 126(61), *174*
Pravikova, N. A., 192(18a), 200(40), 230(18a), 231(18a), 232, 246, *247*, 248, 249, 471(342), 473(342), *495*
Preaux, G., 126(137), *176*
Pregaglia, G., 371(213), *378*
Press, E. M., 126(38), *174*
Prevorsek, D., 362(41e, 41i), *372*
Price, F. P., 471(41), *488*
Protasova, M. S., 370(187), *377*

Pudovik, A. N., 478(309), *494*
Pugh, D. W., 326(33), 331(53), 332(58, 60c), *337, 338*, 467(294), *494*
Purdy, W. C., 470(378), *495*
Purnell, G. G., 141(207), *178*
Purves, C. B., 341(2), *371*
Pyrkov, L. M., 281(77), *284*, 368(154a), *376*

Q

Quitzsch, K., 181(1), 182(7), 184(1), 185(1), 186(1), 187(1), 188(1, 7), *189*, 472(256, 262), 474(256,258), *493*

R

Rabel, W., 196, 228, 230, 231, 232, *247*
Rabinowitz, J. L., 126(53), *174*
Radford, L. E., 280(73), *284*
Radock, R. M., 438(26), *459*
Raetz, R. W., 43(3), *65*, 471(285), *493*
Raff, R. A. V., 86, *93*
Rafikov, S. R., 365(103), *375*, 479(343), *495*
Raman, N. K., 466(344), *495*
Ramien, A., 126(87), *175*
Ranby, B. G., 264, 265, 266, *283*, 309, *316*, 483(345), *495*
Rao, V. S. R., 5(19), *41*
Rapp, N. S., 85, *93*, 479(346), *495*
Rarre, A., 368(155), *376*
Rasmussen, E., 369(173), *377*
Rasmussen, H., 126(18, 74), *174, 175*
Rathmann, G. B., 476(347), *495*
Rauch, B., 184(15), 185, *190*
Ravenhill, J. R., 166(253), *179*
Redlich, O., 78, 79, *92*, 468(348), *495*
Ree, T., 330, *338*
Rehfeld, S. J., 269(43), 272, *283*, 473(437), *497*
Rehner, J., Jr., 463, *464*
Reichard, P., 126(127), *176*
Reichenberger, D., 139(204), *178*
Reichmann, M. E., 320, *337*
Reimschüssel, W., 370(203), *378*
Reiner, M., 318(1), 322(1), *336*
Reisfeld, R. A., 126(90), *175*
Rempp, P., 363(43a), 366(115a, 115b, 126), 367(126), *372, 375*
Renkin, E. M., 137, *178*
Renner, G., 476(376), *495*
Reynolds, W. W., 156(239), *179*
Rich, A., 485(51), *489*

Richards, G. N., 367(140, 141), 368(141), *376*
Richards, H. H., 126(105), *175*
Richards, R. B., 63, 64, *66*
Richardson, M. J., 24(48), *42*, 461, *464*
Rinde, H., 264(23), 279, *283, 284*
Rinderknecht, H., 126(29), *174*
Rinfret, M., 363(45), *373*
Ringertz, N. R., 126(127, 128), *176*
Riou, M., 485(349), *495*
Ritchey, W. M., 362(41j), *372*
Rivest, R., 363(45), *373*
Robb, J. C., 341(4), *371*, 462, *464*
Robb, J. D., 231, *248*
Robbins, K. C., 126(91), *175*
Roberts, R., 366(127), 367(127), 368(127), *375*, 487(191), *491*
Robertson, A., 427, 431(11), *459*
Robertson, R. F., 363(48a), 370(188), *373, 377*
Robins, A. B., 257(9), 273(9), *282*
Rocaboy, F., 367(134), *376*, 471(218), *492*
Rochat, H., 145(214), 163(214), *178*
Rodén, L., 126(25, 26), 127(25), *174*
Rodriguez, F., 172, *179*, 326, *337*
Roe, R. J., 431(16), *459*
Rogers, L. N., 482(310), *494*
Rogovin, S. A., 365(87), *374*
Rolfson, F. B., 402(28), *413*
Rolland, M., 126(35), *174*
Romatowski, J., 98(13), *120*, 472(189), *493*
Rondier, A. J., 366(112e), *375*
Roovers, J., 367(142), 368(142), *376*
Rosenberg, J. L., 462, *464*
Rosenkranz, H. S., 484(34), *488*
Rosenthal, A. J., 351(12), 365(12), *371*
Roseveare, W. E., 483(350), *495*
Rosik, L. S., 363(48), *373*, 486(351), *495*
Roth, L. J., 126(114), 129(114), *176*
Rothman, S., 193(28), *247*
Rothstein, F., 162, 167, *179*
Rouse, B. P., 483(363), *495*
Rouse, P. E., 326, *337*
Rovery, M., 126(43), *174*
Rubina, Kh. M., 126(40, 109), 127(40), 129(109), *174, 176*
Rudd, J. F., 4(6), *41*, 331, 336, *338, 339*, 473(353, 354), *495*
Runge, F., 467(155), *491*
Ruska, H., 461, *464*
Ruskin, A. M., 463, *464*

Russell, C. A., 468(355), *495*
Rutherford, P. P., 70, 84(2), *92*, 98(8), 102(8), 107, 108, 110(35), *120, 121,* 467(329), 472(333), *494*
Ruthven, C. R. J., 127(167), 128, 152(167, 186), *177*
Rutowski, B. N., 470(356), *495*
Ryabova, L. G., 192, 200(40), 230, 231(40), 232(40), *247, 248*
Ryabova, Y. U., 471(342), 473(342), *495*
Rybnikar, F., 479(357), 480(358), *495*
Rylov, E. E., 201(47), 203(47), 204(47), 205(47), *248*
Rymo, L., 126(120), 169, *176, 179*

S

Sabia, R., 320, 332(60a), *337, 338*
Sacerdote, P., 335(77a), *339*
Sachs, A. P., 466(12), *488*
Sadron, C., 335(78), *339*, 366(121), *375*, 470(239), 482(359), 485(92), *489, 492, 495*
Saegusa, T., 356(32), 364(32, 75, 76), 365(79c), 366(123), 369(176), 370(200), *372, 374, 375, 377,* 486(147), *491*
Saha, A. N., 365(81), *374*
Saito, G., 145(210), *178*
Saito, T., 65(47), *66*
Sakaguchi, Y., 365(101, 103a), *374, 375*
Sakiz, E., 126(49), *174*
Sakurada, I., 365(101, 103a), 367(148), *374, 375, 376,* 469(314), *469*
Salmon, J. E., 155(231), *179*
Salyer, I. O., 72, 74, 83(36a), 84(6), 86(6), 88, 89(6), *92, 93,* 106(25), 120(46), *121,* 466(229), 467(228, 229), 472(228), *492*
Samuels, R. J., 62(42), *66*, 476(77), *489*
Samuelsson, G., 126(20), 159(245), *174, 179*
Sasaguri, K., 476(237), *492*
Sasaki, K., 364(71a), *374*
Sather, G. A., 111(32), 113, *121*, 479(180), *491*
Sato, H., 371(208), *378*
Saunders, W. M., 266(35), 270(48), *283*
Savitskaya, M. I., 477(360), *495*
Sayre, E. V., 26(56), 39(56), *42*
Scarso, L., 200(35), 239, *247*
Schachman, H. K., 256(8), 260(10), 275(61), 277(10), *282, 284,* 287(3), *305*
Schäfer, K. H., 328(38, 39, 41), *337*, 473(361), *495*

Schaffhauser, R., 334(70), 335(70), *339*, 427(8), *459*
Schally, A. V., 126(83), *175*
Scheibling, G., 287(5), 288(5), 294(5), 296, 301(5), 304(5), 305(5), *305*
Schenck, H. J., 479(11), *488*
Schen Van, A. G., 365(78), *374*
Scheraga, H. A., 127(173, 174), *177*, 263(18, 20), *283*, 335(82), *339*, 483(208), *492*
Scherer, P. C., 336, *339*, 365(94), *374*, 463, *464*, 482(362), 483(363), 484(364), *495*
Schildknecht, E., 470(365), *495*
Schimiza, H., 356(35), 369(35), *372*
Schlick, S., 366(119), *375*
Schliebener, C., 85(47), *93*
Schmidt, H., 479(230), *492*
Schmieder, W., 349(9), *371*, 462(12), *464*
Schnecko, H. W., 356(34), 369(34), *372*
Schneider, C., 368(151), *376*
Schneider, N. S., 71, 84(41), 85, 86(41), 87(3, 41), 88(41), 90, *92*, *93*, 97(4), 98, 102(4), 106, 107(28), 108, 110(4, 33), 113, 115, *120*, *121*, 193(28a), *247*, 472(366, 367, 368, 369), *495*
Schneider, W. C., 328(45), *338*, 364(67), *373*, 486(89), *489*
Schölner, R., 369(179a), *377*
Scholtan, W., 192, 200(16), 220, 226, 228, 235, 236, 239, 240, *247*, *248*, 474(370, 371), *495*
Scholz, A. G. R., 41(73), *42*, 71, *92*, 112(38), *121*
Schoon, T. G., 461, *464*, 485(372), *495*
Schramek, W., 387, 399, *412*
Schreiber, H. P., 4(5), *41*, 331, 332(60i), *338*, 467(373), *495*
Schultz, A. R., 15(35), 17, *41*, 469(374), *495*
Schulz, G. V., 26(55), 28, 39(55), 41, *42*, 53, 54, 56(34), *66*, 71, *92*, 112, *121*, 192, 193(1), 220, *246*, 266, *283*, 293(18), 298(18), *305*, 318, *336*, 357(40), *372*, 383, 386, 388, 390, *412*, 461, *464*, 471(72), 483(375), 484(124), *489*, *490*, *495*
Schulz, R. C., 371(215), *378*, 476(376, 377), *495*
Schulz, W. W., 470(378), *495*
Schumaker, V. N., 260(10), 277(10), *282*
Schurz, J., 319(11), 320(13), 322(22), 325, 327(34, 35, 36), 328(35, 36, 37, 38, 39, 40, 43, 44), 332(60), *337*, *338*, 363(61), *373*, 468(380), 484(381, 382), 486(379), *495*

Schwaben, R., 322(22), 326(27), *337*
Schwarzl, F., 318(1), 322(1), 332(61), *336*, *338*
Schwindt, H., 476(80), *489*
Scott, K. W., 463, *464*
Scott, R. L., 9(23), 13, 17(23), 19(38), 21, *41*, 42, 52, 53(27), *66*, 146(216, 218), *178*
Scott, R. P. W., 166(253), *179*
Sebastiano, F., 84(44), 88(44), *93*, 468(330), *494*
Sebban-Danon, J., 111(31), *121*, 367(132, 133), *375*, *376*, 487(76), *489*
Sebille, B., 474(383), *495*
Sehon, A. H., 126(140), 129(140), 150, *176*
Sekikawa, K., 26, 27, 39(64), *42*
Sela, M., 126(115), *176*, 431(15), *459*
Selikman, S. G., 363(60), *373*
Semanza, G., 126(145), *176*
Semenova, A. S., 332(60i), *338*
Semple, R. E., 462(10), *464*
Senti, F. R., 484(384), *496*
Sergeev, V. A., 366(117a), *375*
Severini, F., 368(154b), *376*
Severy, E. T., 332(59), *338*
Shahani, K. M., 126(84), *175*
Shalaeva, L. F., 332(60i), *338*
Shalitin, I., 431(14), *459*
Shapiro, B., 126(53), *174*
Sharp, J. O., 90(54), *93*, 113(40), 117(40), 118(40), *121*, 271(51), *283*, 467(303), *494*
Shaw, R. W., 126(130), *176*
Shen (Kwei), K. P., 200(36), 240, 241(36), *247*, 249
Shepherd, G. R., 126(125), 127(125), *176*
Sherman, J. R., 126(92), *175*
Shigatsch, K. F., 365(96), *374*
Shihanda, M., 494(313), *494*
Shima, S., 365(101), *374*
Shimosaka, Y., 364(73), *374*
Shimura, J., 487(385), *496*
Shimura, Y., 363(55a), *373*
Shirayama, K., 369(166a), *377*
Shockley, W., 483(199), *492*
Shooter, E. M., 282, *284*
Shooter, K. V., 257(9), 273(9), *282*, 484(386), *496*
Shostakooskii, M. F., 370(194), *377*
Shtarkman, B. P., 201(47), 203(47), 204(47), 205(47), *248*
Shtern, V. Y., 362(41k), 363(44, 44a), *372*, 486(267), *493*

Shu, T. W., 5(16), 23(16), *41*
Shubtsowa, I. G., 351(16), 366 (16, 108), *372, 375*
Shultz, A. R., 172, *179*
Shulyatikova, N. V., 482(387), *496*
Shyluk, S., 71, 72, 80, 81(4), 83(4), 84(4), 85(4), 86(4), 87(4), 88(4), 89(4), *92*, 468(388), *496*
Sieber, H., 200(45), 239(45), *248*
Siegel, B. M., 461, *464*
Signer, R., 267(38), 268, *283*, 300, *306*, 335(80), *339*, 473(389), 485(390), *496*
Sigwalt, P., 366(116), 370(199), *375, 378*
Sihtola, H., 351(14), 365(14, 89), *371, 374*, 471(392), 482(391), 483(2), *488, 496*
Silina, L., 200(32), 237(32), *247*
Simha, R., 310(14), *316*, 365(80), *374*
Simionescu, C. C., 482(393), *496*
Simon, R. H. M., 4(8), *41*, 332(60e), *338*
Simonazzi, T., 370(193), *377*
Simpson, D. W., 148(224), *178*
Singer, S., 470(106), *490*
Sirianni, A. F., 433(22), *459*
Sirlibaev, T., 236(66b), *248*
Sitenberg, B., 464(30), *464*
Sittel, K., 326(32), *337*
Sjövall, J., 154(230b), *179*
Skrube, H., 126(129), 145(129), *176*
Skvortsova, H., 370(194), *377*
Slonaker, D. F., 84(38, 43), 87(38), 89(38), *93*, 102(18), 104(18), 105(18), 106(26), 107(18), 108(18), 109(18), 110(18), 111(26), 116(18, 26), 117(18), 118(18), *121*, 467(83, 170), 468(83), *489, 491*
Sluyser, M., 126(70), 127(70), *175*
Smets, G., 365(78), 366(128), 367(128, 142, 149), 368(128, 142, 149, 154), *374, 375, 376*, 471(179), 487(240, 394, 395), *491, 492, 496*
Smith, D. G., 139(203), 148(203), 155(203), 156(203), *178*
Smith, H., 25, 26(53), *42*, 109(36), *121*, 369(166), *377*, 466(190), *491*
Smith, K. L., 478(18), *488*
Smith, M. H., 126(143), *176*
Smith, P., 75(15), *92*, 466(252), *493*
Smith, R. G., 4(9), *41*, 331(55), 333(62), *338*, 467(265), *493*
Smith, T. G., 332, *338*
Snell, J. B., 371(214), *378*
Sober, H. A., 366(110), *375*

Soboleva, T. A., 475(331), *494*
Sobue, H., 334, *339*, 365(93), *374*
Sollner, K., 148(223), *178*
Solms, J., 127(182), 145(182), *177*
Solomon, D. H., 477(119), 479(133), *490*
Solomon, O., 363(55c), *373*
Sookne, A. M., 482(396), *496*
Sorge, G., 48(14), *66*, 466(422), *496*
Sorm, F., 127(147), *176*, 366(107), *375*
Sormova, Z., 127(147), *176*, 370(198), *378*
Sorokin, M. F., 236, *248*
Spangler, R. D., 334(74), *339*, 477(439), *497*
Spasskowa, A. I., 364(74), *374*
Spencer, R. S., 39, *42*, 308, 309 *315*
Spiegelman, P. P., 474(397), *496*
Spiegels, M. C., 74, 79(12), 86, 89, *92*, 115(42), *121*, 466(97), *490*
Spinner, I. M., 362(41g), *372*
Spitzy, H., 126(129), 145(129), *176*
Sponar, J., 366(112b), 370(198), *375, 378*
Spurlin, H. M., 341(3), 362(3), 365(98), *371, 374*
Squire, P. G., 266(35), *283*
Stacy, C. J., 484(398), *496*
Stahl, F. W., 280(74), 281(74), 282(74), *284*
Stainsby, G., 484(399), *496*
Stannett, V., 367(140a), *376*
Starck, W., 43(1), *65*
Staudinger, H., 43, 59(2), *65*, 470(400), 482(401), *496*
Staverman, A. J., 318(1), 322(1), 332(61), *336, 338*, 462, *464*
Stead, B. D., 192(13, 14), 201, 205(14), 229, 231, 232, 235, 236(13, 14), *247*, 366(117), 369(117), *375*, 473(282), *493*
Stearne, J. M., 192, 201(21), 203(21), 205(21), 206(21), 207(50), 208, 209, 216(50), 231(50), 232(50), *247, 248*, 487(425), *496*
Steele, R. E., 402(27), *413*
Steelman, S. L., 126(90), *175*
Steere, R. L., 127(170, 171), 137, 142(199), 152(170, 171), 153, *177, 178*
Stefani, R., 476(402), *496*
Stein, W. H., 126(116), *176*
Steiner, T., 328(39), *337*, 363(61), *373*, 468(380), 486(379), *495*
Steinwehr, H., 479(230), *492*
Stemmler, H. D., 463, *464*
Stepanov, V., 126(31), 127(31), *174*
Stinson, E. E., 126(56), *174*

Author Index

Stockmayer, W. H., 12(33), 40, *41*, *42*, 263(22), *283*, 308, *316*, 363(43), *372*, 477(42), 487(403), *488*, *496*
Stokrova, S., 366(112b), *375*
Stolka, M., 187, *190*, 475(244), *493*
Storgårds, T., 128, *177*
Stouffer, J. E., 166(255), *179*
Strain, D. E., 365(100), *374*
Streitzig, H., 328(37, 44), *337*, *338*, 363(61), *373*, 486(379), *495*
Streuli, H., 126(117), 127(117), *176*
Stuart, H. A., 214, *248*
Stumpf, W., 126(55), *174*
Styk, B., 126(50), *174*
Sueoka, N., 282, *284*
Suminoe, T., 364(71a), *374*
Sumitomo, H., 367(138, 139), 368(152), *376*
Summaria, L., 126(91), *175*
Summers, J. T., 84(43), *93*, 106(26), 111(26), 116(26), *121*, 467(83), 468(83), *489*
Suprun, A. P., 475(331), *494*
Suzuki, A., 483(14), *488*
Svedberg, T., 181, *189*, 254, 256(6), 264(23, 24), 277(63), *282*, *283*, *284*, 288(8), *305*
Swanson, D. L., 482(404, 436), *496*, *497*
Synge, R. L. M., 126(103), 128(103), *175*
Szwarc, M., 432(19), 438(27), *459*

T

Tablino, V., 335(77a), *339*
Tada, K., 364(75, 76), *374*
Tager, A. A., 320(18), *337*
Takahashi, A., 76(21), *92*, 468(451), *497*
Takamisawa, K., 65(47), *66*
Takamura, S., 367(139), *376*
Takashima, K., 365(103a), *375*
Takayama, G., 349(10), 364(68a), 365(10), *371*, *373*
Takemori, S., 126(54), *174*
Takenaka, H., 53(28), *66*
Talamini, G., 471(405), 472(405), *496*
Talmud, S. L., 370(195), *377*
Tamblyn, J. W., 50, *66*, 192, 201(2), 205(2), 219, 222, 239, *246*, 365(88), *374*, 482(308), 483(306, 307, 406), *494*, *496*
Tan, M., 126(46, 97, 119), *174*, *175*, *176*
Tanaka, S., 200(40a), 242, *248*
Tanaka, T., 369(159), *276*
Tanenbaum, S. W., 126(64), *175*
Tanimura, M., 369(159), *376*
Tasman, J. E., 309, *316*, 483(407), *496*
Tavani, L., 155(234), *179*
Tavazzani, C., 368(154b), *376*
Taylor, D. L., 473(408), *496*
Taylor, E. W., 197(31), 234(31), *247*, 469(305), *495*
Taylor, G. B., 480(409), *496*
Taylor, W. C., 192, 196, 200(24), 243, *247*, 466(410), *496*
Testerman, M. K., 336, *339*, 463, *464*
Thielen, L., 155(235), *179*
Thoma, F. A., 126(16), *173*
Thomas, B. B., 483(411), *496*
Thompson, J. O., 473(435), *497*
Thompson, R. B., 482(362), *495*
Thomson, J., 485(21), *488*
Thormann, F., 483(438), *497*
Thureborn, E., 126(57), *174*
Thurmond, C. D., 54, 55, *66*, 363(50), *373*, 486(412), *496*
Tikhomirov, B. I., 371(207), *378*
Timasheff, N., 366(112), *375*
Timell, T. E., 365 (92, 97, 98), *374*, 483(413), *496*
Timochin, I. M., 365(96), *374*
Tiselius, A., 98(10), *120*, 124, *173*
Tobias, R. L., 78, 79(29), 86(29), 89, *92*, 389, 390, *412*, 468(93), *489*
Tobolsky, A. V., 4(10), *41*, 333(64), 334(70), 335(70), *339*, 427, *459*, 463, *464*, 473(414), *496*
Tombs, M. P., 484(415), *496*
Tomescu, M., 363(55c), *373*
Tompa, H., 6, 7, 17, 21, *41*, 79, *92*, 146(217), *178*, 401, 413, 422(2), 431(2), *459*
Tompkins, V. N., 126(77), *175*
Topchiev, A. V., 362(41h), 363(44, 44a), 370(184), *372*, *377*, 481(416), 486(267), *493*, *496*
Townend, R., 366(112), *375*
Träxler, G., 217, *248*
Trautman, R., 260, 275, 277(10), *282*, *283*
Tremblay, R., 363(45), *373*
Trementozzi, Q. A., 369(165), *377*
Trommsdorff, E., 477(417), *496*
Tschebotarwski, W. W., 470(356), *495*
Tschoegl, N. W., 331, *338*
Tsuruta, T., 364(73), 369(170), *374*, *377*, 478(226), *492*
Tuijnman, C. A. F., 397, *413*

Tulane, V. J., 99(15), *120*
Tung, L. H., 4(7), 26, *41*, *72*, 78, *92*, 173, *179*, 192, 196, 200(24), 243, *247*, 383, 388, 394, 395, 399(12), 402(12), 403, 408, 409, *412*, 466(410, 418, 419), 467(418), *496*
Turley, S. G., 367(135), *376*
Turner, D. T., 487(10), *488*
Turska, E., 479(420), 480(420, 421), *496*

U

Udvarhelyi, K., 365(83), *374*, 478(63, 64), *489*
Udy, D. C., 486(130), *490*
Ueberreiter, K., 48, *66*, 196, 228, 230, 231, 232, *247*, 466(422), 479(423), *496*
Ulfendahl, H. R., 126(60), *174*
Umstätter, H., 322(22), 326, *337*
Unbehend, M., 462, *464*
Ungerer, E., 127, *177*
Urquhart, A. R., 365(86), *374*
Urwin, J. R., 192, 201(21), 203(21), 205(21), 206(21), 207(50), 208, 209, 216, 231, 232, 235(65), *247*, *248*, 366(113), *375*, 473(424), 487(425), *496*
Ushakov, S. N., 364(71), *374*, 486(426), *496*
Usmanov, Kh. U., 236, *248*
Utracki, L., 479(420), 480(420), *496*

V

Vainryb, M., 200(32), 237(32), *247*
Valan, K. J., 371(204), *378*
Vale, R. L., 369(163), *376*
Vallee, B. L., 126(122), *176*
Valvassori, A., 364(69b, 69c), *373*
Valyi, I., 471(427), *496*
van Dam, A. F., 126(66), *175*
van de Brande, J., 366(111), *375*
van der Bie, G. J., 461, *464*, 485(372), *495*
van der Heyden, A., 470(95), *489*
van der Wende, G., 126(70), 127(70), *175*
van Eijk, H. G., 126(52), *174*
van Hoang, D., 126(43), *174*
van Holde, K. E., 260(11), 271(52), 278(65), 279(65), *282*, *283*, *284*, 401(23), *413*, 467(434), *497*
van Hoorn, H., 75(16), *92*
van Humbeck, W., 367(142), 368(142), *376*
van Schooten, J., 25(52), *42*, 75, 87(52), 89(52), *92*, *93*, 468(447), *497*
van Velden, P. F., 480(196), *492*

Varadaiah, V. V., 5(19), *41*
Varoqui, R., 287(7), 288(7), 294(7), 296(7), 301(7), 304(7, 26, 30), 305(7, 26, 30), *305*, *306*
Vasil'eva, N. N., 126(40, 109), 127(40), 129(109), *174*, *176*
Vaughan, D. J., 370(183), *377*, 469(255), 470(255), *493*
Vaughan, G., 89(53), 90(53), *93*, 104(21), 106(23), 111(21), 116(23), *121*, 367(145, 146, 147), 369(147), *376*, 475(84, 86, 428), 487(85), *489*, *496*
Vaughan, M. F., 98, 118, *120*, 127(154, 155, 156), 129(154, 155, 156), 130, 147, 153(155), 154(154, 155), *177*, 462(11), *464*, 471(166), 472(166), *491*
Ve, T., 371(210), *378*
Veatch, F., 364(74a), *374*
Veerkamp, T. A., 355(26), 366(26), *372*
Veermans, A., 355(26), 366(26), *372*
Venet, A. M., 484(340), *495*
Verhaar, G., 485(429), *496*
Vermeulen, T., 134, *178*
Veselovskaya, L. N., 332(60i), *338*
Vidotto, G., 471(405), 472(405), *496*
Vink, H., 135, *178*
Vinograd, J. R., 280(72, 74), 281(74), 282(74), *284*
Vinogradov, G. V., 320(17), *337*
Virkola, N., 471(392), *496*
Vodrazka, Z., 366(112a), *375*
Vogel, R. E., 478(430, 431), *496*
Vogler, K., 56(35), *66*, 469(101), 474(102), *490*
Volkova, G. I., 192, 201(21a), 204(21a), 206, *247*
von Hofsten, B., 126(75, 85), *175*
Voronov, B. Ya., 192, 201(21a), 204(21a), 206, *247*
Vrij, A., 23(44), *42*
Vyrskii, P., 471(342), 473(342), *495*
Vyrskii, Yu., 200(40), 231(40), 232(40), *248*

W

Wada, S., 363(49b), 366(112d), *373*, *375*
Wagner, H. L., 470(106), *490*, 485(432), *496*
Wagner, M., 126(80), *175*
Wagner, R. H., 469(433), 483(406), *496*, *497*
Wake, R. G., 366(106), *375*

Author Index

Wales, M., 193(28), *247*, 260(11), 269(43), 272, 278(65), 279(65), *282*, *283*, *284*, 401, *413*, 438(26), *459*, 467(434), 473(435, 437), 482(436), *497*
Walker, W. E., 402(27), *413*
Wall, L. A., 356(37), *372*
Wallenius, G., 192, 200(10), 204, 228, 240, *247*
Walsh, E. K., 466(222, 223), *492*
Walton, K. W., 126(73), *175*
Wannow, H. A., 483(438), *497*
Ward, S. G., 320, *337*
Watanabe, H., 356(35), 369(35), *372*
Waters, J. L., 145, *178*
Watkins, J. M., 333(65), 334, *339*, 477(439), *497*
Watson, G. N., 425, *459*
Watson, W. F., 366(114), 367(114), *375*, 431(13), *459*, 485(440), *497*
Watts, D. C., 366(104), *375*
Waugh, G. P., 197(31), 234(31), *247*, 469(305), *494*
Weakley, T. J. R., 102, 104(19), 110(19), 111(29), 113(19), 117, *121*, 477(441, 442), 485(339), *495*, *497*
Weaver, W., 279, *284*
Webb, T., 126(51), *174*
Webber, R. W., 471(178), *491*
Webster, M. E., 126(24), *174*
Weiss, F. J., 466(216), 472(216), *492*
Weiss, F. T., 90(55), *93*, 97, 98(2), 102(2), 104, 106, 110(2), 111(2), *120*
Weiss, G., 438(28), *459*
Weissberg, S. G., 193(28), *247*, 278(66), *284*
Weissermel, K., 365(79a, 79b), *374*
Wenediktow, S. P., 365(87), *374*
Werner, H., 479(241), *493*
Wesslau, H., 65(49), *66*, 74, *92*, 385, 395, *412*, 466(443, 444), 467(444), *497*
Wessling, R. A., 112(37), *121*
West, D. C., 331(51, 52), *338*, 467(373), *495*
Westenbrink, H. G. K., 126(52), *174*
Weston, N. E., 276(62), *284*
Wheaton, R. M., 128, 139(184), 148(184, 225), *177*, *178*
Whitaker, J. R., 127(172), *177*
White, B. B., 351(12), 365(12), *371*
White, E. F. T., 200(34), 237, *247*, 366(122), *375*
White, G., 471(212), *492*
Whitehead, E. P., 366(104), *375*

Whitmore, F. C., 184, *190*
Whitmore, R. L., 320, *337*
Widström, G., 126(41), *174*
Wiederhorn, N. M., 477(445), *497*
Wiegner, G., 127, *177*
Wieland, T., 127(162), 129(162), 133(162), *177*, 472(100), 485(446), *490*, *497*
Wiener, O., 286, *305*
Wijga, P. W. O., 25, *42*, 87, 89, *93*, 468(447), *497*
Wilander, O., 192(9), 204(9), 228(9), 240(9), *247*
Wilcox, P. E., 126(93), 129(93), *175*
Wiles, D. M., 477(88), *489*
Wilkie, K. C. B., 462, *464*
Williams, D. E., 126(90), *175*
Williams, H. L., 366(120), *375*, 486(324), *494*
Williams, J. W., 264, 266, 269, 270(48), 278(67), *283*, *284*, 473(435), 482(404), *496*, *497*
Williams, L. C., 4(9), *41*, 278(66), *284*, 331(55), *338*, 467(265), *493*
Williams, M. L., 334(67), *339*
Williams, R. J. P., 40(71), *42*, 83, 90, *93*, 95, 98(1, 10), 102(1), 103(19), 104(19), 106, 110(1, 19), 111(29), 113(19), 117(19), *120*, *121*, 169, *179*, 472(19), 477(441, 442), 481(22), 485(339), *495*, *488*, *497*
Willits, C. O., 126(56), *174*
Wiloth, F., 479(25, 448), 480(449), *488*, *497*
Wilson, J. D., 102(19), 104(19), 110(19), 113(19), 117(19), *121*, 477(441, 442), *497*
Windisch, K., 328(40, 43), *337*, *338*
Winkler, C. A., 463, *464*
Winzor, D. J., 126(173, 174), *177*
Wipke, W. T., 192(25), 201(25), 204(25), 243(25), *247*, 364(69), *373*, 486(151), *491*
Wishman, M., 364(68), *373*
Wisingerová, E., 126(61), *174*
Witkop, B., 126(144), 127(144), *176*
Witte, J. J., 126(52), *174*
Wittenberger, U., 219(52c), *248*
Wittman, G., 126(67), *175*
Wolf, K. A., 321(21), *337*
Woltersdorf, O. W., 309(12), *316*, 483(345), *495*
Wong, P. S. L., 332, *338*
Wood, H. L., 485(28), *488*
Wood, J. W., 369(172), *377*
Woodward, A. E., 480(295), *494*

Woof, J. B., 126(82), *175*
Worsfold, D. J., 433(22), *459*
Wosesson, M. W., 352(20), 366(20), *372*
Wright, B., 332(60d), *338*
Wunderlich, B., 5(16), 23(16), *41*
Wyman, D. P., 332(60g), *338*

Y

Yabe, K., 366(112d), *375*
Yacmazaki, N., 364(71a), *374*
Yakubchik, A. I., 371(207), *378*
Yakushina, T. A., 246(77a), *249*
Yamada, A., 371(205), *378*
Yamada, K., 351(13), *371*
Yamada, N., 473(450), *497*
Yamaguchi, I., 356(35), 369(35), *372*
Yamaguchi, K., 65(47), *66*, 76, *92*, 468(451), *497*
Yamakawa, H., 5(20), *41*, 263(21), *283*
Yamamoto, N., 370(185), *377*
Yamamuro, Y., 371(205), *378*
Yamashita, Y., 370(186), *377*
Yamazaki, E., 126(49), *174*
Yamin, M., 45(5), *65*, 474(36), *488*
Yanagisawa, K., 363(51b), *373*
Yanagita, M., 371(205), *378*
Yanko, J. A., 363(47), 364(66), *373*, 486(453), 486(452), *497*
Yannas, J. B., 363(49a), *373*
Yardley, J., 89(53), 90(53), *93*, 106(23), 116(23), *121*, 475(86), *489*
Yasumoto, N., 355(29), 367(29), *372*
Yeh, S. J., 472(454), *497*
Young, B. G., 369(172), *377*
Youngson, M. A., 126(103), 128(103), *175*
Yphantis, D. A., 280, *284*
Yuki, H., 363(51c), *373*
Yul'chibaev, A. A., 236(66b), *248*

Z

Zadražil, S., 127(147), *176*
Zambelli, A., 75(20), 85(20), 91(20), *92*, 364(69a, 69b, 69c), *373*
Zapf, F., 483(455, 456), *497*
Zapunnaya, K. V., 370(194), *377*
Zelikman, S. G., 58, *66*, 470(291), *494*
Zeman, R., 431(18), *459*
Zettelmoyer, A. C., 486(457), *497*
Zhdanov, A. A., 365(85), *374*
Zief, M., 484(458), *497*
Ziegler, K., 385, *412*
Zilka, A., 367(148a), 368(148a), *376*
Zimm, B. H., 54, 55, *66*, 75, *92*, 182, 184, 185, *190*, 298, *306*, 330(47), *338*, 363(50), *373*, 471(41), 472(182, 201), 486(412), *488*, *491*, *492*, *496*
Zimmerman, J. M., 365(80), *374*
Zimmerman, R. L., 363(53), 364(53), *373*
Zipp, O., 126(33), *174*
Zlatina, S. A., 362(41l), *372*
Zuber, H., 126(37), *174*
Zwan, J., 126(66), *175*

Subject Index

A

Adsorption, 85
Apparent viscosity, 321
Archibald method, 274–276
Asymmetry factor, 215

B

Beall's method, 395–398
Bio-gel, 149
Birefringence, 335
Branching, 4, 173, 422
Brownian diffusion, 285–305
Bruns' series, 302

C

Centrifugal potential, 253
Chain folding, 24
Channel formation, 85
Chemical inhomogeneity, 341–362
  average value, 357–358
  categories, 342–344
  characterization of fractions, 354
  choice of fractionation conditions, 345–352
  comparison of methods, 361
  determination by density gradient centrifugation, 355
    by electrophoresis, 352
    by fractionation, 344–355
    by gel permeation chromatography, 352–353
    by infrared spectroscopy, 355
    by mechanical measurements, 355
    by microscopy, 356
    by nuclear magnetic resonance, 356
  fractionation of mixtures, 353–354
  literature survey, 362–371
  quantitative description, 356–362
Chemical partition ratio, 358
Chemical potential, 11–12, 18, 20, 22–23, 253
Choice of solvent–nonsolvent, 48–52, 60, 83, 107, 195–197
Chromatographic fractionation, 95–120
  comparison with other techniques, 114–118
  effect of temperature gradient, 114–118
  specific fractionations, 110–111

Chromatographic fractionation apparatus, 96
Claesson grid, 223
Coacervate extraction, 78
Cohesive energy densities, 49
Column construction, 97
Column efficiency, 71
Column elution, 21, 79
  apparatus, 81
Column support materials, 98
Concentration effects, 52
Convolution sums, 417
Countercurrent distribution, 464
Critical composition, 9
Critical concentration, 12, 16, 19, 28
Critical point, 7
Critical temperature, 9
Cross-linking, 22
Crystalline polymers, 5, 9–10, 23–25, 63–65, 75

D

Data smoothing, 439
Degassing, 102
Degradation, 73, 85, 87
Density differences of polymers, 281–282
Density gradient centrifugation, 280–282, 355
Dextran gels, 150
Dielectric dispersion, 463
Dielectric measurements, 336
Differential molecular weight distribution, 380–381
  application of Beall's method, 409–412
  direct calculation, 395–398
  numerical example, 406–407
Differential refractometer, 159, 165
Diffusion, 72
Diffusion cells, 288
Diffusion coefficient, 182
  concentration dependence, 299–301
  distribution, see Distribution of diffusion coefficients
Diffusion constant, 286
  averages, 291–293
  determination, 287
  distribution, see Distribution of diffusion constants

Diffusional equilibrium, 134
Distribution coefficient, 133, 138–141
Distribution of diffusion coefficients, 301
 moments, 302
Distribution of diffusion constants, 289–297
 conversion to molecular weight distribution, 295
 moments, 295
Distribution of sedimentation coefficients, 267–274
 concentration dependence, 270
 conversion to molecular weight distribution, 270–274
 extrapolation to infinite time, 269–270
 hydrostatic pressure correction, 268–269
Double-sector cell, 256
Drying, 57, 88

E

Elastic effects, 4
Elastic energy, 22
Electron microscopy, 461–462
Elution curve, 169
Elution volume, 131, 134, 157
Entanglement points, 21
Entropy of mixing, 12
Equipartitioning of fractions, 56, 60
Error function, 435
Extraction, 74

F

False cloud point, 49
Fick's law, 286
Film extraction, 76
Flory–Huggins theory, 10–12, 69–71
Flow birefringence, 335
Flow control, 102–103
Flow curve, 321–325
 derivation of polydispersity parameters, 325–326
 empirical analysis, 327–329
 evaluation, 321–325
 influence of polydispersity, 322–325
 theoretical analysis, 329–331
Flow rate, 87, 108
Fourier transformation, 302
Fraction collection, 103–104
Fraction recovery, 105

Fractional precipitation, 44–66
 at elevated temperature, 47
 cooling method, 61–64
 nonsolvent addition method, 44–57
 solvent evaporation method, 58–60
Fractional solution, 67–91
 by coacervate extraction, 78
 by column elution, 79–81
 by direct extraction, 74–75
 by film extraction, 76–77
 column design, 82
 comparison between methods, 89
 theoretical considerations, 69–73
Fractionation
 analytical, 68
 preparative, 68
Fractionation efficiency, 27–35, 52, 107, 113
Fractionation flask, 46–47
Fractionation schemes, 53–56
Fractionation theory, 10–25
Free energy of mixing, 11
Frictional coefficient, 254

G

Gamma function, 383, 441
Gaussian integration, 441–443
Gel filtration, *see* Gel permeation chromatography
Gel permeation chromatography, 98, 123–172
 adsorption, 145
 applications, 126
 assessment of optimal conditions, 143–145
 calibration, 171
 column design, 159–161
 electric analog, 136
 evaporation, 166
 history, 127–130
 in aqueous solutions, 127–129
 in nonaqueous systems, 129–130
 incompatibility, 146
 monitoring of column effluent, 165
 packing of columns, 162–163
 partitioning, 146
 preparative separations, 166–169
 principle, 125
 sample load, 164
 selection of gel, 139
 solvents, 163
 temperature, 164
 theory, 130–146
 treatment of data, 172–173

Subject Index

Gel structure, 132, 156
Generating functions, 417
Gralén method, 264
Graphical differentiation, 404

H

Hermite polynomials, 401
Herrington–Robertson distribution, 427
Histogram, 380

I

Inhomogeneity index, 386, 398–400
   calculation, 398–400
Initiation, 421
   first order, 430
   general mechanisms, 433–437
   pulse, 432–433
   second order, 431, 438
   thermal, 449–451
   zero order, 430, 437
Integral molecular weight distribution, 381–382, 439–440
   calculation from fractionation data, 387–398
   elimination of graphical integration, 391–395
   numerical example, 405–406
Interaction coefficients, 11, 44, 70
Interstitial volume, 125, 131
Intrinsic viscosity, 319
   measurement in two solvents, 319, 400–401
Inversion, 186
Ion exchangers, 148

J

Jamin interference optics, 287

K

Kinetic scheme,
   tabulation of results, 429–431

L

Laguerre polynomials, 401, 440–441
Lamm differential equation, 255, 261–263
Lansing–Kraemer distribution, 266, 293, 385
   numerical example, 407–409
Large-scale fractionation, 118–120
Living polymers, 434
   number distribution curve, 436–438

Logarithmic gradient, 102
Log-normal distribution, *see* Lansing–Kraemer distribution

M

Mass flux rate, 182
Melt viscosity, 3, 331
Membrane fractionation, 462
Moebius function, 330
Molecular sieve, 124
Molecular weight,
   number average, 2
   viscosity average, 3
   weight average, 2
   z average, 2
   $(z+1)$ average, 2
Molecular weight averages, 2–3, 277, 383–385, 448
Molecular weight distributions, 400–401
   calculation from two averages, 400–401
   comparison of methods, 402–404
   condensation polymers, 419–421
   cumulative, *see* Integral molecular weight distribution
   differential, *see* Differential molecular weight distribution
   free radical polymerizations, 421–431
   integral, *see* Integral molecular weight distribution
   ionic polymerizations, 431–438
   kinetic interpretation, 415–451
   matrix method, 438–451
      numerical example, 444–451
      tables, 452–458
   numerical analysis, 415–451
   prediction from kinetic schemes, 418–438

N

Newtonian liquid, 321
Non-Newtonian flow, 319, 321
Nuclear magnetic resonance, 463
Nucleation, 24
Number distribution curve, 381
   moments, 416–417

O

Optical birefringence, 335
Overlapping, 39, 389, 392
   correction, 390–391
Oxidation, 73

P

Partition coefficient, 132, 138
Phase relations, 5–9
Poisson–Charlier series, 440
Poisson distribution, 433
Poly-, for information on specific polymers see appendix to Chapter G, 465–487
Polyacrylamide gel, 150–153
Polydispersity, 1–3
  fractionation and, 1
  polymer properties and, 3–5
Polyelectrolytes, 51
Polymer deposition, 72, 80, 85–87, 106
Polymer support, 84
Polystyrene gel, 155–158
  permeability, 158
Porous glass, 154
Precipitation fractionation, see Fractional precipitation
Precipitation turbidimetry, see Turbidimetric titration
Preparatory-scale fractionation, 118–120
Propagation, 422

R

Rayleigh interference optics, 287
Recycling chromatography, 161
Refractionation, 40, 53, 113, 402
Relaxation measurements, 332–335
Relaxation time distribution, 332–335
  conversion to molecular weight distribution, 333–335
Retardation time distribution, see Relaxation time distribution
Retention volume, 134
Reverse-order fractionation, 50, 52, 88
Rheological methods, 317–336
  dynamic measurements, 322
  flow curve, 317–318
  steady state measurements, 322
Rouse model, 331

S

Salting-out, 51
Santocel A, 153
Schlieren optics, 257, 287
Schulz distribution, 383
Sedimentation, 251–282
  theory, 252–255

Sedimentation coefficient, 258
  averages, 271–272
  combination with intrinsic viscosity, 263
  concentration dependence, 260
  diffusion effects, 260
  standard deviation, 266–267
Sedimentation equilibrium, 276–282
  time required to reach equilibrium, 279–280
Sedimentation equilibrium experiments, 251
Sedimentation velocity, 251, 258, 260
Semicrystalline polymers, 63–65
Sephadex, 129, 149
Shear rate, 321
Shear stress, 321
Sol-gel ratio, 463
Solubility, 25–27
Solution fractionation, see Fractional solution
Solvent gradient, 83
Solvent gradient controller, 99
Solvent gradient production, 98–102
Soret coefficient, 182
Spread factor, 29
Standard deviation, 386
  calculation, 398–400
  of number distribution, 386
  of weight distribution, 387
Steady state approximation, 422–428
Stress birefringence, 336
Stress relaxation, 463
Structural fractionation, 24
Summative fractionation, 307–315
  experimental results, 311–315
  mathematical interpretation, 309–311
Svedberg equation, 254, 263, 277
Synthetic boundary cell, 256
Synthetic time, 435

T

Tail effect, 53
Temperature gradient, 109–133
Termination, 422
Theoretical considerations, 1–41
Theoretical plate concept, 141–142
Thermal diffusion, 181–189
  cascade of thermal diffusion columns, 188
  concentration dependence, 184
  convectionless cells, 182
  history, 181–189
  molecular weight dependence, 185

Subject Index

multiple reservoirs, 189
solvent flow, 189
summary of systems studied, 187
theory, 182
thermal diffusion columns, 183
Thermal diffusion coefficient, 182
Thermodynamic equilibrium, 6
Theta conditions, 260, 278, 281–282, 304
Transfer, 422
Treatment of data, 379–412
Triangle fractionation, 55
Tung distribution, 383
  numerical example, 405–406
Turbidimetric titration, 191–246
  apparatus, 199–210
  efficiency, 193
  evaluation of data, 218–228
  limitations, 193
  operating temperature, 198
  polymer concentration, 197, 213
  principle, 193
  rate of addition of precipitant, 197
  rate of stirring, 198
  summary of results, 228–246
Turbidity titration, *see* Turbidimetric titration

U

Ultracentrifugation, 251
Ultracentrifuge, 255–258
Ultrafiltration, 462
Ultraviolet optics, 287
Ultraviolet photometer, 165

V

Viscosity, 330
  normal coordinate theory, 330
  shear dependence, 324
Void volume, 125

W

Watson's Lemma, 425
Weaver's rule, 279
Weissenberg equation, 321
Wesslau distribution, 385, 395

Z

Zimm model, 331
Zone melting, 462–463

DATE DUE

OC